Chemical Engineering Plant Design

McGRAW-HILL SERIES IN CHEMICAL ENGINEERING

Max S. Peters, *Consulting Editor*

EDITORIAL ADVISORY BOARD

Charles F. Bonilla. Professor of Chemical Engineering, Columbia University

John R. Callaham. Editor-in-chief, *Chemical Engineering*.

Cecil H. Chilton. Editor-in-Chief, *Chemical Engineering*.

Donald L. Katz. Chairman, Department of Chemical and Metallurgical Engineering, University of Michigan

Sidney D. Kirkpatrick. Consulting Editor, McGraw-Hill Series in Chemical Engineering, 1929–1960

Walter E. Lobo. Consulting Chemical Engineer

Robert L. Pigford. Chairman, Department of Chemical Engineering, University of Delaware

Mott Souders. Associate Director of Research, Shell Development Company

Richard H. Wilhelm. Chairman, Department of Chemical Engineering Princeton University

BUILDING THE LITERATURE OF A PROFESSION

Fifteen prominent chemical engineers first met in New York more than 30 years ago to plan a continuing literature for their rapidly growing profession. From industry came such pioneer practitioners as Leo H. Baekeland, Arthur D. Little, Charles L. Reese, John V. N. Dorr, M. C. Whitaker, and R. S. McBride. From the universities came such eminent educators as William H. Walker, Alfred H. White, D. D. Jackson, J. H. James, Warren K. Lewis, and Harry A. Curtis, H. C. Parmelee, then editor of *Chemical & Metallurgical Engineering*, served as chairman and was joined subsequently by S. D. Kirkpatrick as consulting editor.

After several meetings, this first Editorial Advisory Board submitted its report to the McGraw-Hill Book Company in September, 1925. In it were detailed specifications for a correlated series of more than a dozen texts and reference books which have since become the McGraw-Hill Series in Chemical Engineering.

Since its origin the Editorial Advisory Board has been benefited by the guidance and continuing interest of such other distinguished chemical engineers as Manson Benedict, John R. Callaham, Arthur W. Hixson, H. Fraser Johnstone, Webster N. Jones, Paul D. V. Manning, Albert E. Marshall, Charles M. A. Stine, Edward R. Weidlein, and Walter G. Whitman. No small measure of credit is due not only to the pioneering members of the original board but also to those engineering educators and industrialists who have succeeded them in the task of building a permanent literature for the chemical engineering profession.

THE SERIES

ANDERSON AND WENZEL—*Introduction to Chemical Engineering*
ARIES AND NEWTON—*Chemical Engineering Cost Estimation*
BADGER AND BANCHERO—*Introduction to Chemical Engineering*
CLARKE—*Manual for Process Engineering Calculations*
COMINGS—*High Pressure Technology*
CORCORAN AND LACEY—*Introduction to Chemical Engineering Problems*
DODGE—*Chemical Engineering Thermodynamics*
GRISWOLD—*Fuels, Combustion, and Furnaces*
GROGGINS—*Unit Processes in Organic Synthesis*
HENLEY AND BIEBER—*Chemical Engineering Calculations*
HUNTINGTON—*Natural Gas and Natural Gasoline*
JOHNSTONE AND THRING—*Pilot Plants, Models, and Scale-up Methods in Chemical Engineering*
KATZ, CORNELL, KOBAYASHI, POETTMANN, VARY, ELENBAAS, AND WEINAUG—*Handbook of Natural Gas Engineering*
KIRKBRIDE—*Chemical Engineering Fundamentals*
KNUDSEN AND KATZ—*Fluid Dynamics and Heat Transfer*
KOHL AND RIESENFELD—*Gas Purification*
LEVA—*Fluidization*
LEWIS, RADASCH, AND LEWIS—*Industrial Stoichiometry*
MANTELL—*Absorption*
MANTELL—*Electrochemical Engineering*
MCADAMS—*Heat Transmission*
MCCABE AND SMITH, J. C.—*Unit Operations of Chemical Engineering*
MICKLEY, SHERWOOD, AND REED—*Applied Mathematics in Chemical Engineering*
NELSON—*Petroleum Refinery Engineering*
PERRY (EDITOR)—*Chemical Business Handbook*
PERRY (EDITOR)—*Chemical Engineers' Handbook*
PETERS—*Elementary Chemical Engineering*
PETERS—*Plant Design and Economics for Chemical Engineers*
PIERCE—*Chemical Engineering for Production Supervision*
REID AND SHERWOOD—*The Properties of Gases and Liquids*
RHODES, F. H.—*Technical Report Writing*
RHODES, T. J.—*Industrial Instruments for Measurement and Control*
ROBINSON AND GILLILAND—*Elements of Fractional Distillation*
SCHMIDT AND MARLIES—*Principles of High-polymer Theory and Practice*
SCHWEYER—*Process Engineering Economics*
SHERWOOD AND PIGFORD—*Absorption and Extraction*
SHREVE—*The Chemical Process Industries*
SMITH, J. M.—*Chemical Engineering Kinetics*
SMITH, J. M., AND VAN NESS—*Introduction to Chemical Engineering Thermodynamics*
TREYBAL—*Liquid Extraction*
TREYBAL—*Mass-transfer Operations*
TYLER AND WINTER—*Chemical Engineering Economics*
VILBRANDT AND DRYDEN—*Chemical Engineering Plant Design*
VOLK—*Applied Statistics for Engineers*
WALAS—*Reaction Kinetics for Chemical Engineers*
WALKER, LEWIS, MCADAMS, AND GILLILAND—*Principles of Chemical Engineering*
WILLIAMS AND JOHNSON—*Stoichiometry for Chemical Engineers*
WILSON AND RIES—*Principles of Chemical Engineering Thermodynamics*
WILSON AND WELLS—*Coal, Coke, and Coal Chemicals*

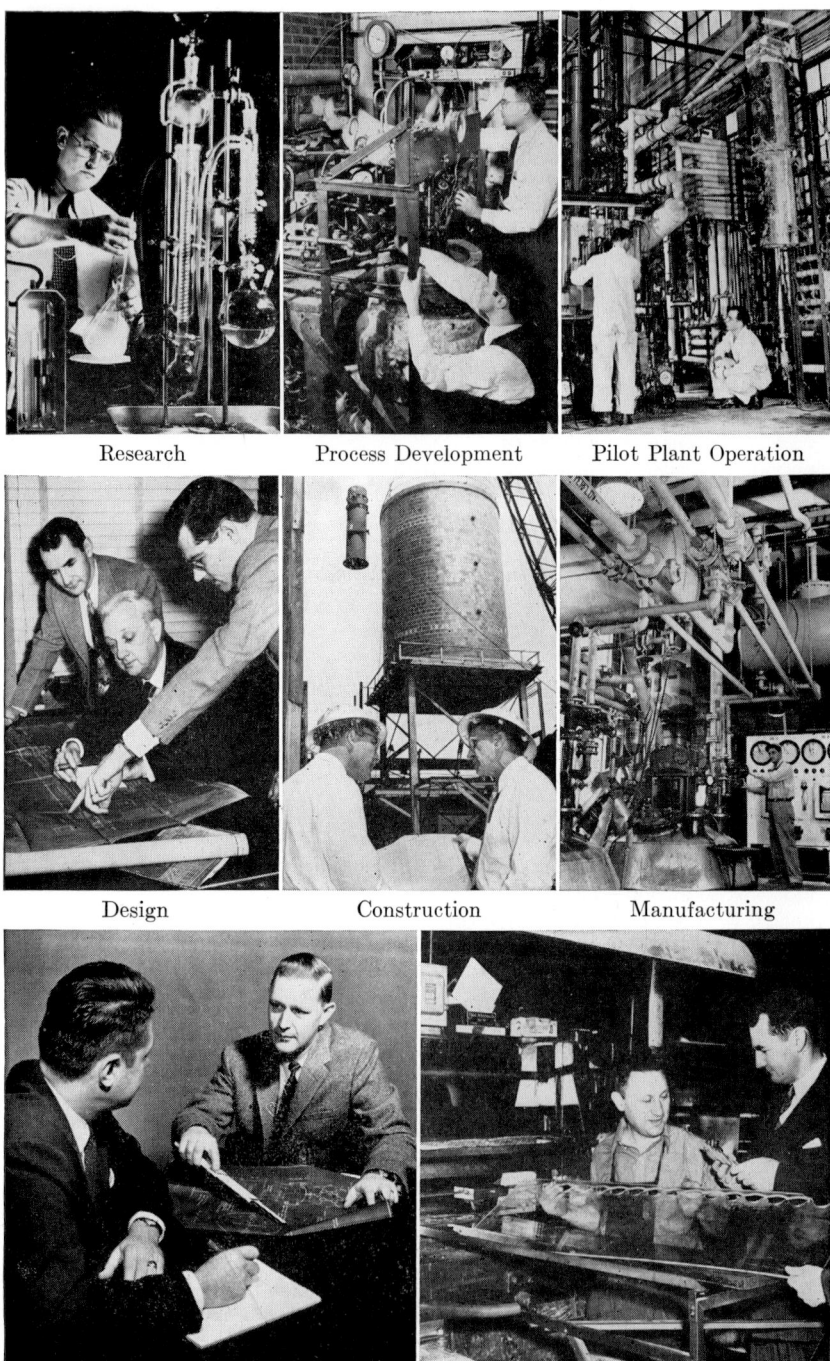

Steps in the commercialization of a chemical process. (*Courtesy of E. I. Du Pont de Nemours & Company.*)

CHEMICAL ENGINEERING PLANT DESIGN

Frank C. Vilbrandt, Ph.D.
PROFESSOR OF CHEMICAL ENGINEERING
VIRGINIA POLYTECHNIC INSTITUTE

Charles E. Dryden, Ph.D.
ASSOCIATE PROFESSOR OF CHEMICAL ENGINEERING
THE OHIO STATE UNIVERSITY

FOURTH EDITION

McGRAW-HILL BOOK COMPANY, INC.

New York   Toronto   London

1959

CHEMICAL ENGINEERING PLANT DESIGN

Copyright © 1959 by the McGraw-Hill Book Company, Inc. Copyright, 1934, 1942, 1949, by the McGraw-Hill Book Company, Inc. Printed in the United States of America. All rights reserved. This book, or parts thereof, may not be reproduced in any form without permission of the publishers. *Library of Congress Catalog Card Number 59-7320*

II

67448

THE MAPLE PRESS COMPANY, YORK, PA.

Preface

Chemical engineering design is divided into equipment design and plant design; it is the purpose of this book to deal only with the latter phase of design as applied to the chemical industries. Chemical engineering plant design is neither a unit operation nor a unit process, but must be considered as one of the tools of the chemical engineering profession. This book presents an analysis of the fundamental principles and factors that are involved in the development of a technically and economically efficient plant process from the laboratory stage through the pilot plant stages to the commercial size unit.

For the student in chemical engineering, this book presents an opportunity for coordinating chemical and engineering information by the application of previously gained or readily available knowledge or facts to the design of a chemical plant; the designed plant is based not only upon the application of accurate fundamental principles and data but also upon the economic phases of the process with emphasis being placed upon costs as an important factor in plant design.

The correlation of the data obtained through laboratory experimentation into a workable basis for designing a plant takes into consideration a thoroughly studied organization of equipment and flow of materials for processing, storage, and future expansion. The writing of specifications for equipment and preconstruction cost estimating are emphasized.

Revision and modernization of the text material from previous editions has been incorporated in a different sequence of presentation. It is assumed that the reader has a sufficient background in process calculations and technology, thermodynamics, unit operations and processes, and elementary engineering economics as taught in an undergraduate chemical engineering curriculum. The first six chapters of the present edition can then be used as a text for course work in process development where the ultimate goal is to arrive at a preliminary cost estimate.

For a more extensive chemical plant design course, detailed information can be used from Chap. 5 on plant layout and from Chap. 7 on plant location. In addition, material from Chaps. 8 and 9, which deal with the design of buildings and piping, control, and power systems, can be incorporated for a well-balanced chemical engineering plant design course in about the same sequence as used in industrial practice. A suggested teaching procedure is given in Appendix A and typical design problems are also incorporated.

Chapter 10 has been added to cover the unique features of nuclear chemical plant design, a subject which will be increasingly useful in the years ahead.

Bibliographic references have been brought up to date; extensive compilations in handbooks, textbooks, and costing journals allowed a reduction of this type of material in this edition by referencing. In particular, frequent reference is made to Perry's "Chemical Engineers' Handbook," 3d ed., McGraw-Hill Book Company, Inc., since students are generally required to purchase this book early in their chemical engineering course work.

The subject matter will be of benefit to professional engineers who would like to keep up to date and also to executives in chemical engineering industries, who have not been trained in chemical engineering, to serve as a guide for their appreciation of the application of chemical engineering principles to plant design.

The authors wish to thank their colleagues in academic and industrial circles for their many contributions to this edition, some of which are directly credited in the text. In addition, it is a pleasure to acknowledge the help of the following in providing valuable experience factors: the Blaw-Knox Co., The Battelle Memorial Institute, E. I. Du Pont de Nemours & Company, General Electric Company, Industrial Nucleonics Co., The M. W. Kellogg Co., Phillips Petroleum Co., E. P. Bartkus, W. T. Butler, H. R. Chope, C. J. Geankoplis, R. P. Genereaux, J. D. Ireland, S. D. Kirkpatrick, T. R. Olive, J. H. Oxley, R. Paffenbarger, E. Pontius, A. Syverson, W. C. Turner, and F. M. Warzell. The support of the College of Engineering at The Ohio State University and the Chemical Engineering Department at Virginia Polytechnic Institute is gratefully acknowledged. In particular, the continued encouragement of Professor J. H. Koffolt, Chairman of the Chemical Engineering Department at The Ohio State University, and the assistance of Professor R. W. Parkinson, OSU Engineering Drawing Department, and of R. J. Shafer in procuring and preparing drawings is very much appreciated.

Frank C. Vilbrandt
Charles E. Dryden

Contents

Preface			vii
Chapter	1.	Introduction	1
	2.	Development of the Project	15
	3.	Process Design	40
	4.	Selection of Process Equipment and Materials	84
	5.	Plant Layout	177
	6.	Economic Evaluation of the Project	189
	7.	Locating the Chemical Plant	265
	8.	Site Preparation and Structures	291
	9.	Process Auxiliaries	340
	10.	Nuclear Chemical Plant Design	427
Additional Selected References			471
Appendix A.	Design Project Procedures		497
Appendix B.	Letter Symbols for Chemical Engineering		504
Appendix C.	Table of Equivalents		516
Name Index			519
Subject Index			525

CHAPTER 1

Introduction

The Role of the Chemical Engineer

The chemical engineer is one who is skilled in development, design, construction, and operation of industrial plants in which matter undergoes a change. Chemical engineers work in four main divisions of the chemical process industries: research and development, design, manufacturing, and sales. This is illustrated in the frontispiece of this book. The chemical engineer prospers because he is versatile; he is well grounded in the fundamental sciences of chemistry, physics, and mathematics, yet knows when to apply empirical engineering know-how to solve problems.

The decisions which make progress possible in engineering (development, design, construction, operation, or management) from an economic necessity are based on inadequate data backed up by experience and sound judgment. Such decisions represent the highest form of expression of engineering. In this manner, the engineer solves the problems which have to be solved.

Chemical Engineering Design

Chemical engineering design consists of process, equipment, and building designs for manufacturing plants to supply the product needs of the customers. More and more, the creative function exemplified in design has become a determining characteristic of the chemical engineer. Since chemical engineering design is a fundamental chemical engineering problem, it is essential that the chemical engineer should recognize design as his responsibility in connection with chemical industries.

Design should follow some prearranged plan based upon space requirements, selections and specifications of process equipment, the layout of process equipment according to processing flows, plant location, plant site selection, and future expansion. Both building and equipment should be designed to give the most efficient production with a minimum of handling of material in process. Provision should be made for storage, for expansion to fit in with the original arrangement without disturbing the flow of work, and for the most favorable and economical conditions of operation of each piece of equipment with respect to all variables. Design is centered about problems of rates of mass and energy transfer and of chemical change. Other factors that should be considered in the design of building and equipment arrangement include possible hazards of fire, explosion, chemical injury, and injury to health, the welfare of the worker, economical distribution of process steam and power, and expansion of production.

All other factors being equal, intelligent and careful design has every advantage over one that has grown up or been put together in a hit-or-miss fashion by alterations, hunches, and additions. The task of the chemical engineer is to calculate quantities and yields, to consider the handling of materials in process and in storage, to apply technical knowledge of material and energy balances, mass and heat transfer by convection, diffusion, and conduction, the flow of fluids, the separation of materials, the thermodynamics and equilibria of reacting systems, the behavior of catalysts, and the kinetics of all types of chemical reactions. In addition, the engineer must develop detailed costs of each unit operation so that even before the plant is in the blueprint stage, he will know not only the cost per ton for processing the raw materials, but also the cost per unit weight of material in each operation, such as grinding, crystallization, filtration, evaporation, drying, etc.

Need for Plant Design

The main factor which dictates the decision to produce a new product or expand or modernize present facilities is generally an economic one as represented by the question "What will be the return on the investment?" The design engineer must be in a position to supply management with preconstruction cost estimates based on a *preliminary plant design* for manufacturing the product so that a sound decision can be made. This plant design analysis includes (1) process design, (2) selection of process equipment and materials, (3) preliminary plant layout and location considerations to estimate labor, building, and land costs, and (4) a manufacturing cost analysis.

It is seen that a good design engineer should have a thorough understanding of chemical economics to make his best contribution to management's problem of making the decision to commercialize the project. If the decision is an affirmative one, then a detailed commercial plant design is required for expediting construction work. The detailed design will probably include optimization and specifications for the process and equipment, in addition to models and working drawings of the building and equipment layout for the construction engineers.

The subject matter required for preliminary and detailed plant designs is given in the subsequent chapters of this book.

Plant Design and Its Relation to Sales

Before a manufacturing plant can be considered, it is necessary to have a product to sell so that profit can be made. This product evaluation job is carried out by a market research group which conducts a scientifically directed study of product design. The fundamental purpose of product design programs[1] is to sell more goods and gain greater profits (1) by keeping the company's products and product lines in a strong competitive position, (2) by diversifying the product lines to serve the industries, (3) by improving or replacing products which, because of market saturation, have shown declining profits, and (4) by advancing by-product or waste products to a profitable status.

Items such as market opportunities, competition, and distribution are studied by the market analysts. Production and economic aspects of product design confront the design engineer with these typical questions:[2]

Production

1. Is the product properly designed from a cost and production viewpoint?
2. Can a necessary new process be integrated with existing plant facilities?
3. What is the best process for producing the product?
4. What new equipment is needed?
5. How much plant space is needed?
6. Can the product be manufactured efficiently?
7. What is the status of raw materials? Should any of these be manufactured?

Economics

1. What is the estimated manufacturing cost per unit?
2. What is the estimated sales and advertising expense per unit?
3. How much capital is required?
4. How much inventory is needed?
5. Can a quality product be produced at a price consumers will pay?
6. How long will it take the product to reach a break-even basis?
7. What is the long-term profit outlook for the product?

[1] *Chem. Eng.*, **57**(9): 129 (1950).
[2] *Chem. Eng.*, **61**(11): 344 (1954).

Questions such as these must be answered by an engineering design group familiar with methods of preconstruction cost estimating based on process and preliminary plant design principles.

Process Design

The question of what is the best process for producing the product bears on the important subject of process design. The process information required for process design comprises the following: (1) written description of the process; (2) notes regarding special safety precautions, possible operating peculiarities, chemical reactions, and properties of materials of construction; (3) knowledge of all raw materials, products, and intermediate process quantities in convenient and appropriate units; (4) knowledge of all process temperatures, pressures, and concentrations; (5) knowledge of physical characteristics and chemical compositions and properties of all raw materials, products, and intermediate process materials under operating conditions; (6) heat, material, and energy balances around all significant operations or pieces of equipment. The complete balances around process equipment as such (a still, a fractionating tower, etc.) should be attempted; but some of the balances, such as energy balances around nonstandard equipment, can be only general and approximate, pending engineering design, procurement of equipment, and information on manufacturers' specified operating efficiencies, etc.; (7) a complete diagrammatic process flow sheet which shows the flow of the process streams and tabulates conditions at the appropriate points on the process streamlines and equipment sketches where convenient. A typical process flow sheet is shown in Fig. 3-4.

Preconstruction Cost Estimation

After the process information has been integrated into one or more flow sheets, the economic aspects of the design are next considered. This involves (1) an estimate of the types and sizes of equipment and materials, buildings, ground area, and utility facilities; (2) a determination of what the process will cost based on physical facilities and construction charges; (3) a cost estimate of utilities consumption (steam, electricity, water, fuel), labor and supervision personnel requirements, maintenance and repairs, raw materials, and finance charges (interest, taxes, insurance, medical benefits, etc.).

On this economic basis, a suitable process design can be chosen and recommendations presented in a preliminary report to company executives for a decision on the project.

Design and Selection of Chemical Engineering Equipment

In the design of all chemical processing equipment, it is important to remember that success depends on continuous performance and the designer should recognize unit operations and unit processes as the basis for selection and design. The designer is concerned primarily in specifying an economical system or piece of equipment suitable for a specific chemical operation. Naturally, this involves problems of temperature, pressure, corrosion, erosion, metal fatigue, and other considerations such as relief from overpressure or vacuum.

Design based upon standard equipment is of primary importance. Requirements for the basic designs of chemical processing equipment are presented in established codes; these are frequently altered in accordance with experiences acquired. Writing of specifications on special equipment for successful operation is equally the task of the designer, should standard equipment not be available to carry on the specific operation. A good design will provide for the processing, handling, and storage of chemical materials in batch and/or continuous systems, which are productive and safe under the conditions involved.

Bases for Good Design

Good designs do not happen; they are founded on well-known, sound principles. To create a good chemical engineering design, it is necessary to possess an interest in and a genuine liking for chemical plant layout and for solving engineering problems, together with a faculty of keen, appreciative observation, and the ability to analyze conditions and data. The chemical engineer accumulates data and determines in minute detail the variables that must be kept under control to ensure economy and success. From these data he makes preliminary designs for the plant and writes specifications for the equipment and the materials needed. He indicates types and sizes of commercial equipment, the feasibility and conditions of economic operation, and supplies information for building and often for designing special equipment. The technical skills required in design involve a comprehensive training in mechanics, engineering drawing, electricity, thermodynamics, materials of construction, materials handling, fluid dynamics, chemistry, physics, and mathematics. In the final analysis, the development through details should be delegated to technicians and draftsmen; but a good engineer must himself be sufficiently skilled so that he can convey his thoughts clearly for others readily to interpret. One of the most important means of communication of design ideas is through the use of engineering graphics skills.

Drawings and Models in Process and Plant Design

The representation of design ideas and the assembly of information for purposes of manufacture, construction, and erection of structures and equipment are done largely through drawings and models. For the average chemical process plant, it is estimated that the cost of preparing these items of design ranges from 3 to 6 per cent of the total erected plant cost.

Sketches, schematic diagrams, engineering drawings of varying degree of detail, and three-dimensional models are used in process and plant design. Typical drawing procedures will be itemized and discussed briefly in the introductory section and references will be made to specific details on drawings and models in the chapters which follow.[1]

Process Design Drawings. The ideas which develop the chemical and physical picture of the process are set down in terms of flow sheets by the chemical design engineer. These vary in degree of complexity from simple block or box diagrams connected by lines to highly complex schematic diagrams showing equipment and process auxiliary requirements. (See Chap. 3 and Figs. 3-1 to 3-10 for explanation and examples.)

Equipment Design Drawings. Usually the chemical plant design group is not vitally concerned with the preparation of engineering drawings required for the manufacture of equipment items. Standard equipment is used wherever possible; drawings required for use in general layout and arrangement, and for erection at the plant site, are supplied by the vendor.

However, there may be times when special equipment is needed and the design engineer must develop the necessary drawings to have the equipment built in his shop or in a vendor's shop. Often it is necessary to assemble a group of standard items to perform an engineering function. A brief review of engineering drawing requirements for these purposes is included.

A series of drawings required to manufacture and assemble the parts to produce an operable piece of equipment would follow this chronological order of development.

1. *Design or Schematic Presentation Sketches.* These first drawings are usually freehand pencil sketches on which original ideas and planning are presented. Calculations must be made to prove feasibility. Often models are constructed at this stage for the same reasons. An example of a schematic diagram, made with drawing tools and not freehand sketched, is shown in Fig. 4-4.

2. *Design Assembly or Layout Drawings.* More details of the design are next worked out on a design assembly drawing, using the design sketches and calculations. The drawing is done in pencil and shows essential dimensions; it is coded for general design specifications of standard parts, materials, finishes, tolerances, and any other information required by draftsmen for producing detail drawings. The specifications can be tabulated on the side of the drawing or on a separate information sheet. The drawings should be sufficiently well presented, using two or

---

[1] For more complete details on drawing procedures, the reader is referred to T. E. French and C. J. Vierck, "Graphic Science," McGraw-Hill Book Company, Inc., New York, 1958.

CHAP. 1] INTRODUCTION 7

three views, so that there can be no confusion about the basic construction of the equipment. An example of this type of work is illustrated in the design of a grit washer pictured in Fig. 1-1. The design assembly drawing in three views required for clarity of design is shown in Fig. 1-2. The code and parts specification listing on the original drawing were not reproduced in this text to save space.

FIG. 1-1. Materials-handling equipment—a grit washer. (*Courtesy of Jeffrey Manufacturing Co., Columbus, Ohio.*)

3. *Detail Drawings.* From a design drawing such as Fig. 1-2, the detail draftsman makes up individual drawings of each part of the equipment so that these can be fabricated. These drawings are accurately to scale and give principal dimensions and manufacturing directions.

4. *Shop Assembly Drawings.* Various classifications of drawings under this heading are used industrially. Referring to the grit-washer example of Figs. 1-1 and 1-2, the assembly of the detailed parts requires a *shop assembly drawing* with necessary code and specifications of parts to allow the final assembly and erection. For complicated machinery such as the grit washer, several *unit or subassembly drawings* of groups of parts are

8   CHEMICAL ENGINEERING PLANT DESIGN   [CHAP. 1

DESIGN NOTES

Capacity – 120 cu ft raw grit per hr
Wash water – 60-70 gpm @ 5# □" pressure
See 200F899F for elevator
Max elevation height to ₵ hd shaft = 50'-0"
Perforated elevator buckets 8 × 4 3/4 × 7 3/4
Spaced 8" on 124 rel. chain – K2ATTS
Jig grit may discharge into side or either end of elevator boot
LH overflow trough may be used on RH jig and vice versa (RH shown)
For outline drawings, see 203F591

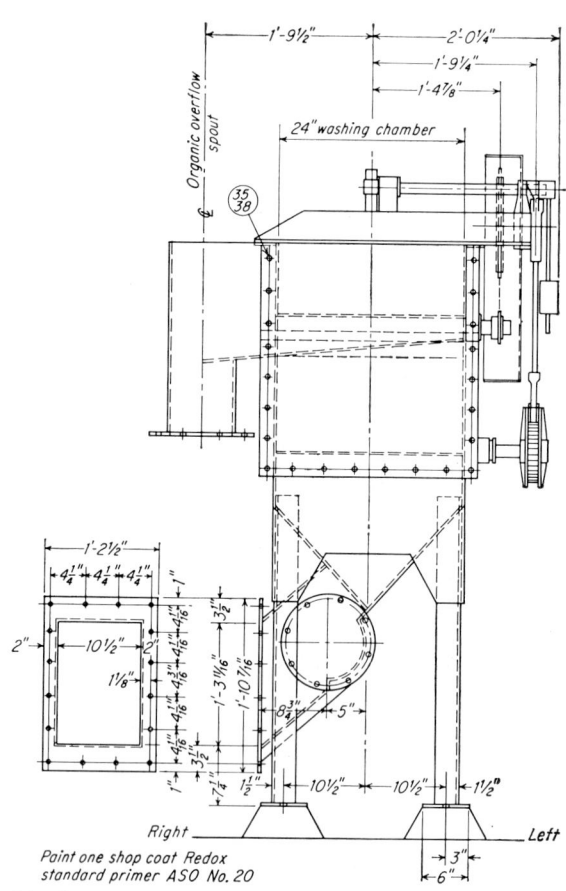

Fig. 1-2. Design assembly drawing of a grit washer (code and parts specification lists Ohio.)

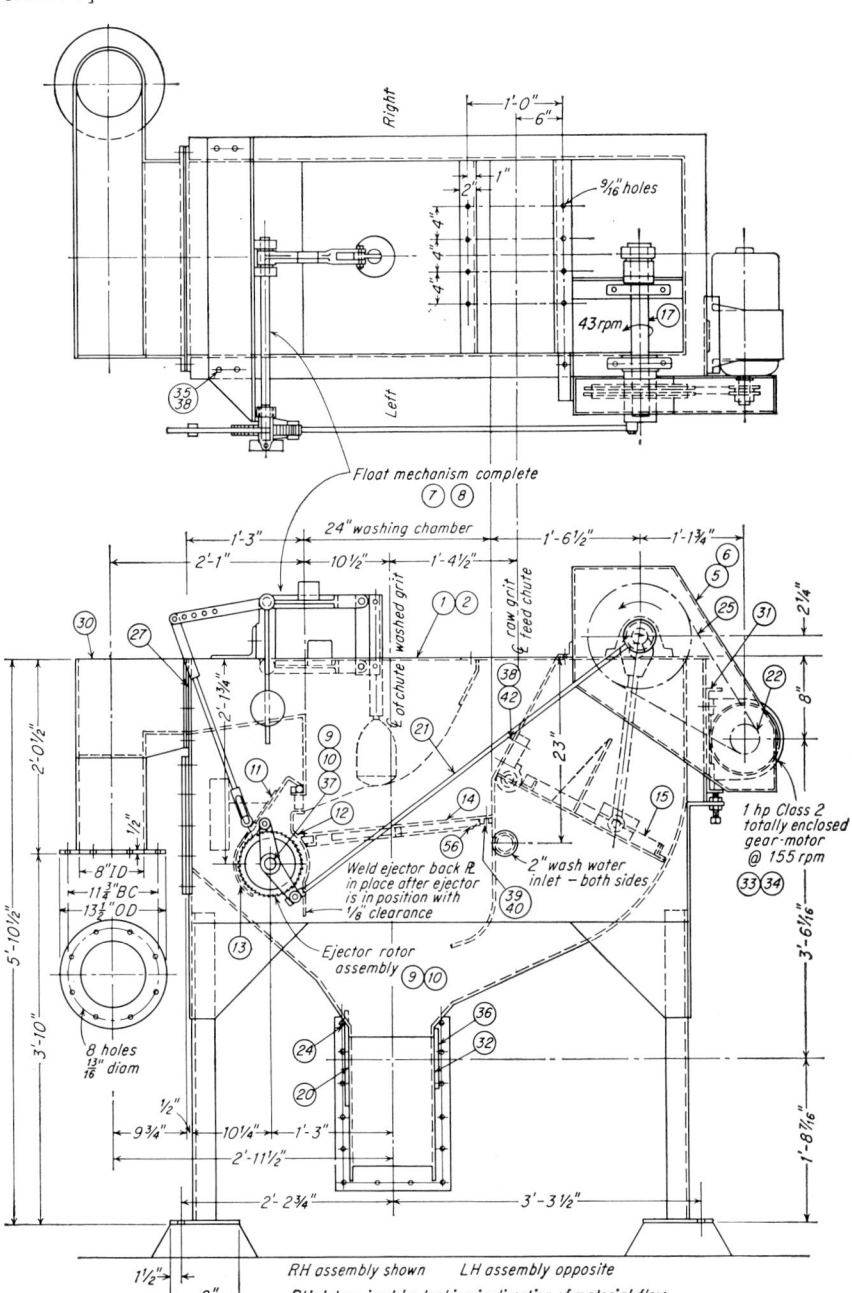

are omitted from original drawing). (*Courtesy of Jeffrey Manufacturing Co., Columbus,*

required to enable the shop mechanic to assemble the complete machine easily. The design and manufacture of complex equipment may require a total of several hundred drawings under classifications 1 through 4.

5. *Outline Assembly or General Arrangement Drawings.* The chemical design engineer must plan for the proper layout of equipment and thus needs a general idea of the exterior shape and principal dimensions of each piece of equipment used in the process. For this purpose, general

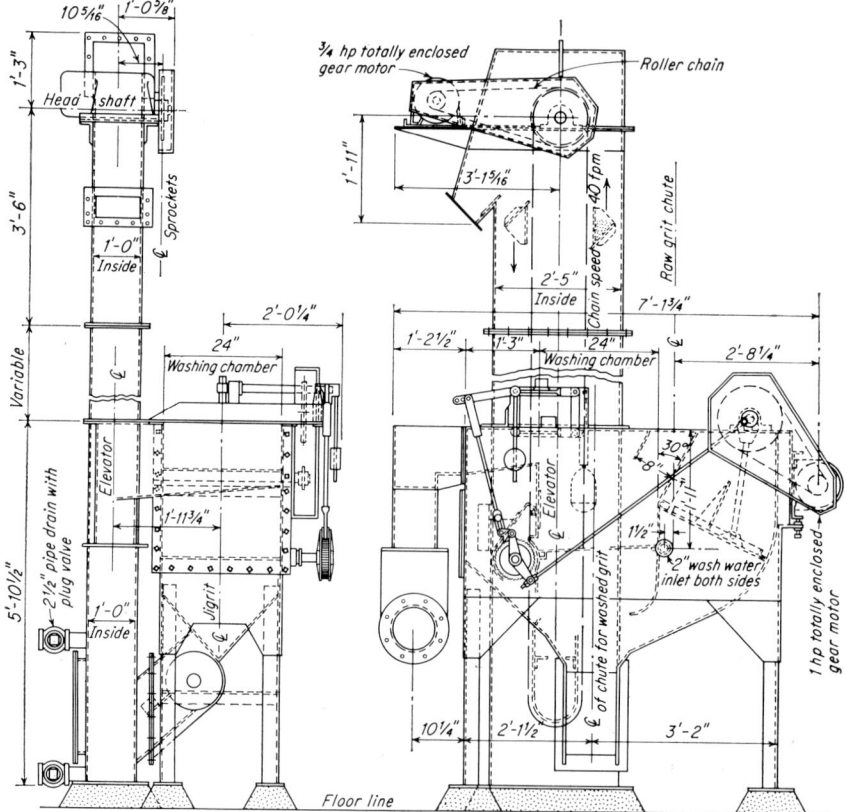

Fig. 1-3. General arrangement drawing of a grit washer. (*Courtesy of Jeffrey Manufacturing Co., Columbus, Ohio.*)

arrangement drawings of the type shown in Fig. 1-3 are available from the vendor or from the equipment design group within the company. Detailed design information such as disassembly and over-all clearance requirements, anchor bolt locations, and piping connections are frequently included.

6. *Equipment Installation Drawings.* Vendor prints which show the details and installation sequencing of purchased equipment are routinely supplied to the design and construction groups. Accurately dimensioned

mounting details for heavy machinery, flow diagrams, and electric power and instrumentation drawings are examples of the drawings required for proper installation and operating performance.

Plant Design Drawings. Chemical plants cannot be put together in random fashion. Careful planning and integration of all process equipment and auxiliaries such as piping, electric power, and instrumentation are done through drawings and models.

1. *Plant Layout Drawings.* After principal equipment has been specified, scale models are usually constructed and arranged by plant layout design for optimum feasibility of operation (Figs. 5-4 and 5-5). Assembly drawings in plan and elevation can be prepared for a permanent record of general layout arrangements of the entire plant from which construction and installation drawings can be detailed by draftsmen.

2. *Construction and Installation Drawings.* All parts of the plant are drawn in considerable detail to scale to show position of equipment and connections, foundations, supports, overhead structures, etc., so that every required item of plant equipment can be fitted into the desired arrangement by a construction crew. Drawings classified under this category are:

  a. Plot plans
  b. Foundation plans, including erection details
  c. Structure plans, including erection details
  d. Piping drawings (Figs. 9-2 to 9-6)
  e. Electrical drawings, largely schematic (see Figs. 8-20 and 9-23 for symbols)
  f. Instrumentation drawings, largely schematic (see Fig. 3-10 for symbols)

Drafting Procedures

Drafting of design requirements by means of drawings is a time-consuming operation and thereby expensive. The use of three-dimensional models of chemical plants has very effectively reduced drawing requirements, but many two-dimensional drawings are still required. A few words about efficient methods of planning and expediting drawings and reproductions would be in order by way of review.

Drawing Instruments. *Drafting machines* which combine the functions of T square, triangles, protractor, and scales are used exclusively in commercial drafting rooms. *Motor-driven rotating erasers* are now employed in combination with shields for correcting errors and making changes. Hand lettering is reduced to a minimum by use of *typing machines*, particularly for drawings where design notes and specifications are included with the drawing on one sheet. Many types of *templates* are available for drawing commonly used symbols.

Drafting Scales. The choice of correct scales and precision of measurement is a matter of experience. The best scale is one that presents a clear, graphic description within the precision of measurement required. Highly congested drawings should be avoided because of the additional time required to produce and to interpret the drawings.

It should be obvious that dimensioning requirements for machine-shop work is much more exact than the requirements for process plant construction drawings. In the latter case, center-line locations of principal process equipment must be specified by scale drawings to give a precision of ± 1 in. This can be achieved by a generally used scale of $\frac{3}{8}$ in. = 1 ft. The scale is increased to $\frac{1}{2}$ in. = 1 ft where more detail is required or where the drawing becomes unusually congested. For an over-all layout the scale is reduced to $\frac{1}{4}$ or $\frac{1}{8}$ in. = 1 ft. Once the equipment is located by this type of center-line scaling procedure, the working dimensions required for installation and repair procedures are shown directly on the drawings.

Reproduction Methods. Pencil drawings on a good-quality tracing paper can be directly reproduced as blueprints or by the newer processes giving black-, red-, yellow-, or blue-on-white prints. The latter method is being adopted throughout the industry since these prints are easier to read, as well as to reproduce and reduce by photographic procedures. The reproduction machines are smaller and portable and employ a fast dry-process procedure. Blueprints are still produced by drafting firms which have made a high investment in blueprint machines and cannot economically afford a change-over.

Use of High-speed Computers

There are many tools used through the entire development of a commercial process but none will contribute as much as high-speed computing machinery when properly utilized. For this reason a special section in this orientation chapter will stress the applications of this relatively new tool in the chemical industry.

High-speed computers consist of electronic circuits and mechanical parts coupled together in such a manner that nearly every conceivable operation in mathematics can be performed at very high speeds. Any process which can be reduced to mathematical language, i.e., addition, subtraction, multiplication, division, integration, and differentiation, is adaptable to and should be considered for high-speed computing. Many problems encountered in process development and design can be solved rapidly with a higher degree of completeness with high-speed computers at less money than was heretofore possible with slide rules and desk calculators. Overdesign and safety factors can be reduced with a substantial savings in capital investment.

Typical Applications. Some of the applications of high-speed computers in the development of a chemical process are listed next with specific reference to chapters in this book where such problems are encountered:

1. Statistical correlation of data—laboratory and pilot plant data for design purposes; plant data for quality control and trouble shooting (Chap. 3).

2. Preparation of process flow sheets—material and energy balances, particularly on recycle operations; ultimately can incorporate many process variables into a general design optimization (Chaps. 3, 4, and 6).

3. Selection of equipment—many equipment design problems require trial and error or complex mathematical solution; examples are absorption, distillation, humidification—dehumidification, evaporation, extraction, fluid flow, and heat exchange (Chap. 4).

4. Economic evaluation—solution of cost equations for optimization of process and equipment selection; this is also a part of items 2 and 3 (Chaps. 3, 4, and 6).

5. Piping design studies—study of piping forces to obtain points of maximum stress; proper location of pipe supports (Chap. 9).

6. Instrumentation and control—study of control loops for proper process control design and operation (Chap. 9).

7. Plant operations—operations research and linear programming of the commercial plant for maximum economic returns; cost accounting, payroll preparation, and other business procedures.

Types of Computers. High-speed computing machinery can be divided conveniently into two types based on their mode of operation. The *digital computer* is a very fast calculator performing basically addition and subtraction with multiplication and division possible by a repetitive series of basic operations. Such machines are expensive to install and program via punch cards, are more precise, capable of repetitive operation on the same type of problem, and produce solutions in tabular form on typed sheets. The cost of the digital computer depends on the speed of operation and the number of its memory units or spaces where numerical information can be filed for future sequential mathematical processing. Small units capable of solving simple problems cost around $50,000, while large machines will require an expenditure of $500,000 or more. In some cases it is possible to rent the larger machines.

Analog computers are essentially electronic amplifying circuits capable of solving differential equations without the use of numerical methods required for digital work. The effect of a complex set of variables on a process can be studied without an expensive card programming. The resultant electronic output data as the answer can be automatically plotted for a permanent record. The inherent accuracy of this type of machine is not as good as that of the digital. It ranges from 0.1 to 2 per cent, depending on the electronic components, but the analog computer is usually less expensive to purchase and operate than a digital computer. For example, a kit can be purchased and assembled for student training for less than $2,000. Analog computers also have the disadvantage of lack of memory storage required for optimization studies.

Engineers should become familiar with the capabilities of these machines and use them to the utmost. It is only then that the large investment in this tool for the chemical industry will pay off. Most colleges now have both digital and analog equipment so that the instructor in plant design should have the class solve a few simple problems by this means. Further study references in this area are listed in the Additional Selected References at the back of the book.

PROBLEMS

1-1. Design of Chemical Stoneware Absorption Tower for SO_2 Absorption System

Specifications
1. Scrubbing length, 45 ft; 30-in.-diameter tower sections
2. Packing:
    a. One-third diaphragm, 4-in. size
    b. One-third spiral, 3-in. size
    c. One-third raschig rings, 1-in. size
3. Stoneware aspirator to pull gases through tower
4. Lantern consisting of Pyrex brand glass-pipe section to be fitted into system at convenient and accessible level, between tower and aspirator
5. Tower saucer on bottom and tower cover on top, distributor on top section
6. Tower to rest on concrete base, built upon floor
7. Sufficient supporting plates to carry the packing rings
8. Tower to be supported by wooden scaffolding
9. Scaffold risers to be made of 1- by 8-in. boards, crosspieces of 2 by 4 in.; supports to be properly cross-braced
10. Pump to transfer water at 25°C to top of tower at rate of 1,200 gph, providing a 1 per cent solution of product at base of tower

Sketches required
1. Sketch layout and setup
2. One detailed sketch of tower showing arrangement of sections of tower; break sections to show nature of packing
3. One detailed sketch of tower showing arrangement of supporting structure

1-2. Design the Installation of a Stainless-steel Sulfonator

Specifications
1. Select spot location for equipment.
2. Draw necessary sketches (8½ by 11-in. paper).
3. Services must be obtained from master valves and switch boxes as exist near spot location.
4. Prepare *complete* bill of materials for the installations, including supports, drainage, and all utilities.

CHAPTER 2

Development of the Project

A chemical engineering plant project may have any one of the following objectives:[1] (1) the design and erection of a new plant, (2) the design and erection of an addition to an existing plant, or (3) the revamping and modernizing of an existing plant. The project is born the moment company executives decide real thought should be given to one of the foregoing objectives and someone is directed to investigate it. A preliminary study is made and then, if it seems warranted, more detailed analyses are completed. This gives management enough facts to make a reliable decision as to whether or not appropriation of time, men, and money for the execution of the project should be authorized.

Authorization for the initiation of a development project is made by management, usually after a study of the recommendations and the development budget submitted by the development group. The exact level at which the authorization is granted differs among companies and is according to the size of the predicted expenditure. Some development departments or steering groups have funds at their disposal with which to pursue general development work or the development of projects within specific areas of the company's activities.

A clear and concise statement of the project with specifications for the plant process, all laboratory and pilot plant data, and any other pertinent chemical or engineering facts must be presented to the design group before the study is begun. The design group must integrate the technical factors with economics. Sometimes, the design study will progress to the point where more data must be obtained to complete a reliable design. The research and development group must then cooper-

[1] D. Gordon, Project Engineering, *Chem. Eng.*, **57**(3): 125 (1950).

ate in getting this data. The development of a project from its inception to a manufacturing operation requires complete integration of facilities and personnel (see Frontispiece). The relationships of various stages in this development are listed next.

EVOLUTION OF A PROCESS

A logical evolution of a process may proceed via the following stages:
1. Process research—library and laboratory work
2. Research evaluation for possible commercialization
3. Process development
4. Preliminary engineering studies
5. Pilot plant
6. Semicommercial plant
7. Commercial plant

Many of these stages may be carried out simultaneously to expedite the work. This is particularly true of the first five items, since the budget for such work is generally small compared to the cost of a commercial plant.

A very important part of successful project development is the organization of personnel within the company to handle the various aspects of the job. Figure 2-1 shows a typical organization chart for industrial research. As the project develops, it is essential to obtain suggestions and criticisms from personnel in various departments from time to time. This can be done by conferences and/or circulation of progress reports. Communications problems must be solved for effective cooperation. The proper method to employ depends on the size and physical setup of the company.[1] As a further illustration of the requirements for cooperation and communications, Table 2-1 shows how each of seven departments was involved during various phases of the development of a chemical product.

Process Research

The conception of an idea may be originated by a chemical engineer, a chemist, a physicist, or any other person. It may be proved sound by the use of existing data, but frequently chemical research must be carried out to provide a more quantitative basis for evaluating the economic feasibility of the process. In general, the object of process research is to find out by library survey and laboratory work if the product can be made and what the yields and rates of conversion are.

The practical objective of process research is to provide scientific data that will permit the rational design of a manufacturing process with the

[1] "Communications," AIChE Symposium Series 49-8, 1953.

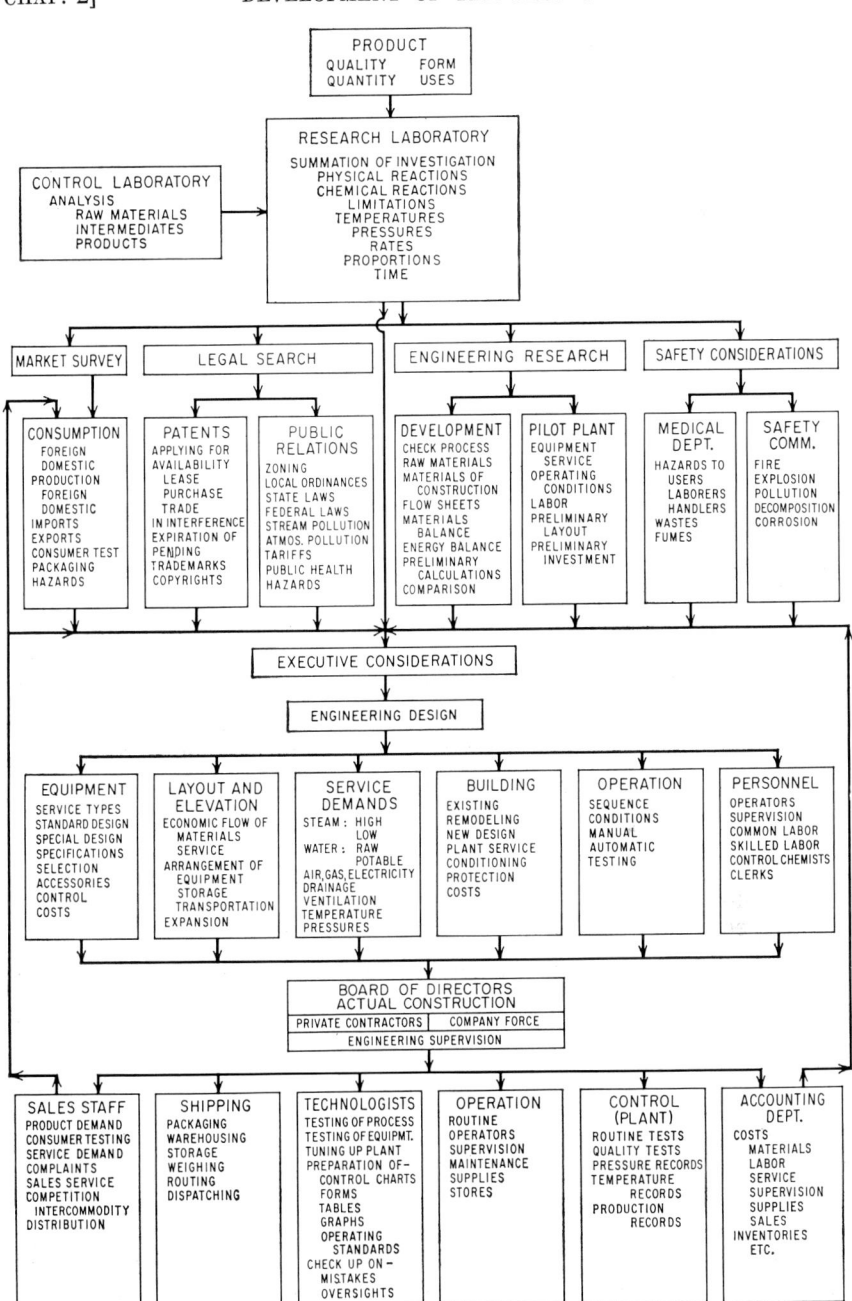

Fig. 2-1. Organization check plan. (*Courtesy of Department of Chemical Engineering, Virginia Polytechnic Institute.*)

TABLE 2-1. DEVELOPMENT OF A CHEMICAL PRODUCT*

Stage of development	Research	Engineering	Patent group	Survey group	Product group	Construction	Production
Literature search	In charge	....	Search art				
Bench scale	In charge	Follow	Patent study	Information	Market study	Cost study	
Intermediate scale	In charge	Follow	Preliminary stage				
Pilot plant design	Advise	In charge	....	Preliminary	Preliminary evaluation	Preliminary estimate	
construction	Advise	Assistant adviser	....	....	....	In charge	
operation	In charge	Advise	....	....	Distribution samples		
Commercial design	Advise	In charge	Final stage	Final survey	Market evaluation	Cost estimate	Follow
construction	Advise	Assistant adviser	....	....	Market program	In charge	Follow
operation	Advise	Assistant adviser	....	....	Assistant adviser	Assistant adviser	In charge

* Nitrogen Division, Solvay Process Co., *Chem. Inds.*, **59**(6): 1000 (1948).

minimum time and equipment spent in pilot plant studies and to operate the final process under the most favorable and economical conditions with respect to all variables such as feed composition, space velocities, recycling, heat and power requirements, and temperature gradients. The objective is to reduce the time and expense of translating a process from an idea to full-size plant construction. Statistical planning of experiments[1,2] in a projected process will greatly assist in the reduction of time.

Research Evaluation

The chemical engineers' contact with the project may begin long before process development and pilot plant work is started. The first stage for many projects is an engineering and economic analysis of available

[1] O. L. Davies, "Design and Analysis of Industrial Experiments," Hafner Press, New York, 1954.

[2] W. L. Gore, "Statistical Methods for Chemical Experiments," Interscience Publishers, Inc., New York, 1952.

data before any laboratory work is attempted. The object of this analysis is to determine the potentialities of a project for further research and development work and for eventual exploitation. *In many cases, a comparison of raw material and finished product prices may rule out a process.*

The second stage is reached after some or all of the initially planned laboratory work has been completed. At this point enough information is available for a satisfactory material balance and for a preliminary analysis of the unit operations and process involved, although several assumptions may be necessary to make up for the data that are lacking. An engineering evaluation of projects at the research stage is designed to (1) make an economic analysis of project, (2) outline unit operations and chemical processes involved in projected manufacturing operations, and (3) show what additional information will be needed to complete design of the process and equipment for a plant. The evaluation completed at this stage should provide the basis for a decision in regard to the advisability of going ahead with additional laboratory work or with process development and pilot plant work.

The principal reason for preparing engineering and economic evaluations of projects at the research stage is to detect, as early as possible, projects that are economically unsound or impractical. In case two or more processes for the manufacture of a chemical are feasible, evaluation may show which of the processes is preferred so that the research work can be concentrated on this process. The engineering and economic analysis should begin as soon as a reasonably reliable material balance is available and the rough details of the processing and purification steps are outlined. Through circulation of the written report on the evaluation, other chemical engineers and chemists may become aware of the engineering and design problems and the combined knowledge of several individuals may be focused on the project or projects. If the engineering and economic evaluations show that the project is an interesting one, items in the manufacturing cost can be ascertained and methods for reducing them suggested. The engineer may find that some additional information needed can be obtained most advantageously before laboratory work is halted.

Research and Development Requirements

The research evaluation has pointed out the necessary process development studies that should be conducted to obtain process design information from the research laboratory and the pilot plant. Reasonably accurate material and energy balances must be available in addition to physical properties of the materials being handled.[1] In general, this process research group will submit values for such properties as:

[1] L. Friend, Tools for Process Design, *Chem. Eng. Progr.*, **44**: 253 (1943).

1. Molecular weight
2. Boiling point
3. Melting point
4. Vapor pressure
5. Critical temperature
6. Critical pressure
7. Specific gravity of solid, liquid, or vapor
8. Enthalpy relationships
9. Liquid-vapor relationships of mixtures
10. Temperature-entropy relationships
11. Thermal conductivities
12. Viscosity

Preliminary Engineering Studies

In order to operate the process being considered, physical equipment and facilities must be specified. The research laboratory can work with glass apparatus or small-scale improvised equipment. The commercial plant will require large-scale equipment capable of operating continuously without undue maintenance. Furthermore, the materials of construction will generally not be glass, but rather metals, ceramics, or glass-lined steels. The engineering studies may be directed toward the development of unusual equipment which is not commercially available and the testing of this equipment as well as standard items in a completely integrated process. This is one of the functions of a pilot plant.

Pilot Plant

Pilot or prototype plants are complete, medium-scale processing units containing all essential product-producing elements, including control.

Pilot-scale equipment falls into either or both of the following two categories: (1) capable of producing results translatable, according to supplied instructions, into full-scale design and operational data, or (2) designed and constructed to permit a much wider range of operating characteristics than is normally available in production models.

The conversion of laboratory data handed down from the research group into plant design data is only one function of the pilot plant. Such a task requires the setting up of a definite program, including a thorough investigation of basic reactions and reactants, time, temperature, concentration, and catalysis factors, a study of raw materials, operations needed, control specifications, and safety and health hazards. The process must be investigated with the thoroughness and zeal of a pure researcher but from the viewpoint of a chemical engineer. The pilot plant in such cases is a research unit. It must be used for the selection of suitable equipment and materials, provide time and labor study information, and enable a study of by-product recovery and waste-disposal problems. After a new plant or process is designed, the pilot plant continues investigations of problems that require the elimination of the earlier compromise acceptance of data.

Operating processes use the pilot plant as the trouble-shooting division; when trouble shooting is not being carried on, the pilot plant is doing development work on lines of alternative raw materials, improvement of products and by-products, lower costs, safety, bringing the plant up to date, etc. The proper personnel in a pilot plant constitutes the most important feature of its success or failure.

A check list[1] is suggested as including all items to be considered in the pilot plant investigation.

1. Flow relations:
   a. Chemical flow diagrams
   b. Breakdown into unit operations
   c. Engineering equipment flow diagrams
   d. Material balance
2. Materials:
   a. Raw materials, availability, substitute raw materials, costs
   b. Impurities in raw materials and in products
   c. Corrosion, erosion, dust, fumes
   d. Solvents
   e. Wastes and recovery
3. Equipment or operation:
   a. Selection of equipment, elimination of obviously unsuitable equipment
   b. Cost of operation
   c. Control specification
   d. Material of construction
   e. Heat transfer
   f. Mass transfer
4. Materials handling:
   a. Proper methods of handling around the plant
   b. Intermediate storage
   c. Industrial hazards (corrosion, fire, erosion, safety, health, pollution, fumes, explosions)
   d. Public nuisances
   e. Storage
5. Labor:
   a. Operators needed
   b. Supervision
   c. Control specifications from operator's viewpoint
   d. Process simplification from operator's viewpoint
   e. Safety from operator's viewpoint
   f. Saving of time and labor

There is usually a change of personnel when the process is transferred from the laboratory to the pilot plant. Chemical engineers often form

[1] F. C. Vilbrandt, *Ind. Eng. Chem.*, **37**: 419 (1945).

an integral part of the group operating and observing pilot plant operations. However, in a few companies the pilot plant group is headed by the research chemist or engineer previously in charge of the laboratory research, since he is most familiar with all the details of the process. The study of the product and process by chemical engineers constitutes an evaluation from a different point of view from that of the research chemists. Many executives emphasize the importance of the pilot plant for this reason.

The ultimate desire is to operate the pilot plant with the assurance that all the risks, both technical and economic, in the full-scale commercial plant have been minimized or, preferably, eliminated. The pilot plant must be capable of operating over relatively long periods under conditions that are not changed frequently, to obtain a fair approximation of labor costs and manufacturing expenses. In a pilot plant the pieces of equipment are selected specifically for the work to be performed. It should not be dismantled until such times as the full-scale commercial plant is in successful operation, because it is the place where quality and manufacturing improvements are worked out.

A practice that is quite prevalent during critical times is for the engineering department to bypass the pilot plant in order to expedite the design of the commercial unit, and sometimes actually to build the commercial unit before the pilot plant is erected. A summary of scale-up practice with or without piloting is listed in Table 2-2. The concept is held that a well-trained engineering force, with experience and chemical engineering pilot plant data available from other processes, is able to translate laboratory data into plant practice and thus avoid the delay attendant upon the completion of a pilot plant study. This is a reasonable procedure where a new process is similar in most respects to an established manufacturing process and one in which the operating data are already sufficient. In such cases the best method for materials handling, illumination, depreciation, maintenance, etc., can be fairly accurately predicted from experience and data obtained in the existing plant. The information needed, however, may be the peculiar difficulties which did not show up in the old process and which require solution before passing to the next stage.

Control and Instrumentation. (See Chap. 9.) When pilot plant design and operation are part of process development, it is particularly important to stress control problems. The requirements of process control are (1) material balance, (2) energy balance, (3) conditions of quality, and (4) conditions of economy. In general, a control engineer may study a flow diagram to see whether the first two conditions are met without too specific knowledge of the process, but to meet the second two conditions requires specialized process operating information.

TABLE 2-2. SUMMARY OF SCALE-UP AND PILOTING PRACTICES

(1) Type of equipment or process	(2) Is large-scale piloting necessary?	(3) Controlling conditions for design	(4) Principal size- or capacity-controlling variable	(5) Average scale-up ratio allowable based on (4)	(6) Average safety factor used in over-design of operating performance, %
Cooling towers*	No	Air humidity, temp. difference	Input rate	>100:1	15
Crystallizers	Yes	Rate of heat transfer, degree of mixing	Volume	10:1	20
Dryers	No	Gas humidity, temperature, surface area:volume ratio	Input rate	>100:1	20
Evaporators	Yes	Latent heat of vaporization, rate of heat transfer	Volume	25:1	20
Fluid-bed columns	Yes	Solids and gas densities	Input rate	10:1	25
Heat exchangers, shell-and-tube*	No	Temperature, fluid velocities	Heat-transfer area	>100:1	15
Mechanical separations:					
Solids from solids	No	Relative physical properties	Input rate	>100:1	10
Solids from liquids					
Centrifuges	No	Particle size, viscosity, and surface tension, relative densities	Input rate	>100:1	15
Filters	No	Cake resistance	Filtering area	>100:1	15
Solids from gases	No	Particle size	Input gas rate	10:1	10
Mixing:					
Solids	No	Particle size	Volume	>100:1	10
Liquids	No	Viscosity, surface tension, solubility, densities	Volume	15:1	20
Liquids-solids	Yes	Particle size, viscosity, surface tension, relative densities	Volume	10:1	20
Packed and plate columns*	No	Equilibrium data Vapor-liquid flow	Input rate Diameter	>100:1 10:1	15
Pumps*	No	Discharge head	Input rate and power Impeller size	>100:1 10:1	10
Reciprocating compressor*	No	Compression ratio	Input rate and power Piston displacement	>100:1 10:1	10
Spray condenser*	No	Latent heat of vaporization	Input rate Height-to-diameter ratio	70:1 12:1	25
Size reduction	Yes	Final particle size	Input rate and power	50:1	20
Materials handling	No	Bulk density	Volume	70:1	20

* Information compiled by M. Laurent, R. D. Beattie, and T. H. Goodgame, *Chem. Eng. Progr.*, **50**(7): 333 (1954).

Procedure for control analysis will vary with the process but the following fundamental steps should be kept in mind: (1) listing all the variables, (2) defining the magnitude, source, and timing of changes in the variables, (3) selecting the significant variables, (4) eliminating the variables subject to measurement, (5) selecting the variables susceptible to control, (6) deciding on primary variables for control, and (7) selecting important secondary variables.

The incorporation of process instrumentation should start at the beginning of process development. The pilot plant serves a very important purpose of testing instrumentation and automatic control. In turn, the instrumentation provides the means for obtaining process design data.

Pilot Plant Costs. Pilot plant costs constitute one of the largest items of expense in research budgets; the total expense of building and running them is usually charged to cost accounts. To find the cost of the pilot plant investigation, the chemical engineer must prepare a preliminary plan of investigation that will provide all the data he needs for detailed process design of a full-scale commercial unit. Also, a process and equipment design and estimate of investment cost for the proposed pilot plant must be prepared. Based on the preliminary plan of pilot plant investigation, an estimate is made of the time that will be required to carry out the projected pilot plant program, the total operating cost of which can then be estimated. This cost, together with the estimated investment cost of the proposed pilot plant, represents the total probable cost of pilot plant investigation. If this figure is relatively high compared with the cost of a commercial unit, the decision may be to avoid completely a pilot plant investigation; instead, a commercial unit may be installed directly on the basis of chemical research data. In this case, the laboratory research work should be more intensive than where a pilot plant investigation is carried out.

Semicommercial Plant

The semicommercial plant, larger than the pilot plant, has as its primary purpose the production of sufficient quantities of the new chemical to permit sales in small lots. However, it is still experimental, and its purposes include all those listed under the pilot plant. This size unit is more often bypassed. A distinction is sometimes made on the basis of new chemicals already made by competitors and those which are entirely new and must be introduced to the market. For the former, a pilot plant just large enough for experimental purposes is used. For entirely new products, a semicommercial plant is designed larger than necessary for a purely experimental pilot plant and yet large enough to permit fairly economical production, so that the product can be sold at the cost of manufacture. This permits the company to "break even" on the costs of operation.

The output of the semicommercial plant is introduced to the market by a staff of market development men who specialize in such work. As the demand grows, the semicommercial plant is operated at an increasing percentage of its capacity. When full capacity is reached, preparations are made to transfer the production to a commercial plant.

Commercial Plant

If the process can survive the foregoing exacting tests of operation and if estimates indicate that the production cost will be sufficiently low, the last and final stage of development—the full-sized commercial plant—may be carried out with the assurance that all the risks, both technical and economic, have been minimized. The size of the plant will depend upon the requirements set forth in the original demands for the design of the plant.

This final step is the coordination of all chemical and engineering data obtained and their translation into a detailed commercial plant design. Access must be had to trade literature and handbooks for selection of types and specific pieces of equipment. Capacities and performance are studied. Preliminary layouts are attempted, and the best flow arrangements obtained. Organization of the equipment by means of a template and study models gives a picture of the possibilities of different layouts. After arriving at the most desirable layout, the actual drawing of the plan and elevation of the assembly is undertaken, followed by accurate preconstruction costing. In order to design a commercial unit, including housing for the production of the specified commodity, the following considerations are important:

1. Specifications of equipment
2. Specifications of materials
3. Selection of commercial equipment
4. Plan
5. Elevation
6. Location of plant
7. Operating instruction for labor
8. Selection of personnel
9. Preconstruction costing
10. Production costs per unit of material

There are numerous factors which must be considered under each item listed above. These are discussed next.

It is of considerable importance to a careful survey to stress the factors that will play an important role, not only in the design itself but also in the construction and operation of the chemical plant. Plant design not only must be technically satisfactory but also must be economically satisfactory; the goal of the design is to secure a workable plant with the maximum return on the necessary investment. Any plant design must

26   CHEMICAL ENGINEERING PLANT DESIGN   [CHAP. 2

also consider the safety factors not only for the sake of its workmen but also for the public at large, the equipment, the plant, and the product. The general wheel outline as presented in Fig. 2-2 brings out an interrelation of the various factors. Each individual design of a chemical plant

FIG. 2-2. Plant design factors.

is usually a highly specialized case, and a detailed generalization scheme cannot be outlined that will be in its entirety a thorough coverage of the requirements of all chemical plant design problems.

TECHNICAL FACTORS

Markets

Market surveys are generally made by the trade or market survey department or division of the company, or they can be obtained from special trade survey consultants or organizations that have been established in various parts of the country to provide complete reports on any commodity, its uses, forms, quality, quantities, availability to markets, import and export data, tariffs, trade agreements, consumer testing consumption, production, and distribution. This applies not only to finished products but likewise to raw materials and by-products.

Flow Diagrams

To the design engineer a flow sheet of equipment and materials in process is considered the first clarifying step. This is a transposition of the research and development laboratory notes and reports into the terminology of the engineer. The flow diagrams present a picture of material and energy flows, process operations, and equipment, materials handling, storage, future plant expansion, and water, power, and fuel requirements. From this picturization the departments, the possible sequence and number of units required, and the distribution of labor are evident. The materials balance and an energy balance are worked out, and the quantitative interrelationships are then presented on the flow sheets (see Chap. 3).

Equipment

Performance and service are demanded from all equipment. Much valuable information for the selection of equipment is available from manufacturers of equipment. Much of the equipment for materials handling and for unit operations and processes is standardized and, whenever such equipment serves the purpose, it is selected in preference to special designs, thus substantially lowering the cost and providing for ready duplication of equipment and availability of repair parts. Large companies have organized their own set of standards to avoid repeated design costs on items of routine nature. One should not hesitate to meet any problem that requires a special design and the use of special materials even if it is considered that a new design is an experiment for the manufacturer and the user. The changes in demands and services for commodities sooner or later lead into pioneer fields of equipment design. Manufacturers hesitate to use a new material on a standard design, since either a change in process equipment to meet a satisfactory compromise with the manufacturer or an exhaustive pilot plant study must be undertaken.

Plant Layout

Plant layout or economic arrangement of equipment follows selection of type of equipment. Arrangements of equipment even for the same process are varied and generally are an expression of some architectural inclinations of the designer, not always resulting in as economical and practical organization as would be possible if layouts and arrangements of similar or allied plants as published in current literature were studied and improvements made upon such layouts, if changes are deemed necessary. Accessories and control devices are essential to effective operation of equipment and flow of process. The operation of a proposed piece of equipment or process is studied carefully. Frequently the proper arrangement of equipment effects a material saving in labor (see Chap. 5).

Buildings

The chemical processing and the materials handled govern the general design requirements of buildings. Careful attention is given to the arrangement and layout of the equipment, and then a building is considered as surrounding all this assembly or only such portions as require housing. In chemical buildings, special attention is given to foundations for building and equipment, sanitation and plumbing, the type of floor, structural frame, walls, roof, fume handling, explosion possibilities, lighting, ventilation, drainage, heating, air conditioning, fire protection, and power-plant orientation. The types of buildings and the service requirements for each can be supplied by the industrial building manufacturers. The building serves as a protective cloak to be used as shelter for equipment or operators (see Chap. 8).

Location

In general, the following items are considered vital in plant location: proximity to market, raw materials, transportation, labor supply, water supply, power supply, economic interrelation with other industries, and specific plant requirements. There are other plant and process location matters that are of real importance to, and exist as real responsibilities of, the chemical plant design engineer of commercial plants for the manufacture of chemicals and chemical formulations, such as land, local ordinances, public improvements, utilities, waste disposal, and climatic conditions. All the factors that enter into the problem of plant or industry location also affect the choice of local sites and must be considered by the design engineer (see Chap. 7).

ECONOMIC FACTORS

Economic Contribution

The success of the chemical engineer is directly related to his economic contribution to an enterprise. The engineer will attain success who frequently effects a significant reduction in cost of production, makes a new design at lower cost, obtains an increase in production at a profit, leads the way to an increase in the yield of the more valuable products, shows how to produce a new product at a profit, makes a profitable trade, or obtains a valuable patent.

The chemical engineer in practice must familiarize himself with the economic status of the enterprise with which he is connected. This will enable him to determine the most likely possibilities for increasing the earnings of the business. If he has knowledge of the processing costs of the manufacturing plant, he will be in a position to determine these costs which, if reduced, would result in a large annual saving to the company.

Preliminary Process Appraisal

The process appraisal usually consists of a preliminary process design of a commercial plant of suitable size and an estimate of investment cost, processing cost, and profit. This is of a preliminary nature, for it involves extrapolation of the results obtained on small laboratory equipment to full-scale commercial equipment. Much judgment, based upon a background of experience, must be exercised by the chemical engineer in regard to what commercial performance can be expected. Experience provides a basis for comparison of commercial performance with process-research results for other processes and is almost essential for this work. Rarely does commercial performance equal that obtained on small laboratory equipment; with increase in size, the optimum operating condition, as determined for the small plant, may not hold true. Consequently, flexibility in design is still desirable for a large unit. In the extrapolation of research data to full-scale commercial plants, it is frequently necessary to set up the process design for a continuous unit on the basis of chemical research data that were obtained with batch equipment. The translation of data from batch equipment to a basis for a continuous unit is indeed dangerous, and unless the chemical engineer stays on the solid foundation of fundamentals, he is likely to arrive at a process design for the continuous unit that is far different from what actually is feasible. If possible, the economic appraisal should be based on pilot plant performance rather than laboratory scale work.

Costs and Profitability Analysis

A compilation of all data relative to the cost of raw materials, land, buildings, labor and supervision, equipment, legal fees, taxes, insurance, interest, etc., should be obtained by the designer on an accurate preconstruction cost estimating basis as a forerunner to actual operation cost accounting if the project goes commercial. A preconstruction cost estimation report can be issued, showing what possibilities exist for profits and earnings under proper management, even before a commercial plant investment is made. Cost data are only current and must be modified as price conditions vary, but in competent hands, current cost data are useful and serve their purpose—that of indicating possible profit or of stopping further expenditure if the venture labels itself as uneconomic. Chapter 6 contains a detailed exposition on preconstruction cost estimation.

The general procedure for preparing economic evaluations consists in estimating the following: (1) total manufacturing cost, (2) total capital investment, (3) estimation of a selling price, and (4) return on investment. These give definite information on possible return on total capital investment after accounting for all costs involved, or conversely, the return that can be realized from selling the product at a fixed price.

Costs in Safety. The hazards present are a direct function of the fire insurance rates that can be obtained. A decrease of the hazards may effect important savings in the insurance items of the overhead costs. In this connection, the cost of a complete automatic sprinkler system may be entirely defrayed by the resulting decrease in fire insurance rates. Such protective schemes and equipment are not so easily paid for, but they frequently can be justified on economic grounds alone. Hazards also involve loss of production and men's services as well as impairment of product quality. Reduction of hazards thus becomes an economic problem.

SAFETY CONSIDERATIONS

It is important that engineers continue to recognize safety and fire hazards of the processes they are operating. From every standpoint, these hazards are liabilities, and their actual and potential costs must be emphasized in process thoughts and plans. The chemical industries today, through their medical, safety, and fire-protection departments, have intensively investigated hazards of all kinds, and much information and data thereon are available. There is every incentive and a real necessity for including a survey of safety and fire hazards in a study of chemical engineering processes. This text emphasizes this approach in every chapter. An excellent treatment of safety and fire prevention is given in Perry's "Chemical Engineers' Handbook," 3d ed., sec. 30, pp. 1847–1884. Some of the important safety considerations in the chemical industries are summarized next.

Building and Process Equipment

Rational plant design is concerned with safety factors and with the need for minimizing such building and equipment hazards as corrosion, fire, explosion, and personal hazards from fume and poison. Process leaks and spillage hazards, hazards due to poor lighting, and reduction of corrosion through the selection of proper materials come under the jurisdiction of safe practices. The relation of equipment hazards to personal hazards is self-evident, and proper design considers not only process flow but the course of action of the operators and other personnel in a plant. Also, safety must be considered when dealing with the disposal of wastes as affecting persons outside the jurisdiction of the plants. The effect of zoning ordinances and other legal restrictions to operations cannot be minimized. No matter how highly satisfactory a plant design may be from the technical and economic viewpoint, disregard of safety, air pollution, and waste-disposal problems will nullify an otherwise sound engineering plant design.

TABLE 2-3. SAFETY REGULATORY GROUPS AND INFORMATION SOURCES

1. American Petroleum Institute, 50 W. 50th St., New York.
2. American Society for Testing Materials, 1916 Race St., Philadelphia.
3. American Society of Mechanical Engineers, 29 W. 39th St., New York.
4. American Standards Association, 70 E. 45th St., New York.
5. Associated Factory Mutual Insurance Companies, 1151 Providence Highway, Boston.
6. Bureau of Explosives, Washington.
7. Interstate Commerce Commission, Constitution Ave. and 12th St., N.W., Washington.
8. Local inspection bureaus.
9. Local and state regulations.
10. Manufacturing Chemists' Association, 1625 Eye St., N.W., Washington.
11. National Board of Fire Underwriters, 85 John St., New York.
12. National Bureau of Standards, Washington.
13. National Fire Protection Association, 60 Batterymarch St., Boston.
14. National Safety Council, 425 N. Michigan Blvd., Chicago.
15. Underwriters' Laboratories, Inc., 207 E. Ohio St., Chicago.
16. U.S. Bureau of Standards, Washington.
17. U.S. Defense Department, Washington.
18. U.S. Department of Labor, Washington.
19. Various material manufacturers' association.

Fire Prevention and Control[1]

The objectives of fire prevention or protection engineering are to minimize opportunities for personal injury, loss of life, property damage, and production interruptions. These objectives are achieved by (1) prevention, (2) control, and (3) extinguishment of fire. The term "fire prevention" applies to that phase of process design which minimizes fire hazards inherent in the process. "Fire control" refers to that phase of process design which seeks to control and protect against fires which have already been started, until available extinguishing forces can become effective.

[1] J. J. Duggan, Designed Fire Prevention, *Chem. Eng.*, **58**(6): 125 (1951).

It is of prime importance in process design to recognize and incorporate the minimum safe practices prescribed by nationally recognized fire protection associations, engineering authorities, and governing bodies. Table 2-3 lists a number of these. Such recommendations and regulations should be followed as closely as possible, and in every case supplemented as indicated by experience (see Chap. 8).

Mechanical Hazards

Some specific sources of literature pertaining to the standard safety codes on machinery and tools are also found in Table 2-3; see items 4, 6, 7, 8, 9, 12, 14, and 18. A well-designed machine must be equipped with safety guards. Visual warnings of danger, such as signs and color schemes, have become extremely useful methods of combating potential hazards. Some of the safety measures which must be taken when using hand tools are (1) nonsubstitution of one tool for another when each has a definite specified function, (2) guarding and sheathing sharp-bladed tools, (3) insulation of electrical hand tools, and (4) use of antispark tools in explosive or inflammable areas. Safe practices for power tools are the same as for machinery.

Electrical Hazards

Accidents attributed to electrical hazards are (1) shocks by alternating current and burns by direct current due to poor indication of and protection from high voltage, (2) faulty or poor wiring, (3) insufficient care of equipment as regards dryness, cleanliness, and operation, (4) fires from overloaded circuits not properly equipped with fuses or circuit breakers, (5) fires from capacitor discharge because of improper maintenance, (6) fires from sparking or arcing of switches or brushes near inflammable materials, and (7) static electricity discharges. Preventive measures against shocks and burns are the proper enclosure of high-voltage equipment and discharge capacitors, indication of accessible voltage sources, proper maintenance of wiring and equipment, capable personnel, and good housekeeping. Fires can be prevented by proper design of electrical and chemical equipment with all electric circuits designed according to an approved code (see Chap. 9).

Chemical Hazards[1]

The most important manuals of safety for chemicals are the Chemical Safety Data Sheets, compiled by the Manufacturing Chemists' Association. Chemical Safety Data Sheets discuss the safe handling of most hazardous chemicals; drawings, tables of data, and graphs are given to aid the reader in understanding the material presented. Safe handling

[1] A. B. Steele and J. J. Duggan, Handling Reactive Chemicals, *Chem. Eng.,* **66**(8): 157 (1959).

of steel drums, fiber drums, paper sacks, glass carboys, and other containers is presented in the data sheet. Unloading and loading of chemicals shipped in tank cars and trucks is also discussed in these sheets. The Interstate Commerce Commission has issued regulations for shipping most of the hazardous chemicals. Also, a handbook of dangerous materials is available with an alphabetical listing of 5,000 dangerous materials.[1] The hazardous properties, toxicity, treatment, antidote, effective extinguishers, storage and handling, personnel safety precautions, physical properties, description, shipping regulations, and labeling instructions of the material are presented. Other material presented is the discussion of explosives, fungus and fungicides, radiation and radiation effects, and ICC requirements for shipping of all types of materials.

One of the most common occupational diseases is lead poisoning; zinc poisoning is also prevalent but of much shorter duration. Recent Atomic Energy Commission activities have added a very dangerous material to the list, namely, beryllium and its compounds.

Ventilation

One of the major functions in most plant surveys is a study of ventilating systems, both local and general, which are used to control the health hazard. By *local exhaust ventilation* the air contaminant is removed at the place of generation before it comes into the worker's breathing zone. Often in general ventilating systems, incorrect types of fans are installed, velocities are too low, the wrong type of hood is used, or there are no air inlets to the room. Sometimes it may be necessary to redesign the whole ventilating system. In any case, industrial hygienists in the Public Health Departments are available to assist with these problems (see Chap. 8).

Labor Relations in Safety

Most public and private enterprises are cognizant of their moral and ethical responsibilities toward those who have placed their health, welfare, and livelihood in their hands. Safety hazards are potential deterrents to attainment of optimum technical efficiencies and product quality. If any hazards are known and if proper safeguards and protection are not provided, the psychology of the operator will frequently be such that his attention will be drawn thereto and, to that extent, withdrawn from his immediate duties. As his attention is taken from his real job and duties, his efficiency and, therefore, the efficiency of his operation will decrease, resulting in the improper discharge of his duties, and, through improper washing, filtration, drying, heating, or any other of the more or less

[1] N. I. Sax (ed.), "Dangerous Properties of Industrial Materials," Reinhold Publishing Corporation, New York, 1957.

essential operations of a chemical process, may result in an unsatisfactory product.

A study of the cause of injuries in 50,000 major and minor industrial injury cases reveals that 30 per cent are due to faulty instruction, 22 per cent to inattention, 14 per cent to unsafe practice, and 12 per cent to poor discipline, whereas 12 per cent are due to inability, physical and mental unfitness of the employee, and only 10 per cent to mechanical hazards.

If the man or men who have been specially trained for a particular job or jobs are temporarily or permanently removed from the operation as a result of ill health, accidents, or fires, the process suffers. Therefore, the costs of an operation or process may be increased appreciably in a large number of ways, particularly as regards the items mentioned above: loss of production, impairment of product quality, insurance costs, disability claims, loss of service of trained men, and the costs of training other men.

Health Hazards

In almost every state there is some sort of board, committee, or bureau set up by the state government to work hand in hand with the industries of that state in combating the numerous health hazards that might exist in the industries. Upon request, surveys and detailed studies are made of industrial plants in the state to determine potential health hazards. Requests for services may be made by plant management, labor unions, local health authorities, and individuals. Industrial hygiene activities prevent occupational health hazards by controlling dusts, gases, vapors, mists, X rays, and radioactive radiation. (Obtain maximum permissible limit values from local or state regulatory boards.[1]) These activities also prevent and control circumstances which produce fatigue, such as excessive noise, inadequate ventilation, poor lighting, and excessive heat and humidity. Unpleasant and excessive temperatures, humidities, noises, and radiation affect workers. Improper lighting can cause eye strain, headaches, and irritability. Excessive noise is objectionable and may cause hearing defects. During the past few years, X-ray radiation has become important. Industrial X-ray machines are now being used which have as much as 2,000,000-volt capacities. Radium is employed to detect defects in metal, and radioactive materials are used for research purposes. These present serious health hazards. While most chemical engineers are aware of some of these hazards, their critical relation to employee productivity is not always apparent.

Radiation Hazards. The sources of radiation hazards are alpha, beta, gamma, and X rays, thermal neutrons, and fast neutrons. Danger comes

[1] For proximate design data, see *ibid.*

from ingestion or external radiation. For a complete discussion of these hazards, see Chap. 10.

Sanitation. Sanitation in the plant as affecting conditions outside the plant is a process problem, and care must be taken in considering (1) the potable water supply and its protection from contamination by the plant or outside sources, (2) sanitary plumbing, (3) drainage, and (4) waste and sewage disposal.

LEGAL PHASES

Patents

The patent situation pertinent to any product, process, equipment, use, or application of any commodity should be considered by the legal department concurrently with the design. The commodity or processing for the commodity may be so involved in patents that one could not proceed with the actual production and distribution. Not only existing patents, but patents in interference and under adjudication and patents pending must likewise be considered by the legal department. Patents available by purchase and lease and by participation in patent pools are considered as safeguards for carrying out a plant design. Trademarks and copyrights are likewise property that must be recognized and properly protected through legal agreements carried out by the legal department.

Infringement

A search in the Patent Office is made during which every feature (even the apparently unimportant details are frequently of real importance) of each logical division or step of the process is studied. In this search, both the expired and the unexpired art must be included, the expired art to indicate what can be done legitimately and the unexpired art to indicate the limitations or restrictions on what can be done legitimately. If unexpired patents are found which would be infringed by the desired process, then a study of their validity and scope must be made with reference to the expired patents. As regards infringement, it should be emphasized that no unexpired patent can cover the material disclosed in an expired patent except by the well-known legal procedure of combining previously known items to make a new contribution to knowledge. If a single expired patent can be found that covers the process desired for use, there is no reason why the process cannot be legitimately and safely used, at least from the patent standpoint. The determination of the scope and the validity of a patent with respect to its infringement is a question of law and, as such, should be undertaken by someone who is conversant with such matters, preferably by one who is an experienced patent attorney, before any definite action is taken. But, the patent attorney should not be relied upon for the technical information involved; the technical aspects belong in the sphere of the technical man. If the proposed process is new, it is desirable that it be protected by patent or patents as completely as possible from a monopolistic standpoint. Such patent protection is essential.

Public Relations

The legal department should acquaint itself with all local, state, and Federal laws that pertain to the manufacture of the commodity, its transportation and application, atmospheric and stream pollution through the disposal of wastes, and the possibility of claims for injury, death, or disabilities in connection with the production and use of the commodity to be manufactured. The design engineer should be familiar with all such legal entanglements, so that he may advise the legal department on probable hazardous conditions and unsafe practices.

Contracts

The design engineer leaves all contracts to the legal department but, for purposes of acquainting himself with limitations that may later arise in the fulfillment of his design project, he should recognize the detailed contractual relations that must be promulgated for actual commodity production.

SOURCES OF INFORMATION

Published Books

The chemical engineer must have a thorough knowledge of the principles of chemical engineering to use the quantitative data obtained from fundamental equations. Standard reference *textbooks* develop these principles in sufficient clarity for study purposes. *Handbooks* furnish information of a fundamental and an applied nature. Compilation of data can be found in the chemistry and physics handbooks in addition to the various engineering handbooks. Consult your technical librarian for a complete list of published books in this area of engineering, the Additional Selected References for each chapter at the back of the book, particularly the process-data section of Chap. 3.

Chemical Engineering Catalogs

The best reference books on manufacturers of chemical engineering equipment and products are the "Chemical Engineering Catalog" and the "Chemical Materials Catalog."[1] These catalogs are published annually. The "Chemical Materials Catalog" is a compilation of chemical

[1] Both published by Reinhold Publishing Corporation, New York.

materials, of chemical and physical properties, specifications, uses and applications, reactions, availability, shipping points and regulations, and structural formulas; it is alphabetically indexed by firm names, product names, by trade names, by uses and applications, and by functional ideas. Essentially it is a producers' directory. The "Chemical Engineering Catalog" is a process industries catalog, alphabetically compiled into the following subdivisions: (1) an index of firms represented, (2) functional index, (3) equipment name index, (4) plants and specialized services index, (5) pilot plant index, and (6) trade name index. Unfortunately these are incomplete inasmuch as not all manufacturers of chemical engineering equipment, supplies, chemicals, and raw materials advertise in them. Nor do they contain the voluminous quantity of data, direct and relative, that are contained in the mass of literature sent directly by manufacturing companies. To include all this information would make the catalogs cumbersome; therefore, the selection of material to be included depends upon the judgment of the advertising companies and of the advisory committee that aids in editing it.

Trade Literature

There is a great profusion of trade literature consisting of pamphlets, circulars, bulletins, and catalogs containing information, such as advertising claims, applications, specifications, and other pertinent information on all types of chemical engineering equipment and materials. That the data and information contained in these pamphlets, circulars, and bulletins are quite valuable will be attested to by anyone who has had occasion to design a plant, a piece of equipment, or a process, or who has wished to purchase a new or replace an old piece of equipment, or to obtain information on chemical products. Many of these pieces of literature are comprehensive treatises on the theory and application of the equipment which their issuers have for sale. To the chemical engineer, the performance data and dimensional drawings contained in some of these pamphlets are of inestimable aid in the realistic solution of his design and process problems. This follows in part from the fact that he depends upon and uses standard designed equipment whenever a satisfactory product is available, resorting to specially designed parts and units only when a standard design does not fit his need.

Chemical Engineering Publications

It should be pointed out that the rapid advances in our chemical industry have created the situation where the designer must keep up with the very latest data that may in any way become useful to him, not only in the present problem but also in any likely future work in his line. In particular, chemical engineering periodicals and other publications devoted to his branch of engineering are in a position to bring him the latest practical results of experimental designs. The *American Institute of Chemical Engineers Journal, Chemical Engineering Progress*, and the magazines *Chemical Engineering, Chemical Week, Industrial and Engineering Chemistry*, and *Chemical and Engineering News* are at present the outstanding American sources of information on progress in the chemical engineering field.

The petroleum and petrochemical industries require a great deal of design information which is published in *Petroleum Processing, Petroleum Refiner, Petroleum Engineer*, and *The Oil and Gas Journal*.

General Periodicals and Publications

In lieu of the strictly chemical engineering books, sources of general information which are used, not only in their own rightful functions but also to compensate for the lack of information, include (1) bulletins of manufacturers of chemicals, (2) bulletins of the manufacturers of special materials and metals, (3) publications of university laboratories and experiment stations, (4) pamphlets from industrial and trade associations and industrial institutes, (5) government publications from various bureaus, such as Chemistry and Agricultural Economics in the Department of Agriculture, the Bureaus of Census, Mines, Patents, and Standards in the Department of Commerce, as well as the Internal Revenue in the Treasury Department, and the U.S. Tariff Commission.

Plant Visits and Expositions

Another group of information services which should be assiduously cultivated for general information but which only rarely renders available specific data to aid in solving a particular problem includes visits to process plants, expositions, and other demonstrations. It is indeed rare that opportunity affords visits to plants making the same products, but there are elements of similarity in most plants that are of considerable value. The exposition is of value principally in the opportunity it affords to study the latest developments in equipment, products, and materials.

Equipment Manufacturers' Laboratories

A service that can be of great value and yet is frequently missed is the equipment manufacturer's demonstration laboratory. The danger lies principally in the attempt to use it instead of the pilot plant. It is much easier if the miniature equipment can be transported from the manufacturer's laboratory to the private plant, used, and returned. Several equipment companies now provide all types and kinds of assembled equipment and process units on a unitized basis.

Personal Experience

The last information service to be specifically mentioned is the engineer's personal experience file. It should be applied last, but it is most important. Attempts have been made to publish certain standardized specifications for mechanical equipment, but so far this has not been done in a general fashion for a large number of types of chemical equipment; thus it is still the fundamental job of the engineer to keep his own check list. This should contain details of the specifications and, in particular, should have underscored the obvious details most frequently omitted.

All these tools must be used with keen judgment. It is rare indeed that an engineer makes an error of commission, i.e., a mistake in a detail of design that he should not have made in light of his experience. The most frequent errors are those of omission, where some vital factor not previously met in the engineer's experience will invariably ruin a design that would otherwise have proved meritorious. Experience, especially the broadening kind in fields other than the particular one containing the problem on which the engineer is at the moment working, is a shield against such dangers.

PROBLEMS

2-1. Chemical and Physical Materials Data

Prepare a tabulated report, using tabulated format, on the physical and chemical properties of the raw materials, intermediate products, by-products, and principal chemicals which will be encountered in the manufacture of (a) maleic anhydric hydrazide; (b) ethylene; (c) uranium hexafluoride; (d) gamma benzene hexachloride; (e) soybeans by solvent extraction; (f) denatured alcohol from potatoes or any selected industrial chemical product.

2-2. Market Price and Marginal Profit Curves

Plot graphs of the semiannual market prices from 1940 on for raw materials, by-products, and principal products of any manufacture given in Prob. 2-1. Prepare marginal profit curves from this information. Sources of information: *Oil, Paint and Drug Reporter; Chemical Markets; Chemical Engineering; Chemical Week; Chemical and Engineering News.*

2-3. Proposed Capacity of Plant Project

Write a brief report answering the following questions: (a) Why do we wish to manufacture this product? (b) Who will buy it? (c) What is the potential market? (d) What working schedule should be proposed in production? (e) What will be the potential daily production? (f) How will by-products be disposed of?

CHAPTER 3

Process Design

A basic step in making a preliminary plant design for cost estimation or for establishing a detailed commercial plant design is to work out a process design. Briefly, one presents the basic *chemical* and *physical* operations of a process. Examples of the latter type include filtration, drying, mixing, and other *chemical engineering operations*. These facts are illustrated in flow-sheet form.[1] It is then possible for the engineer to apply industrial stoichiometry principles to the process as outlined to obtain material and energy balance flow sheets. After all these facts are available, the designer is ready to specify the type of equipment required to do the job. By knowing the cost of such equipment, the start of a plant cost estimate has been made.

Choice of Process

Nearly all chemical plants employ processes where the raw materials undergo chemical changes to produce a more valuable product. Frequently, the choice of process cannot be made at the library or laboratory research level. It is necessary that the chemical engineer become thoroughly familiar with all aspects of the proposed processes in order to aid in the best selection. A rough cost estimation based on laboratory results and engineering experience can sometimes settle the decision in favor of one particular process. At other times, a more thorough process design and cost estimation will be required before any decision can be rendered. Consultation among the designers, the research and development staff, and the legal department is of prime importance in getting the best over-all evaluation of a particular process.

[1] J. P. O'Donnell, How Flowsheets Communicate Engineering Information, *Chem. Eng.*, **64**(9): 245 (1957).

Selection of Process Cycles

Assuming that the annual production requirement has been established, one of the first problems faced in process design is to choose a process cycle so that material and energy balances on a time basis can be worked out and all of the necessary flow sheets prepared. Questions related to a 24-hr or 8 hr/day operation and production by sequential batch operation or on a steady-state continuous basis must be resolved.

Continuous vs. Batch Processing.[1] It is the usual rule in process design to choose continuous processing in preference to batch processing based purely on economic reasons. By operation on a continuous 24 hr/day basis, smaller-size, less expensive processing equipment is used. Process operation is steady state and easier to control by automatic instrumentation than is batch operation. Thus, capital investment, fixed charges, and labor requirements are minimized. However, batch processing is not a completely obsolete method in the chemical industry. It is feasible in such cases as:

1. Small-volume output of relatively expensive products when sales demand is not steady, or when the same equipment can be used for several processes of this nature.

2. When batch equipment from an abandoned process is available at a low transfer cost to the current project.

3. When continuous process equipment has not been satisfactorily developed and batch-process equipment has been satisfactorily demonstrated; this is a short-range solution when emergency deadlines must be met.

4. When yields and quality of products cannot be achieved by continuous processing because of such things as very low reaction rates and correspondingly long residence times in the processing equipment.

Often a few steps in a process may better be accomplished by batch operation and the remainder on a continuous basis with provision for sufficient buffer storage facilities between steps.

Shifts and Operating Time Schedules. Although "around the clock" operation is the usual rule in the chemical industry, other production schedules listed below are sometimes set up, depending on labor conditions and volume of sales:

1. *Around the clock operation.* Three shifts of operation per day, seven days a week, using four shifts of labor, with or without scheduled maintenance shutdowns.

2. *Five-day operation.* One, two, or three shifts of operation and labor; stand-by conditions on weekends and holidays; overtime if production

---

[1] J. R. Donovan, Batch vs. Continuous. Don't Write Off the Batch Process, *Chem. Eng.*, **64**(11): 241 (1957).

demands must be temporarily increased; maintenance done during periods of nonoperation.

3. *Scheduled shutdown.* Scheduled shutdowns or "turn around" after operating for periods up to 2 years, or annually for 2 weeks with operating labor on vacation and maintenance labor doing inspection, repair, and replacement work on the plant.

4. *Unscheduled shutdowns.* Interruption in scheduled operation due to lack of raw materials, labor strikes, failure of utilities such as power and water, fire and explosions, and extreme weather conditions, or most probably equipment failure. In starting up new plants, this latter factor should be taken into account as an estimated per cent onstream time for at least the first year.

The names of shifts encountered in the chemical industry are:
1. First or day shift (e.g., 7 A.M. to 3 P.M.)
2. Second, evening or swing shift (e.g., 3 P.M. to 11 P.M.)
3. Third, midnight or graveyard (e.g., 11 P.M. to 7 A.M.)

Increased pay is allowed for the second and third shifts (see Table 6-15). Operations can be on a permanent or rotating shift basis, depending on plant policy and local labor conditions. Shift rotation usually occurs every 2, 3, or 4 weeks.

Example of a Process Design Project

The market analysis staff of Company X, a medium-sized organic products company, has recommended to its management group that there was some chance to diversify their product line by developing a pesticide. The research and development staff suggest that benzene hexachloride (BHC) might be a good possibility. It can be sold in the so-called crude form which contains 12 to 15 per cent of the active pesticide, γ isomer of BHC, or it can be marketed as the concentrated product Lindane (99 per cent γ isomer of BHC). Market analysis showed that an annual production of 1,200,000 lb of crude BHC or up to 250,000 lb of Lindane could be easily sold without undue competition. Management would like to compare the economics of crude BHC and Lindane production before making the final decision to invest money in a manufacturing plant. This requires process design study followed by a preconstruction cost estimate and profitability analysis.

For the purpose of teaching process design, only the complete development of the process for producing crude BHC will be illustrated in this text. A similar procedure can be used for the Lindane production problem (see Process Design Project 6-1). Enough of the economics of both projects will be considered in Chap. 6 to show the relative merits of the two projects.

A statement of the chemical process for the production of crude benzene hexachloride is specified by the research and development group as follows:

Chemical Process Considerations.[1] The production of benzene hexachloride is to involve the reaction of gaseous chlorine with liquid benzene at the refluxing temperature in the absence of actinic light and substitution catalysts but in the presence of water. The reaction is to be carried out on a continuous basis in a glass-lined reactor fitted with a refluxing system; conversion of the chlorine is expected to be 99 per cent and the reactor product should consist of about 33 per cent benzene hexachloride dissolved in chlorobenzenes and unreacted benzene. The crude product is to be continuously removed for further processing. The accompanying 21.5 per cent muriatic acid layer should be decanted and further concentrated to the salable 36 per cent acid. The concentration is to be accomplished in a Karbate stripping column which produces a bottoms of 20 per cent HCl solution. Since this bottoms product is not salable as such and since the HCl cannot be stripped farther without the use of excessively high pressures, it is to be recycled into the chlorinator for reconcentration.

The organic layer is to be neutralized with 1 per cent NaOH solution and then concentrated to 56 per cent BHC by evaporation of some of the volatile ingredients in a flash still. The overhead from this still is largely benzene and monochlor benzene which are separated by fractionational distillation. The benzene is recycled and the monochlor benzene is to be sold as a by-product. A large fraction of the benzene hexachloride is to be crystallized out in a double-pipe "chiller" and the crystals removed by means of a continuous centrifuge. The filtrate is to be recycled to a flash still for reconcentration.

The crystalline BHC product is to be dried in a rotary vacuum dryer equipped with a solvent recovery system. The recovered solvents from various operations are to be combined and fractionated. The overhead distillate consisting primarily of benzene is to be reused in the process whereas the chlorobenzol solvents in the bottoms are to be stored for sale as crude monochlorobenzene.

Laboratory and Pilot Plant Data and Specifications

1. Raw materials specifications:
    Benzol, purified grade, considered as 99.5 per cent benzene, 0.5 per cent toluene
    Chlorine considered to be 100 per cent pure
    Water catalyst, demineralized process water
    Sodium hydroxide, flake, 98 per cent NaOH

[1] F. L. Gradishar, Dissertation, Department of Chemical Engineering, Virginia Polytechnic Institute, Blacksburg, Va.

2. Chlorinator conditions:
    Reaction temperature: 70°C
    Chlorine conversion: 99 per cent
3. Characteristics of organic product layer from chlorinator:

Composition	Weight per cent
Benzene	58.85
Water	0.31
Monochlorobenzene	7.08
Benzene hexachloride (BHC fraction):	
Alpha	23.44
Beta	2.04
Gamma	4.22
Delta and cogeners	3.55
Heptachlorocyclohexane	0.26
Trichlorobenzene	0.15
Hydrogen chloride	0.10
Total	100.00

Specific gravity at 70°C: 1.034

4. Characteristics of aqueous layer from chlorinator:

Composition	Weight per cent
Acid (as HCl)	24.60
C_6H_6	0.22
Cl_2	0.13
H_2O	75.05
Specific gravity at 70°C	1.0975

5. Neutralizer conditions:
    Purity of caustic.................................... 96.9% NaOH
    % excess of caustic used............................ 100.0%
    Concentration of NaOH in neutralizing solution...... 1.0%
    Spent caustic characteristics:
        C_6H_6 .... 0.2%
6. Solvent fractionation still conditions:

Overhead composition	Weight per cent
C_6H_6	99.48
H_2O	0.52

Bottoms composition	Weight per cent
C_6H_6	0.10
C_6H_5Cl	99.64
$C_6H_6Cl_6$	0.23
$C_6H_3Cl_3$	0.03

7. Solubility data for chiller-centrifuge calculations:
    At 30°C: 28.3 g of crude BHC in 100 g total mixture
    At 70°C: 76.0 g of crude BHC in 100 g total mixture
8. Dryer data:
    Volatile content of solids from centrifuge: 3%

Analysis of volatiles | Weight per cent

C_6H_6	57.0
C_6H_5Cl	29.3
$C_6H_5CCl_3$	13.0
H_2O	0.7

Analysis of recovered solvent from dryer | Weight per cent

C_6H_6	67.5
C_6H_5Cl	31.7
H_2O	0.8
Volatile content of solids leaving dryer	0.5%

9. General specifications of BHC product:
   Grade: technical
   Guaranteed gamma isomer content: 12.5%
   Color and form: slightly discolored crystals
   Odor: minimum
   Volatiles: less than 0.5%
   Containers: 100-lb fiber drums
   Average analysis of product:
   BHC fraction
   Benzene hexachloride

Alpha	69.4%
Beta	6.1%
Gamma	12.6%
Delta and cogeners	10.6%
Heptachlorocyclohexane	0.8%
Monochlorobenzene	0.1%
Trichlorobenzene	0.4%
Total	100.0%

10. Heat of chlorination:

    $C_6H_6 + 3Cl_2 \rightarrow C_6H_6Cl_6 + 107.4$ kcal/g mol
    $C_6H_6 + Cl_2 \rightarrow C_6H_5Cl + HCl + 24$ kcal/g mol

11. Average specific heat of BHC = 0.211 cal/(g)(°C)
12. Vapor-pressure data:
    Benzene hexachloride[1]

Temp., °C	323.4	240.5	176.2	113.0
Pressure, mm Hg	760	100	10	0.36

13. Temperature at critical points:
    Chlorinator: 70°C
    Decanter: 65°
    Neutralizer: 60°C
    Flash still equilibrium operating temperature: 100°C
    Liquor from chiller: 30°C
    Raw materials input: 30°C
    Reflux from chlorinator overhead condenser: 30°C

---

[1] R. R. Dreisbach, "Physical Properties of Chemical Substances," sec. 13-15, Dow Chemical Co., 1953.

46   CHEMICAL ENGINEERING PLANT DESIGN   [CHAP. 3

    Dryer: 60°C
    Cooling water available at 25°C
    Room temperature: 25°C
14. Pressure conditions:
    Essentially atmospheric conditions for all process equipment except vacuum dryer which operates at about 20 in. Hg vacuum
15. Available steam-line pressures:
    15-lb gauge (121°C), latent heat: 945 Btu/lb
    75-lb gauge (160°C), latent heat: 895 Btu/lb
16. Other necessary data:
    Data not listed above are available in handbooks

Qualitative Block-type Process Flow Sheet

The next step for the design engineering group is to translate the written description of the chemical process into a working pattern. It might be well at the outset to explain that this working pattern can be developed in a number of ways. Each industry, each company, and each individual designer has a somewhat different approach to process design, particularly when it comes to pictorial representation of the design data. We can present in this text only a few of the possible ways of working out a process design. The main points that a student should realize are that the design should be (1) technically sound, (2) on a reasonable basis, and (3) clearly pictured.

An examination of this process description for manufacturing benzene hexachloride and the supporting data shows that the principal chemical reaction is that of chlorine addition.

$$C_6H_6(l) + 3Cl_2(g) \xrightarrow[\text{Presence of H}_2\text{O, 70°C}]{\text{Absence of actinic rays and substitution catalysts}} C_6H_6Cl_6$$

Benzene   Chlorine   Benzene hexachloride or cyclohexane, 1,2,-3,4,5,6-hexachlorine

The working pattern for this process is put down on paper by drawing rectangular boxes and inserting the types of physical and/or chemical operations that take place, together with a description of the streams entering and leaving the box. The result is a qualitative block-type process flow sheet or flow diagram. Figure 3-1 illustrates this principle. The process engineer now has a better idea of the over-all process and obtains an insight as to further data he might require.

Material Balance

Before any significant design progress can be made, a quantitative approach to the process design must be made in terms of material and energy flows and balances. A time basis is required since the annual production has been specified. A proposed operating schedule was prepared as follows:

Proposed Operating Schedule. The process as planned and developed is essentially continuous; hence, it would be highly desirable to operate with as few shutdowns as are consistent with good employee relations.

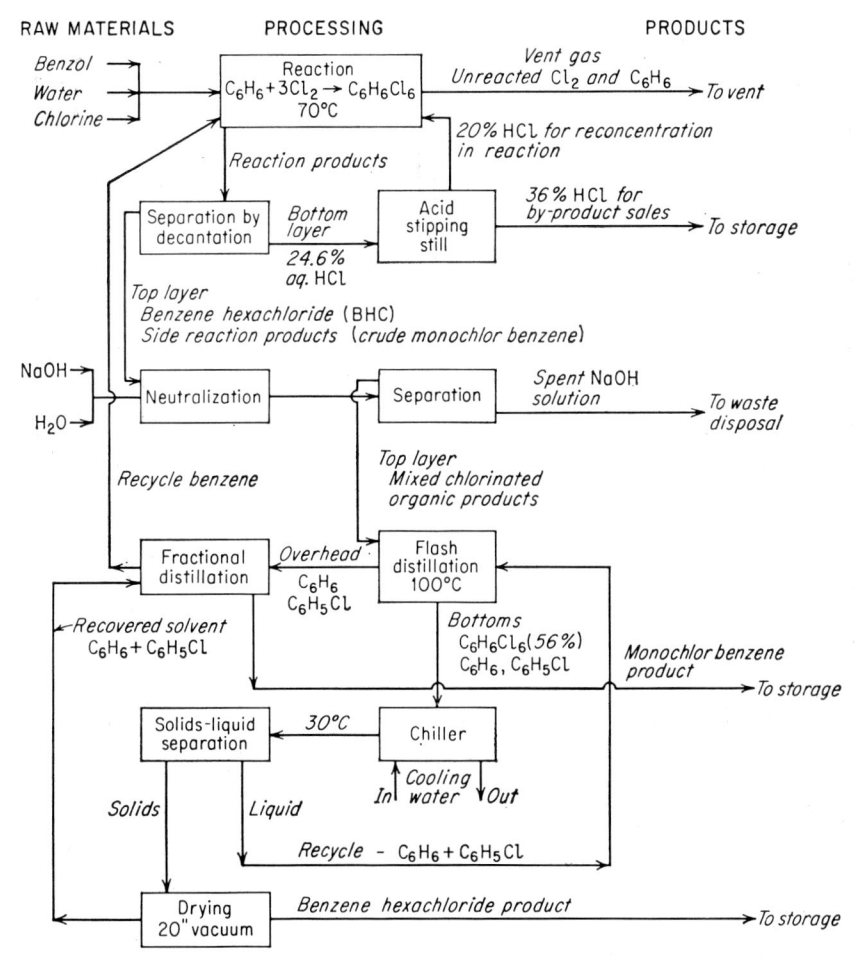

FIG. 3-1. Qualitative block-type process flow sheet for continuous production of benzene hexachloride. (*Note:* This shows both the chemical and physical operations of the process.)

It is deemed best, therefore, to operate the plant 24 hr per day, 5 days per week. All employees will work an 8-hr day, 5-day week with time-and-a-half pay for all necessary overtime and holiday duty. A 2-week unitized vacation period will be allotted the employees during the 2-week period

48   CHEMICAL ENGINEERING PLANT DESIGN   [CHAP. 3

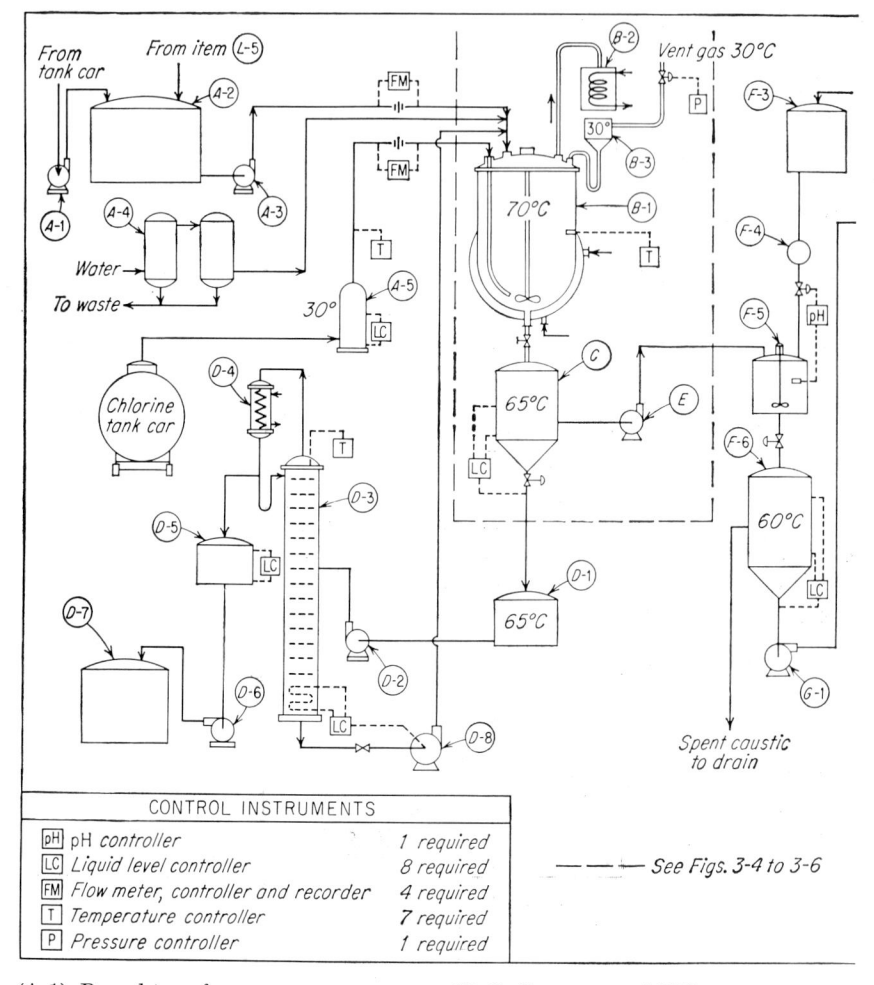

	CONTROL INSTRUMENTS	
pH	pH *controller*	1 required
LC	Liquid level controller	8 required
FM	Flow meter, controller and recorder	4 required
T	Temperature controller	7 required
P	Pressure controller	1 required

— — — — See Figs. 3-4 to 3-6

(A-1) Benzol transfer pump
(A-2) Benzol storage tank
(A-3) Benzol feed pump
(A-4) Demineralizer
(A-5) Chlorine evaporator
(B-1) Chlorinator
(B-2) Reflux condenser
(B-3) Vent gas separator
(C) Decanter
(D-1) Dilute acid receiver
(D-2) Acid still feed pump
(D-3) Acid stripping still
(D-4) Acid still condenser

(D-5) Concentrated HCl receiver
(D-6) Concentrated HCl pump
(D-7) Concentrated HCl storage
(D-8) Acid recycle pump
(E) Crude product pump
(F-1) Caustic make-up tank
(F-1a) Caustic make-up agitator
(F-2) Caustic transfer pump
(F-3) Caustic storage tank
(F-4) Caustic feeder
(F-5) Neutralizer
(F-6) Spent caustic separator
(G-1) Flash still feed pump

FIG. 3-2. Equipment flow sheet

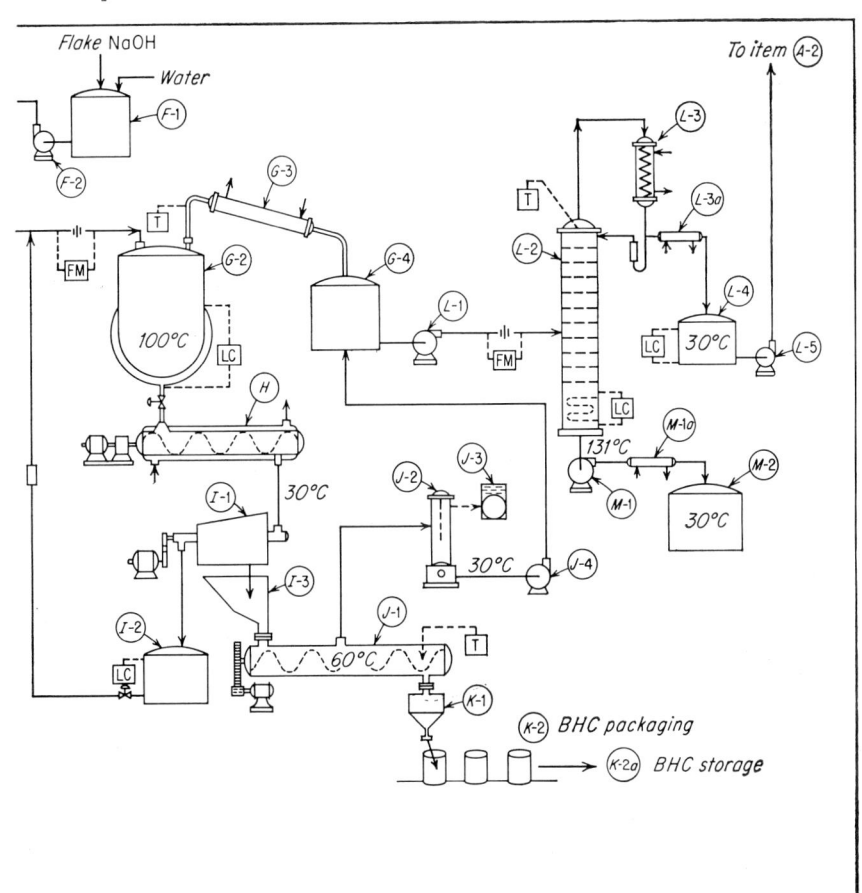

(G-2) Flash still
(G-3) Flash still condenser
(G-4) Solvent receiver
(H-1) Crystallizer feed pump
(H-2) Crystallizer
(I-1) Continuous centrifuge
(I-2) Recycle liquor receiver
(I-3) Wet crystal hopper
(J-1) Rotary vacuum dryer
(J-2) Vacuum condenser and receiver
(J-3) Vacuum pump
(J-4) Recovered solvents pump

(K-1) Dry BHC hopper
(K-2) BHC packaging
(K-2a) BHC storage
(L-1) Solvent still feed pump
(L-2) Solvent still
(L-3) Solvent still condenser
(L-3a) Condensate cooler
(L-4) Wet benzol receiver
(L-5) Wet benzol pump
(M-1) Monochlor pump
(M-1a) Monochlor cooler
(M-2) Monochlor storage

of the benzene hexachloride plant.

that includes Christmas and New Year's Day. Major repairs to the plant will be performed during the vacation period by the maintenance group.

On the basis of the above decisions, a summary of the operations and production schedule is presented below. (Start-up and shutdown time are neglected in the calculations.)

> Annual output: 1,200,000 lb of BHC
> Operation: 3 shifts (8 hr each) per day; 40-hr week; 10-day vacation, unitized
> Operating days per annum: 250 days
> Designed daily production: 4,800 lb of BHC
> Designed hourly production: 200 lb of BHC

A convenient time basis for the material flow is *1 hr*, unless the main flow quantities prove too small or an accurate balance on impurities is to be carried out. Such was the case with the example problem and so a *24-hr basis* was chosen. The main requirement is to set a reasonable time basis which gives meaningful values for all flow quantities.

To obtain a systematic organization of the material balance for a complex process of this type, a tentative equipment flow sheet was prepared (Fig. 3-2). This was coded by letter for each key operation, for example, B for chlorination, D for acid concentration, etc. All the equipment conveniently associated with the key operation was numbered, for example, B-1, B-2, etc. A study of this diagram will be helpful in following this particular plan of execution.

The equipment pictured may not represent the final choice as subsequent material and energy balances could reveal that a certain type of equipment was not technically feasible to carry out the desired process step.

The material balance calculations were carried out and listed in the pages to follow under the various equipment code numbers as flow-entering and flow-leaving quantities. The calculation of these values is not given in this text, but is listed as a problem for the process design class. The students are presumed to have had adequate background courses in industrial stoichiometry and process calculations.

Material Balance (basis: 4,800 lb of technical BHC per 24-hr day). All data are given in avoirdupois pounds.

A-1. Benzol transfer pump

Entering		*Leaving*	
From benzol tank car		To benzol storage tank, A-2	
C_6H_6	2,019	C_6H_6	2,019
C_7H_8	10	C_7H_8	10
Total	2,029	Total	2,029

A-2. Benzol storage tank

Entering		*Leaving*	
From benzol transfer pump, A-1		To benzol feed pump, A-3	
C_6H_6	2,019	C_6H_6	10,429
C_7H_8	10	C_7H_8	10
Total	2,029	Water	44
From wet benzol pump, L-5			
C_6H_6	8,410		
H_2O	44		
Total	8,454		
Total	10,483	Total	10,483

A-3. Benzol feed pump

Entering		*Leaving*	
From benzol storage tank, A-2		To chlorinator, B-1	
C_6H_6	10,429	C_6H_6	10,429
C_7H_8	10	C_7H_8	10
Water	44	Water	44
Total	10,483	Total	10,483

A-4. Demineralizer

Entering		*Leaving*	
From process water supply		To chlorinator B1	
Water	589	Water	589
Impurities	0.2	To waste	
		Impurities	0.2
Total	589.2	Total	589.2

A-5. Chlorine vaporizer

Entering		*Leaving*	
From chlorine tank car		To chlorinator, B-1	
Cl_2	4,228	Cl_2	4,228

B-1. Chlorinator

Entering		*Leaving*	
From benzol feed pump, A-3		To decanter, C	
C_6H_6	10,429	Product layer	
C_7H_8	10	C_6H_6	8,427
H_2O	44	C_6H_5Cl	1,013
Total	10,483	$C_6H_6Cl_6$	
From chlorine vaporizer, A-5		Alpha	3,355
Cl_2	4,228	Beta	293
From demineralizer, A-4		Gamma	605
H_2O	589	Delta and cogeners	509
Recycle from vent gas separator, B-3		$C_6H_5Cl_7$	38
Cl_2	10	H_2O	44
C_6H_6	10,152	HCl	14
C_6H_5Cl	222	$C_6H_3Cl_3$	21
$C_6H_6Cl_6$	13	Total	14,319
H_2O	650	Aqueous layer	
HCl	587	H_2O	2,338
Total	11,634	HCl	778
From acid recycle pump, D-8		C_6H_6	7
HCl	447	Cl_2	4
H_2O	1,749	Total	3,127
Total	2,196	To reflux condenser, B-2	
		C_6H_6	10,160
		C_6H_5Cl	222
		$C_6H_6Cl_6$	13
		H_2O	650
		HCl	587
		Cl_2	52
		Total	11,684
Total	29,130	Total	29,130

B-2. Reflux condenser

Entering		*Leaving*	
From chlorinator, B-1		To vent gas separator, B-3	
Cl_2	52	As vapor	
C_6H_6	10,160	Cl_2	42
C_6H_5Cl	222	C_6H_6	8
H_2O	650	Total	50
HCl	587	As liquid	
$C_6H_6Cl_6$	13	Cl_2	10
		C_6H_6	10,152
		C_6H_5Cl	222
		$C_6H_6Cl_6$	13
		H_2O	650
		HCl	587
		Total	11,634
Total	11,684	Total	11,684

B-3. Vent gas separator

Entering

From reflux condenser, B-2
As vapor
Cl$_2$	42
C$_6$H$_6$	8
Total	50

As liquid
Cl$_2$	10
C$_6$H$_6$	10,152
C$_6$H$_5$Cl	222
C$_6$H$_6$Cl$_6$	13
H$_2$O	650
HCl	587
Total	11,634

Total ... 11,684

Leaving

As vent gas
Cl$_2$	42
C$_6$H$_6$	8
Total	50

To chlorinator, B-1
Cl$_2$	10
C$_6$H$_6$	10,152
C$_6$H$_5$Cl	222
C$_6$H$_6$Cl$_6$	13
H$_2$O	650
HCl	587
Total	11,634

Total ... 11,684

C. Decanter

Entering

From chlorinator, B-1
Product layer
C$_6$H$_6$	8,427
C$_6$H$_5$Cl	1,013
C$_6$H$_6$Cl$_6$	
Alpha	3,355
Beta	293
Gamma	605
Delta and cogeners	509
C$_6$H$_5$Cl$_7$	38
H$_2$O	44
HCl	14
C$_6$H$_3$Cl$_3$	21
Total	14,319

Aqueous layer
H$_2$O	2,338
HCl	778
C$_6$H$_6$	7
Cl$_2$	4
Total	3,127

Total ... 17,446

Leaving

To dilute acid receiver, D-1
H$_2$O	2,338
HCl	778
C$_6$H$_6$	7
Cl$_2$	4
Total	3,127

To crude product pump, E
C$_6$H$_6$	8,427
C$_6$H$_5$Cl	1,013
C$_6$H$_6$Cl$_6$	
Alpha	3,355
Beta	293
Gamma	605
Delta and cogeners	509
C$_6$H$_5$Cl$_7$	38
H$_2$O	44
HCl	14
C$_6$H$_3$Cl$_3$	21
Total	14,319

Total ... 17,446

D-1. Dilute acid receiver

Entering

From decanter, C
H$_2$O	2,338
HCl	778
C$_6$H$_6$	7
Cl$_2$	4
Total	3,127

Leaving

To acid still feed pump, D-2
H$_2$O	2,338
HCl	778
C$_6$H$_6$	7
Cl$_2$	4
Total	3,127

D-2. Acid still feed pump

Entering		*Leaving*	
From dilute acid receiver, D-1		To acid stripping still, D-3	
H₂O	2,338	H₂O	2,338
HCl	778	HCl	778
C₆H₆	7	C₆H₆	7
Cl₂	4	Cl₂	4
Total	3,127	Total	3,127

D-3. Acid stripping still

Entering
From acid still feed pump, D-2
- H₂O 2,338
- HCl 778
- C₆H₆ 7
- Cl₂ 4

Total 3,127

Leaving
To acid recycle pump, D-8
- H₂O 1,749
- HCl 447
- Total 2,196

To acid still condenser, D-4
- H₂O 589
- HCl 331
- C₆H₆ 7
- Cl₂ 4
- Total 931

Total 3,127

D-4. Acid still condenser

Entering		*Leaving*	
From acid stripping still, D-3		To conc. HCl receiver, D-5	
H₂O	589	H₂O	589
HCl	331	HCl	331
C₆H₆	7	C₆H₆	7
Cl₂	4	Cl₂	4
Total	931	Total	931

D-5. Conc. HCl receiver

Entering		*Leaving*	
From acid still condenser, D-4		To conc. HCl pump, D-6	
H₂O	589	H₂O	589
HCl	331	HCl	331
C₆H₆	7	C₆H₆	7
Cl₂	4	Cl₂	4
Total	931	Total	931

D-6. Conc. HCl pump

Entering		*Leaving*	
From conc. HCl receiver, D-5		To conc. HCl storage, D-7	
H₂O	589	H₂O	589
HCl	331	HCl	331
C₆H₆	7	C₆H₆	7
Cl₂	4	Cl₂	4
Total	931	Total	931

CHAP. 3] PROCESS DESIGN 55

D-7. Conc. HCl storage

Entering
From conc. HCl pump, D-6
H$_2$O	589
HCl	331
C$_6$H$_6$	7
Cl$_2$	4
Total	931

Leaving
Storage
H$_2$O	589
HCl	331
C$_6$H$_6$	7
Cl$_2$	4
Total	931

D-8. Acid recycle pump

Entering
From acid still, D-5
H$_2$O	1,749
HCl	447
Total	2,196

Leaving
To chlorinator, B-1
H$_2$O	1,749
HCl	447
Total	2,196

E. Crude product pump

Entering
From decanter, C
C$_6$H$_6$	8,427
C$_6$H$_5$Cl	1,013
C$_6$H$_6$Cl$_6$	
Alpha	3,355
Beta	293
Gamma	605
Delta and cogeners	509
C$_6$H$_5$Cl$_7$	38
H$_2$O	44
HCl	14
C$_6$H$_3$Cl$_3$	21
Total	14,319

Leaving
To neutralizer, F-5
C$_6$H$_6$	8,427
C$_6$H$_5$Cl	1,013
C$_6$H$_6$Cl$_6$	
Alpha	3,355
Beta	293
Gamma	605
Delta and cogeners	509
C$_6$H$_5$Cl$_7$	38
H$_2$O	44
HCl	14
C$_6$H$_3$Cl$_3$	21
Total	14,319

F-1. Caustic make-up tank

Entering
Flake caustic storage
NaOH	31
Impurities	1
Total	32

Process water supply
H$_2$O	3,069
Total	3,101

Leaving
To caustic transfer pump, F-2
NaOH	31
Impurities	1
H$_2$O	3,069
Total	3,101

F-2. Caustic transfer pump

Entering
From caustic make-up tank, F-1
NaOH	31
Impurities	1
H$_2$O	3,069
Total	3,101

Leaving
To caustic storage, F-3
NaOH	31
Impurities	1
H$_2$O	3,069
Total	3,101

F-3. Caustic storage

Entering		*Leaving*	
From caustic transfer pump, F-2		To caustic feeder, F-4	
NaOH	31	NaOH	31
Impurities	1	Impurities	1
H₂O	3,069	H₂O	3,069
Total	3,101	Total	3,101

F-4. Caustic feeder

Entering		*Leaving*	
From caustic storage, F-3		To neutralizer, F-5	
NaOH	31	NaOH	31
Impurities	1	Impurities	1
H₂O	3,069	H₂O	3,069
Total	3,101	Total	3,101

F-5. Neutralizer

Entering		*Leaving*	
From crude product pump, E		To spent caustic separator, F-6	
		Product layer	
C_6H_6	8,427	C_6H_6	8,421
C_6H_5Cl	1,013	C_6H_5Cl	1,013
$C_6H_6Cl_6$		$C_6H_6Cl_6$	
Alpha	3,355	Alpha	3,355
Beta	293	Beta	293
Gamma	605	Gamma	605
Delta and cogeners	509	Delta and cogeners	509
$C_6H_5Cl_7$	38	$C_6H_5Cl_7$	38
H₂O	44	H₂O	44
HCl	14		
$C_6H_3Cl_3$	21	$C_6H_3Cl_3$	21
Total	14,319	Total	14,299
From caustic feeder, F-4		Aqueous layer	
H₂O	3,069	H₂O	3,076
NaOH	31	NaOH	16
Impurities	1	NaCl	22
Total	3,101	C_6H_6	6
		Impurities	1
		Total	3,121
Total	17,420	Total	17,420

CHAP. 3] PROCESS DESIGN 57

F-6. Spent caustic separator

Entering		*Leaving*	
From neutralizer, F-5		To flash still feed pump, G-1	
Product layer		Product layer	
C_6H_6	8,421	C_6H_6	8,421
C_6H_5Cl	1,013	C_6H_5Cl	1,013
$C_6H_6Cl_6$		$C_6H_6Cl_6$	
Alpha	3,355	Alpha	3,355
Beta	293	Beta	293
Gamma	605	Gamma	605
Delta and cogeners	509	Delta and cogeners	509
$C_6H_5Cl_7$	38	$C_6H_5Cl_7$	38
H_2O	44	H_2O	44
$C_6H_3Cl_3$	21	$C_6H_3Cl_3$	21
Total	14,299	Total	14,299
Aqueous layer		To waste	
H_2O	3,076	H_2O	3,076
NaOH	16	NaOH	16
NaCl	22	NaCl	22
C_6H_6	6	C_6H_6	6
Impurities	1	Impurities	1
Total	3,121	Total	3,121
Total	17,420	Total	17,420

G-1. Flash still feed pump

Entering		*Leaving*	
From spent caustic separator, F-6		To flash still, G-2	
C_6H_6	8,421	C_6H_6	8,421
C_6H_5Cl	1,013	C_6H_5Cl	1,013
$C_6H_6Cl_6$		$C_6H_6Cl_6$	
Alpha	3,355	Alpha	3,355
Beta	293	Beta	293
Gamma	605	Gamma	605
Delta and cogeners	509	Delta and cogeners	509
$C_6H_5Cl_7$	38	$C_6H_5Cl_7$	38
H_2O	44	H_2O	44
$C_6H_3Cl_3$	21	$C_6H_3Cl_3$	21
Total	14,299	Total	14,299

G-2. Flash still

Entering		*Leaving*	
From flash still feed pump, G-1		Bottom to crystallizer feed pump, H-1	
C_6H_6	8,421	C_6H_6	3,415
C_6H_5Cl	1,013	C_6H_5Cl	1,847
$C_6H_6Cl_6$		$C_6H_6Cl_6$	
Alpha	3,355	Alpha	3,817
Beta	293	Beta	299
Gamma	605	Gamma	1,413
Delta and cogeners	509	Delta and cogeners	1,188
$C_6H_5Cl_7$	38	$C_6H_5Cl_7$	89
H_2O	44	H_2O	24
$C_6H_3Cl_3$	21	$C_6H_3Cl_3$	61
Total	14,299	Total	12,153
From recycle liquor receiver, I-2		Vapor to flash still condenser, G-3	
C_6H_6	3,331	C_6H_6	8,337
C_6H_5Cl	1,803	C_6H_5Cl	969
$C_6H_6Cl_6$		$C_6H_6Cl_6$	24
Alpha	486	H_2O	43
Beta	6	$C_6H_3Cl_3$	3
Gamma	808	Total	9,376
Delta and cogeners	679		
$C_6H_5Cl_7$	51		
H_2O	23		
$C_6H_3Cl_3$	43		
Total	7,230		
Total	21,529	Total	21,529

G-3. Flash still condenser

Entering		*Leaving*	
From flash still, G-2		To solvent receiver, G-4	
C_6H_6	8,337	C_6H_6	8,337
C_6H_5Cl	969	C_6H_5Cl	969
$C_6H_6Cl_6$	24	$C_6H_6Cl_6$	24
H_2O	43	H_2O	43
$C_6H_3Cl_3$	3	$C_6H_3Cl_3$	3
Total	9,376	Total	9,376

G-4. Solvent receiver

Entering

From solvent pump, J-4

C_6H_6	83
C_6H_5Cl	39
H_2O	1
Total	123

From flash still condenser, G-3

C_6H_6	8,337
C_6H_5Cl	969
$C_6H_6Cl_6$	24
$C_6H_3Cl_3$	3
H_2O	43
Total	9,376
Total	9,499

Leaving

To solvent still feed pump, L-1

C_6H_6	8,420
C_6H_5Cl	1,008
$C_6H_6Cl_6$	24
$C_6H_3Cl_3$	3
H_2O	44
Total	9,499

H-1. Crystallizer feed pump

Entering

From flash still, G-2

C_6H_6	3,415
C_6H_5Cl	1,847
$C_6H_6Cl_6$	
Alpha	3,817
Beta	299
Gamma	1,413
Delta and cogeners	1,188
$C_6H_5Cl_7$	89
H_2O	24
$C_6H_3Cl_3$	61
Total	12,153

Leaving

To crystallizer, H-2

C_6H_6	3,415
C_6H_5Cl	1,847
$C_6H_6Cl_6$	
Alpha	3,817
Beta	299
Gamma	1,413
Delta and cogeners	1,188
$C_6H_5Cl_7$	89
H_2O	24
$C_6H_3Cl_3$	61
Total	12,153

H-2. Crystallizer

Entering

From flash still, G-2

C_6H_6	3,415
C_6H_5Cl	1,847
$C_6H_6Cl_6$	
Alpha	3,817
Beta	299
Gamma	1,413
Delta and cogeners	1,188
$C_6H_5Cl_7$	89
H_2O	24
$C_6H_3Cl_3$	61
Total	12,153

Leaving

To continuous centrifuge, I-1

C_6H_6	3,415
C_6H_5Cl	1,847
$C_6H_6Cl_6$	
Alpha	3,817
Beta	299
Gamma	1,413
Delta and cogeners	1,188
$C_6H_5Cl_7$	89
H_2O	24
$C_6H_3Cl_3$	61
Total	12,153

I-1. Continuous centrifuge

Entering		*Leaving*	
From crystallizer, H-2		To wet crystal hopper, I-3	
C_6H_6	3,415	C_6H_6	84
C_6H_5Cl	1,847	C_6H_5Cl	44
$C_6H_6Cl_6$		$C_6H_6Cl_6$	
Alpha	3,817	Alpha	3,331
Beta	299	Beta	293
Gamma	1,413	Gamma	605
Delta and cogeners	1,188	Delta and cogeners	509
$C_6H_5Cl_7$	89	$C_6H_5Cl_7$	38
H_2O	24	H_2O	1
$C_6H_3Cl_3$	61	$C_6H_3Cl_3$	18
		Total	4,923
		To recycle liquor receiver, I-2	
		C_6H_6	3,331
		C_6H_5Cl	1,803
		$C_6H_6Cl_6$	
		Alpha	486
		Beta	6
		Gamma	808
		Delta and cogeners	679
		$C_6H_5Cl_7$	51
		H_2O	23
		$C_6H_3Cl_3$	43
		Total	7,230
Total	12,153	Total	12,153

I-2. Recycle liquor receiver

Entering		*Leaving*	
From continuous centrifuge, I-1			
C_6H_6	3,331	C_6H_6	3,331
C_6H_5Cl	1,803	C_6H_5Cl	1,803
$C_6H_6Cl_6$		$C_6H_6Cl_6$	
Alpha	486	Alpha	486
Beta	6	Beta	6
Gamma	808	Gamma	808
Delta and cogeners	679	Delta and cogeners	679
$C_6H_5Cl_7$	51	$C_6H_5Cl_7$	51
H_2O	23	H_2O	23
$C_6H_3Cl_3$	43	$C_6H_3Cl_3$	43
Total	7,230	Total	7,230

CHAP. 3] PROCESS DESIGN 61

I-3. Wet crystal hopper

Entering
From continuous centrifuge, I-1
  $C_6H_6Cl_6$
    Alpha.................... 3,331
    Beta..................... 293
    Gamma.................. 605
    Delta and cogeners......... 509
  $C_6H_5Cl_7$.................... 38
  C_6H_6....................... 84
  C_6H_5Cl..................... 44
  $C_6H_3Cl_3$.................... 18
  H_2O........................ 1
  Total.................... 4,923

Leaving
To rotary vacuum dryer, J-1
  $C_6H_6Cl_6$
    Alpha.................... 3,331
    Beta..................... 293
    Gamma.................. 605
    Delta and cogeners......... 509
  $C_6H_5Cl_7$.................... 38
  C_6H_6....................... 84
  C_6H_5Cl..................... 44
  $C_6H_3Cl_3$.................... 18
  H_2O........................ 1
  Total.................... 4,923

J-1. Rotary vacuum dryer

Entering
From wet crystal hopper, I-3
  $C_6H_6Cl_6$
    Alpha.................... 3,331
    Beta..................... 293
    Gamma.................. 605
    Delta and cogeners......... 509
  $C_6H_5Cl_7$.................... 38
  C_6H_6....................... 84
  C_6H_5Cl..................... 44
  $C_6H_3Cl_3$.................... 18
  H_2O........................ 1

Total....................... 4,923

Leaving
To dry BHC hopper, K-1
  $C_6H_6Cl_6$
    Alpha.................... 3,331
    Beta..................... 293
    Gamma.................. 605
    Delta and cogeners......... 509
  $C_6H_5Cl_7$.................... 38
  C_6H_6....................... 1
  C_6H_5Cl..................... 5
  $C_6H_3Cl_3$.................... 18
  Total.................... 4,800
To vacuum condenser and receiver, J-2
  H_2O........................ 1
  C_6H_6....................... 83
  C_6H_5Cl..................... 39
  Total.................... 123
Total....................... 4,923

J-2. Vacuum condenser and receiver

Entering
From rotary vacuum dryer, J-1
  C_6H_6....................... 83
  C_6H_5Cl..................... 39
  H_2O........................ 1
  Total.................... 123

Leaving
To recovered solvent pump, J-4
  C_6H_6....................... 83
  C_6H_5Cl..................... 39
  H_2O........................ 1
  Total.................... 123

J-3. Vacuum pump[1]

Entering
Air leakage (estimated)......... 10 cfm

Leaving
Air leakage................... 10 cfm

[1] An atmospheric condenser may be required beyond the vacuum pump.

J-4. Recovered solvent pump

Entering		Leaving	
From receiver, J-2		To solvent receiver, G-4	
C_6H_6	83	C_6H_6	83
C_6H_5Cl	39	C_6H_5Cl	39
H_2O	1	H_2O	1
Total	123	Total	123

K-1. Dry BHC hopper

Entering		Leaving	
From rotary vacuum dryer, J-1		To BHC packaging and storage, K-2	
$C_6H_6Cl_6$		$C_6H_6Cl_6$	
Alpha	3,331	Alpha	3,331
Beta	293	Beta	293
Gamma	605	Gamma	605
Delta and cogeners	509	Delta and cogeners	509
$C_6H_5Cl_7$	38	$C_6H_5Cl_7$	38
C_6H_6	1	C_6H_6	1
C_6H_5Cl	5	C_6H_5Cl	5
$C_6H_3Cl_3$	18	$C_6H_3Cl_3$	18
Total	4,800	Total	4,800

K-2. BHC packaging and storage

Entering		Leaving	
From dry BHC hopper, K-1		Packaged and stored	
$C_6H_6Cl_6$		$C_6H_6Cl_6$	
Alpha	3,331	Alpha	3,331
Beta	293	Beta	293
Gamma	605	Gamma	605
Delta and cogeners	509	Delta and cogeners	509
$C_6H_5Cl_7$	38	$C_6H_5Cl_7$	38
C_6H_6	1	C_6H_6	1
C_6H_5Cl	5	C_6H_5Cl	5
$C_6H_3Cl_3$	18	$C_6H_3Cl_3$	18
Total	4,800	Total	4,800

L-1. Solvent still feed pump

Entering		Leaving	
From solvent receiver, G-4		To solvent still, L-2	
C_6H_6	8,420	C_6H_6	8,420
C_6H_5Cl	1,008	C_6H_5Cl	1,008
$C_6H_6Cl_6$	24	$C_6H_6Cl_6$	24
$C_6H_3Cl_3$	3	$C_6H_3Cl_3$	3
H_2O	44	H_2O	44
Total	9,499	Total	9,499

CHAP. 3] PROCESS DESIGN 63

L-2. Solvent still

Entering
From solvent still feed pump, L-1
C_6H_6 8,420
C_6H_5Cl 1,008
$C_6H_6Cl_6$ 24
$C_6H_3Cl_3$ 3
H_2O 44

Total 9,499

Leaving
Tops to solvent still condenser, L-3
C_6H_6 8,410
H_2O 44
Total 8,454

Bottom to monochlor pump, M-1
C_6H_6 10
C_6H_5Cl 1,008
$C_6H_6Cl_6$ 24
$C_6H_3Cl_3$ 3
Total 1,045

Total 9,499

L-3. Solvent still condenser

Entering
From solvent still, L-2
C_6H_6 8,410
H_2O 44
Total 8,454

Leaving
To wet benzol receiver, L-4
C_6H_6 8,410
H_2O 44
Total 8,454

L-4. Wet benzol receiver

Entering
From solvent still condenser, L-3
C_6H_6 8,410
H_2O 44
Total 8,454

Leaving
To wet benzol pump, L-5
C_6H_6 8,410
H_2O 44
Total 8,454

L-5. Wet benzol pump

Entering
From wet benzol receiver, L-4
C_6H_6 8,410
H_2O 44
Total 8,454

Leaving
To benzol storage tank, A-2
C_6H_6 8,410
H_2O 44
Total 8,454

M-1. Monochlor pump

Entering
From solvent still, L-2
C_6H_6 10
C_6H_5Cl 1,008
$C_6H_6Cl_6$ 24
$C_6H_3Cl_3$ 3
Total 1,045

Leaving
To monochlor storage, M-2
C_6H_6 10
C_6H_5Cl 1,008
$C_6H_6Cl_6$ 24
$C_6H_3Cl_3$ 3
Total 1,045

64 CHEMICAL ENGINEERING PLANT DESIGN [CHAP. 3

RAW MATERIALS · PROCESSING · BY-PRODUCTS

BENZOL	
C_6H_6	2,019
C_7H_8	10

WATER	
H_2O	589

CHLORINE	
Cl_2	4,228

CHLORINATOR	
C_6H_6	10,429
C_7H_8	10
H_2O	2,382
HCl	447
Cl_2	4,228

VENT GAS		
Cl_2	42	To vent
C_6H_6	8	

DECANTER	
C_6H_6	8,434
C_6H_5Cl	1,013
BHC	4,800
$C_6H_5CCl_3$	21
H_2O	2,382
HCl	792
Cl_2	4

ACID RECYCLE	
H_2O	1,749
HCl	447

ACID STILL	
H_2O	2,338
HCl	778
C_6H_6	7
Cl_2	4

ACID BY-PROD.		
H_2O	589	To storage
HCl	331	
C_6H_6	7	
Cl_2	4	

CAUSTIC	
NaOH	31
Impur.	1

WATER	
H_2O	3,069

NEUTRALIZER	
C_6H_6	8,427
C_6H_5Cl	1,013
BHC	4,800
H_2O	3,113
HCl	14
$C_6H_5CCl_3$	21
NaOH	31
Impur.	1

SEPARATOR	
C_6H_6	8,427
C_6H_5Cl	1,013
BHC	4,800
H_2O	3,113
HCl	14
$C_6H_5CCl_3$	21
NaOH	31
Impur.	1

SPENT NaOH		
H_2O	3,076	
NaOH	16	To waste
NaCl	22	
C_6H_6	6	
Impur.	1	

RECYCLE	
C_6H_6	8,410
H_2O	44

SOLV'T STILL	
C_6H_6	8,420
C_6H_5Cl	1,008
BHC	24
$C_6H_5CCl_3$	3
H_2O	44

FLASH STILL	
C_6H_6	11,752
C_6H_5Cl	2,816
BHC	6,830
$C_6H_5CCl_3$	64
H_2O	67

MONOCHLOR.		
C_6H_6	10	
C_6H_5Cl	1,008	To storage
BHC	24	
C_6H_5CCl	3	

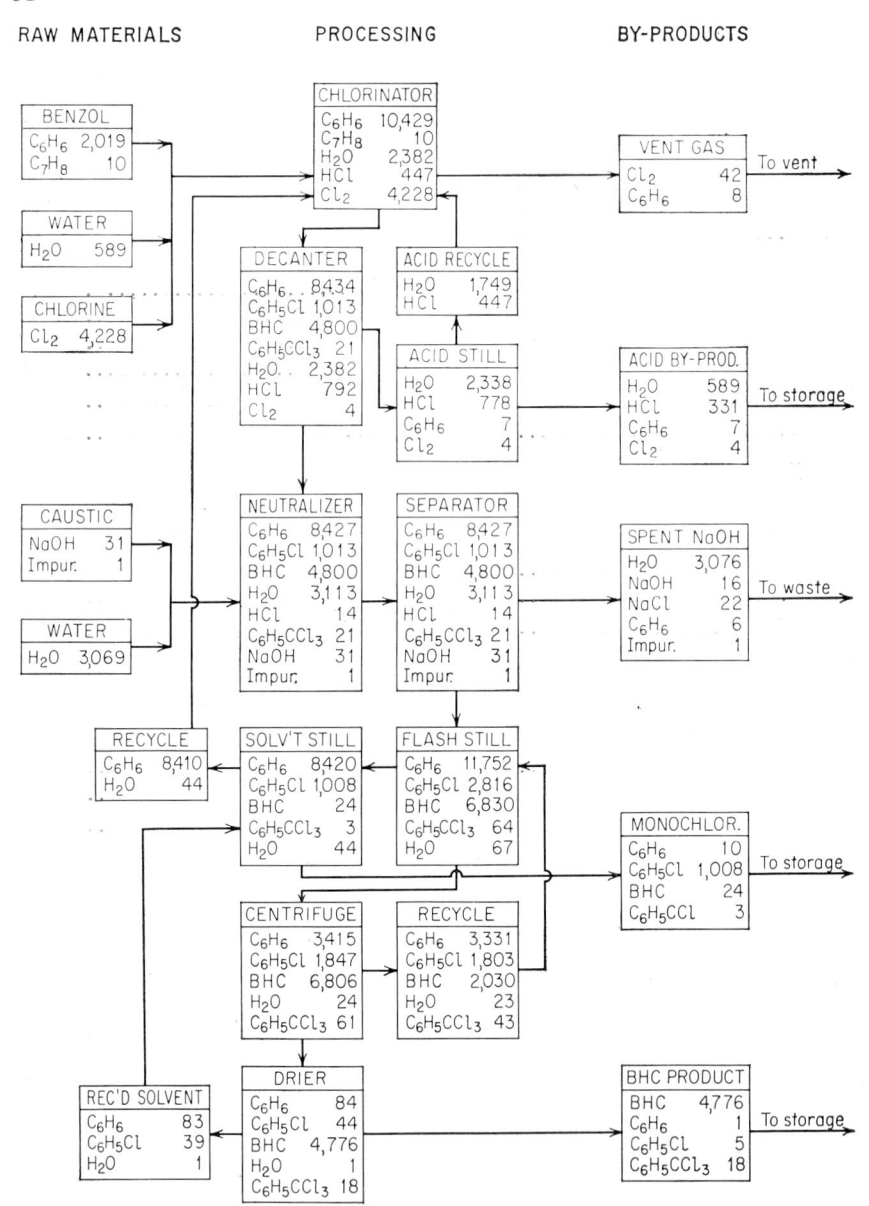

CENTRIFUGE	
C_6H_6	3,415
C_6H_5Cl	1,847
BHC	6,806
H_2O	24
$C_6H_5CCl_3$	61

RECYCLE	
C_6H_6	3,331
C_6H_5Cl	1,803
BHC	2,030
H_2O	23
$C_6H_5CCl_3$	43

REC'D SOLVENT	
C_6H_6	83
C_6H_5Cl	39
H_2O	1

DRIER	
C_6H_6	84
C_6H_5Cl	44
BHC	4,776
H_2O	1
$C_6H_5CCl_3$	18

BHC PRODUCT		
BHC	4,776	
C_6H_6	1	To storage
C_6H_5Cl	5	
$C_6H_5CCl_3$	18	

FIG. 3-3. Material balance flow sheet. Production of 4,800 lb of technical benzene hexachloride per 24-hr day. All values given in pounds.

M-2. Monochlor storage

Entering		Leaving	
From monochlor pump, M-1		Stored	
C_6H_6	10	C_6H_6	10
C_6H_5Cl	1,008	C_6H_5Cl	1,008
$C_6H_6Cl_6$	24	$C_6H_6Cl_6$	24
$C_6H_3Cl_3$	3	$C_6H_3Cl_3$	3
Total	1,045	Total	1,045

Material Balance Flow Sheet

A summary of the information obtained from the preceding calculations has been entered on the quantitative flow sheet of Fig. 3-3.

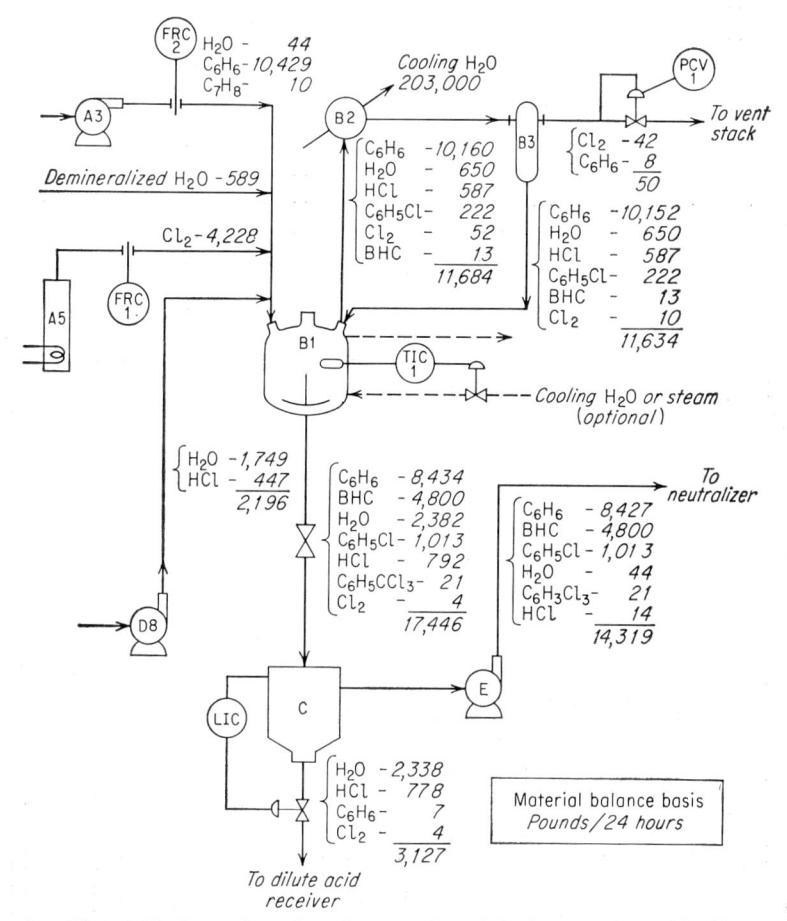

FIG. 3-4. Material balance flow sheet for a section of the benzene hexachloride process—includes equipment and instrumentation.

Location →	A3 → B1	A4 → B1	A5 → B1	D8 → B1	B3 → B1	B1 input
Material →	Lb/24hr					
C_6H_6	10,429				10,152	20,581
C_7H_8	10					10
H_2O	44	589		1,749	650	3,032
Cl_2			4,228		10	4,238
C_6H_5Cl					222	222
BHC					13	13
$C_6H_5CCl_3$						
HCL				447	587	1,034
Total in	10,483	589	4,228	2,196	11,634	29,130
Total out						
Gph	59.8	3.0		10.4		0

Location →	B1 → C or C input	B1 → B2	B1 output	B1 output −input	C → D1	C → E	C output	C output −input
C_6H_6	8,434	10,160	18,594	−1,987	7	8,427	8,434	0
C_7H_8		650	3,032	−10				
H_2O	2,382	650	3,032	0	2,338	44	2,382	0
Cl_2	4	52	56	−4,182	4		4	0
C_6H_5Cl	1,013	222	1,235	1,013		1,013	1,013	0
BHC	4,800	13	4,813	4,800		4,800	4,800	0
$C_6H_5CCl_3$	21		21	21		21	21	0
HCL	792	587	1,379	345	778	14	792	0
Total in								
Total out	17,446	11,684	29,130	0	3,127	14,319	17,446	0
Gph						87.8		

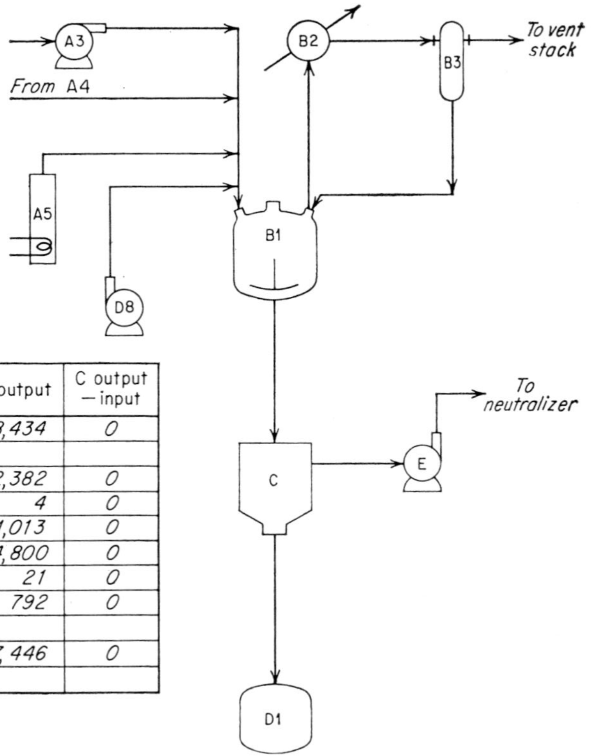

FIG. 3-5. Material balance flow sheet for a section of the benzene hexachloride process—shows tabular presentation of stream components.

CHAP. 3] PROCESS DESIGN 67

Several alternative methods for presenting material balances in graphic form will be mentioned. Figure 3-4 shows a method where all the stream components and weights are placed at the proper location on the flow diagram. There is only a section of the entire process shown here. This method has utility for very simple process flow sheets, but it can be seen that a rather cumbersome method of presentation results when the number of stream components is large.

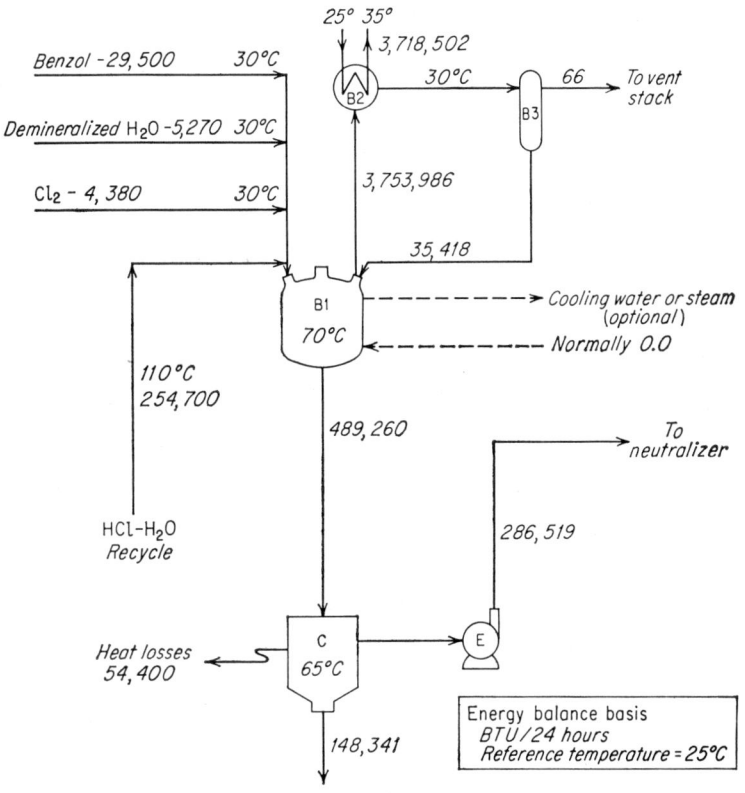

FIG. 3-6. Energy balance flow sheet for a section of the benzene hexachloride process.

To avoid this difficulty and to provide for easier checking of the material balance at intermediate points in the plant, a third method is presented in Fig. 3-5. The equipment and flow streams are coded by the diagram drawn on the right-hand side with components listed in vertical columns on the left-hand side. Each stream is identified by two equipment code numbers. For example, A5 → B1 means the flow from the chlorine vaporizer A5 to the chlorinator B1. Balancing of the material flows can be made by summing and comparing input and output data at

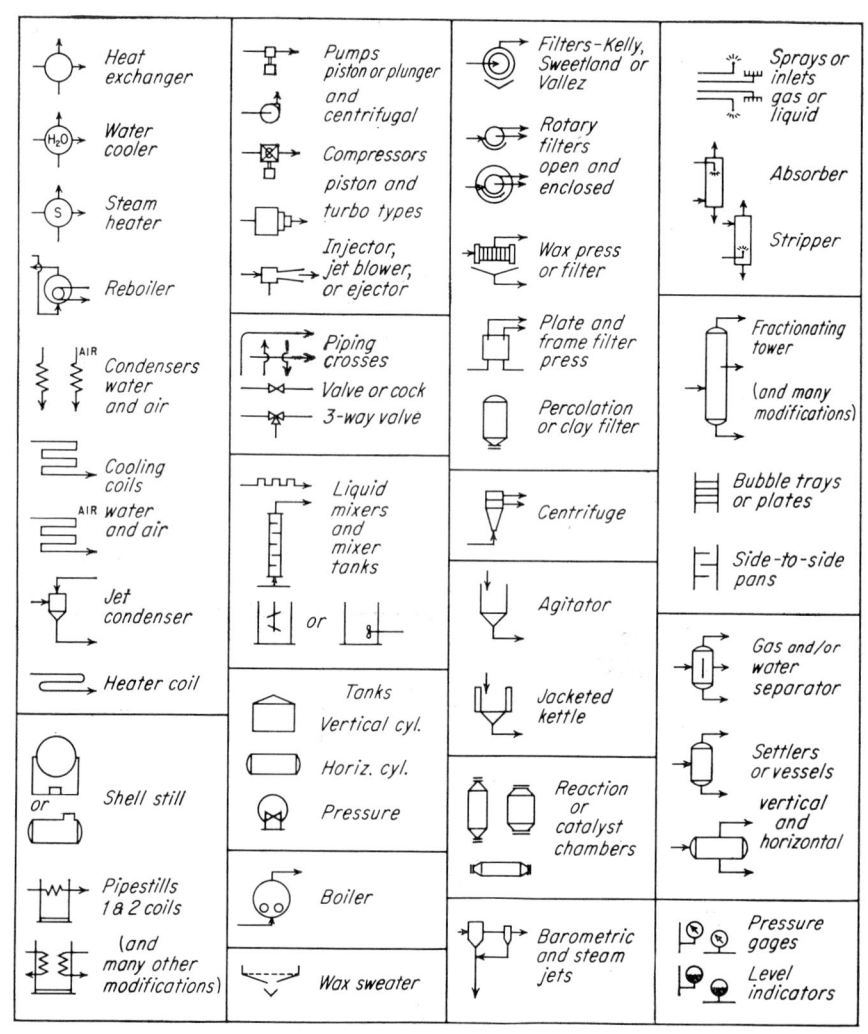

FIG. 3-7. Equipment symbols. (*Courtesy of W. L. Nelson, Oil Gas J., Dec. 9, 1944.*)

each point in the process flow sheet. In some cases, volumetric flow tabulations can be incorporated.

Supplementary reference material can be shown on the same flow sheets. These include density, molecular weight, and specific and latent heat.

Energy Balance

After a thorough material balance has been worked out, the mass quantities can be used to compute an energy balance. Temperature and pressure levels at various key points in the process, particularly at each major piece of equipment, will serve as the guide in making the heat balance. The results of these heat balances, usually in terms of Btu per

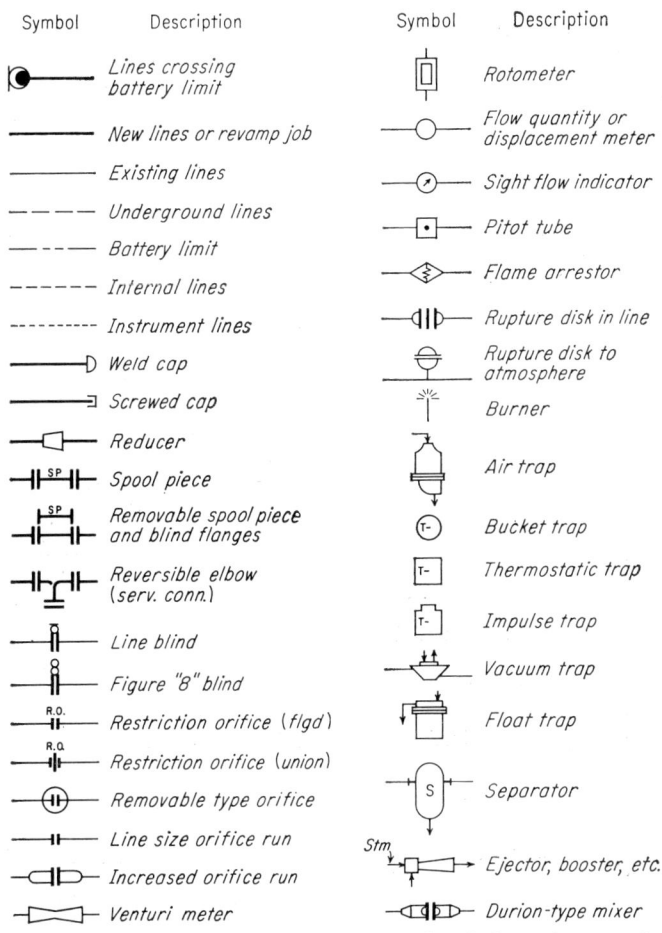

Symbol	Description	Symbol	Description
	Lines crossing battery limit		Rotometer
	New lines or revamp job		Flow quantity or displacement meter
	Existing lines		Sight flow indicator
	Underground lines		Pitot tube
	Battery limit		Flame arrestor
	Internal lines		Rupture disk in line
	Instrument lines		Rupture disk to atmosphere
	Weld cap		Burner
	Screwed cap		Air trap
	Reducer		
	Spool piece		Bucket trap
	Removable spool piece and blind flanges		Thermostatic trap
	Reversible elbow (serv. conn.)		Impulse trap
	Line blind		Vacuum trap
	Figure "8" blind		Float trap
	Restriction orifice (flgd)		
	Restriction orifice (union)		Separator
	Removable type orifice		
	Line size orifice run		Ejector, booster, etc.
	Increased orifice run		
	Venturi meter		Durion-type mixer

Fig. 3-8. Flow-sheet symbols, particularly for detailed equipment flow sheets. (*Courtesy of The M. W. Kellogg Co., New York.*)

unit time, can be made up in various types of flow-sheet forms as demonstrated for the material balance. Where convenient, major energy requirements are sometimes listed on the material balance flow sheets.

The complete energy balance calculations for the benzene hexachloride process are left as an exercise for the students. A typical portion of the process is shown in Fig. 3-6. Here, an energy balance around an exothermic chlorinator is described with heat removal in a reflux condenser.

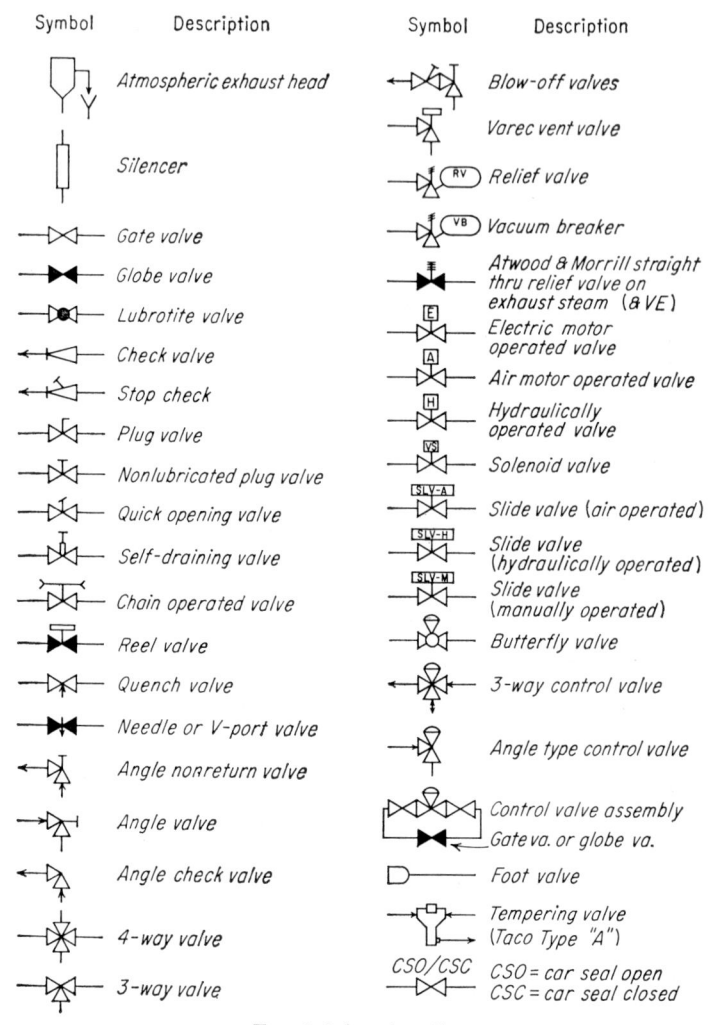

Symbol	Description	Symbol	Description
	Atmospheric exhaust head		Blow-off valves
			Varec vent valve
	Silencer		Relief valve
	Gate valve		Vacuum breaker
	Globe valve		Atwood & Morrill straight thru relief valve on exhaust steam (&VE)
	Lubrotite valve		Electric motor operated valve
	Check valve		Air motor operated valve
	Stop check		Hydraulically operated valve
	Plug valve		Solenoid valve
	Nonlubricated plug valve		Slide valve (air operated)
	Quick opening valve		Slide valve (hydraulically operated)
	Self-draining valve		Slide valve (manually operated)
	Chain operated valve		Butterfly valve
	Reel valve		3-way control valve
	Quench valve		Angle type control valve
	Needle or V-port valve		
	Angle nonreturn valve		Control valve assembly Gate va. or globe va.
	Angle valve		Foot valve
	Angle check valve		Tempering valve (Taco Type "A")
	4-way valve	CSO/CSC	CSO = car seal open CSC = car seal closed
	3-way valve		

FIG. 3-8 (*continued*).

Engineering Flow Diagrams

The engineering flow diagrams consist of various types.

1. *Simplified Equipment Flow Sheets.* An example is given on the right-hand side of Fig. 3-5 in symbol form. The interrelationships of major items of process equipment are shown by line connections.

Useful equipment symbols are shown in Fig. 3-7. Other types of symbols are illustrated in Fig. 3-2. Drawing templates for some of these symbols are available.[1] As mentioned previously, flow-sheet symbols are not as yet standardized within the industry. Hence, symbols should

[1] Compass Instrument and Optical Co., New York.

CHAP. 3]   PROCESS DESIGN   71

be chosen from the standpoint of clarity and simplicity with a fairly close resemblance of the actual equipment employed in the field.

2. *Detailed Equipment Flow Sheets.* These may be required but are not generally necessary for a preconstruction cost estimating design.

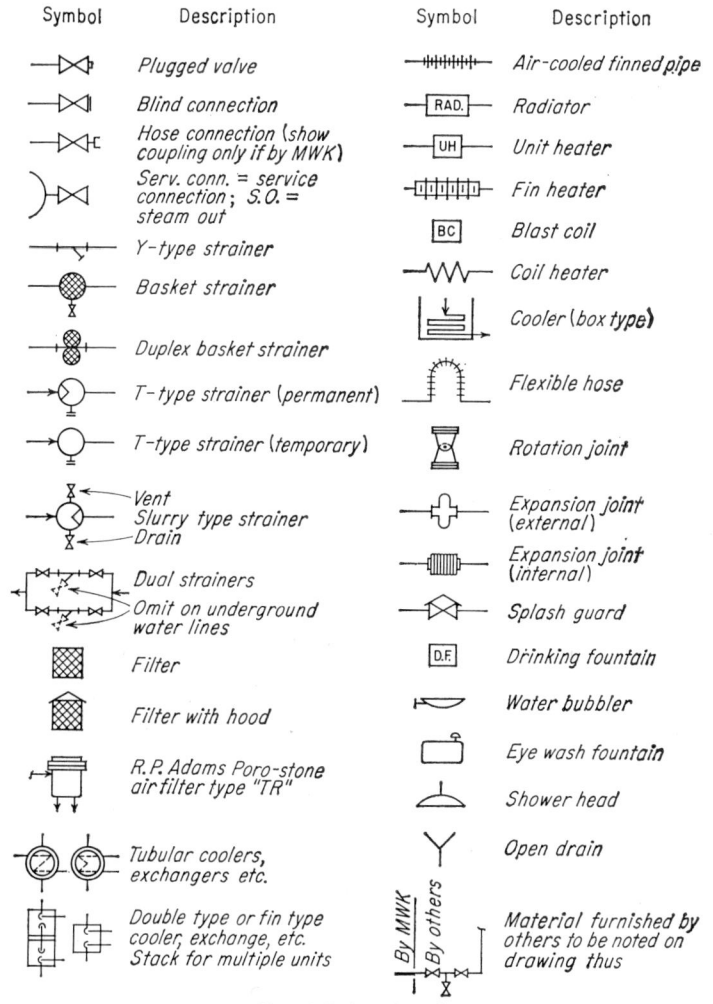

Symbol	Description	Symbol	Description
	Plugged valve		Air-cooled finned pipe
	Blind connection	RAD.	Radiator
	Hose connection (show coupling only if by MWK)	UH	Unit heater
	Serv. conn. = service connection; S.O. = steam out		Fin heater
	Y-type strainer	BC	Blast coil
	Basket strainer		Coil heater
	Duplex basket strainer		Cooler (box type)
	T-type strainer (permanent)		Flexible hose
	T-type strainer (temporary)		Rotation joint
	Vent / Slurry type strainer / Drain		Expansion joint (external)
	Dual strainers / Omit on underground water lines		Expansion joint (internal)
	Filter		Splash guard
	Filter with hood	D.F.	Drinking fountain
	R.P. Adams Poro-stone air filter type "TR"		Water bubbler
	Tubular coolers, exchangers etc.		Eye wash fountain
	Double type or fin type cooler, exchange, etc. Stack for multiple units		Shower head
			Open drain
		By MWK / By others	Material furnished by others to be noted on drawing thus

FIG. 3-8 (*continued*).

These would include process piping, valving, drains, bypasses, vents, etc., as well as the process equipment requirements. Such flow sheets are useful for plant construction work.

American Standards Association graphical symbols are available for piping detail work (see Fig. 9-2). Other useful symbols are shown in Fig. 3-8. One type of detailed equipment flow sheet is shown in Fig. 3-9.

72   CHEMICAL ENGINEERING PLANT DESIGN   [CHAP. 3

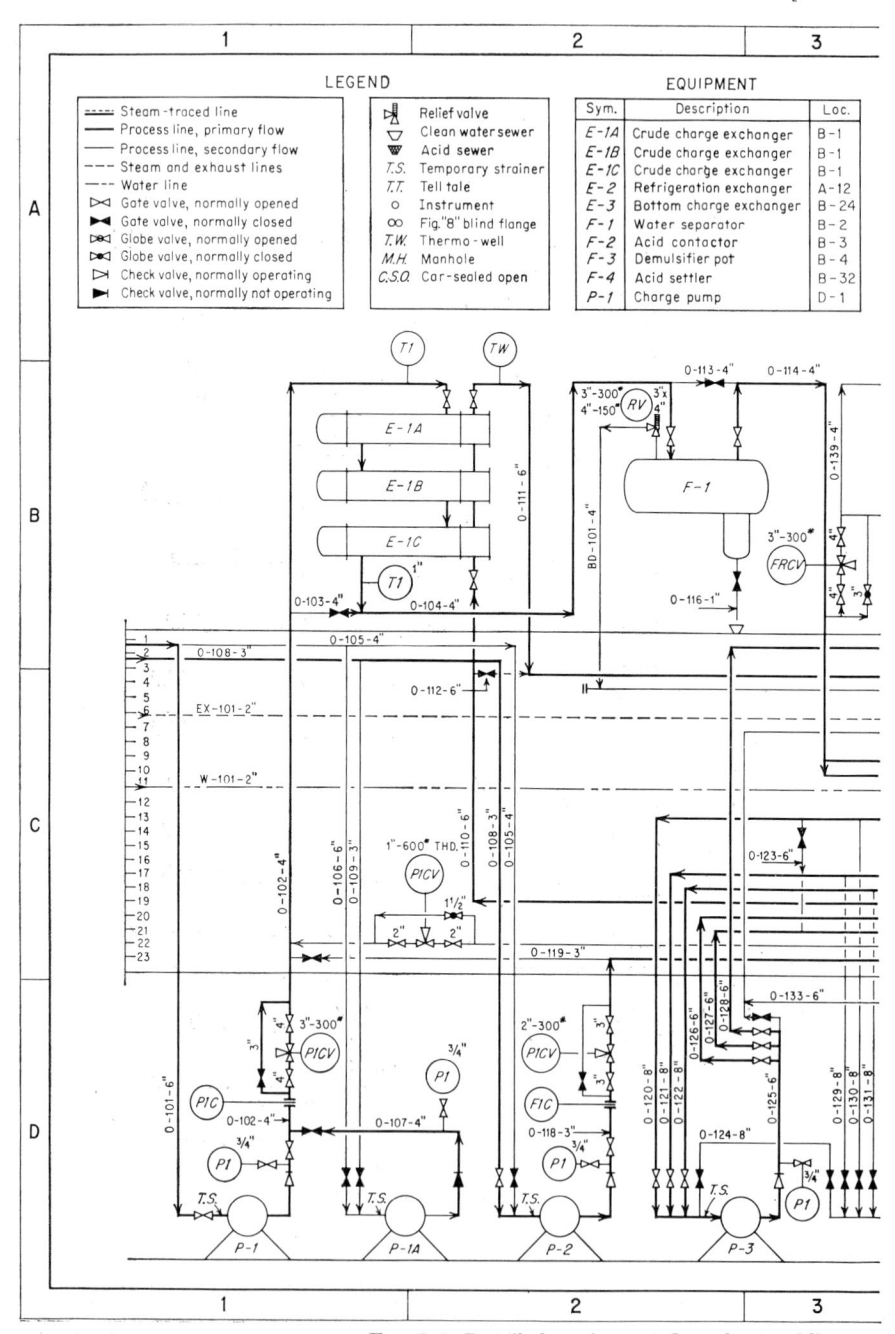

Fig. 3-9. Detailed equipment flow sheet. [*Courtesy*

PROCESS DESIGN

EQUIPMENT

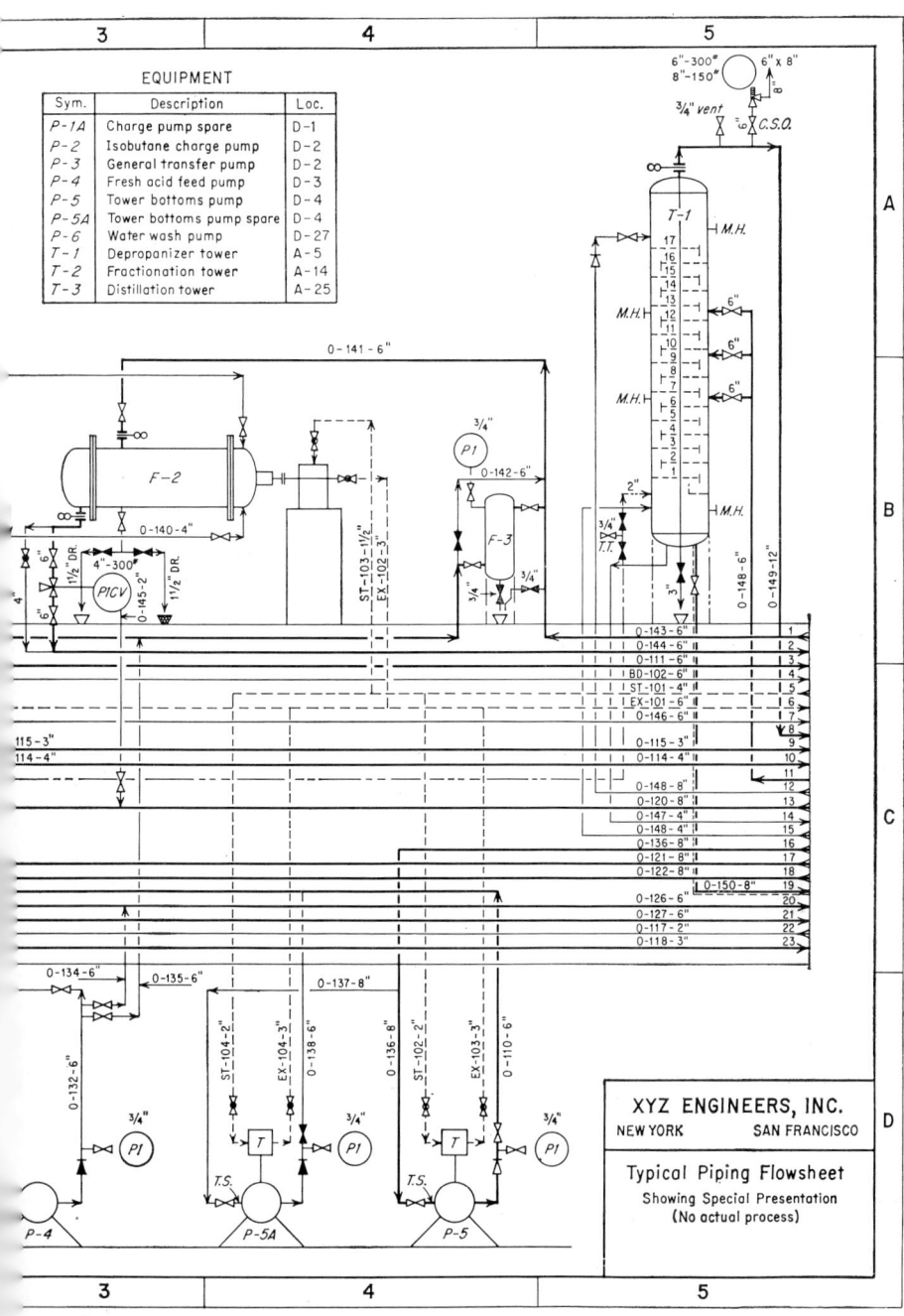

Sym.	Description	Loc.
P-1A	Charge pump spare	D-1
P-2	Isobutane charge pump	D-2
P-3	General transfer pump	D-2
P-4	Fresh acid feed pump	D-3
P-5	Tower bottoms pump	D-4
P-5A	Tower bottoms pump spare	D-4
P-6	Water wash pump	D-27
T-1	Depropanizer tower	A-5
T-2	Fractionation tower	A-14
T-3	Distillation tower	A-25

XYZ ENGINEERS, INC.
NEW YORK   SAN FRANCISCO

Typical Piping Flowsheet
Showing Special Presentation
(No actual process)

[. P. Romeo, *Chem. Eng.*, **64**(2): 256 (1957).]

3. *Instrumentation flow sheets* are useful in determining the requirements for process control and instrumentation. The Instrument Society of America has pioneering standardization of instrumentation flow plan symbols. These systems of nomenclature are illustrated in Tables 3-1

TABLE 3-1. INSTRUMENTATION: LETTERS OF IDENTIFICATION*
Definition and Permissible Positions in Any Combination

Upper-case letter	First letter: process variable or actuation	Second letter: type reading or other function	Third letter: additional function
A	——	Alarm	Alarm
C	Conductivity	Control	Control
D	Density	——	——
E	——	Element (primary)	——
F	Flow	——	——
G	——	Glass (no measurement)	——
H	Hand (actuated)	——	——
I	——	Indicating	——
L	Level	——	——
M	Moisture	——	——
P	Pressure	——	——
R	——	Recording (recorder)	——
S	Speed	Safety	——
T	Temperature	——	——
V	Viscosity	——	Valve
W	Weight	Well	——

* Reproduced from ISA-RP5.1 by permission of the Instrument Society of America.
Note 1: When required the following may be used optionally as a first letter for other process variables:
 a. "A" may be used to cover all types of analyzing instruments.
 b. Readily recognized self-defining chemical symbols such as CO_2, O_2, etc., may be used for these specific analysis instruments.
 c. The self-defining symbol "pH" may be used for hydrogen ion concentration.
Note 2: Although not a preferred procedure, when considered necessary it is permissible to insert a lower-case "r" after "F" to distinguish flow ratio. Likewise, lower-case "d" may be inserted after "T" or "P" to distinguish temperature difference or pressure difference.

and 3-2 with recommended symbols shown in Fig. 3-10. Copies of the complete listing can be obtained as Bulletin ISA-RP5.1 from the Instrument Society of America, Pittsburgh, Pennsylvania. For a further discussion on the subject of instrumentation and control, see Chap. 9.

In many cases, the instrumentation requirements can be incorporated in simplified engineering flow sheets, as shown in Figs. 3-2 and 3-4. If instrumentation is more complex, a separate flow diagram, bringing into bold relief all instruments and controls, is required. Along these lines, a successful scheme has recently been brought to the attention of the

TABLE 3-2. INSTRUMENTATION: COMPLETE GENERAL IDENTIFICATIONS*
(Combinations of Letters)

		Second and third letters (type of device)								
		Controlling devices			Measuring devices				Alarm devices		
		Separate controllers		Self-actuated (integral) regulating valves	Safety (relief) valves	Record-ing	Indi-cating	Glass devices for observation only (no measure-ment,	Record-ing	Indi-cating	Blind	Primary element	Wells
	Record-ing	Indi-cating	Blind										
Letter	-RC	-IC	-C	-CV	-SV	-R	-I	-G	-RA	-IA	-A	-E	-W
Temperature... T-	TRC	TIC	TC	TCV	TSV	TR	TI	—	TRA	TIA	TA	TE	TW
Flow... F-	FRC	FIC	LC	LCV		FR	FI	FG	FRA	FIA	LA	FE	—
Level... L-	LRC	LIC	PC	PCV	PSV	LR	LI	LG	LRA	LIA	PA	PE	—
Pressure... P-	PRC	PIC	DC			PR	PI	—	PRA	PIA			—
Density... D-	DRC	DIC	HC	HCV		DR	DI	—	DRA	DIA			—
Hand... H-		HIC	MC					—					—
Moisture... M-	MRC	MIC				MR	MI	—	MRA	MIA	MA	ME	—
Conductivity... C-	CRC	CIC				CR	CI	—	CRA	CIA	CA	CE	—
Speed... S-	SRC	SIC	SC	SCV	SSV	SR	SI	—	SRA	SIA	SA	—	—
Viscosity... V-	VRC	VIC				VR	VI	VG	VRA	VIA			—
Weight... W-	WRC	WIC				WR	WI		WRA	WIA		WE	—

First letter—process variable (or actuation)

Note: The optional additional process variables given in footnotes of Table 3-1, when used, shall be combined with second and third letters as per above. Dashes indicate impossible combinations; blanks improbable combinations.
* Reproduced from ISA-RP5.1 by permission of the Instrument Society of America.

75

BASIC INSTRUMENTATION SYMBOLS

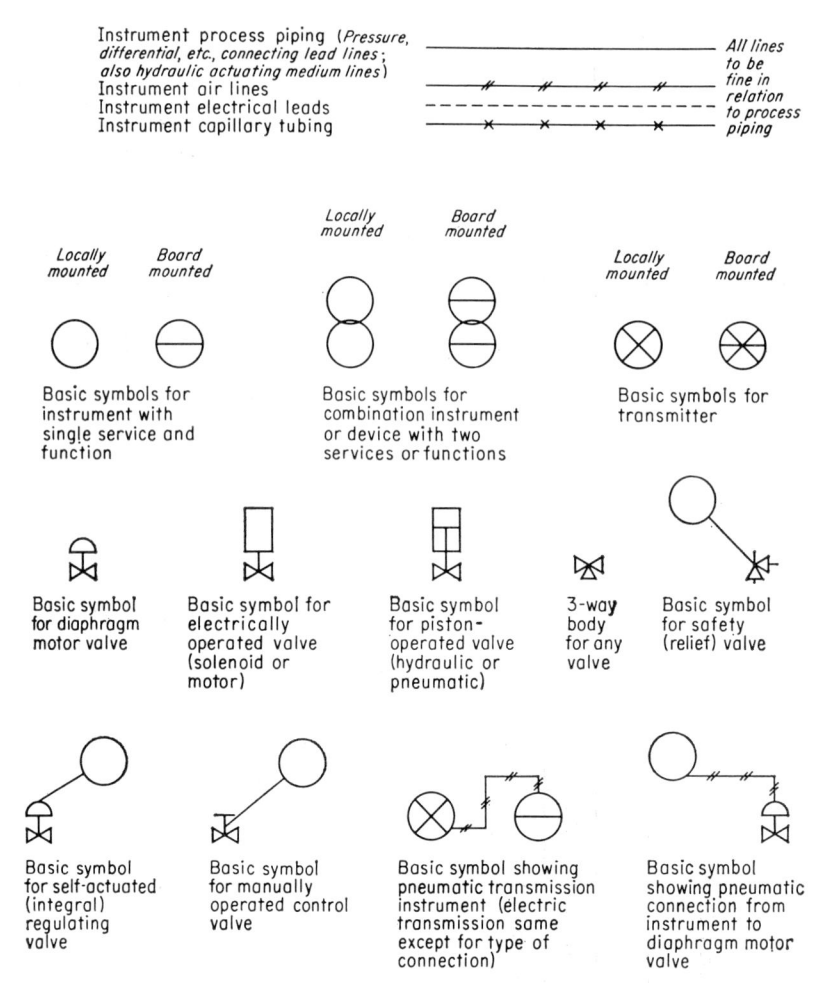

Fig. 3-10. Recommended basic instrument symbols. (*Reproduced from* ISA-RP5.1 *by permission of the Instrument Society of America.*)

authors.[1] A simplified equipment flow sheet is made up in black and white. Mounted on top of these are transparencies which contain one on each sheet, the detailed piping, the instrumentation, and the auxiliaries. Each transparent sheet is lined with different color ink. In this manner any or all of the flow sheets required for a complete engineering flow diagram can be observed separately or together.

4. *Auxiliary flow sheets* cover the auxiliary process requirements such as steam, water, fuel, air, and other utilities. As with instrumentation,

[1] W. T. Butler, Kennecott Copper Corp., private communication, 1956.

CHAP. 3] PROCESS DESIGN 77

TYPICAL INSTRUMENTATION SYMBOLS FOR FLOW

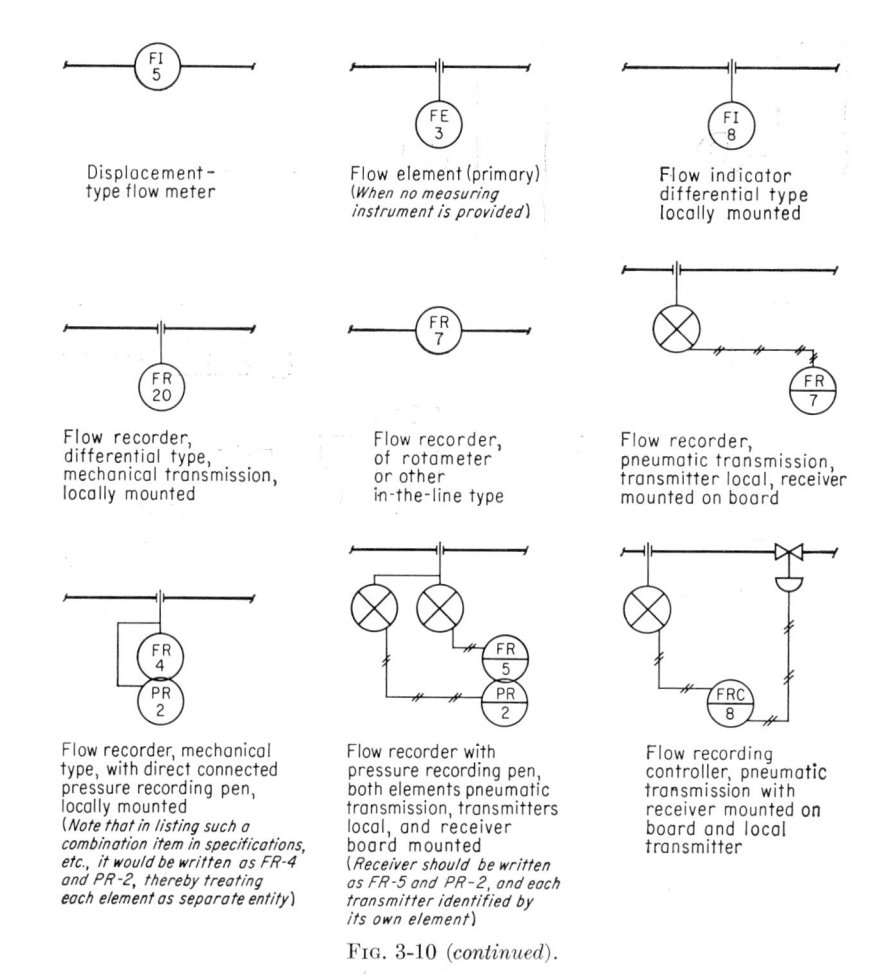

FI 5 — Displacement-type flow meter

FE 3 — Flow element (primary) (*When no measuring instrument is provided*)

FI 8 — Flow indicator differential type locally mounted

FR 20 — Flow recorder, differential type, mechanical transmission, locally mounted

FR 7 — Flow recorder, of rotameter or other in-the-line type

FR 7 — Flow recorder, pneumatic transmission, transmitter local, receiver mounted on board

FR 4 / PR 2 — Flow recorder, mechanical type, with direct connected pressure recording pen, locally mounted (*Note that in listing such a combination item in specifications, etc., it would be written as FR-4 and PR-2, thereby treating each element as separate entity*)

FR 5 / PR 2 — Flow recorder with pressure recording pen, both elements pneumatic transmission, transmitters local, and receiver board mounted (*Receiver should be written as FR-5 and PR-2, and each transmitter identified by its own element*)

FRC 8 — Flow recording controller, pneumatic transmission with receiver mounted on board and local transmitter

FIG. 3-10 (*continued*).

it is possible on some jobs to combine auxiliaries on the same flow sheet as the simplified equipment symbols. This was done by dashed lines in Fig. 3-4. One or more auxiliary flow sheets will be needed for more complex jobs.

Equipment Selection

As a result of coordinating material, energy, and engineering flow diagrams, the design engineer is now in a position to select possible types of equipment which will do the job and obtain a cost estimate for each major item. The selection of process equipment will be discussed next in Chap. 4.

78 CHEMICAL ENGINEERING PLANT DESIGN [CHAP. 3

TYPICAL INSTRUMENTATION SYMBOLS FOR LEVEL

Blind level controller, internal type

Gage glass

Level recorder, pneumatic transmission, with board mounted receiver, external type transmitter

Level indicating controller and transmitter combined with board mounted level indicating receiver

Level recording controller, external type, pneumatic transmission

Level alarm, internal type

Level recording controller and level recorder, pneumatic transmission combined receiver board mounted

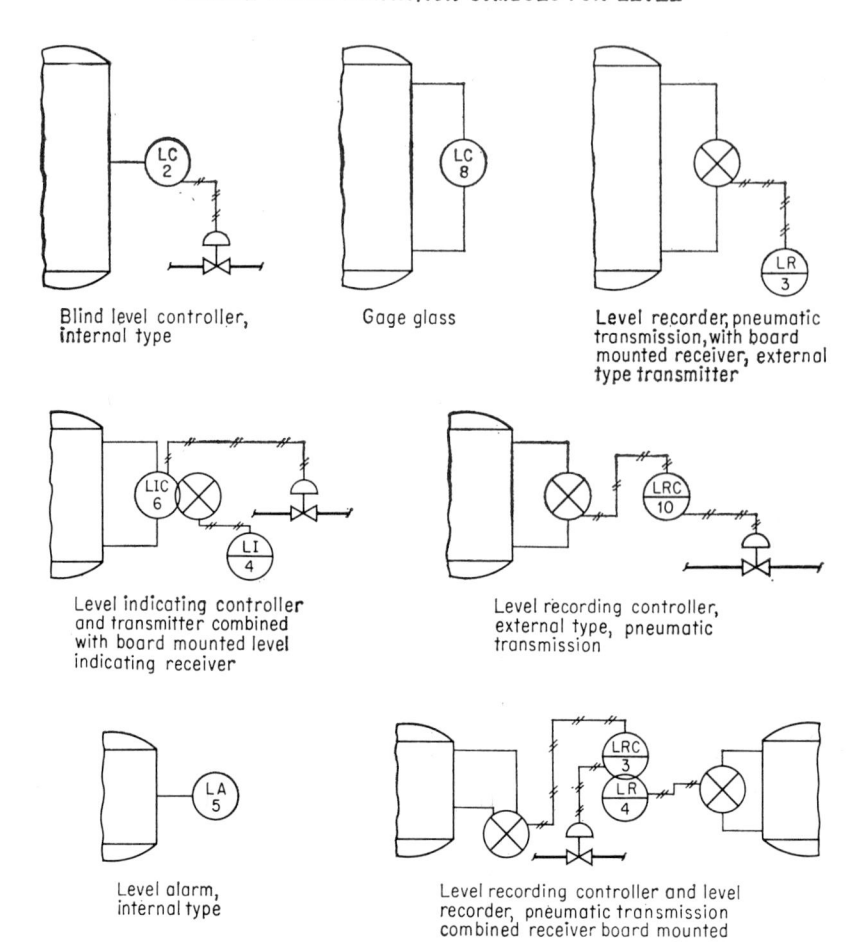

FIG. 3-10 (*continued*).

CHAP. 3] PROCESS DESIGN 79

TYPICAL INSTRUMENTATION SYMBOLS FOR PRESSURE

Pressure indicator, locally mounted

Pressure recorder board mounted

2-pen pressure recorder, board mounted, 1-pen pneumatic transmission

Pressure recording controller, pneumatic transmission, with board mounted receiver

Pressure alarm local

Self-actuated (integral) pressure regulating valve

Pressure controller, blind type
(*Show controller directly above diaphram if so mounted*)

Pressure recording controller (differential), pneumatic transmission; with pressure recorder, combined instrument board mounted

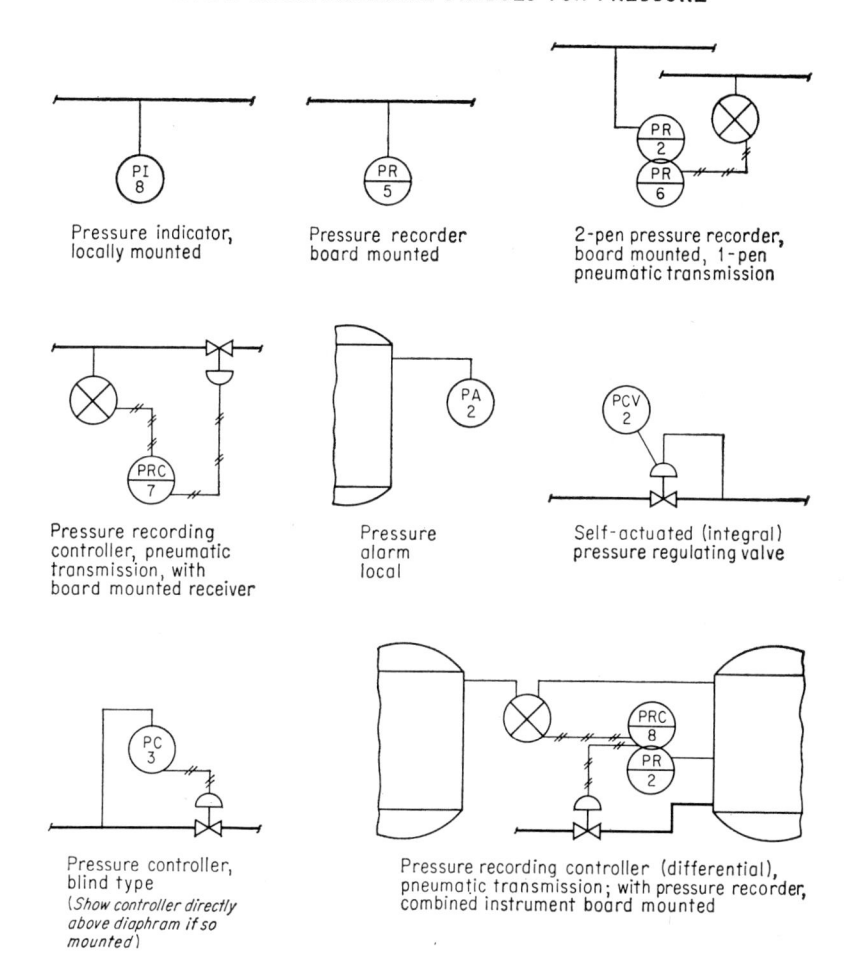

FIG. 3-10 (*continued*).

TYPICAL INSTRUMENTATION SYMBOLS FOR COMBINED INSTRUMENTS

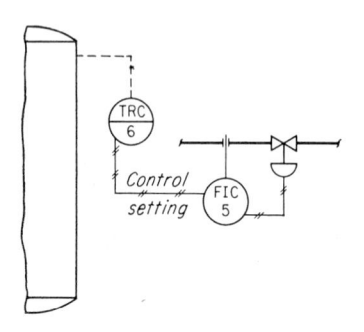

Temperature recording controller, board mounted, resetting locally mounted flow indicating controller
(*Note that "Control setting" should be shown alongside air line to indicate cascade control*)

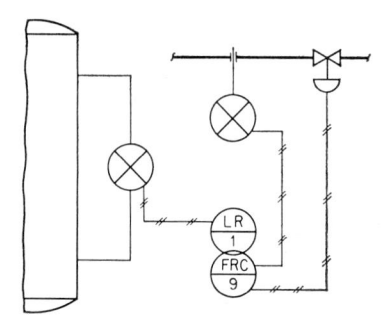

Flow recording controller with level record. Both elements pneumatic transmission. Level transmitter external type. Combined receiver board mounted

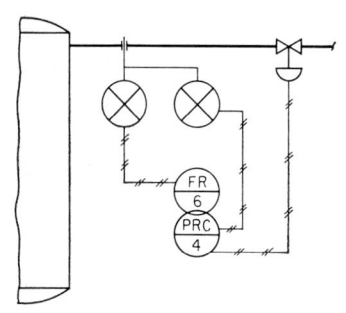

Pressure recording controller with flow record. Both elements pneumatic transmission, combined receiver board mounted

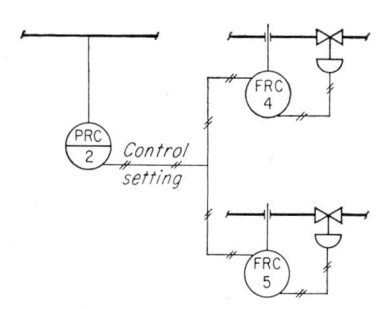

Pressure recording controller, board mounted, resetting locally mounted flow recording controllers

FIG. 3-10 (*continued*).

TYPICAL INSTRUMENTATION SYMBOLS FOR TEMPERATURE

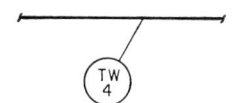

Temperature well (TW 4)

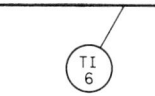

Temperature indicator or thermometer (local) (TI 6)

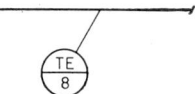

Temperature element without connection to instrument (TE 8)

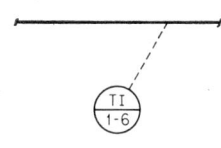

Temperature indicating point connected to multipoint indicator on board (TI 1-6)

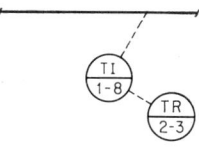

Temperature indicating and recording point connected to multipoint instruments on board (TI 1-8, TR 2-3)

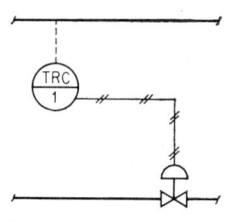

Temperature recording controller, board mounted (electric measurement) (TRC 1)

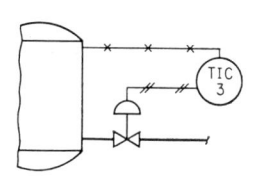

Temperature indicating controller, filled system type, locally mounted (TIC 3)

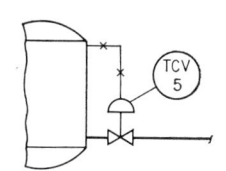

Temperature controller of self-actuated type (TCV 5)

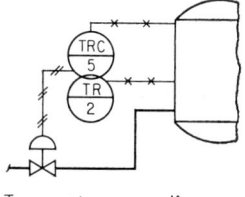

Temperature recording controller and temperature recorder, combined instrument board mounted (TRC 5, TR 2)

FIG. 3-10 (*continued*).

TYPICAL INSTRUMENTATION SYMBOLS (MISCELLANEOUS)

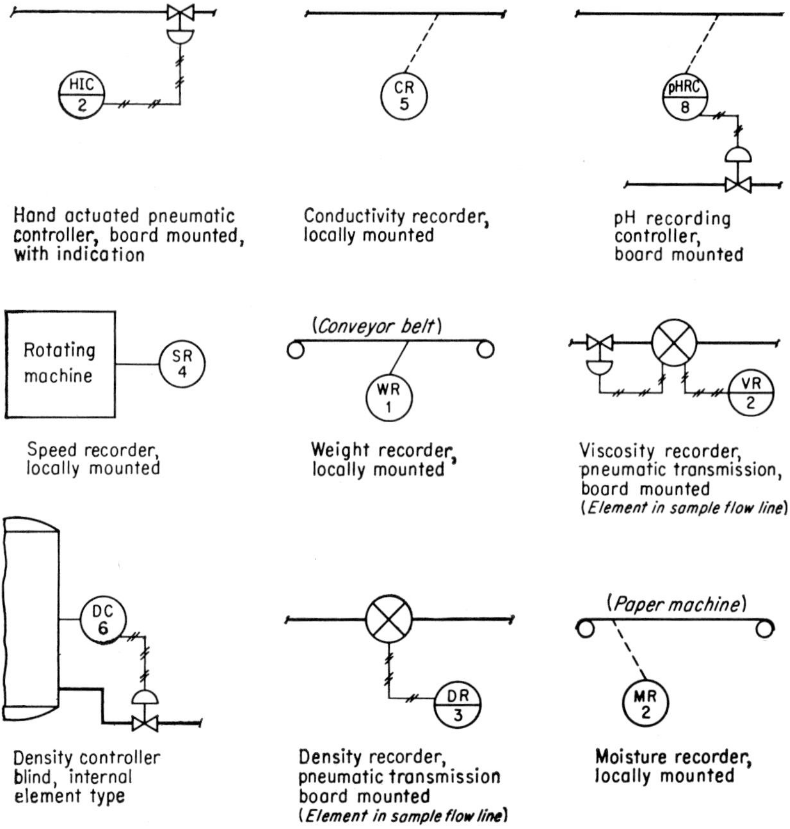

Hand actuated pneumatic controller, board mounted, with indication

Conductivity recorder, locally mounted

pH recording controller, board mounted

Speed recorder, locally mounted

Weight recorder, locally mounted

Viscosity recorder, pneumatic transmission, board mounted
(*Element in sample flow line*)

Density controller blind, internal element type

Density recorder, pneumatic transmission board mounted
(*Element in sample flow line*)

Moisture recorder, locally mounted

FIG. 3-10 (*continued*).

PROJECT

3-1. BASIC

Draw flow diagrams of assigned units or areas of maleic acid hydrazide production, based upon literature search: (*a*) production of maleic anhydride; (*b*) purification of maleic anhydride; (*c*) reaction for preparation of maleic acid hydrazide; (*d*) production of sodium hypochlorite; (*e*) production of hydrazine hydrate; (*f*) purification of hydrazine hydrate; (*g*) purification of maleic acid hydrazide.

PROBLEMS

3-1. Blocked Qualitative Process Flow Sheet

Write all general equations involved; then sketch a simplified qualitative process flow sheet with blocks for major unit operations, naming each block and including chemical formulation at each station.

3-2. Simplified Equipment Flow Sheet

Repeat Prob. 3-1, except that equipment symbols at each station be substituted for blocks, and that chemical compositions of each flow line be identified, and the chemical composition, with temperatures and pressures, be identified at each piece of equipment.

3-3. Basic Material Balance

Organize a material balance based upon fundamental chemical and equipment sequence; assume only pure reactants entering into the balance according to chemical equations. Place information on flow sheets as shown in Fig. 3-4 or Fig. 3-5.

3-4. Basic Energy Balance

Using information developed in Prob. 3-3, make an energy balance and make up a flow sheet such as shown in Fig. 3-6, or use tabular form as shown in Fig. 3-5.

3-5. Expanded Material Balance

Enlarge upon the material balance made in Prob. 3-3, including all closed-circuit or loop flows, all impurities, all side reaction products, and new operations equipment to dispose satisfactorily of all elements of flow and composition.

3-6. Expanded Energy Balance

Repeat Prob. 3-4, but use the information developed in the expanded material balance in Prob. 3-5.

3-7. Instrumentation Flow Sheet

Fill in the type of instrumentation for each piece of equipment on the flow sheet of Prob. 3-2.

3-8. Auxiliaries Flow Sheet

Draw a flow sheet basically similar to the flow sheet of Prob. 3-2, but substitute utilities used instead of chemicals or materials in each unit.

3-9. Detailed Engineering Flow Sheets

After process and plant design studies have been made (Chaps. 3 through 9), construct a model of the plant, including piping, instrumentation, equipment, and building layouts. Then draw detailed instrumentation, piping, and equipment layout flow sheets. (See Fig. 3-9 as a possible method.)

CHAPTER 4

Selection of Process Equipment and Materials

In Chap. 3, the methods used in selecting and pictorially representing a chemical process were discussed. The next step in arriving at a preconstruction cost estimation and the eventual construction of a chemical plant is the selection of (1) process equipment and (2) suitable materials of construction for all parts of the plant. The important decisions to be made in this type of work depend to some degree on engineering field experience. For any large-scale plant design, the final choice is made after critical review by seasoned process engineers. However, any well-trained engineering student can apply his own background of knowledge and common sense to make a reasonable selection. This chapter will point out some of the procedures which can be employed to carry out the job of selecting the best materials and equipment.

SELECTION OF MATERIALS

The design engineer who is responsible for selecting materials of construction must have a thorough understanding of all the basic process information available. He must be assured that the final choice of process raw materials has been made and that he can work with these materials for laboratory corrosion testing. Unless this choice has been definitely established, the materials of construction chosen for certain parts of the plant may be entirely unsuitable because of impurities which might be present in a different source of raw material. The selection of materials of construction can be logically planned with these points in mind.

Plan for Selection of Materials

Selection of materials is intimately connected with the design and selection of proper equipment. A brief plan for studying materials of construction is:

A. Preliminary selection:
 1. Experience
 2. Manufacturers' data
 3. Special literature
 4. General literature
 5. Availability
 6. Safety; mechanical and physical properties
 7. Preliminary tests by standard laboratory methods as check on deductions from experience, literature, and opinion

B. Laboratory testing:
 1. Revaluation of apparently suitable materials, with test pieces included in laboratory runs of the proposed processes

C. Application of data and final selection:
 1. Interpret laboratory results and other data in terms of plant operation, giving consideration to
   a. Presence of air in equipment
   b. Possibility of impurities
   c. Segregation of alloy constituents
   d. Fabrication method
   e. Avoidance of electrolysis
   f. Effect of temperature
   g. Effect of method of heating
   h. Effect of agitation
 2. Compare economic features of apparently suitable materials
   a. Material cost
   b. Production cost
   c. Probable life
   d. Lost-time costs
   e. Cost of product degradation
   f. Liability to special hazards
 3. Determine need for semiworks check of data

Some of the items listed in the above plan warrant discussion in more detail.

Preliminary Selection

From the standpoint of the student in a plant design course, the preliminary selection of materials based on published literature references is about the only recourse he has to solving the problem. To assist in this selection, a Bibliography is listed in the Additional Selected References at the end of the book. The most complete handbook on the subject of materials is Mantell, "Engineering Materials Handbook," McGraw-Hill Book Company, Inc., New York, 1958. A suitable source of information is Perry's "Chemical Engineers' Handbook," 3d ed. The tables appearing in this handbook are indexed below for ready reference.

Table 1. Chemical Resistance of Construction Materials; pp. 1461–1526.
Table 2. Resistance of Metals and Alloys to Corrosive Gases; p. 1526.
Table 3. Directory of Materials for Construction of Chemical Equipment; pp. 1527–1538.
Table 4a. Materials of Construction—Base Metals; Physical Properties and Methods of Fabrication; pp. 1539–1543.
Table 4b. American Iron and Steel Institute Standard Steels. Chemical Compositions; pp. 1543–1548.
Table 5. Precious Metals and Their Alloys; p. 1548.
Table 6a. Glass, Glass-lined, and Fused Silica Equipment; p. 1548.
Table 6b. Glass-lined and Enameled Steel; p. 1549.
Table 7. Chemical Stoneware, Porcelain, and Cements; p. 1549.
Table 8. Refractory Materials; p. 1549.
Table 9. Structural Carbon and Graphite; p. 1550.
Table 10. Physical Properties of Synthetic and Natural Rubber; p. 1551.
Table 11. Vulcanized Fiber; p. 1551.
Table 12. Wood for Chemical Equipment; p. 1552.
Table 13. Chemical Resistance of Gasket Materials; pp. 1554–1558.
Table 14. Concrete for Chemical-tank Construction; p. 1558.

Glass-lined equipment is used extensively by the chemical industry because of its inertness to a great many chemicals. For this reason, Table 4-1 is included in this text. As a general rule, glass-lined equipment is of the same order of magnitude costwise as the alloy or alloy-clad equipment. It is particularly suitable for pharmaceutical processes where trace impurities of metals in alloy construction may give trouble in the final product. Extreme care in installing and maintaining glass-lined equipment is required because of its poor resistance to mechanical shock.

A word of warning is necessary in the use of these tables. The choice of materials by this method is a tentative one and should by no means be considered as the final selection. The presence of impurities or galvanic corrosion effects can give entirely different results than predicted by the tables. It is for this reason that the outlined plan shown on page 85 stresses laboratory testing of possible materials of construction in contact with process materials. A procedure for testing can be found in Perry's "Chemical Engineers' Handbook," 3d ed., p. 1458.

In some instances for an *industrial* design problem, the manufacturer of the equipment may have sufficient experience with the type of process materials to render assistance in proper selection of materials.

Final Selection

Economic considerations have some part in the final selection of materials and such things as initial cost, maintenance, and probable life of the materials should not be overlooked. As an example of initial cost, the relative prices of a selected group of fabricated equipment using various materials of construction are: steel, 1.0; copper, 1.4; aluminum, 1.5; lead, 1.6; type 304 stainless steel, 2.0; type 316 stainless steel, 2.3; monel or nickel, 2.5; Inconel, 2.9; Hastelloy, 3.2. More detailed prices for equipment are found in Figures 6-3 to 6-30.

In making a choice on this basis, the engineer is often faced with the problem of where to use claddings or coatings over relatively cheap base materials such as steel or wood. For example, a vessel requiring an expensive alloy-steel surface in contact with the process liquid may be built with the alloy itself or with a cladding of the alloy on the inside of a carbon-steel structural material.

The data presented in Fig. 4-1 give a criteria for economic selection of stainless-steel-clad construction. Similar data for other metals can be obtained from cladding manufacturers.

Other examples of commercial coating for chemical process vessels

---

In the fabrication of pressure vessels, any comparison of clad steels versus the corresponding solid high alloy must take into consideration the allowable stress figures permitted in design. These stress allowances are established and published under the ASME Code.

The comparisons shown on Charts 1 and 2 are based on the 1956 Revision of Section 8 of the ASME Code for unfired pressure vessels. The clad-steel stress values used are based on the use of the full thickness of the composite plate as permitted by this Code. A factor of safety of 4 is used in all calculations.

Comparisons are made in the range of minus 20 to 100° and at 650, 800, and 900°F, and show the comparative merits of type A-285 Grade C flange quality, A-212 Grade B firebox quality, and A-204 Grade C firebox quality, as backing steels for the clad plate at each temperature. Comparisons are shown in each case for both 10 and 20 per cent clad steels. The extra charge for the A-212 and A-204 backing steels is included in the clad-steel calculations where applicable.

All data on the charts is expressed as savings in dollars per ton in use of clad steel of the types shown and with the backing steels shown at each temperature.

In the temperature range of 100 to 650°F, the allowable design stress values for the steel backing plates remain constant, while those for the solid high alloy material drop off gradually. Thus the comparative savings in clad steels over the solid high-alloy materials in this temperature range will lie between the values shown on the charts for minus 20 to 100° and for 650°.

In addition to the base price savings possible in the use of clad steel, these economies are also obtainable:

Standard overweight tolerances are substantially less for clad steel than the comparable plate of solid high alloy, including stainless steel and Inco alloys. This can mean especially important savings on large plates where high overweight can represent significantly increased cost.

Mill forming of clad steels is less expensive than solid high alloy for such items as heads and cylinders. The difference in forming cost for large-diameter heads represents a considerable saving with clad steels.

Note: All solid stainless-steel grades include the base extra of 6 cents per pound for annealing and pickling. This is done to make the solid base price comparable to that of stainless-clad steel since the annealing and pickling operations are included in stainless-clad base prices.

include baked ceramic or glass coatings, flame-sprayed metal, hard rubber, and many of the organic plastics. The durability of coatings is sometimes questionable, particularly where abrasion and mechanical wear conditions exist. As a general rule, if there is little to choose between a coated type versus a completely homogeneous material, a selection should favor the latter material, mainly on the basis of better mechanical stability.

SELECTION OF PROCESS EQUIPMENT

The selection of the types and sizes of equipment for the process plant requires considerable experience in this field to do the best job, particularly if the process is partially or completely new. If the process is an established one or in operation elsewhere, then the task is chiefly one of comparative calculations, scaling the equipment and accessories up or down, and incorporating pertinent innovations and improvements that past experience suggests. Any new process requires a complete study of the unit operations and unit processes involved and then follows a selection of the types and sizes of equipment required for guaranteed performance. It is true that the selection of the best equipment is a difficult yet surmountable job for a student taking a process design course where the process engineering is completely new. He will generally be without benefit of a thorough pilot plant study or be seasoned by field experience. However, he has the advantage of being freshly acquainted with the theory and calculations of unit operations. He can set up cost equations and determine optimum design,[1] a procedure frequently omitted for preliminary cost estimations. By suitable background reading in this chapter and other equipment books plus good design calculations, he can do an excellent job of equipment selection. The engineering student should bear in mind that this procedure of design and selection with lack of complete pilot data and experience is common practice for any new process of industry. Competence and success in this area of engineering is a goal worth achieving.

Equipment Selection Procedures

After the engineer has listed the equipment requirements from engineering flow sheets and made the necessary design calculations, he prepares a

---

[1] J. Happel, "Chemical Process Economics," John Wiley & Sons, Inc., New York, 1958.

J. O. Osburn and K. Kammermeyer, "Money and the Chemical Engineer," Prentice-Hall, Inc., Englewood Cliffs, N.J., 1958.

M. S. Peters, "Plant Design and Economics for Chemical Engineers," McGraw-Hill Book Company, Inc., New York, 1958.

H. E. Schweyer, "Process Engineering Economics," McGraw-Hill Book Company, Inc., New York, 1955.

specification form for each major piece of equipment using standard equipment if possible. If outside bids are required, detailed specification sheets must be presented to suppliers, as discussed later in this chapter. Standard forms are frequently available from individual vendors or from manufacturers' associations.

If the design is to be used only for preconstruction cost estimation and plant layout work, a standard specification sheet as shown in Table 4-2 is useful, either in part or as a whole. It is particularly suited for summarizing process equipment calculations in a plant design course.

TABLE 4-2. EQUIPMENT SPECIFICATION SUMMARY SHEET FOR PRECONSTRUCTION COST ESTIMATING

1. Code No. _____ on Flow Sheet No. _____   2. Date _____
3. Name of equipment: _____
4. Type: _____
5. Number required: _____
6. Process materials handled (type, composition): _____
7. Operating conditions: Temp. _____ Pressure _____
   Design throughput (mass or volume/unit time): _____
8. Volumetric capacity (gal or ft^3): _____
9. Dimensions: Ht. _____ Width or diam. _____ Length _____ Floor area _____
10. Principal design dimension (filtering, heat transfer, on screening area, conveyor length, etc.) _____
11. Recommended materials of construction: _____
12. Piping requirements: Inlet size (NPS) _____ Outlet size (NPS) _____
    Other fittings: _____
    Special piping hardware (relief and check valves, snubbers, etc.): _____

    Materials of construction: _____
13. Instrumentation requirements: _____
    Estimated cost (installed) _____
14. Utility requirements:
    Electric motors: type _____ hp _____ kva _____
    Other electrical equip.: type _____ kva _____
    Steam: _____ psi _____ lb/hr
    Gas: _____ ft^3/hr   Compressed air _____ ft^3/hr
    Cooling water: _____ °F max temp. _____ gph
15. Construction details: _____
16. Possible suppliers:
17. Estimated operating labor required:
18. Cost estimation summary [see Eq. (6-1)]
    Reference source _____
    Date of reference _____   Price index type _____
                                            Price index value (I_k) _____
    Their cost (C_k) _____   Basis–purchased or installed
    Present cost calculation:
      Date computed _____   Price index value (I_x) _____
      Your computed cost _____   Basis–purchased or installed
      Installation cost _____   Total installed cost _____
19. Remarks

The cost estimating section of Table 4-2 will be explained in Chap. 6. Use of published costs and adjusted prices is recommended for student design problems since it avoids antagonizing equipment manufacturers with student inquiries on equipment for hypothetical plants.

Some of the material which follows points out industrial methods of equipment selection where manufacturer contact and bidding are involved. This is inserted for completeness and to orient young engineers just starting out in this work.

Standard versus Special Equipment

The value of using standard equipment such as pumps or stock-size heat exchangers is well recognized in the chemical engineering field. Performance and service are demanded from all equipment; mistakes in judgment are hazardous—and inexcusable—if service data on like equipment for a similar or related process are available. The experience of others is quite valuable and should be used as fully as possible. Since the good will of the chemical engineer is the aim of all equipment manufacturers, they are desirous of giving service. Much valuable information for the solution of problems is ready for the asking, available from manufacturers who see possibilities of placing orders for equipment. However, they are equally anxious not to enter a field or process wherein they find that their equipment will not give satisfactory service.

Although it is a chemical engineering axiom to select standard equipment whenever possible, oftentimes the engineer is confronted with the situation in which his problem requires a special design and probably the use of special materials. In such cases he must draw upon his training and experience to design the requisite equipment. To do this need not awe any designer; he has his specifications; he understands the rules of machine designing; all he needs to do is to apply himself to the task of converting his specifications into a line picture or workshop drawing which the shopmen can convert into a three-dimensional piece of equipment. Much of the equipment for materials handling and for unit processes is standardized and, whenever such equipment will serve the purpose, it should be selected in preference to special designs. Not only will the first cost be substantially lower, but the duplication of equipment and the making of repairs on old equipment will be made much easier.

One should assure himself that he has completely exhausted the trade literature for his requirements before he embarks on the design of special equipment. Standard equipment has been tried out and has stood the rigorous test of service. It has produced results and gone through long periods of experimentation. Usually it is a result of many modifications of its original design. Standardization means not only a minimum cost in manufacturing but also that a machine built according to standard methods and in standard sizes has usually been given the best of thought in its designing. Under such circumstances, the manufacturers can and do deliver equipment under a guarantee of satisfactory performance. A new design is as much an experiment for the user as for the designer; it must stand up under use to acquire recognition for the giving of service. But when the engineer finds himself in a situation demanding the design of new equipment, he should have no hesitancy about executing the commission.

Specifications

Before one makes a search of the "Chemical Engineering Catalog" and the trade literature files, or before he corresponds with manufacturers

TABLE 4-3. FRACTIONATING COLUMN SPECIFICATIONS

REQUIRED: Two (2)
TYPE AND SIZE:
Supply ...., one still-pot, fractionating column, complete with dephlegmator condenser, all accompanying piping, gauges (pressure and vacuum), and thermometers. The unit is to be used for pilot plant study and shall conform to the following specifications:
1. No. of plates: 20
2. Plate spacing: 8 to 10 in.
3. Type plates: bubble cap—skirt
4. Column diameter (inside): 12 in.
5. Still-pot heating surfaces: approx. 16 sq ft
6. Still-pot volume: approx. 30 gal

The unit shall be built to stand operation under a maximum vacuum of 15 in. Hg. The still pot shall be jacketed for steam heating where pressures will not exceed 150 psig.

All the plates shall be built to stand operation equipped with sampling cocks for withdrawal of samples, and affording complete drainage when desired.

A weir or other suitable flow-measuring device shall be provided for measure of reflux ratio, in the range of from 2 to 6 gph (liquid sp gr, 1.12). The weir will be calibrated by the purchaser.

All necessary supporting structures shall be supplied by manufacturer, along with two receiving tanks, each of approx. 20 gal capacity.

MATERIALS:
Of construction:
1. All metals contacting liquids or vapors shall be Type 304 stainless steel.
2. All gasketing shall be neoprene.
Handled:
1. $C_6H_6Cl_6$ and other chlorinated hydrocarbons. Approx. 80 lb/hr is to be handled as product; bottoms are to be approx. 10 lb/hr.

INSTALLATION:
Unit is to be installed by purchaser. Assembly by the manufacturer is to be as nearly complete as possible, and shipment shall contain instructions.

DELIVERY:
Immediate to 30 days delivery.

ALLOWABLE VARIATION:
One-half to two times specifications as to capacity; any substitutions, including any pilot plant fractionating column on hand, is acceptable, pending approval of purchaser.

TABLE 4-4. COMPARISON OF EVAPORATOR BIDS FROM FOUR MAJOR EQUIPMENT VENDORS

Vendor	1	2	3	4
Operating service requirements				
Steam, lb/hr (psig)	99,500 (110)	106,000 (65)	106,000 (75)	92,000 (110)
Main hp	450	300	375	375
Water, gpm (°F)	7,600 (88)	7,200 (90)	7,200 (88)	6,700 (80)
Operating hr/day	21.6	22.0	22.0	22.0*
Principal evaporation components:				
Type evaporator	Triple effect, backward feed	Triple effect, backward feed	Triple effect, backward feed	Triple effect, backward feed
	Horizontal	Horizontal	Horizontal	Vertical
Heaters	1st—3,500 ft², † 2d—8,300 ft², 3d—8,300 ft², total 20,100 ft²	1st, 2d, and 3d avg 6,500 ft² each, 19,500 ft², total	1st, 2d, and 3d each 7,200 ft², total 21,600 ft²	1st, 2d, and 3d each 6,350 ft², total 19,050 ft²
Main circ. pumps	Centrifugal (3), 150 hp each	Axial flow (3), 100 hp each	Centrifugal (3), 125 hp each	Mixed flow (3), 125 hp each
Separators	Centrifix units top each evaporator body 9 ppm	Centrifugal type in each body	Centrifix unit top each body, less than 9 ppm	Open-end pipe design; no centrifugal element. Do not know if 9 ppm can be reached, 25 ppm max
Barometrics	Included	Included	Included	Included
Salt crystal	Reasonable for centrifugal filtration	Majority plus 60 mesh, uniform	Satisfactory for filtering on centrifugal unit	Special emphasis on growing large crystal 60 mesh average
Units operating	Yes	Yes	No	No
Guaranteed	Yes	Yes	Yes	Yes

94

Equipment furnished and materials of construction:				
Evaporator bodies:				
1st effect	10 dia. × 33 ft × 7/16 in.	17 dia. × 27 ft × 1/8 in. min. nickel thickness on all effects	11 dia. × 15 ft × 3/8 in.	18 dia. × 9 ft × 3/4 in.
2d effect	10½ × 33 ft × 7/16		14½ × 15 ft × 7/16	18 × 9 ft × 3/4
3d effect	11½ × 25 ft × 7/16		20 × 15 ft × 7/16	20 × 9 ft × 3/4
	All 20% Ni-Clad		All 20% Ni-Clad	All 10% Ni-Clad
Circulating pumps				
1st effect, 2d effect, 3d effect	13,100 gpm each, solid nickel, Ingersoll-Rand Included	Gpm not given, solid nickel, Bingham ‡	11,000 gpm each, solid nickel, Ingersoll-Rand ‡	16,000 gpm each, solid nickel, Ingersoll-Rand Included
Motors				
Salt tanks	Incorporated in evaporator body	1st—10 × 10 ft nickel-clad steel, 1/8 in. min. Ni	1st—8 dia. × 4 ft, 2d—10 × 4 ft, 3/16 in. solid nickel wall, w/60° cone	3, 18 ft dia. tapered to 15.75 × 10 ft × 3/4 in., 10% Ni-Clad
Overhead vapor piping	All carbon steel, w/1/8 in. corrosion allowance, prefab.	Nickel where in contact with liquid, balance carbon steel, prefab.	Monel from 1st effect, carbon steel from 2d and 3d, monel from flash tank to 3d effect, prefab. except for one field weld each	Nickel elbow at evaporator, remainder carbon steel, prefab.
Large liquor piping	Nickel, 3/16 in. wall, prefab.	Nickel, prefab.	Nickel, 10 ga. wall, prefab.	Nickel, prefab.
Small process piping	Liquor piping connecting effects and pumps of nickel, valves and fittings, control valves of nickel. Furnished prefab.	Nickel piping and fittings. Control valves butterfly type, others plug type. Piping furnished random lengths for fabrication by purchaser	Nickel piping w/monel valves, control valves butterfly type. Pipe prefab.	Nickel Sch. 10 in. random length for fabrication by purchaser
Instrumentation	Complete	Minimum requirements	Complete	Very complete
Miscellaneous by vendor	Structural-steel supports (no platforms, walks, stairways)	Mountings (not supports)	Mountings (not supports)	Supports not included

TABLE 4-4. COMPARISON OF EVAPORATOR BIDS FROM FOUR MAJOR EQUIPMENT VENDORS (*Continued*)

Vendor	1	2	3	4
Engineering drawings	Equipment—genl. assembly, plans, elevations, flow sheets, piping	Equipment—genl. assembly, plans, elevations, flow sheets, piping	Equipment—genl. assembly, plans, elevations, flow sheets, piping	Equipment—genl. assembly, plans, elevations, flow sheets, piping
Code construction	ASME, heaters only to be stamped	Not given	Not given	ASME-U-69, heaters only to be stamped
Cost comparison:§				
Equipment and material by vendor	1.00	0.997	0.875	1.18
Extra rigging for field assembly	None	1.5	None	1.8
Piping erection	1.00	Included above	1.00	Included above
Structural steel	Included	4.0	3.5	3.5
Foundation extra, due to weight and size	None	2.5	2.5	2.5
Extra for spare pipe and equipment required by purchaser	1.00	0.402	0.402	None
Extra instruments	1.00	1.50	1.00	None
Extra engineering	None	0.90	0.80	1.00
Freight	1.00	1.18	Included	1.27
Final ratios (expected erected cost)	1.00	1.08	0.927	1.21

* Not given, assumed from other information.
† First effect is considered unit being fed primary steam, thus producing the highest concentration of product.
‡ Not furnished.
§ Cost ratios are on an erected basis referenced to vendor 1. The ratios given hold relatively within a given row and are not additive in any column.

From E. E. Ludwig and A. F. Shockey, *Chem. Eng.*, **62**(1): 183 (1955).

of equipment, he should formulate a carefully written specification in which ranges of performance and other requirements have been carefully worked out. The writing of specifications must not be considered a special art, but rather a requisite[1] of every chemical engineer. The specifications should contain all information deemed essential, including composition, physical and chemical characteristics of materials handled, kind and quality of service available, service requirements on the equipment, packing, and marking of containers, delivery requirements, and quotations. Manufacturers of equipment ordinarily supply a form in which are included the questions that the individual manufacturer deems sufficient, if answered, to supply him with the information he needs to satisfy the demands. However, as excellent as this service is, the time that would be lost in correspondence may often be saved by sending a well-written specification to the manufacturer (see Table 4-3).

Specifications for Competitive Bidding. For practically all large-scale equipment purchases, competitive bidding is required for economic reasons. Ludwig and Shockey[2] show one example (Table 4-4) of how the design engineer can tabulate bids for comparative study. This system makes for easy review of the major items covering service requirements, principal components for the proposed flow cycle, accessories furnished, together with materials of construction and cost ratio on an installed basis. Delivery time, experience, and reliability of the supplier and total costs are important items to be considered in the final selection of the supplier.

For sake of completeness, a more detailed discussion of a typical selection problem is justified. Take the evaporator problem on which comparative bids were prepared and listed in Table 4-4. The design engineer's analysis is shown in Table 4-5. He must clearly understand the problem to analyze what is known, unknown, and desired. Each new application of a unit operation, or equivalent piece of equipment designed for carrying out a unit operation, should receive individual attention. It is oftentimes better, when dealing in the area of uncertainties, to supply to equipment manufacturers only the process data rather than exact equipment specifications as visualized by the design engineer. On the theory that two or more heads are better than one, several equipment manufacturers should be permitted to propose independently equipment and cycles of operation. In this manner a better performance and relative cost selection can be made.

[1] D. T. Canfield and J. H. Bowman, "Business, Legal, and Ethical Phases of Engineering," 2d ed., McGraw-Hill Book Company, Inc., New York, 1954.
[2] E. E. Ludwig and A. F. Shockey, Guide for Picking Bids, *Chem. Eng.*, **62**(1): 183 (1955).

TABLE 4-5. How a Problem Breaks Down*

Known

    Production rate and analysis of product.
    Feed flow rate, analysis, temperature, approximate quantity available.
    Services available to serve as design basis for steam, water, gas, etc.
    Disposal of condensate (location) and its purity.
    Probable materials of construction.

Unknown

    Evaporation cycle, pressure, temperature, solids capacities, and concentration features of various evaporation cycles.
    Number of evaporator effects.
    Best type of evaporator body and heater arrangement.
    Practice of competitors in same and related products.
    Best instrumentation.
    Experience record of manufacturer.

Desired

    Answers to unknown above.
    Operational guarantee.
    Material and workmanship guarantee.
    Performance record of evaporator manufacturer in same or related process.
    Filtering characteristics of any solids or crystals.
    Details on:
        Equipment dimensions, arrangements, fabrication details.
        Heat-transfer surfaces, velocities of fluids, coefficients.
        Separator elements for purity of overhead vapors.
        Types and details of circulating pumps, if any.
    Utility requirements:
    Steam
    Power
    Water
    Air

Review These Design Features

Evaporator body

Is vapor velocity above the boiling solution reasonable for reducing entrainment? Refer to separator manufacturer for optimum and limiting velocities.

Is the body adequately baffled to reduce upward entrainment yet maintain low pressure drop?

Are there limiting (minimum or maximum) operating conditions, and what are they? This refers to flexibility in operation.

A cone of about 60° is required to prevent salt bridging in the body and consequent plugging.

Growth of salt crystals must be controlled to secure good crystal size. Too high a degree of supersaturation yields fine, hard-to-filter crystals.

Heaters

Are heaters horizontal or vertical? In either case, check relative arrangement to ensure liquor head covering tubes to avoid dry tubes in heater.

TABLE 4-5. How a Problem Breaks Down (*Continued*)

Is liquor flowing through tubes at about 6–10 fps average velocity? This appears to be reasonably good velocity for operation.

What is pressure drop through liquor side of heater?

A reasonable "U" value for over-all heat transfer appears to be 400 to 500 Btu/(hr)(°F)(ft^2). Some units may run as high as 700 to 800, but check to determine if this value can be maintained under dirty tube conditions.

Entrainment separators

Is the evaporator body provided with auxiliary means for reducing liquor carryover with the overhead vapor?

What is the required outlet vapor specification and what is the guarantee on the separating device?

* E. E. Ludwig and A. F. Shockey: *Chem. Eng.*, **62**(1): 181 (1955).

SELECTION OF EQUIPMENT FOR CHEMICAL ENGINEERING UNIT OPERATIONS AND PROCESSES

It is obvious from the above discussions that the design engineer must be thoroughly familiar with unit operations and processes of chemical engineering to select the required equipment with possible alternatives. A student or inexperienced engineer will require background material to aid in his selection. He has already had thorough study and calculation training in the conventional unit operations, including absorption, distillation, fluid flow, and heat transfer. He should also be familiar with the unit processes as given in Groggins.[1] The information compiled in the next sections of this chapter and the References serve to supplement this basic knowledge, particularly from an equipment selection standpoint. There are other sources which provide additional knowledge for a better over-all choice of process equipment. For convenience of the reader, the basic chemical engineering equipment is listed in Table 4-6, together with page references from this book and other recommended references.

Equipment Costs

The cost of equipment in many instances has an important bearing on the final selection. For example, it would be foolish to select an automatic basket centrifuge costing $15,000 in preference to a wooden filter press costing $1,000 for a small-scale batch operation where ample labor for cleaning the press was already available. A complete discussion of equipment costs is given in Chap. 6 (page 195). Table 4-6 lists other published sources for estimating the cost of equipment. These costs must be brought up to date by the method of cost index ratioing, as described in Chap. 6 (page 198).

[1] P. H. Groggins (ed.), "Unit Processes in Organic Synthesis," 5th ed., McGraw-Hill Book Company, Inc., New York, 1958.

Table 4-6. Chemical Process Equipment and Cost References

Equipment	Description—page no.					Cost—page no.		
	Vilbrandt and Dryden[a]	Riegel[b]	Rase and Barrow[c]	Perry[d]	Peters[e]	Vilbrandt and Dryden[a]	Aries and Newton[f]	Zimmerman and Lavine[g]	Peters[e]
Absorbers	117–119	254		667	367	219	70–71		414
Compressors and fans	142–143	200	297	1258, 1439	294	213	22, 25		297
Crystallizers	119–122	423	381	1050		203	30–31	71–82	414
Distillation towers	117–119	484	187	561	367	219	70–71	83–110	418
Dryers	122–124	441	382	799	418	204	33–35	57–64	343
Evaporators	124–127	390	384	499	147		39–41	49–56	
Fluidization	127–132			1618–1620		132			
Heat exchangers	113–117	544	214	390, 464	306	116, 207	49–50	35–48	93, 341
High-pressure equipment	136, 414	622		1243	255, 301	217–218	96–97		300
Instrumentation	402–415	637	497	1265	187	415–417	27–29	10, 11, 31	95
Materials handling	103–106	86, 179	389	1343	257	200–203			257
Mechanical separations:							58		
Solids from solids	107–108	44		955, 1091		198, 206, 219	23–24, 44–46	111–143	437
Solids from liquids	108–113	308	375, 386	964, 992	424	199	36–37	201–209	
Solids from gases	106–107	227		1013		208–211	53–54		439
Mixing	132–135	264	390	1195	195, 290	212	103–105	363–367	297
Motors and turbines	137–141		348	1767, 1645	272	208, 398–402	77–95	10, 31	95
Process piping	340–375		394	389, 413	284	215	55–56	220–233	296
Process pumps	144–150	138	249	1414		232–233	109, 111, 112	236	
Refrigeration	232–233	588	347	1675		216	59	145–199	
Size reduction	101–102	7	379	1107		216–218	62–68		300
Tanks and reaction vessels	135–137	112	187		300	209, 386–389	97–101	369–370	94
Thermal insulation	375–389		473	480	135, 283	213	57	259–273	
Vacuum producing equipment	142	510	322	990, 1439					

[a] F. C. Vilbrandt and C. E. Dryden, "Chemical Engineering Plant Design," 4th ed., McGraw-Hill Book Company, Inc., New York, 1959. (Refers to this text.)
[b] E. R. Riegel, "Chemical Process Machinery," 2d ed., Reinhold Publishing Corporation, New York, 1953.
[c] H. F. Rase and M. H. Barrow, "Project Engineering of Process Plants," John Wiley & Sons, Inc., New York, 1957.
[d] J. H. Perry (ed.), "Chemical Engineers' Handbook," 3d ed., McGraw-Hill Book Company, Inc., New York, 1950.
[e] M. S. Peters, "Plant Design and Economics for Chemical Engineers," 1st ed., McGraw-Hill Book Company, Inc., New York, 1958.
[f] R. S. Aries and R. D. Newton, "Chemical Engineering Cost Estimation," McGraw-Hill Book Company, Inc., New York, 1955.
[g] O. T. Zimmerman and I. Lavine, "Chemical Engineering Costs," Industrial Research Service, Dover, N.H., 1950. (This reference also contains ample background discussion of equipment selection.)

In addition to the cost references listed in this table, the magazine *Cost Engineering*,[1] published specifically on the subject of equipment specifications and costs, should be available to the design engineer.

Size Reduction[2]

Size reduction is a general term, encompassing a multitude of specific operations, such as crushing, grinding, cutting, cracking, shearing, shredding, pulverizing, granulating, rubbing, defiberizing, and hulling. Generally speaking, different mills do different jobs, and usually one particular type is best for a particular problem. Selection should be in terms of performance and of cost, especially operating cost. The purchase price itself has little effect on the over-all cost of the machine during its operating life; low maintenance costs and reliability are the marks of proper equipment.

The chemical engineer should know the particle size of the feed, especially the maximum value, the tonnage per hour required, and the particle size and other characteristics desired in the product. He should then look for the size-reduction machine that produces a maximum amount of that product with a minimum of power and a minimum of wear on the working parts.

Size-reduction equipment is often a mill with multiple functions. In some units, drying accompanies the grinding; in others dry feeds are blended, or a solid is dispersed in a liquid. Still other units break pieces of ore away from worthless rock and, in conjunction with accessory machines, separate the feed into component parts.

Size-reduction machines, of whatever design, could not do the jobs they do without the help of accessory equipment. Detailed descriptions of the accessories are given by Riegel.[3] They include screens and air classifiers for separating dry particles; elutriators, drum and rake classifiers, and centrifugal classifiers for wet grinding. Magnetic separators, another accessory, aid in keeping tramp iron from entering the mill, and sometimes in removing magnetic particles from the crushed product.

Selection of Size-reduction Equipment. The following factors are determinants in the selection of equipment for size reduction of materials:

1. Physical properties of materials:
    a. Hardness
    b. Mechanical structure, i.e., whether the material is brittle or fibrous, tough or soft, or thermoplastic

[1] O. T. Zimmerman and I. Lavine (eds.), "Cost Engineering," Industrial Research Service, Dover, N.H.

[2] J. C. Smith, Size Reduction, Chemical Engineering Report, *Chem. Eng.*, **59**(8): 151 (1952).

[3] E. R. Riegel, "Chemical Process Machinery," 2d ed., Reinhold Publishing Corporation, New York, 1953.

    c. Moisture content
    d. Specific gravity
 2. Size of feed and product
 3. Tonnage to be ground
 4. Speed of the mill
 5. Physical properties of grinding of equipment:
    a. Shape and character of lining
    b. Shape and character of grinding medium

An extensive compilation of information and data on the construction, design, capacity, and horsepower requirements of grinding and milling equipment of all sorts has been published in Perry's "Chemical Engineers' Handbook," 3d ed., sec. 16. This book has also compiled information on

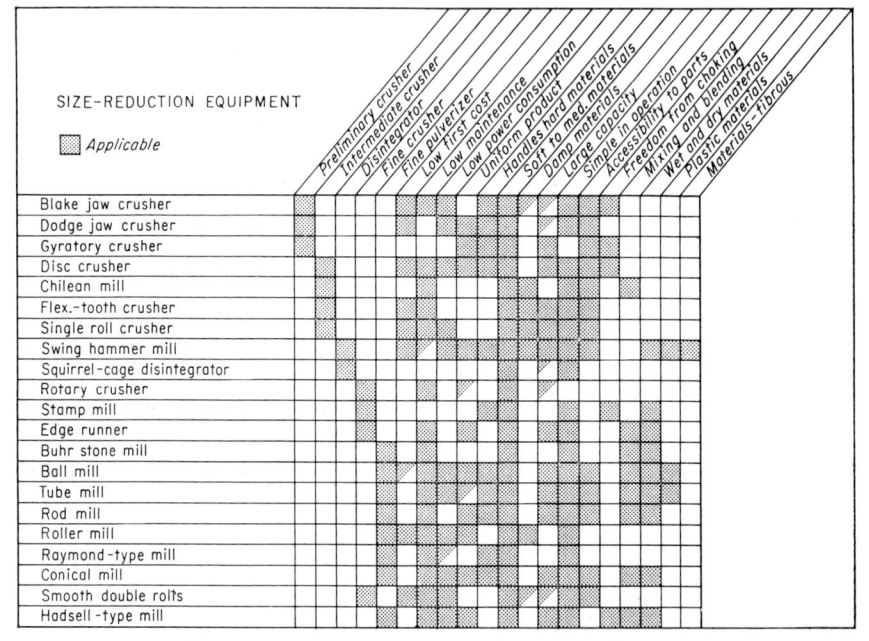

FIG. 4-2. Application of size-reduction equipment.

the industrial applications of milling equipment, giving operating characteristics on a large number of products. The use of air as a conveying medium in grinding mills, for removing the product and for accomplishing drying simultaneously with size reduction, is also covered in a discussion of closed circuiting of mills by means of air separators.

Figure 4-2 is an application chart for size-reduction equipment, showing classification of types and characteristics of several examples in each type. Figure 6-25 lists prices for typical size-reduction equipment.

Materials-handling Equipment

Materials-handling equipment may be either manual or mechanical. Mechanical handling equipment is the better coordinator of processes; not only does it eliminate manual work, but it also serves to pace the process, tie together various pieces of equipment, and frequently convert batch to continuous operation. To ensure the lowest cost of operation, mechanical handling should be substituted for manual handling whenever it can be justified. Materials-handling equipment is logically divided into continuous and batch types, and into classes for the handling of gases, liquids, and solids. Liquids and gases are handled by means of pumps and blowers; in pipes, flumes, and ducts; and in containers such as drums, cylinders, and tank cars. This unit operation is so highly specialized that the chemical engineer would do well to consult with competent mechanical engineers in selecting the equipment. The latter cannot do the job alone because special materials of construction will frequently be required, because special hazards, including corrosion, fire, heat damage, explosion, pollution and poison, together with special service requirements, must generally be met in design. In the main, materials-handling problems in chemical engineering industries do not differ widely from those in other industries except that the existence of these six hazards will frequently influence design.

Corrosion is often the most difficult of these hazards to surmount, and its solution will generally be based on (1) the cheapest type of equipment available, (2) the use of a high-first-cost, corrosion-resistant material in the best type of handling equipment, or (3) the use of containers which adequately protect the equipment. Fire and explosion hazards are reduced by grounding the handling equipment where static electricity is likely to develop, by ventilation to reduce dust concentration, by handling materials in containers to eliminate dust scattering, by the use of low-oxygen-content gases in conveying systems, by jarproof conveyances, and by screens to avoid contact with sparks or fire. Poison hazards are reduced by remote handling or closed container conveyances. Where food products are handled, sanitary requirements to prevent pollution demand sealed containers, frequently of special materials of construction, and the employment of easily cleanable equipment with moisture-proof bearings.

Selection of Materials-handling Equipment. The selection of materials-handling equipment depends upon (1) the cost and (2) the work to be done.

Factors that must be considered in choosing materials-handling equipment include:

1. Chemical nature of the material to be handled

2. Physical nature of the material to be handled

3. Character of the movement to be made, whether horizontal, vertical, or a combination of the two

4. Distance of movement

5. Quantity moved per hour or other unit of time, such as weight, number of pieces, or volume

6. Nature of feed to handling equipment

7. Nature of discharge from handling equipment

8. Nature of flow—continuous or intermittent

Conveyors. A *belt conveyor* consists of a continuous canvas, rubber, leather, composition, or wire screen belt supported on idler pulleys, generally arranged to trough the belt, and driven by application of power to a head pulley. Most belts are in the range of width from 12 to 60 in. In the large sizes, belt speeds may be as high as 600 fpm, depending on the type of material and the loading; for packaged materials, flat belts are used with speeds up to 200 fpm. Very high capacity is possible with belt conveyors. They offer the further advantage of relatively low maintenance, reasonable power consumption, relatively low cost, and continuous discharge. Belt conveyors may be used on horizontal or inclined runs for handling practically all sorts of solids, ranging from fine powders, through grains and crystals, to large lumps such as coal, ore, or stone.

In *chain conveyors*, chains are dragged through shallow trenches or troughs and serve to convey such materials as hot ashes and hot cement clinker. By the attachment of dogs, blocks, or plates, chains and cables may be used for pulling cars up inclines, pulling logs up chutes in paper mills, or dragging bulky material such as coal or stone through troughs. When a chain is supported from trolleys run on an overhead track, it is useful for handling packages and other bulky materials such as tires. Since the track may run both horizontally and inclined, chain trolley conveyors are very flexible.

When attached to *platforms, slats, aprons, and pans,* chains serve to move both packaged and bulk materials at speeds from 30 to as high as 100 fpm. Apron conveyors are frequently used as feeders for handling coarse material to and from crushers. Such equipment, at low speeds, reduces breakage to a minimum.

For the transportation of bulk materials over paths that may vary anywhere from horizontal to vertical, buckets supported on chains and rollers are often used. Because of the flexibility of the *bucket conveyor,* both in path and in bucket material, this conveyor is particularly adapted to the handling of abrasive and otherwise difficult materials. *Roller conveyors* are used when gravity can supply the motive power or for those which are driven and used for the transportation of boxes, packages, etc.

Consultation with manufacturers is particularly necessary in the selection of *screw conveyors*. The cut-flight type is used for materials which tend to pack, or, when placed in a trough having a perforated lining, for removing foreign materials from grain. This type of conveyor also is supplied with mixing paddles and is used for mixing materials during conveying. A similar type, the cut-and-folded conveyor, will thoroughly stir material which passes through it. The ribbon type is particularly adapted to handling sticky materials which would tend to collect in a standard conveyor at the point where the flights join the shaft.

Drag-line scrapers employ bucketlike scoops or disks which are moved back and forth by steel cables to drag loose materials from a large storage area, usually outdoors, toward a central elevator or conveyor hopper for subsequent delivery to the plant. Such equipment is used for storing and reclaiming materials like coal or stone.

The use of air, or occasionally of inert gases, for the sweeping of comparatively light or powdered materials through ducts has come into widespread use in the chemical industry. *Pneumatic conveyors* are used for handling materials as coarse as shavings and ashes, but their greatest application is for nonflammable materials such as soda ash, phosphate rock, and other free-flowing chemicals and chemical raw materials. The *solids pump* (Fuller-Kinyon pump) is used to a considerable extent for the handling of pulverized materials such as feldspar and portland cement. It uses screw feeder pressure in addition to aeration. The solids pump is more costly than the pneumatic conveyor but employs smaller pipe, requires no cyclone for receiving discharged material, and offers increased flexibility in many uses.

Many forms of *interfloor elevators* are available for the handling of pans, barrels, packages, trays, and other material. These generally employ continuous chains, operating vertically, which carry platforms or arms to support the packages. Some of them are devised for automatic pickup and discharge.

As in the case with other manufacturers of process equipment, materials-handling-equipment makers supply information blanks to assist in the making of recommendations.

Costs. Cost curves are presented in Figs. 6-5 to 6-9 for several of the conveyor types of materials-handling equipment.

Solid Feeders.[1] A solids feeder is any device that will maintain a reasonably uniform flow of bulk material, implying a metering function. It deals with solids in bulk, solid-liquid, and solid-gas mixtures which may be free-flowing, lumpy, sticky, corrosive, erosive, fluidizable, hot, plastic, or pasty. It deals with feeding against pressure or vacuum, with feeding where precision is required, as well as with more usual conditions. Finally, since many materials tend to hang up in bins and hoppers, and

[1] T. R. Olive, Solids Feeders, *Chem. Eng.*, **59**(11): 163 (1952).

since feeders can function only if the material reaches them, bin and bin-discharge problems enter the picture.

Almost any sort of materials-handling device that can move bulk loads can be adapted to feeding service; many of them have belts, aprons, screws, flights, and vibrating conveyors. These belong to the class of volumetric feeders, which meter their loads as more or less constant volumes per unit of time. But the weight per unit volume of bulk materials can vary widely, as much as 15 per cent or more. If precision in the order of ± 1 to 2 weight per cent is needed, gravimetric feeders are required.

Feeder troubles are often actually bin troubles, since many materials tend to bridge, arch, and hang up in bins and hoppers, thus failing to reach the feeder at a uniform rate. Such troubles are solved by the design of the bin itself, by the use of agitators within the bin, by jetting air into the bin contents, by vibrating or pulsing of the bin walls, and finally by suitable built-in dischargers.

Bin design is still the subject of controversy between those who believe in the use of one or more straight sides all the way down to the discharge and the proponents of full conical or four-sloped hopper bottoms. The least slope in a hopper will occur in the valleys between adjacent sides; such valleys should generally not have less than a 45-deg slope and preferably should be a 60-deg slope. Certain materials will tend to hang up almost regardless of bin design.

An excellent discussion of feeders and feeding mechanisms is found in Perry's "Chemical Engineers' Handbook," 3d ed., pp. 1370–1376.

Mechanical Separation

A large number of unit operations are included under the general term of *mechanical separation*. This group is one of the most important employed in the chemical industry. The separations may be grouped into five headings as follows:

1. Separation of solids from gases
2. Separation of solids from solids
3. Separation of solids from solids in liquids
4. Separation of solids from liquids
5. Separation of liquids from liquids

Separation of Solids from Gases. Solids are separated from gases by a variety of methods including settling, centrifugal force, filtration, impingement on particles of liquid or on wetted or sticky surfaces, and by means of electrostatic precipitation. These methods are covered in Perry's "Chemical Engineers' Handbook," 3d ed., pp. 1013–1050. Such separation methods are employed to eliminate waste solids from contaminated air, to separate valuable solids from conveying air in materials-handling equipment, and to separate pulverized materials from conveying air in air-swept pulverizers. Costs of various types of separators in the dust collection classification are presented in Fig. 6-4.

Separation of Solids from Solids. Solids may be separated from solids in the dry state by a variety of methods including:

1. Grizzlies
2. Trommel screens
3. Fine screens:
    a. Shaking and gyrating screens
    b. Vibrating screens
    c. Rotating screens
    d. Impact screens
    e. Sieving and bolting
4. Magnetic separation
5. Electrostatic separation
6. Air separation

Screen Cloth. Many thousands of different meshes, sizes, grades, and weights of screen cloth are available, manufactured from a variety of different materials. Although the first consideration is the size of opening, corrosive and abrasive characteristics of the material are also of paramount importance. In determining a suitable screen cloth, the following characteristics of the process and screen should be considered:

Product	*Screen*
Size	Size of mesh or opening
Weight and abrasive nature	Wire diameter or thickness
Capacity or output	Size of screen deck
Material dry or wet	Plain or twilled weave
Chemical characteristics	Kind of metal or alloy

In specifying the size of screen opening, it should be noted that a square opening has a larger area than a round opening when the diameter of the latter is equal to the side of the square.

Separation of Solids from Solids in Liquids. The resistance offered to the fall of particles through a liquid is used in several methods of separation of particles. Some of these methods depend principally on differences in particle size and others on differences in specific gravity. All these schemes were developed in the metallurgical industry, where they are used to a much greater extent than in the chemical industry. The following arbitrary classification is offered for this equipment:

1. Basis of particle size:
    a. Settling basins
    b. Classifiers
    c. Elutriators

2. Basis of specific gravity:
    a. Hydraulic jigs
    b. Concentrating tables
    c. Flotation machines
    d. Cyclones

Of these several methods, only the classifiers have been used to any considerable extent in the chemical industry, although the equipment of the second group is being used to an increasing extent in the concentration of nonmetallic materials for chemical processes.

A classifier developed by the Hardinge Company also finds application in the chemical industry. This type of classifier consists of a revolving cylinder with a conical bottom and containing a spiral flight, which, when the cylinder rotates, tends to convey all material that settles to the central discharge. Fines overflow with the water from the lower end.

Separation of Solids from Liquids. A wide variety of equipment is employed in the separation of solids from liquids. In general, however, there are five principal divisions of this field as indicated in the following classification:

A. Pressing:
  1. Expellers
  2. Curb presses
B. Draining:
  1. Natural draining
  2. Drag conveyors
  3. Classifiers
C. Filtration:
  1. Gravity:
    a. Sand filters
    b. Bags
    c. Nutsches
  2. Vacuum:
    a. Intermittent
    b. Continuous
  3. Pressure:
    a. Plate filter presses
    b. Leaf pressure filters
    c. Continuous
D. Centrifugal:
  1. Batch
  2. Semicontinuous
  3. Continuous
E. Settling and decanting:
  1. Thickeners:
    a. Nonmechanical
    b. Mechanical
  2. Liquid separators:
    a. Swing pipe
    b. Multiple drawoff

Filters. Since filtration is one of the most commonly used operations, both in chemical plants and in metallurgy, a great variety of filters have been developed. Filtration is the operation of separating a solid from a liquid by means of some form of membrane, usually a wire or fabric filter cloth. The membrane retains the solid, while the liquid passes through under whatever pressure is being used to bring about the filtration.

The continuous type of *vacuum filters* has three principal representatives:

1. Drum-type vacuum filters in which the filter membrane covers the outer periphery of the drum, example Oliver.

2. Drum-type filters in which the filter membrane covers the inner periphery of the drum, example Dorco.

3. Vacuum filters with disk- or spindle-shaped filter elements, example American, Oliver horizontal filter.

With any sort of vacuum filter, the filter pressure is less than $14.7/\text{in.}^2$ Consequently, this method is suited mainly to free-filtering materials, which will rapidly build a thin, unbroken cake on the filter surface. A precoat filter, using filter aid as a base, can be successfully adapted where only the filtrate is valuable.

Filtration rates on a previously untested problem cannot be reliably estimated. Small-scale test units, employing a single dip leaf, can be utilized in the laboratory to give a more accurate estimate.

Filters of the continuous type require little labor and have the advantage over batch-type pressure filters by giving a continuous discharge. Furthermore, in certain filters, the cake can be dried to a greater extent after washing by permitting warm air to be sucked through the cake before discharge. Perry's "Chemical Engineers' Handbook," 3d ed., p. 968, describes a test procedure to be used for estimating filter areas. To establish some of the factors affecting a filtration problem and to give the reader some idea of possible filtration rates, Table 4-7 is included. More exact capacities, however, should be determined by laboratory tests. *Costs* for vacuum filters as well as other types are found in Fig. 6-13.

The principle of *pressure filters* differs from that of vacuum filters only in the fact that a positive rather than a negative pressure is used to force the filtrate through the filter membrane. On this account, pressures as high as feasible can be attained. Consequently, materials not filterable on vacuum filters may be handled by this means. *Filter presses* are of three general types: those employing both plates and frames, those using recessed plates, and continuous rotary pressure filters.

Leaf pressure filters are of several types, but the principal variations are (1) filters containing stationary leaves and (2) filters containing rotating leaves. The latter give better uniformity of cake but are more expensive. The first type is exemplified by the standard Sweetland filter and the Kelly filter. The rotating-leaf type is exemplified by the Vallez filter which uses either pancake-type elements or radial leaves rotating within a cylindrical casing.

Filter Cloths. Success in filtering depends largely on the suitability of the membrane chosen for the separation in question. Filter cloths are of numerous materials including cotton, nylon, natural and synthetic fibers, glass fibers, wool, formed rubber, metallic and nonmetallic fibers, and metals and alloys. No general-purpose filter cloth, suitable for all materials, has ever been developed. To assist in filtration, it is often necessary to add a filter aid to the sludge before passing it onto the filter. Filter aids are generally composed of diatomaceous earths, which are inert toward most of the substances filtered. They assist materially in clarification by depositing a noncompressible solid on the cloth and presenting a much less pervious filter surface. Filter aids are particularly useful for the removal of very finely divided materials.

TABLE 4-7. FACTORS AFFECTING SELECTION OF TYPE OF FILTER AND CHARACTER OF PULPS HANDLED*

Typical materials	Character	In. Hg vacuum or lb pressure	Approx. filter capacity, psf/day	Plate and frame	Shell type	Continuous vacuum
Cyanide slime	Finely ground quartz ore	18–25 in.	400–2,000			X
Flotation concentrates	Minerals, finely ground	18–25 in.	400–1,800			X
Gravity concentrates and sand	Metallic and nonmetallic minerals almost free from slime	2–6 in.	10,000–70,000			X
Cement slurry	Finely ground limestone and shale, or clay, etc.	18–25 in.	400–2,000			X
Pulp and paper	Free-filtering fibers	6–20 in.	200–1,200 and 1½–20 gal water/ft²-min			X
Crystals, salt, etc	Granular, crystalline	2–6 in.	3,000–12,000			
Cane-sugar-liquor clarification, beverages, etc.	Sirups and solution with small percentage of solids with filter aid	40–50 lb	36–1,400 gal/ft²-day	X	X	
Pigments	Smeary, sticky, finely divided, noncrystalline	20–27 in. 40–50 lb	200–500 Batch operation	X	X	X
Sewage sludge	Colloidal and slimy	22–24 in.	25–250			X
Varnish	Cloudy viscous liquid, filter aid used for clarification; filtered hot	15–16 lb	5 gal/ft²-hr	X		
Mineral oils with or without wax	Removal of bleaching clay from petroleum products; 1 to 20% clay used	50 lb max pressure	3–30 gal/ft²-hr (lubricating oils); 25–75 gal/ft²-hr (gasoline)		X	
Cane mud	Vegetable fiber and cane juice					X

* J. H. Perry (ed.), "Chemical Engineers' Handbook," 3d ed., McGraw-Hill Book Company, Inc., New York, 1950.

Thickeners and Centrifugals. The simplest form of *thickener* is an ordinary settling tank arranged for the decantation of the clear liquid after settling. Nonmechanical, continuous settling tanks have conical bottoms from which thickened slurries high in solids are continuously discharged. The most widely used thickener is the mechanical type, of which the Dorr is typical. Cost estimates for continuous thickeners are shown in Fig. 6-29.

A *centrifugal separator* consists of a rotating bowl or basket into which the materials to be separated are fed. One type, operating by filtration, uses a perforated basket on which the solids are retained while the liquid passes through. The other uses a solid basket or bowl against which the solids deposit, while the liquid remains close to the center and is withdrawn over a dam. This latter type operates by accelerated settling and decantation and is also used for the separation of immiscible liquids.

The volume capacity of a basket centrifugal is only one factor in determining time capacity. To estimate the size required it is also necessary to know the operating cycle that can be maintained. Table 4-8 lists a

TABLE 4-8. TYPICAL CAPACITIES OF BASKET CENTRIFUGALS*
(Basis: 40-in. basket diameter; 7.8 ft^3 per load)

Product	Cycle time, min	Basket load, lb	Capacity, tons/hr
Top-suspended machine:			
Sugar................	2–5	350–500	2–5
Inorganic salts.........	10–15	350–700	0.7–2.5
Organic crystals........	10–20	250–350	0.4–1.0
Fine powders..........	30	350	0.35
Link-suspended machine:			
Waste...............	10	80	0.4
Textile piece goods......	15	70–120	0.15–0.25

* J. C. Smith, *Chem. Eng.*, **59**(4): 140 (1952).

few typical cycle times and weight capacities for 40-in. machines in a variety of industries. In general, bottom-discharge machines with mechanical unloaders have shorter cycles than underdriven machines. Free-draining crystals in a slurry of 50 per cent or more crystal concentration (e.g., sugar massecuite) can be handled at very high production rates, with 2- to 5-min cycles for charging, washing, spinning, and unloading. Special large valves and feeder lines must be used for such short cycles.

In general chemical service, it is well to assume tentatively four to six loads per hour for fairly free-draining crystals in concentrations of 25 per cent or more of solids. For fine powders or slurries containing less than 10 per cent solids, two charges per hour may be assumed. With fines, slimes, or compressible products, or flat crystals, cycles of 45 min or longer may be expected, especially if washing is needed (chloride removal particularly) or if the solid-liquid ratio is low. Where the solids are so fine that they might be lost through fine screens or through strips of filter cloth held in place by expander rings, it is desirable to use an underdriven machine with a filter bag, which can easily be inserted in such a machine and removed for washing. With an underdriven machine the slower hand unloading is required but this may be of little consequence where high-priced products are handled or fines loss may be avoided. Filter bags are convenient to use when a variety of products must be handled in the same machine. Two to four loads per hour (with two a safe average estimate) should be figured for underdriven centrifugals, depending on crystal size, slurry consistency, and washing conditions. With either type of centrifugal it is necessary to increase the cycle times when slimy or fine-grained products are handled.

The *horizontal-shaft ter Merr automatic centrifugals* are of high-tonnage batch type with automatic cycle control of charging, spinning, washing, and discharging. Capacity varies with the material handled, the capacity given being based on a cake discharged with 13.5 per cent moisture, feeding a slurry of 20 per cent solids having a particle size of 60 per cent on 200 mesh and 89 per cent on 400 mesh. The *continuous* type of machine is used for separating free-draining solids from slurries of high solids concentration (in general, 40 per cent or more solids of crystalline, granular, or fibrous character, with at least 90 per cent retained on 100 mesh). Capacity varies widely with the material handled.

The *rotating-conveyor-discharge* type utilizes the conical horizontal revolving chamber, throwing the solids against the inner face of the cone, from which they are scraped and moved forward to the small discharge end, while the liquid flows out at the larger end of the cone. Assuming an ordinary crystalline material, a discharge capacity of 30 cfh corresponds to about 1 ton of solids per hour for the 18- by 20-in. machine, with other sizes in proportion to the indicated output volumes. The maximum feed rate should not be exceeded. The relative clarifying capacity depends on the settling characteristics of the material handled. In a typical application, a 20 per cent solids suspension of NaCl crystals (mostly plus 200 mesh) is separated from a 25 per cent electrolytic NaOH solution discharging solid-free caustic and solid salt of 4 to 5 per cent moisture and less than 0.3 per cent NaOH, using 0.3 lb wash water per pound of crystals.

High-speed centrifugals include tubular-bowl and disk-bowl machines, used for both liquid-liquid and liquid-solid separations. Capacities are particularly difficult to anticipate for application of the types carried out in such machines. Approximate capacities for several materials are shown in Table 4-9. Apparently small changes in operating conditions may make large differences in capacity. For example, a substantial increase in capacity occurs in certain liquid-liquid extraction operations in penicillin manufacture when a very small amount of wetting agent is added to help break emulsions.

TABLE 4-9. POWER LOADS AND TYPICAL CAPACITIES OF HIGH-SPEED CENTRIFUGALS*

Type	Material handled	Power load, hp	Typical throughput capacity, gph
Tubular	Transformer oil	....	1,200
	Diesel lube oil	2	350
	Olive oil and water	2	200
	Organic chemicals	....	40–400
Disk	Turbine oil	....	1,000
	Vegetable oil	....	450
	Lube oil	1½	100–150 (small)
		2	225–250 (large)
	Organic chemical liquids	....	40–70 (small)
			300–500 (large)
Three-way separator (Sharples Autojector)	Wool grease from scouring liquor	18	1,500

* J. C. Smith, *Chem. Eng.*, **59**(4): 140 (1952).

High-speed centrifuges are of two types: (1) the Sharples type, which employs a long hollow bowl of small diameter and is rotated at a very high speed, and (2) the DeLaval type, employing a short disk bowl of large diameter. Such equipment is used largely for clarification and for separation of immiscible liquids.

A variation of the high-speed type, for intermittent discharge of the solids, is the rotojector which uses hydraulic pressure generated by the rotation of the bowl to uncover ports for occasional discharge of the solids.

Installed costs of typical centrifugal separators are graphed in Fig. 6-3.

Heat-transfer Equipment

Heat-transfer equipment in the form of heat exchangers and condensers is a vital part of the chemical process industries. Heat exchangers are used for the cooling or heating of all sorts of process materials, while condensers are used largely for the condensation of vapors from evaporators and consequent production of vacuum, and for the recovery of materials volatilized from stills.

Heat Exchangers. Heat exchangers are built in a great number of designs, but three types are most common: (1) coils submerged in liquid; (2) tubular heat exchangers consisting of tubes supported within a shell by means of tube sheets, one of which may float to provide for differential expansion between tubes and shell; (3) double-pipe heat exchangers consisting of two concentric pipes, one for each fluid. They may be of the bare tube or extended (fin) surface design if the heat-transfer coefficient for one fluid is very low, e.g., in the case of gases or viscous liquids.

Recommended standard heat-exchange design practice has been set up by standards committees.[1,2] For example, stock lengths of tubes are 8, 12, 16, and 20 ft; tubing should be specified to match one of these lengths. Shell sizes are NPS up to 24 in. with rolled plate used for larger-diameter requirements. Minimum baffle spacing is 2 in. or one-fifth the ID of the shell, whichever is greater, while the smallest allowable pitch is $1\frac{1}{4}$ times the tube OD.

It is thus necessary to consider certain practical limitations when a heat exchanger is designed. Selections should be made from standard items if possible. This will prove to be less expensive than using nonstandard items from both an initial investment and maintenance standpoint. Exchanger surface should be divided into enough separate bundles so that no single one is too heavy for the maintenance crew to handle.

Condensers. Many condensers are constructed along the lines of one of the heat exchangers mentioned above, principally types 1 and 2. In these cases they are known as *surface condensers* and are used where it is necessary to avoid mixing the condensed phase with the cooling fluid. Condensers for use on evaporators are generally of the jet type, in which the cooling water mixes with and condenses the vapor. *Jet condensers* are of two types: one in which the flow of vapor and cooling water is parallel and the other in which the flows are countercurrent. Jet condensers are generally provided with a barometric leg consisting of a tail pipe extending beneath the surface of water in the hot well which serves as a barometric seal. By this means, water may be removed from the condenser at any vacuum possible with the existing cooling-water temperature and without the use of a vacuum pump. However, it is generally necessary to provide a small dry vacuum pump or ejector for the removal of noncondensable vapors.

Still a third form of condenser is the *eductor* type, consisting of a venturi-type jet compressor in combination with a surface condenser. This type requires large volumes of cooling water but has the advantage of operating at a higher cooling-water temperature by reason of the compressor action of the jet and of operating without a vacuum pump.

[1] Tubular Exchanger Manufacturers' Association, Inc., 53 Park Pl., New York.
[2] AIChE-ASME Joint Committee on Heat Exchanger Testing Procedures.

Design Procedures. There is ample material in the engineering literature describing design methods for heat exchangers.[1] In general, they are based on economic optimum design equations in which basic variables of fluid velocity and heat-transfer area are counterposed. Increasing fluid velocities means higher heat-transfer coefficients, a correspondingly lower heat-exchanger size, and *lower annual fixed costs* for a given heat duty. The resulting higher pressure drop *increases the pumping cost* so that an economic design can be worked out in terms of cost equations. A corollary case can be introduced into the cost picture if one of the fluid streams is costly and used only for sensible heat-transfer pickup, e.g., cooling water as a utility, not a process fluid. Increasing the water velocity in a heat exchanger of fixed design would increase the hourly water rate and operating cost.

Calculations on the above basis are complex, tedious, and time-consuming. Simplifying assumptions can often be made, such as (1) power costs on the shell and/or tube side negligible, (2) flow rate of one or both fluids fixed, (3) constant temperature of one fluid throughout the entire exchanger, e.g., in a condenser. However, the more rigorous equations lend themselves to easy solution on digital computers (see Chap. 1). The calculation procedure is usually the same and only the substituted numbers vary in each case.[2] It is recommended that the instructor develop the basic equations in class and then use a digital computer in class demonstration for rapid solution, if time permits and the necessary equipment is available.

The following general procedures should be followed in designing heat exchangers:

1. Specify material and enthalpy balances in the exchanger, specifying inlet and outlet temperatures of all fluids where possible. (See Perry's "Chemical Engineers' Handbook," 3d ed., fig. 16, p. 479, for a cooling-water case.)

2. Specify all fluid properties: density, bulk and wall viscosities, thermal conductivity, specific heat, and latent heat.

3. Calculate and estimate heat-transfer coefficients and fouling factors.

4. Pick a standard heat-exchanger layout and hence a trial area: length, pitch, and size of tubes, baffle arrangement.

5. Route fluids to shell-and-tube sides: use tube side for corrosive

[1] D. Q. Kern, "Process Heat Transfer," pp. 239–252, McGraw-Hill Book Company, Inc., New York, 1950.
W. H. McAdams, "Heat Transmission," 3d ed., pp. 431–441, McGraw-Hill Book Company, Inc., New York, 1954.
J. H. Perry (ed.), "Chemical Engineers' Handbook," 3d ed., sec. 6, particularly p. 479, McGraw-Hill Book Company, Inc., New York, 1950.

[2] R. E. Githens, Jr., Let Computers Pick Your Exchangers, *Chem. Eng.*, **65**(3): 43 (1958).

and/or high-pressure conditions; use the shell side for condensing fluids or low-pressure-drop requirement.

6. Calculate the heat-transfer area and pressure drop. Repeat the procedure until calculated and assumed areas match and pressure drop is an economical value, generally below 25 psi.

Specification Sheets. Certain basic information is required for preparing heat-exchanger designs and quotations. The Tubular Exchanger Manufacturers' Association issues specification sheets which serve as both a request form by the purchaser and a final bid form by the supplier. Copies are available on request. Important information to be made available to the designer is:

1. Fluids handled
   a. Name and chemical formula
   b. Physical properties for heat-transfer equations[1]
   c. Per cent liquid, vapor, and noncondensables
   d. Corrosion allowances[1]
   e. Type of cooling water if required—raw or treated
2. Flow rates: average and/or maximum, lb/hr
3. Temperatures in and out
4. Pressures: operating pressures and allowable pressure drop[1]
5. Quantity of fluid vaporized or condensed, lb/hr
6. Heat-exchange duty, Btu/hr[1]
7. Fouling factors[1]
8. Available space[1]

From this information the supplier develops the optimum design for bid purposes. In large chemical companies the complete mechanical design is worked out by their own design group. Further information sent to the fabricator for construction bidding then includes:

1. Tube-side construction details
   a. OD, length, wall thickness (BWG), and pitch
   b. If fins are required, number, thickness, and height of fin
   c. Number of passes
   d. Material of construction
2. Shell-side construction details
   a. ID, OD, and thickness
   b. Number of passes
   c. Baffle arrangements
   d. Material of construction
3. Type of tube sheet and shell construction, i.e., floating or fixed head
4. Code requirements and test pressure
5. Remarks

Cost. Rubin[2] presents extensive cost data on varieties of heat exchangers with detailed specifications on sizes and materials of construction.

[1] Optional—may be suggested to the supplier.
[2] F. L. Rubin, *Chem. Eng.*, **60**(10): 202 (1953).

These are based on 1953 purchased costs and should be multiplied by 1.5 to be put on an ENR 750 (1958) basis. A composite graph of heat-exchanger costs on an installed basis is presented in Fig. 6-14.

Mass-transfer Equipment

The transfer of mass as well as heat from one material phase to another is quite commonly encountered in chemical process flow sheets. The same physical laws, rate equations, and design principles can be applied to mass-transfer operations as occurring in *absorption, adsorption, crystallization, distillation, drying, extraction, fluidization,* and *humidification.*[1] Equipment is designed to obtain intimate contact between phases, in either a stagewise or continuous manner, and many special types of equipment have been developed for any given operation. This discussion will be limited to the conventional types of equipment.

Contacting Columns. A tall cylindrical column or tower can be filled with packing for continuous contact of two or more phases or fabricated with a number of trays at fixed distances apart for stagewise contact operation. Such columns, either alone or in series, are commonly specified for separations in gas absorption, distillation, extraction, and humidification.

Trays are designed with the following types of gas-liquid contactors:

1. Bubble caps: round or rectangular cups with serrated edges inverted over nozzles spaced uniformly over the plate area; require downcomers on each tray.

2. Sieve plates: numerous small holes of $\frac{1}{8}$ to $\frac{3}{16}$-in. diameter spaced on uniform triangular pitch to cover about 15 per cent of the plate area; require liquid downcomers on each plate; sieve trays are cheaper to construct and give lower pressure drop for the same plate efficiencies.

3. Turbogrid trays:[2] use rectangular slots instead of circular holes; no liquid downcomers required; have the lowest cost and best operating performance.

Packing for towers consists of rings, berl saddles, Fiberglas pads, and helices (see Perry's "Chemical Engineers' Handbook," 3d ed., p. 685). Because of uncertainties of scale-up and ease of flooding, packed towers are seldom used for large-scale operations, being limited to diameters less than 2 ft.

A modification of the packed tower has been made by Schiebel[3] in

[1] R. E. Treybal, "Mass-transfer Operations," McGraw-Hill Book Company, Inc., New York, 1955.
J. H. Perry (ed.), "Chemical Engineers' Handbook," 3d ed., secs. 8–14, McGraw-Hill Book Company, Inc., New York, 1950.
[2] Engineering Staff, Shell Development Co., *Chem. Eng. Progr.*, **50**(1): 57 (1954).
[3] E. G. Schiebel and A. E. Karr, *Chem. Eng. Progr. Symposium Ser.*, **50**(10): 73 (1954).

which a vertical column is designed with a series of alternate mixing and calming sections in stagewise operation. Each mixing section has a paddle-type stirrer turning on a common axial shaft to mix intimately the two counterflowing phases. The calming section is packed with wire mesh to coalesce small droplets and separate the two phases. Stage efficiencies are higher than in plate columns, giving lower column height and improved throughput.

The stage efficiency of sieve-plate towers for extraction can be increased by use of a pulse wave generated at the base of the column. The discontinuous phase rises or falls to the next plate in its passage through the column and is forced through the sieve-plate holes by the pulsing energy wave. The finely divided jets of discontinuous phase create additional surface area and improved agitation within the vicinity of the plate. This type of column is particularly well suited for fission-product separation in the nuclear field,[1] since column height and shielding requirements can be reduced by factors of 2 to 4 over packed or sieve-plate designs. The pulse column has the disadvantage of high pulse power requirements and scale-up difficulties beyond 3-ft-diameter columns because of lack of suitable pulse generators.

Cooling Towers. Since the operation of condensers depends upon adequate cooling water of sufficiently low temperature, economy sometimes dictates the use of recirculation and of atmospheric evaporative cooling. For this purpose, either spray ponds or cooling towers are employed. Spray ponds are rarely used except in large installations, whereas cooling towers have the advantage of ease of operation and relatively small size. In either case, advantage is taken of the evaporative cooling of the water in contact with air. Cooling towers are of two types: natural draft and forced draft. In either type, the water is distributed over a large surface, usually of wood grids, so as to facilitate evaporation.

Design Methods. Continuous contactors are rated on the basis of a height of transfer unit (HTU) or a height equivalent to a theoretical plate or stage (HETP) with equilibrium data, material and energy balances used in evaluating the number of transfer units or equilibrium stages. Design methods for binary systems and simple ternary systems have been handled in previous chemical engineering courses by the McCabe-Thiele graphical method and by graphical or analytical integration of approach to equilibrium differential equations. Multicomponent distillations by key component calculation methods are described in Perry's "Chemical Engineers' Handbook," 3d ed., pp. 622–629. High-speed computers are applicable to the problem and a key code is now available from computer manufacturers.

[1] G. Sege and F. W. Woodfield, *Chem. Eng. Progr. Symposium Ser.*, **50**(13): 179 (1954).

The economic balance enters the plans for contacting equipment design. In distillation calculations,[1] the pivot point in design is the reflux ratio, which can vary between minimum and total reflux values. Higher reflux ratios require greater quantities of steam and cooling water and a larger column diameter, but the column height requirements are lowered. The economic reflux ratio is usually 1.1 to 1.2 times the minimum for most cases. For gas adsorption and humidification calculations[2] there is an optimum gas velocity calculated on the basis of a balance between tower fixed charges, which decrease as gas velocity increases, and power costs for pumping the gas through the tower. A second optimum for this equipment involves tower height as a function of solute value lost and the cost of removing the solute from the solvent.

Costs. Preconstruction cost estimates can be made for packed and plate towers, using Fig. 6-30. For more accurate costs, an assembly drawing should be prepared and sent out for bid.

Crystallizers. Crystallization involves, generally, the evaporation and subsequent cooling of a solution to the point of supersaturation, whereupon the formation of crystals takes place. Much of the work that has been done in the evolution of crystallization apparatus has been pointed toward the control of crystal size, since trade demands frequently are rigorous in this regard.

The phenomenon of a salt coming out of solution is especially complex because it involves diffusion, formation of nuclei, and crystal growth, all of which may take place simultaneously. At present, it is not possible to calculate the rates of any of these exactly and so crystallizer design remains empirical. Many different types of crystallizers are produced to meet the various demands.[3]

Since supersaturation is the important prerequisite of crystallization, crystallizers can be conveniently classified according to the primary methods by which supersaturation is brought about or released by (1) supersaturation by cooling, (2) supersaturation by evaporation of solvent, (3) supersaturation by adiabatic evaporation (cooling plus evaporation by vacuum), (4) circulation of solution over crystal bed to release supersaturation produced by one of the above methods, and (5) salting out.

The oldest and simplest representative is the *tank crystallizer*. It consists of an open tank, either rectangular or circular in section, exposed to the atmosphere, which provides the necessary cooling. Frequently, ropes, rods, or lead strips are suspended in the bath to provide a base

---

[1] A. P. Colburn, *Ind. Eng. Chem.*, **28**(5): 526 (1936).

[2] T. K. Sherwood and R. L. Pigford, "Absorption and Extraction," 2d ed., p. 246, McGraw-Hill Book Company, Inc., New York, 1952.

[3] D. E. Garrett and G. P. Rosenbaum, Crystallization, Chemical Engineering Report, *Chem. Eng.*, **65**(16): 125 (1958).

upon which crystals can grow. Crystals tend to build to a very large size on these strips and on the walls of the tank. They are removed by hammers or crushers in the case of the strips. This method also gives some crystals of various sizes all the way down to a fine sludge. Impurities are frequently occluded in the product from this crystallizer, especially in the case of tank bottom material. The obvious defects in this method are the large floor space it requires, the high labor cost, the large quantities of material in process, the lack of control, and the consequent poor quality of the product.

The *agitated batch crystallizer* provides agitation and artificial cooling. The water or other coolant is circulated through cooling coils and the solution is agitated by the propeller blades on the central shaft. The agitation performs several useful functions: (1) it increases the rate of heat transfer and keeps the solution temperature more nearly uniform; (2) it produces a large number of nuclei so that the number of small crystals increases; (3) it provides a better opportunity for the crystals to grow uniformly instead of forming agglomerates. The over-all result is the production of comparatively small but uniform crystals. Coil cooling design gives rise to the deposition of solids on the cooling coils, an action that rapidly reduces the rate of heat transfer. This necessitates frequent cleaning by dumping or dissolving the adhering crystals which may result in the introduction of excessive water into the system. The *Acme crystallizer* employs two oppositely directed helical coil sections. Because of greater coil length, it is customary to arrange multiple decks with countercurrent flow of solution and cooling water. A crystallizer that has been used to a considerable extent in Europe is the *Wulff-Bock* which consists of a shallow, inclined trough set on rollers so that it can be rocked from side to side. This crystallizer has small capacity but can make unusually large crystals. The *Jeremiassen*, or *Oslo*, crystallizer has recently been introduced into the United States. It is made in various designs for multiple-stage evaporation, evaporation with recompression of the vapor, vacuum cooling, or cooling to low temperatures with cooling liquids. However, all these forms control crystal growth by causing the supersaturated solution, cooled to crystallization temperature, to pass upward through a perforated plate above which crystal nuclei are kept in suspension. Positive circulation is maintained by a centrifugal pump. Crystals are removed continuously by a salt elevator or some other suitable means. *Double-pipe crystallizers* are found in the chillers used in the petroleum industry to separate wax from oil, represented by the *Vogt oil chiller*, the *Buflovak crystallizer*, the *Worthington inclined chiller*, and the *dual-worm crystallizer*.

Salts that exhibit little increase in solubility with temperature (e.g., sodium chloride) can frequently be crystallized in a conventional evaporator provided with some form of salt separator. It is also crystallized in the *long-pan open grainer*. The *Buflovak grainer* is used chiefly for the conversion of ammonium nitrate into a granular product. The solution of ammonium nitrate first is concentrated to about 97 per cent solids in open pans heated with coils and is poured into the previously heated graining kettle; cooling water is then circulated while the solidifying mass is granulated by means of the agitator which is set close to the bottom and to the side wall of the kettle. As the material cools down, it reaches a transition point at which it changes from one crystal form into another. During this transition some heat is developed which is sufficient to evaporate most of the residual moisture. The sugar vacuum pan which is designed especially for the graining of sugar comes under the classification of a *crystallizer evaporator*. It is designed to provide the highest possible rate of evaporation and economy of operation. The Lafeuille rotary rapid-cooling crystallizer is widely used in the American and European beet sugar industry and in many cane sugar factories.

In a *vacuum crystallizer*, the solution is exposed to a pressure below its corresponding vapor pressure, resulting in both evaporation and adiabatic cooling. Vacuum crystallization may be accomplished batchwise or continuously. In principle, the vacuum crystallizer is essentially a vessel which may be evacuated to extremely low pressure, usually by steam-jet ejectors, into which the feed may be introduced. When the desired final temperature is reached, the vacuum is broken and the charge dumped for filtering or centrifuging. In vacuum crystallization, (1) heat transfer is not inhibited by salted-up or corroded media; (2) the end temperature is not limited by the temperature of the available cooling water; (3) operating costs are low; and (4) relatively easy control and regulation of conditions are possible. The *circulation vacuum crystallizer* maintains the temperature of the solution just within the supersaturated region such that no new crystals are formed and crystal growth takes place only on crystals already present in solution. The *Krystal* vacuum crystallizer requires no heater, the feed being mixed with the circulating mother liquor and conducted to the vaporizer where the sensible heat of the feed and heat of crystallization are utilized to evaporate the solvent. The *Zaremba* crystallizing evaporator employs circulation of the solution but does not have a separate container for the crystal bed. The *Swenson* evaporator comprises mainly a cooling system of either the flash or heat-exchange type and a separate classification and magma control zone wherein the dissolved material is precipitated under controlled conditions. It is possible in this design to separate the supersaturation system from the precipitation portion of the cycle and control each one independently of the other.

Crystallization by Salting Out. Thompson and Molstad[1] suggested that crystallization using organic precipitants has definite commercial possi-

[1] A. R. Thompson and M. Molstad, *Ind. Eng. Chem.*, **37**: 1244–1248 (1945).

bilities. They also pointed out that the use of organic solvents offered improved fractional crystallization in many cases. The proposed process and equipment can be found in the reference cited.

Costs of Crystallizers. Representative installed cost data for several types of crystallizers are shown in Fig. 6-10.

Dryers. According to Marshall and Friedman (Perry's "Chemical Engineers' Handbook," 3d ed., p. 800) drying refers to the removal of liquid, usually water, from a solid. There is no hard-and-fast distinction between drying and evaporation, except that the former usually concerns solids that are not in solution, whereas the latter deals with the concentration of solutions. A further distinction, necessarily, is in the type of equipment employed. Drying may be accomplished by various means, but the only one to be considered here is the means employing evaporation of the water. The mechanical forms of drying, including centrifuging, pressing, filtering, and draining, are sometimes used in advance of thermal drying in order to reduce the moisture content and decrease drying costs.

In all types of dryers, some means must be provided for supplying the heat required to evaporate the moisture present, and for removing the vapor. A rough classification may be based on the method of removing the moisture during evaporation, thus dividing dryers into two types: (1) dryers in which the moisture is swept away from the material by air or other gas and (2) dryers in which the moisture is removed by condensation in a separate condenser, with the material placed within a vacuum chamber. In the first type, heat is generally conveyed to the material by the same air which removes the moisture. On the other hand, occasionally the air may be supplied at atmospheric temperature and the drying accomplished simply by increasing its degree of saturation. In the second type, heat is generally supplied to the material indirectly by contact with heated (generally steam-heated) surfaces.

A more extensive classification of dryers is based upon the type of material to be dried. This is logical, in view of the fact that dryer form is largely determined by material form:

A. Materials in sheets or masses carried through on conveyors or trays:
   1. Batch dryers:
      a. Atmospheric cabinet dryers
      b. Vacuum chamber and shelf dryers
   2. Continuous dryers:
      a. Continuous conveyor and tray dryers
      b. Roll dryers

B. Granular or loose materials:
   1. Rotary dryers:
      a. Atmospheric, direct heat, countercurrent
      b. Atmospheric, direct heat, parallel current
      c. Atmospheric, direct heat, two-pass

       d. Atmospheric, indirect-direct heating
       e. Vacuum rotary
C. Paste and sludges or caking crystals:
    1. Agitator dryers:
       a. Atmospheric
       b. Vacuum
D. Materials in solution:
    1. Drum dryers:
       a. Atmospheric
       b. Vacuum
    2. Spray dryers:
       a. Air
       b. Superheated steam

For the drying of materials in sheets or masses, dryers capable of supporting the material in the desired atmosphere for an adequate period of time are necessary. In batch dryers, material is held in place and subjected to the desired conditions or cycle of conditions. In continuous dryers, the material is conveyed through the dryer, generally through a number of zones where different conditions are maintained.

Practically all *atmospheric cabinet dryers* fall within two general classifications: (1) those dryers in which the material is supported on trays placed by hand on shelves within the drying cabinet and (2) those in which the trays are wheeled into the cabinet on trucks. Both types are used extensively for handling such products as pigments, dyes, and other granular and pasty materials. In the truck type of tray dryer, operations either may be on the batch system or may be semicontinuous, in that one or more trucks may be run into the dryer at intervals, while an equal number of trucks are removed at the same time from the other end. By this method of operation, the material progresses through the dryer in a semicontinuous manner. *Vacuum drying* is required for many products which would be injured by the higher temperature of atmospheric drying. Furthermore, in some cases, vacuum drying is more economical because its heat can be supplied at low temperature by means of exhaust steam. Frequently, drying under vacuum is much more rapid than atmospheric drying. Vacuum dryers for solids are similar to the atmospheric cabinet dryers mentioned above, except that the cabinet is of much heavier construction, adapted to be held under vacuum, and provided with a condenser and vacuum pump. *Conveyor dryers* include nontilting and tilting pans, belts of various sorts, traveling buckets, rolls for handling web and sheet materials, festoon carriers, and chain conveyors for hauling tray trucks. A *rotary dryer* consists of a slightly sloping cylinder, open at both ends and supported on rolling wheels; the material is fed at one end, and fuel (gas, oil, or powdered coal) is burned at the other end. Numerous variations are possible, but the fundamental differences depend only on whether the heating medium supplies its heat to the material undergoing drying by direct or indirect contact. *Agitator dryers* generally consist of a vertical cylindrical shell 5 to 6 ft in diameter, provided with a steam jacket and a heavy slow-speed agitator for scraping the bottom. They are useful for the handling of pastes and sludges as well as caking crystals and various noncaking materials such as wood flour. Material is fed through an opening in the top and raked from a door in the side. These dryers may be operated either under atmospheric pressure or under vacuum.

When materials are to be carried from solution to the dry state in one operation, the *drum dryer* is generally employed. This consists of either one or two drums on the surface of which the material to be dried is sprayed in a thin film. The heating medium is conducted to the inside of the drum. As soon as the material has been dried, it is scraped from the surface of the drum. The principal variations of construction in *atmospheric drum dryers* relate to the number of drums and to the method of applying the solution.

When drying must be carried from solution to solid under low-temperature conditions, *vacuum drum dryers* are sometimes employed. Such equipment is similar to the single-drum dryer mentioned above, except that the drum must be enclosed in a chamber capable of evacuation. Furthermore, means must be provided for discharging the solids without breaking the vacuum.

Spray Dryers. Spray dryers generally consist of a chamber through which heated air passes upward, countercurrent to the fall of finely divided droplets of the material to be dried. The spray of material is produced either by conducting the solution under pressure to spray heads or by turbine-type dispersers. The bottom of the drying chamber ordinarily contains some form of conveyor for removing the dried material. Such dryers are employed in the manufacture of soap powder, milk powder, and similar materials.

Dryer Costs. Representative dryer costs on an installed 1958 price basis are given in Fig. 6-11.

Evaporators.[1] Evaporation is empirical and a lot of the information is available only in manufacturers' files; it is also true that there are a wide variety of problems and solutions. Each problem is affected by a variety of variables and must be treated individually. The best solution depends on the local conditions, properties of material, plant heat balance cost, and the availability of utilities like power, steam, and cooling water.

Evaporation is the removal of vapor (solvent) from a relative nonvolatile solute, which is usually a solid. Usually, the solvent is not

[1] E. Lindsey, Evaporation, Chemical Engineering Report, *Chem. Eng.*, **60**(4): 227 (1953).

completely removed and the concentrated product remains a liquid, although sometimes a very viscous one.

Basic considerations in any evaporation process are:

1. To supply the heat necessary: sensible, latent, heat of solution, and heat of crystallization. (The last two are sometimes improperly omitted in heat-balance calculations.)

2. To separate the vapor from the concentrate.

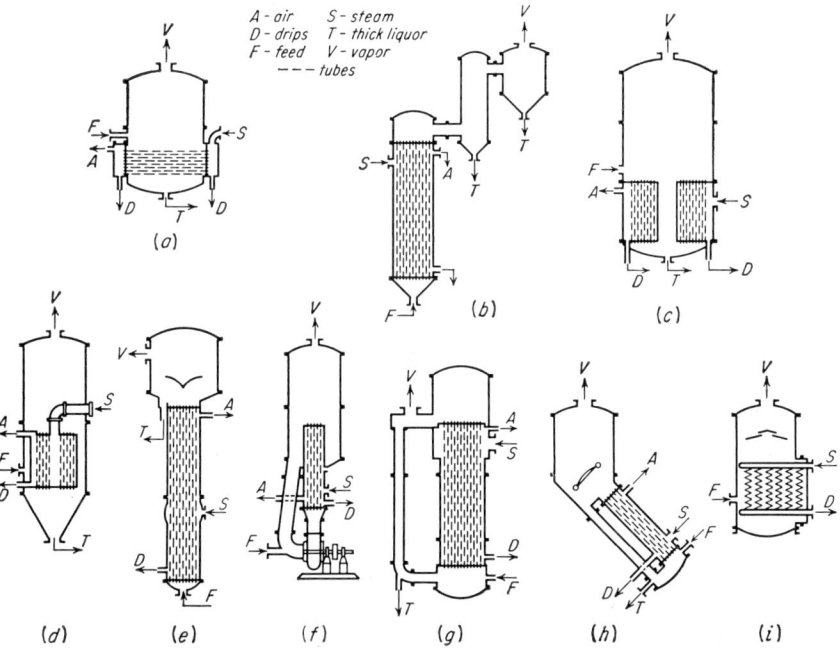

Fig. 4-3. Typical types of evaporators. (a) Typical horizontal-tube evaporator; (b) long-tube evaporator without vapor head; (c) standard vertical-tube evaporator; (d) basket-type evaporator; (e) long-tube natural-circulation evaporator without downtake; (f) forced-circulation evaporator; (g) long-tube natural-circulation evaporator with downtake; (h) Buflovac inclined-tube evaporator; (i) Griscom-Russell evaporator. [J. H. Perry (ed.), by permission from "Chemical Engineers' Handbook," 3d ed., fig. 15, p. 506, McGraw-Hill Book Company, Inc., 1950.]

3. To minimize (usually) any chemical change, thermal decomposition, or growth of organisms that might tend to occur simultaneously.

The preferred types in use today are (1) climbing-film long-tube vertical type, (2) forced-circulation type, and (3) falling-film long-tube vertical type. Examples of these types are shown in Fig. 4-3. Mechanical compression and multiple effect are favored where high evaporation temperatures and relatively high fuel costs exist.

The steam-jet and heat-pump cycles are favored when the evaporation is at such a low temperature the vapors cannot readily be condensed with the cooling water available and when there is an upper limit set on the heating-medium temperature. Conditions in the Florida citrus industry favor the jet and heat pump. Here cooling water runs 75 to 85°F. Evaporation must be done about 60°F—certainly no higher than 80°F—in the falling-film type used and 104°F is about the upper safe limit of the heating medium to avoid a cooked flavor. Many citrus evaporators combine the jet or heat pump with multiple-effect operation.

Classification of Evaporators. According to Badger (Perry's "Chemical Engineers' Handbook," 3d ed., p. 505) evaporators can be classified as follows:

A. Apparatus using solar heat
B. Apparatus heated by direct fire
C. Apparatus with heating medium in jackets, double walls, etc.
D. Steam-heated evaporators with tubular heating surfaces:
   1. Tubes horizontal:
      a. Steam inside tubes
      b. Steam outside tubes
   2. Tubes vertical:
      a. Standard (calandria) type
      b. Basket type
      c. Long-tube type
      d. Forced-circulation type
   3. Tubes inclined
   4. Specially shaped tubes

In certain industries, single-effect evaporation cannot be used as an economy measure because of the comparatively high temperature maintained in the first effect. In the case of heat-sensitive materials such as concentrated milk products, fruit juices, and pharmaceuticals, it is often necessary to use multiple-effect vacuum evaporators and accept high operation cost in order to preserve the quality of the product.

Small water stills, caustic dehydration pots and, of course, steam boilers are the principal representatives of direct-fired evaporators. Much small-scale evaporation is accomplished in *jacketed kettles* and similar apparatus. Such evaporation is generally atmospheric, although, if the kettle construction is suitable, it may be carried out under vacuum. *Horizontal-tube evaporators* usually have the steam inside the horizontal tubes. This type is best suited for nonscaling, noncrystallizing, and nonviscous solutions. The bodies of horizontal-tube evaporators vary between 5 and 10 ft with tube lengths from 3 to 16 ft. Standard and basket-type evaporators are essentially similar in that they employ a *vertical* cylindrical *evaporator* body of diameter less than the height, containing a number of vertical tubes with the steam outside. These are constructed in sizes with diameters from 6 to 24 ft. In the standard type, the steam space is formed by tube sheets which extend horizontally across the shell, with space allowed for a central downtake. In the basket type, the heating element is a separate unit and the downtake is an annular ring between the shell and the heating element. The *long-tube natural-circulation (Kestner type)* single-pass and the *Webre forced-recirculation* type are additional types. The standard vertical-tube evaporator is especially adaptable to solutions that deposit scale or crystals. The application of the basket type is similar. The long-tube type is not suitable for scaling and salting liquids, while the forced-circulation type, by the addition of a salt separator, may treat either salting or clear liquors. This last type is especially adaptable to the evaporation of viscous materials and liquors requiring expensive materials for the heating surface. Because of the high velocity, the coefficient of heat transfer is especially high in this type. Heating surfaces from 35 to 8,000 ft^2 are available.

Inclined-tube evaporators are similar in construction to the long-tube type except that, by reason of the slope of the tubes, these need not be so high. This type employs recirculation and, because of the high velocity attained, gives a high coefficient of heat transfer. Such evaporators are generally not suitable for salting and scaling liquids, although in some modifications the ready removal of scale from the tubes is accomplished.

Coiled tubes, or slightly bent tubes, are employed in strike pans for crystallizing second and third sugars in sugar mills with capacities of 25 to 120 tons, varying from 8 to 18 ft in diameter, and also distilled-water evaporators. The design is such that temperature changes cause movement of the coils and thus serve to crack off scale.

Evaporator Costs. With so much variation in evaporator construction, materials, and accessories, manufacturers are reluctant to give standard cost figures. The most reliable method seems to be to estimate the unit cost as cylindrical shells, plates, heads, tubes, fittings, etc., based on the weight of material and the unit material and labor costs. This is the method generally used by fabricators. A published set of cost curves for preconstruction cost estimating is given in Fig. 6-12.

Fluidization[1]

A *fluid bed* is a relatively stable mixture of a fluid, generally a gas, and finely divided solids which is intermediate between a fixed bed and the pneumatic transport condition. It resembles a boiling liquid in behavior. Fluid velocities are in the range which will lift and stir the suitably sized particles without carrying the particles completely out of the containing vessel. Solid particle sizes in the range from $\frac{3}{8}$ in. to 10 microns (μ)

---

[1] D. F. Othmer (ed.), "Fluidization," Reinhold Publishing Corporation, New York, 1956.

M. Leva, "Fluidization," McGraw-Hill Book Company, Inc., New York, 1959.

M. Sittig, Fluidized Solids, *Chem. Eng.*, **60**(5): 219 (1953).

can be handled satisfactorily with some gradation in size preferred for smooth fluidization.

Fluid-bed techniques for producing intimate contact between small solid particles and fluids are standard design practice where these characteristics are desirable:

1. High rates of reaction between solids and gases under controlled conditions: exposed reaction surface area is inversely proportional to particle size so that a large reaction surface per unit volume of reactor is possible; violent agitation which occurs between the carrier gas and the solids creates high rates of mass and heat transfer at very uniform temperature conditions.

2. Uniform reactor temperatures: fluidized beds have high thermal conductivities as compared to the carrier gas alone, leading to very uniform bed-temperature profiles when compared to fixed-bed operation.

3. High rates of heat transfer: heat-transfer rates from the fluosolids mixtures to any transfer surface within the agitated zone are quite large as compared to gases alone, with values as high as 200 Btu/(hr)(ft^2)(°F) being reported; thus, minimum heat-transfer surface is required when using fluid-bed reactors.

4. Ease of transport of solids: the fluidized solids behave as a liquid, so they can be transported from one reactor to another for adding or removing heat, or for carrying out another step in the process.

Basic Operation of a Fluid-Solids System. The principal components of a fluidized solids system are shown in Fig. 4-4a as ①—the fluid-bed reactor, ②—the solids transfer lines, and ③—the solids recovery system. Various combinations of these elements can be used as shown in Fig. 4-4 and in Perry's "Chemical Engineers' Handbook," 3d ed., fig. 21, p. 1619, and fig. 22, p. 1620. A qualitative explanation of fluid-bed operation is given next; quantitative evaluations are found in the design section which follows.

Fluid-bed Reactor. Finely divided solid particles are transferred from a fixed bed to a fluidized bed when the gas velocity creates a pressure drop sufficiently great to overcome the total weight of the solids in the bed. The gas is introduced via a bottom grid plate. Increases in gas velocity will cause the bed to expand in volume and the bed density to decrease in a manner depending on the species of gas, solids particle size and shape, system pressure, and temperature. Fluid beds are inherently unstable as gas tends to separate from the solids and form bubbles which grow in size as they rise, causing slugging. This condition can be minimized by even distribution of gas at the bottom entrance. Grids should be preferably oriented concave upward with more hole area toward the outer edge of the grid to balance the pressure drop uniformly across the entire grid area.

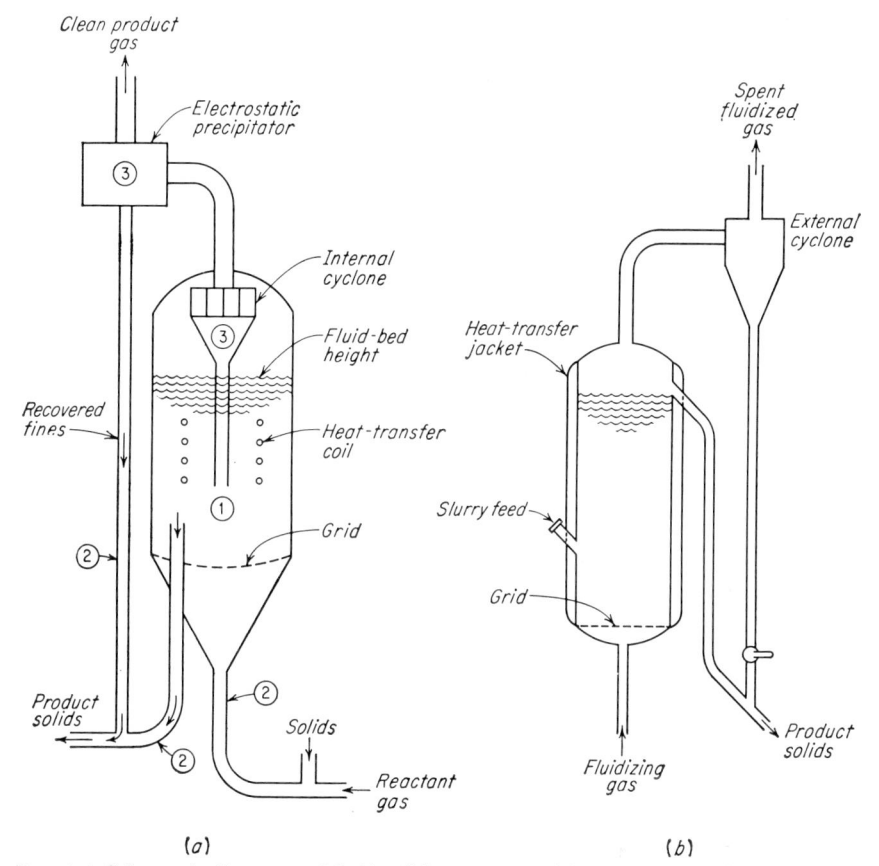

CHAP. 4] SELECTION OF EQUIPMENT AND MATERIALS 129

Fig. 4-4. Schematic diagrams of fluid-solids systems. (a) Reactor with bottom solids discharge and internal cyclone. (b) Reactor with external cyclone and side slurry feed.

Typical particle size ranges employed in fluid-bed reactors are:

Type of process	Particle size range for major percentage of the solids
Fluid catalytic cracking	30–80 microns
Fluid coking	20–100 mesh
Phthalic anhydride production	20–200 mesh
Shale retorting	8–100 mesh
Roasting	¼ in. to 80 mesh

Transfer Lines. Pneumatic conveying of solids is an important factor in successful fluid-bed operation just as pumping of liquids is vital in other process operations. Lines are divided into two types: *standpipes* in which solids flow downward concurrently with the gas and *risers* in which the action is countercurrent. The mixed density must be controlled so that hydrostatic balances are maintained. For rapid transport of solids, a "dilute phase" or low density (less than 10 lb/ft^3) is maintained, using superficial gas velocities of 25 to 35 fps.

Solids Recovery Equipment. Gas take-off from the surface of the fluid bed is accompanied by entrainment of solids, particularly of the fine and intermediate size range created in part by attrition. A hindered settling or disengaging zone above the bed is provided and the fine particles with low free-fall velocities are carried over into dust-recovery equipment consisting of one or a combination of dry or wet cyclones, multiclones, and electrostatic precipitators. Bag filters are prone to plug but must sometimes be used where expensive solids are involved. Cyclones may be placed within the reactor shell (Fig. 4-4a) or external to it (Fig. 4-4b), depending on the freedom of the cyclone operation from mechanical troubles.

Application of Fluidization in Process Industries. Commercial fluidization had its greatest growth with catalytic cracking processes in the petroleum industry, later expanding within this industry to include desulfurization, hydroforming, and fluid coking. Fluidization cannot be used for processes employing expensive solids because of the high cost of solids losses, e.g., platinized catalyst processes.

The nonpetroleum applications of fluidization have been expanding rapidly with such processes being included as manufacture of phthalic anhydride by oxidation of naphthalene, calcining of limestone, roasting of ores, and production of high-purity metals by thermal decomposition of metallic salts.

Design Methods. Important design variables in fluidized-bed calculations are gas velocity and pressure drop, rates of reaction, and heat transfer. In nearly all cases, a low scale-up ratio from pilot plant fluidized-bed reactors should be the basis for accurate designs, since production rates in terms of lb/hr-ft^2 of cross section are uncertain for new processes and often decrease with increasing size of reactor. However, some preliminary design estimates can be made on the basis of past performances with this type of equipment.

Rates of Reaction. The following production figures for several types of processes are typical:

Chemical reaction	Processing rate, lb/hr-ft^2
Fluid catalytic cracking of petroleum	250–400
Roasting of zinc sulfide or iron pyrites concentrates	20–50
Calcining limestone	50–100

Reactor Size and Shape. On the basis of flow-sheet material balance requirements and rate of reaction, the total square feet of reactor bed required to do the job can be computed. The number of units required depends on the allowable maximum diameter of a single unit. Reactors as large as 60 ft in diameter have been operated successfully.

The total shell height to diameter ratio varies from 2 to 5, being larger for small-size units, for multiple-bed operation, or for solids separations within the reactor proper. If the height-diameter ratio is too low, considerable bypassing of reactants, low conversion, and higher entrainment result. The actual fluo-solids bed height is often as low as 20 per cent of the actual shell height to provide an adequate disengaging zone.

Gas Velocity and Pressure Drop. The minimum gas velocity in feet per second ($v_{\min}$) required to fluidize a fixed bed of solid particles has been correlated by Leva et al.:[1]

$$v_{\min} = \frac{0.005 d_p^2 \epsilon_0^3 (\rho_p - \rho_f) g}{(1 - \epsilon_0) \mu_f} \tag{4-1}$$

where $d_p \cong 6/S_p$ and S_p is the surface area per unit vol, ft^{-1}
ϵ_0 = fixed-bed void volume (usually from 0.3 to 0.5)
ρ_p = density of solids, lb mass/ft^3
ρ_f = density of fluidizing gas, lb mass/ft^3
μ_f = viscosity of fluidizing gas, lb mass/ft-hr
g = gravitational acceleration, ft/sec^2

Practical operating velocities range from 0.5 to 3 fps; higher rates cause excessive carry-over to the solids-gas separating system.

The pressure drop through the bed is equal to the weight of solids it contains for bed expansions of 20 per cent or less. It can be calculated from the following equation:

$$\Delta p = L(1 - \epsilon)(\rho_p - \rho_f)g/g_c \tag{4-2}$$

where Δp = pressure drop in the fluid bed, lb force/ft^2
L = total fluid-bed height, ft
g_c = gravitational constant, lb mass–ft/lb force–sec^2
ϵ = expanded-bed void volume

The relationship between the percentage of bed expansion (X) and ϵ is given by

$$\epsilon = \frac{\epsilon_0 + 0.01X}{1 + 0.01X} \tag{4-3}$$

where X is a complex function of the gas velocity.[2]

Actual pressure drops at bed expansions greater than 20 per cent are higher than calculated by Eq. (4-2), sometimes as much as 50 per cent

[1] M. Leva, M. Grummer, N. Weintraub, and M. Pollchik, *Chem. Eng. Progr.*, **44**: 511, 619 (1948).
[2] R. F. Benenati, "Fluidization," pp. 12–13, Reinhold Publishing Corporation, New York, 1956.

higher, depending on reactor geometry, and also increasing with fluid Reynolds numbers.

Heat Transfer. Fluid-bed reactors are characterized by high heat-transfer coefficients on the side walls and on internal heat-transfer devices such as coils or axial bayonet-type tubes, with heat-transfer coefficients ranging from 50 to 125 Btu/(hr)(ft²)(°F). Coefficients between the gas and solid particles of high surface area are somewhat higher than attained under natural convection conditions. For example, a natural convection coefficient for 50-μ particles in air is around 200 and, in a fluidized system, this value would be even higher.

Correlation equations of Gamson[1] can be used for estimating heat transfer within the agitated zone of the reactor.

For transfer to side walls,

$$j_h = \left(\frac{h}{c_p G}\right)\left(\frac{c_p \mu_f}{k_f}\right)^{2/3} = 2.0 \left(\frac{d_p G}{\mu_f}\right)^{-0.69} (1 - \epsilon)^{-0.3} \qquad (4\text{-}4)$$

For transfer to internal heating and cooling devices,

$$j_h = 2.5 \left(\frac{d_p G}{\mu_f}\right)^{-0.80} (1 - \epsilon)^{-0.3} \qquad (4\text{-}5)$$

where c_p = specific heat of fluidizing gas, Btu/lb
k_f = thermal conductivity of fluidizing gas, Btu/(hr)(ft²)(°F/ft)
G = mass velocity of fluidizing gas, lb/hr-ft²

Heat transfer to the side walls often suffices for handling many mildly exothermic reactions. If highly exothermic reactions are being considered, designs might incorporate one or more of these methods: (1) add internal coils, (2) recycle the bed through an external cooler, (3) use the sensible heat in the fluidizing gas, and (4) reduce the reactor diameter to increase the surface area to volume ratio. The same considerations apply to endothermic processes but these are much more difficult to handle designwise.

Costs of Fluidization Systems. Published cost information on these systems is lacking. As a method of approach for preconstruction cost estimation of installed fluidization equipment, use the cost figures from Fig. 6-28 for the reactor, increasing these by 35 per cent if heat-transfer coils or jacketing is required. Add to this the cost of a dust-collection system of the required gas-handling capacity as obtained from Fig. 6-4.

Mixers and Blenders

Mixing, to a greater extent than any other of the chemical unit operations, retains its status as an art, for it still has very little scientific foundation. On this account, a great number of types of mixers have been

[1] B. W. Gamson, *Chem. Eng. Progr.*, **47**(1): 19 (1951).

developed, many of which are far from satisfactory. Furthermore, each industry has developed its own particular form of mixer, whereas it is probable that, with a better scientific basis, a comparatively smaller number of types would serve for all industries.

Valentine and MacLean (Perry's "Chemical Engineers' Handbook," 3d ed., sec. 17) have stated that the practical aims of mixing are four:

1. To produce simple physical mixtures, such as that of two or more miscible fluids, two or more uniformly divided solids, or a mixture of phases where no reaction or changes of particle size take place.

2. To accomplish physical change, such as the solution of one component in another, the formation of crystals from a supersaturated solution, the selective adsorption of minor constituents by adsorbents such as fuller's earth, and the flocculation or deflocculation of particles.

3. To accomplish dispersion, wherein a quasi-homogeneous product is produced from two or more immiscible fluids, or one or more fluids with finely divided solids.

4. To promote a reaction. This latter is perhaps the most important use of mixing in the chemical industries, since intimacy of contact between reacting phases is necessary as a condition of proper reaction.

The requirements of a satisfactory mixer, according to Valentine and MacLean, are (1) that it yield a desired degree of mixing at the point of most intense agitation, (2) that a satisfactory rate and direction of motion of the entire body of material, however remote from the mixing element, must be established and maintained, and (3) that it require the minimum expenditure of power and in the shortest, most economical period of time. Whether a particular type of mixer will meet these criteria in a given problem can often be determined only by experiment. Valentine and MacLean have divided some 40 types of mixers into the following:

1. Flow mixers
2. Paddle or arm mixers
3. Propeller or helical mixers, including screw conveyors
4. Turbines or centrifugal impeller mixers
5. Colloid mills and homogenizers
6. Miscellaneous types including slurry, mass, solid, and drum mixers

It will be noted that this classification is based upon equipment rather than materials.

In *flow mixers* the material is pumped and the mixing effect produced by interference with the flow. Mixers of this type are used in continuous or circulating systems, generally for miscible fluids or occasionally for the mixing of two phases. This principle of mixing is employed in mixers where one jet impinges upon another, in injector mixers where a second ingredient is injected into the main stream, in baffle and orifice columns, in air-lift and long draft-tube mixers, in mixers using centrifugal pumps with or without recirculation, and in towers for the absorption of gas in liquids.

Paddle or arm mixers include a great number of types ranging from simple paddles to combinations of stationary and movable paddles and double-motion agitators consisting of two paddles operating in opposite directions; horseshoe type of scraper agitator, traveling paddle agitators, off-center paddle, mulling wheels within a rotating pan, epicyclic course paddles, and heavy double-arm dough mixers with impellers of the Z or S type for the handling of heavy, doughy, gummy, and plastic masses.

Propeller mixers operate with peripheral speeds in the range between 1,000 and 2,000 fpm. The portable types generally use two propellers with blades set to propel in opposite directions (push-and-pull type). The propeller may be driven from the top or through the side of the tank, with or without a draft tube surrounding the propeller.

Turbine mixers use impellers similar to a centrifugal pump impeller, submerged in the material to be mixed and rotated at moderately high velocity. The impeller may or may not be provided with stationary deflecting rings. One patented form of the turbine mixer, *the turbomixer*, is made in a number of variations for both batch and continuous mixing of liquids ranging from low viscosity up to that of paints and for the contacting of liquids and gases.

The action of *colloid mills* is a combination of fluid shear and impact caused by high centrifugal force. Although there are many different types of construction, all colloid mills operate by forcing the materials to be dispersed or emulsified between surfaces placed very close together and having a high relative velocity with respect to each other. Colloid mills have very high power requirements, the commercial units varying from 5 to 100 hp at 3,600 rpm.

The *homogenizer* consists of a high-pressure hydraulic pump in combination with a spring-loaded valve through which the pressure is suddenly released so as to give a very high degree of impact of the components being emulsified against a plate or ring. Pressures in excess of 1,000 psi are generally employed. Homogenizers are used in the production of certain pharmaceuticals and cosmetics, and to a considerable extent in the dairy and ice-cream industries.

The tumbling barrel, ball mill, rake mixer, Vee-cylinder blenders, paper beater, mixing or compounding roll, and putty chaser are among the miscellaneous types of mixing equipment. The compounding of heavy, semidry masses has led to the introduction of the rotating pan mill, which has a revolving pan as well as a circular motion of the muller.

Power Consumption of Mixers. Valentine and MacLean in Perry's "Chemical Engineers' Handbook," 3d ed., sec. 17, give theoretical horsepower requirements for propellers. They state that the power consumption of colloid mills and homogenizers, for production of 100 gph, will vary between 20 and 50 hp. These authors also give figures on power consumption for occasional specific problems.

Choice of Process Mixers. A great number of specific pieces of process equipment incorporate mixers of one sort or another, usually for the promotion of reaction. Among these pieces of equipment may be mentioned autoclaves, bleaching equipment, cookers, chlorinators, digesters, dissolvers, emulsifiers, extractors, kettles, nitrators, percolators, retorts, reducers, and sulfonators.

As has been indicated, mixer choice frequently is a matter of experience or experiment. Consequently, the tabulated data of Valentine and MacLean, in which specific recommendations for certain mixing ranges are made, should be of great value. Practically any material of construction may be used in a mixer. Mild steel is the most common material, but almost all the special metals and alloys, as well as nonmetallic coatings, can be used.

Costs. Installed cost data on practically all types of mixers and blenders are given in Figs. 6-15 to 6-21.

Tanks and Reaction Vessels

Storage, transportation, and processing of materials, particularly gases and liquids, require the use of tanks and reaction vessels for a wide variation of temperature and pressure conditions. A few principles employed in this type of materials handling will be presented.

Storage of Liquids. Bulk storage of liquids is generally handled by closed tanks to prevent escape of volatiles and contamination. In some instances, such as water storage, where contamination and dilution are not a factor, large open reservoirs can be employed. Natural terrain, concrete-walled excavations, or concrete tanks without tops are typical construction. Reinforced-wall design is required and the concrete must be waterproofed with a suitable paint to prevent any possibility of leaking.

Storage of liquid materials in a typical process industry is carried out in tanks classified as spherical or vertical and horizontal cylindrical. Since safety is an important consideration in storage tank design, the National Fire Protection Association[1] and the American Petroleum Institute[2] publish rules for safe design and operation. *Vertical tanks* are most commonly used for outdoor storage for such materials as petroleum products. Water towers are typical of elevated vertical tank outdoor construction. They are used for maintaining a uniform head of water to store water for temporary emergencies and for fire protection. *Hori-*

---

[1] "Storage, Handling and Use of Flammable Liquid," NFPA No. 30-L, National Fire Protection Association, 60 Batterymarch St., Boston, Mass., 1954.

[2] Standard 12c, American Petroleum Institute, 50 W. 50th St., New York.

zontal tanks are most frequently found inside buildings where floor loading and allowable headroom are prime considerations. A series of horizontal pressure vessels of 2 to 6 ft in diameter are used for bulk storage of liquids under pressures greater than 250 psig. Typical piping includes individual pressure relief valves and a common header. *Spherical tanks* using the thin-steel construction of vertical storage tanks are economically designed for pressures ranging between 2 and 250 psig.

Venting of storage tanks is an important design consideration. In the low vapor-pressure, nonflammable category, an inverted U or downcomer cap of the open type should be specified. The venting capacity in terms of scf air should be twice the pumping rate. Storage of flammable liquids requires flame arrestors on all vapor openings. Flame arrestors consist of a number of thin slotted plates in a small chamber of 40- to 60-mesh wire screen of such a design that rapid heat dissipation prevents sustained combustion. Tanks for conserving valuable fluids of significant vapor pressure are vented via pressure-vacuum controlled openings. These are commercially available and work on an open-close pressure range tolerance. This avoids loss by normal tank breathing caused by ambient temperature cycling.

Gas Storage. Atmospheric pressure gas can be stored in vertical cylindrical tanks commonly known as wet- or dry-seal gas holders. The wet gas holder maintains a liquid seal of water or oil between the top movable inside tank and the stationary outside vertical tank. A dry-seal holder maintains a seal between the inner and outer tanks by means of a flexible rubber or plastic sheet. It is specified where dry gas storage is a necessity.

Recent developments in bulk natural gas or gas product storage show that pumping the gas into porous underground strata is the cheapest method available. High-pressure gas must be stored in a bank of spherical or horizontal cylindrical pressure vessels.

Transportation of Liquids. Tank cars, consisting of horizontal cylindrical tanks made of riveted- or welded-steel plate construction, permanently mounted on railway-car chassis, are used for bulk shipment of liquids and liquefied gases. Construction must conform to specifications of the ICC.[1] Tank-car capacities for general-service conditions range from 4,000 to 12,000 gal. Materials of construction of the tank must conform to corrosion specifications. Liquids which attack carbon steel are handled in lined tanks. Alloy cladding, rubber or plastic coatings, and glass or lead linings are prevalent. Materials which are normally solid at room temperature but melt at elevated temperatures are handled by use of internal heating coils. During loading and unloading,

[1] Pamphlet No. 9, Interstate Commerce Commission, Bureau of Explosives, 30 Vesey St., New York.

steam, Dowtherm, or hot oil is circulated to liquefy the tank contents, which can then be removed by pumping, siphoning, or blowing.

Tank cars can be either rented from railroads or purchased outright from fabricators. Rental rates vary from $50 per month for steel tank cars to $250 per month for stainless-steel tank cars with special fittings. Purchased price of tank cars is 40 to 60 per cent higher than that of the tank itself (see Figs. 6-26 and 6-27).

Tank trucks moving on highways furnish a second method of transportation, generally for short-radius haulage. Design considerations are similar to those for tank cars. The maximum tonnage is limited by state and ICC regulation with a top loading of 15 tons in most states.

In recent years, pipeline pumping has been used for moving large-volume raw materials, such as petroleum, for thousands of miles across country at a cost lower than for tank cars and competitive with water transportation.

Transportation of Gases. Insulated tank cars are available for shipping liquefied oxygen and nitrogen. Where high-pressure gas is transported locally, trucks carrying racks of horizontal cylindrical pressure vessels are available. Smaller volume deliveries are made in vertical gas cylinders. All designs must conform to ICC regulations.

Transportation of gases, particularly from petroleum operations, via pipelines over a range of several thousand miles is a common practice in the gas industry. (See pages 142 to 144.)

Processing of Materials. Tanks, towers, and reactors of all types are common in chemical processing. Open-head tanks of thin-gauge steel to autoclaves of 3,000 psia are typical. For comments on the design of vessels of this category, see the following section on mechanical design and fabrication (pages 150–151).

Cost of Tanks and Reaction Vessels. Installed cost data for typical equipment discussed in the previous sections are given in Figs. 6-26, 6-27, 6-28, and 6-30. Where jacketed tanks are required, add 35 per cent to the installed cost prices.

Electrical Equipment

Motors. Certain fundamental considerations must be taken into account in selecting the correct motor and control for a given job. The first factor to consider is the size or horsepower rating of the motor required to handle the job; with standard drives, most machinery builders have established the exact brake horsepower demanded by their particular machine, so that a motor of the nearest higher standard horsepower rating is recommended. On drives of an intermittent nature, where peak loads or momentary overloads and a frequent number of starts, stops, and/or reversals occur, the motor rating is based not only on torque capabilities but also on thermal capacity sufficient to handle the average or root-mean-square loads.

Some process equipment such as reciprocating pumps, compressors, and conveyors must be started while fully loaded; other drives such as centrifugal pumps, fans, and agitators have light loads at starting. Many applications require considerably more torque to start and accelerate the load than to run it; the reverse is also true. Starting torque, maximum or breakdown torque, inertia of the load, and the accelerating time must be considered.

The operating speed characteristics of the process equipment dictate the type of motor and control to be applied. Most drives operate at a speed lower than that of the motor, thus requiring some form of speed reduction. The gearing may be via direct-connected coupled motor or a speed reducer may be used. Variable or adjustable speed performance must be definitely established as to speed range, degree of speed adjustment, and load requirements at all speeds. Constant-torque or constant-horsepower drives both require variable-speed or multispeed motors with suitable control equipment.

Classification of Motors. Motor and control selection is strongly influenced by the ambient conditions under which they will be operating. The National Electric Manufacturers Association[1] has classified motors according to mechanical features, particularly cooling methods as related to surrounding conditions:

1. Open machine: windings directly exposed to cooling air
    a. Dripproof: material cannot enter the motor if the approach angle is not greater than 15° from the vertical
    b. Splashproof: material cannot enter the motor if the approach angle is not greater than 100° from the vertical
2. Totally enclosed machine: cooling is not done by open, direct exchange with ambient gas, but motor is not gastight
    a. Nonventilated: no method provided for cooling
    b. Fan-cooled: forced circulation by an integral fan unit
    c. Explosion-proof: motor built to prevent ignition of combustible gases by sparks, flashes, or explosions and to withstand an explosion within the motor housing
    d. Dust-explosion proof: same as (2c) except it deals with combustible dust-laden atmospheres
    e. Weatherproof: prevents water, even in a direct stream, from entering the motor; leakage which may occur around the shaft is prevented from entering the oil reservoir by automatic draining

If the location is indoors, the temperature and availability of clean ventilating air must be investigated. For outdoor operation, consideration should be given to prevailing weather, rain, snow, sand, or dust storms, and especially extremely high or low temperatures such as above

[1] NEMA Standards for Motors and Generators, 1955; National Electric Manufacturing Association, 155 E. 44th St., New York.

40°C or below 10°C. In explosive atmospheres, the liquid, vapor, gas, or dust involved must be clearly established so as to specify suitable electrical equipment. Special pumps and agitators are specified to operate with submerged motors. Hazards such as corrosive fumes, acid spray, excessive moisture, oil vapor, salt air, abrasive dust, steam, and fungus growth must be given serious attention to make sure motors and controls are selected with proper protection, or are located away from the contaminated area. Provision must be made to take care of any abnormal vibration, shock, or tilting that may be transmitted from the driven machine to the motor, or controlled through shafts, couplings, and mountings.

Motors and controls are rated for satisfactory operation at rated load without exceeding the nameplate temperature rise, with an ambient temperature not exceeding 40°C, at any elevation from sea level up to 3,000 ft; the motor temperature rise will increase 1 per cent of the nameplate rise for each 330 ft above 3,300 ft. For operation above 6,000 ft, continuous duty resistors, autotransformers, and control circuit transformers must be dropped to 75 per cent of their normal kva rating.

Power Supply. A few plants may be located in areas where only direct current is available from the local power supply; the combined demand for large amounts of low-voltage d-c power and low-pressure steam for process work may well be chosen for all drives throughout the plant, particularly if materials-handling equipment such as cranes, hoists, and car dumpers is being operated or if a number of machines require speed adjustment. If a great many constant-speed motors and/or if large horsepower ratings are involved, economic considerations may justify bringing in a new a-c power supply or installing a-c generating equipment to permit the use of squirrel-cage or synchronous driving motors. Alternating-current motors are designed to operate successfully at rated load with a variation in voltage of plus or minus 10 per cent of rated voltage or a variation in frequency of plus or minus 5 per cent of rated frequency. Any deviation from rated frequency and voltage will result in changes in power factor, torque, speed, and efficiency.

Metal-clad switchgear equipment with air or oil circuit breakers should be selected for (1) starting and controlling motors for operation above 5,000 volts or (2) where the motor rating exceeds the capacity of industrial control contractors and circuit breakers. For low-voltage circuits a separate starter, with or without a combination air circuit breaker for short-circuit protection, might be specified for each motor. Where a number of motors are operating near a central location, the choice might be a group control or control center with air circuit breakers and contactors for each motor mounted in separate compartments of a common metal-enclosed structure.

TYPICAL APPLICATIONS ▨ – Satisfactory	AC MOTORS									DC		
	Normal-torque squirrel cage	High-torque squirrel cage	High-slip squirrel cage	Wound rotor	Synchronous	Low-torque single phase	Medium-torque single phase	High-torque single phase	Shunt wound	Compound wound	Series wound
Agitators – 1/2 to 15 hp	▨		▨			▨		▨			
Attrition and ball mills – 20 to 900 hp		▨			▨					▨	
Banbury mixers – 200 to 900 hp		▨									
Beater – up to 200 hp			▨								
Blowers – up to 500 hp	▨				▨		▨		▨		
Bucket elevators – 5 to 25 hp		▨						▨			▨
Chippers – up to 1,500 hp			▨								
Compressors – up to 600 hp				▨	▨					▨	
Conveyors – 3 to 100 hp		▨						▨			▨
Cranes and hoists – 3 to 150 hp				▨							▨
Crushers – 5 to 300 hp		▨		▨						▨	
Extractors – 3 to 100 hp											
Fans – up to 150 hp	▨					▨			▨		
Grinders (pulp) – 1,000 to 4,000 hp					▨						
Grinders and granulators – 1/4 to 30 hp	▨						▨				
Hammer mills – 20 to 200 hp		▨	▨							▨	
Jordans – up to 400 hp	▨										
Kilns – 20 to 100 hp		▨		▨					▨		
Mixers – 2 to 200 hp	▨							▨	▨		
Pulverizers – 10 to 250 hp		▨									
Pumps (centrifugal) – up to 1,000 hp	▨				▨				▨		
Pumps (reciprocating) – up to 200 hp		▨						▨		▨	
Shredders – 5 to 300 hp		▨							▨		
Stokers – 5 to 50 hp	▨										

STARTER TYPES

Squirrel cage	Magnetic full or reduced voltage.
Wound rotor	Manual, semimagnetic or full magnetic with secondary resistance.
Synchronous	Magnetic full or reduced voltage.
Single phase	Manual or magnetic full voltage.
Direct current	Manual or magnetic with field and armature resistance.

RECOMMENDED VOLTAGES

Motor hp	Voltage
Up to 150	550 or less
200 to 2,000	2,300
2,000 to 6,000	4,160
Above 6,000	13,800

FIG. 4-5. Typical motor specifications and applications. Applications and starter types due to V. J. Kropf, *Chem. Eng.*, **58**(7): 124 (1951).

Power Factor. Most public utilities have a clause in their power contracts requiring a plant with low power-factor loads to pay a penalty charge or an increased rate; the power factor of a motor represents the percentage of the load or power current to the total line current. The line current is made up of magnetizing or reactive current and power or active current. This reactive current represents just as real a burden to the power system as the active or load current, even though it does not represent any actual work accomplished at the driven machine. The limiting factor of both generators and transmission lines as well as plant distribution systems is current- or ampere-carrying capacity. The power company not only charges process plants for the kilowatthours used but also imposes a severe penalty on any plant with excessive reactive loads. Attention is being given to raising plant power factor in an effort to keep power costs and equipment investment to a minimum. Many plants use overexcited synchronous motors with 80 per cent leading power factor for constant-speed drives to compensate for the lagging power factor of other machines. Power companies usually permit some lagging current without penalty so that the power factor need only be brought up to the minimum allowable value or a little above it to obtain the most favorable power rates. Consideration must also be given to plant-wiring capacity and transformer ratings in determining exactly how far to go in correcting plant power factor to gain the maximum benefits.

Demand Charges. Power companies usually include a so-called demand charge in their rate structure based on the maximum load carried during a given period of time; the power cost is sometimes computed on maximum demand alone (see Table 6-13). Thus a peak load for only 1 hr or 1 day, regardless of how light the normal load may be, can result in electrical charges of the same amount as though the plant had actually used the peak power every day for the entire billing period. In most process plants, motor-driven machinery represents a continuous power load rather than an intermittent one. There are many applications, however, requiring various motor operations from time to time as the occasion demands. The plant operator or process engineer may save his company a good many dollars in demand charges by scheduling additional motor loads at off-peak periods and by operating the plant for a majority of the time or for as much time as is practical at or near the maximum peak power required.

Application of Types of Motors and Typical Costs. A classification of types of motors applicable to specific processing equipment has been prepared by Kropf in Fig. 4-5 with costs of motors given in Table 6-2 and Fig. 6-22.

Gas Facilities

Equipment for storage and handling of heating gases, air, and process gas in a liquefied or compressed gas state must be provided in most chemical processes. Tank storage and transportation of gases have been discussed in the section on tanks and reaction vessels. Some of the design requirements for chemical process and gas-moving systems will be discussed next.[1]

---
[1] Staff report, Handling Compressible Fluids, *Chem. Eng.*, **63**(6): 175 (1956).

Gas Requirements. Modern chemical processes require one or more gases under a wide range of pressures and temperatures, as shown in Fig. 4-6. Air-handling equipment is still in the largest demand, with such requirements as:

Type and use	Pressure range, psig	Requirements
Instrument air..................	25–35	Extremely dry, oil-free
Utility air for motors, hoists, solids transport, fluid jet pumps, blow cases, and cleaning	100–125	Condensed phase oil and water removed by centrifuge or filter
Process air for oxidation........	15–3,000	Same as for utility air
Liquid air separation...........	100–600	Extremely dry, oil-free

Gas-moving Equipment. The principal types of equipment available for gas pumping are blowers, compressors, ejectors, fans, and vacuum pumps. The following is a general guide for selection of equipment based on pressure and capacity requirement:

Type	Max discharge pressure commercially usable, psig	Inlet capacity range, cfm
Compressors:		
Reciprocating...............	40,000	10–10,000
Centrifugal.................	3,000	500–100,000
Rotary displacement.........	150	100–5,000
Axial flow..................	100	100,000–5,000,000
Blowers:		
Rotary......................	30	20–70,000
Centrifugal turboblower......	30	1,000–100,000
Fans.........................	1	100–30,000

Vacuum equipment	Min suction pressure, mm Hg
Rotary displacement........................	3
Steam jet ejectors...........................	5×10^{-2}
Mechanical rotor or vane, oil seal..............	5×10^{-3}
Diffusion....................................	1×10^{-7}

The reader is referred to Perry's "Chemical Engineers' Handbook," 3d ed., pp. 1258–1261, 1439–1456, for a general description and for calculation methods on gas-moving equipment.

Selection of Compressor Equipment. Compressors for such gas processes as liquid-air separation, hydrogenation, and desulfurization must be nonlubricated to avoid possible explosive reaction with lubricating oils. Graphite rings are used in reciprocating compressors, and labyrinth or

CHAP. 4] SELECTION OF EQUIPMENT AND MATERIALS 143

mechanical seals are designed for centrifugals to provide the nonlubrication features.

In general, reciprocating compressors are more economical than centrifugals for capacities below 10,000 cfm, but accurate total annual cost comparisons should be made, starting at 1,000 cfm.

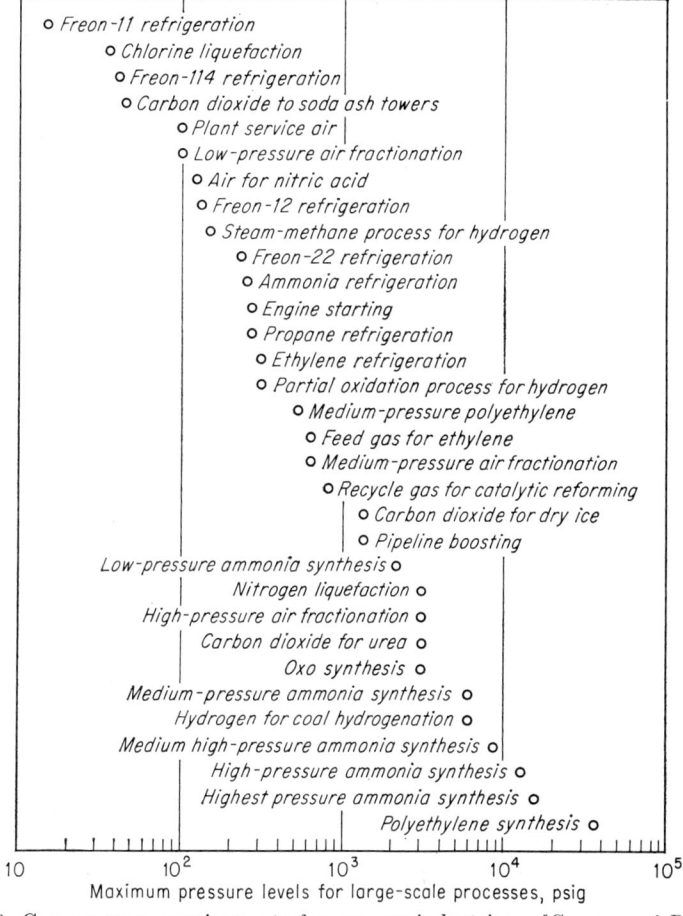

```
  o Freon-11 refrigeration
    o Chlorine liquefaction
     o Freon-114 refrigeration
      o Carbon dioxide to soda ash towers
        o Plant service air
         o Low-pressure air fractionation
           o Air for nitric acid
            o Freon-12 refrigeration
             o Steam-methane process for hydrogen
              o Freon-22 refrigeration
               o Ammonia refrigeration
                o Engine starting
                 o Propane refrigeration
                  o Ethylene refrigeration
                   o Partial oxidation process for hydrogen
                    o Medium-pressure polyethylene
                     o Feed gas for ethylene
                      o Medium-pressure air fractionation
                       o Recycle gas for catalytic reforming
                        o Carbon dioxide for dry ice
                         o Pipeline boosting
Low-pressure ammonia synthesis o
          Nitrogen liquefaction o
High-pressure air fractionation o
          Carbon dioxide for urea o
                Oxo synthesis o
Medium-pressure ammonia synthesis o
Hydrogen for coal hydrogenation o
Medium high-pressure ammonia synthesis o
       High-pressure ammonia synthesis o
      Highest pressure ammonia synthesis o
                Polyethylene synthesis o
```
10   10^2 10^3 10^4 10^5
Maximum pressure levels for large-scale processes, psig

FIG. 4-6. Gas pressure requirements for process industries. [*Courtesy of P. R. Des Jardins, Chem. Eng.*, **63**(6): 178 (1956).]

Pipeline Transmission. Transportation of large volumetric rates of compressible fluids is most economically accomplished by pipelines. For example, gas pipelines are used to transmit gases, such as natural gas, acetylene, ethylene, and LPG (propane, butane), over distances up to several thousand miles. Booster compressors, operating at 300 to 600 psi suction and 800 to 1,200 psi discharge, are stationed along the pipeline to maintain gas-flow rates. Pipeline design is discussed in Chap. 9.

Costs of Gas-moving Equipment. Figure 6-23 gives installed costs of gas-moving equipment based on rated volumetric input. These costs are for preconstruction cost estimating only. The maintenance cost picture should also be examined when installed costs of several different types of compressors are about equal. For example, centrifugals require less maintenance than reciprocating or rotary machines.

Pumps and Pumping

The severe pumping demands in modern chemical plant service require continuous heavy duty for long periods with freedom from forced shutdowns, flexible operating characteristics, ease of control, availability in wide choice of materials of construction to meet a wide range of operating conditions (head, capacity, temperature, viscosity, etc.), maximum interchangeability of pumps and parts, handling of solids and abrasives in suspension, and a design that can tolerate some erosion and corrosion.

Pumps are used in chemical plants for a great number of purposes in transferring liquids, colloidal solutions, or solids suspended in gases or liquids from one point to another. Pump transportation covers both long and short distances, horizontal and vertical, under pressure heads ranging from subatmospheric to very high pressures. Pumps are also used to produce both high and low pressures in equipment to aid physical or chemical processing reactions. Siphons, air-lift systems, acid eggs, and barometric legs are but rarely used in chemical plants for limited applications; reciprocating pumps are next in extent of usage with centrifugal pumps being the major choice for chemical plants.

Reciprocating Pumps. The reciprocating pump is the oldest and best-known form of pump. The delivery of liquid is effected by the displacement of a piston or plunger. As a general rule, reciprocating pumps are used for comparatively small capacities against high heads. Ordinarily they will operate on higher suction lifts and will handle fluids of higher viscous consistency than centrifugal pumps. Since they will keep their prime, they are particularly suitable for installations that require automatic control. The capacities depend on the displacement of the plunger and the speed with which it is operated. By increasing the speed, the capacity may be changed without affecting the head except for increased friction in the pipes caused by the increased velocity of the liquid.

The variety of reciprocating pumps is indicated by the specific applications for which pumps are used. Among the variety of pumps available can be included deep-well, boiler-feed, condensation, dry-vacuum, wet-vacuum, compound, filter-press, hydraulic-press, creamery, milk, oil, proportioning, fuel-oil, lime, and magma pumps. Some data on the capacity of reciprocating pumps are found in Perry's "Chemical Engineers' Handbook," 3d ed., pp. 1426–1436.

The limitations of reciprocating pumps are (1) costly design to construct in special alloys, (2) rubbing contact and danger of seizure limit choice of materials, (3) large space requirements, (4) pulsating flow, (5) not generally suitable for dirt or abrasive-laden liquids, (6) relatively inflexible operating characteristics, (7) most types require protection against overpressure and power overload, and (8) reciprocating rod difficult to seal against leakage.

Diaphragm Pumps. This is the only pump that is immune from clogging and abrasive wear in handling pulps, sludges, and other non-homogeneous materials. The suction type is an open pump and is not designed to work against pressure heads. Diaphragm pumps have also been developed to fill the recognized need in the chemical, metallurgical, and sanitary fields for a pump that has all the wearing and operating advantages of the open-suction type, with the additional property of forcing pulps and sludges to high elevations or through horizontal pipes for long distances. Capacities of diaphragm pumps can be varied by changing the speed, length, and diameter of the piston diaphragm and cylinder.

Centrifugal Pumps.[1] The flow produced by a centrifugal pump is free from pulsations, and its advantages are numerous. Such pumps are compact, rugged, dependable, and simple to operate. They may be operated manually or employ combined manual and automatic operation. Auxiliary stand-by gas-engine drives may be used with either turbine or motor drive, ensuring constant service. They will operate against a closed discharge without building up dangerous pressures and will deliver an increased capacity at reduced heads. The centrifugal pump has no valves to stick, no reciprocating parts that must constantly be kept in motion to prevent corrosion, and no close tolerances. Centrifugal pumps are more dependable, easier to maintain, and cost less to install than reciprocating types.

The centrifugal pump must be primed each time it is started. It will not develop so high a suction lift as the reciprocating pump, and it shows a decreased capacity when handling viscous liquids. Although a single-stage pump has delivered at a head of 1,000 ft, practical designs limit the head to not over 300 ft per impeller. For the higher heads, two or more impellers are connected in series, the discharge from one impeller being the suction of the next; the total head is the head of each impeller multiplied by the number of impellers. For the sake of economy such an arrangement is built in one case and is known as a *multistage* pump, each impeller constituting a stage. Ordinarily not more than six stages can be built in

[1] Perry, "Chemical Engineers' Handbook," 3d ed., pp. 1417–1419.

one pump casing on account of the length of shaft. Room for a large shaft is difficult to get because of the space required to lead the liquid into the eye of the impeller, and a long shaft must be rigid in order to prevent undue vibration. Where more than six stages are required, the usual practice is to split the requirements into two pumps placed on both ends of the motor shaft, and back to back to neutralize end thrust by having one pump oppose the thrust of the other. As the capacity of the centrifugal pump varies with the speed and the head varies with the square of the speed, it is important that the pump be operated at the proper rate. No installation should be made without checking speed, capacity, horsepower, etc., with the tables given for the pump by its manufacturer.

Specification of Centrifugal Pumps. Typical specifications of the all-metal chemical centrifugal pumps are (1) deep stuffing box; (2) plenty of room for mechanical seal; (3) rugged construction of casing, bearings, and shaft; (4) provision for water cooling of bearings; (5) extra erosion-corrosion allowance for longer life; (6) external bolting often used on casings to permit use of special alloys; (7) simplified castings; (8) impeller nut integral with impeller; (9) a solid corrosion-resistant shaft at the pump end; (10) maximum interchangeability of pumps and parts, through simplification and standardization in design; (11) flanged connections and elimination of drains and vents to reduce corrosion (of threads, and at sharp corners) and leakage; (12) easy adjustment of impeller clearance in casing; (13) means of collecting leakage; (14) protection of bearing housing, bearings, and base plates from corrosion; (15) availability with either open, semienclosed, or closed impellers for different conditions; and (16) special designs such as self-priming and submerged.

Rotary Gear Pumps. These employ two meshing gears within a close-fitting case. Liquid is trapped by the gear teeth and carried from intake to discharge. The meshing of the gears seals the pump against backflow. The *screw pump* is a special type of gear pump, employing two meshing screws in a figure-of-eight casing. Such pumps are built to handle any liquid or semiliquid that will flow through a suction pipe, such as molasses, brine, water, heavy and light grease, oils, and acid sludges. They are built in capacities ranging from 2 to 4,200 gpm, against pressures up to 1,000 psi or more. Such pumps have been used for extruding cellulose nitrate solutions at pressures as high as 2,500 psi. The *sliding-vane pump* embodies an eccentric casing; either the rotor and shaft are eccentric to the casing or the casing is elliptical in shape. In the sliding-vane or ring types, the rotor carries the slides in and out because they rest on a stationary ring. Because of their simplicity of design and ruggedness, most rotary gear and vane pumps are particularly adapted to the pumping of liquids more viscous than water, such as molasses, tar, soap, and oil. Such thick, slow-flowing liquids cannot always be handled satisfactorily by either piston or centrifugal pumps. Rotary pumps usually depend for their lubrication on the material being pumped and, if long life is desired, should not be used for liquids that do not have some lubricating qualities. They are built with close clearances and consequently should not be used for handling liquids containing grit or solids.

Materials of Construction for Pumps. The materials used in pump construction depend upon the service demanded. Although all-purpose, corrosion-resisting materials do not exist, there are materials that have great resistance to specific corrosive reagents, and the knowledge of the chemical reactivity between various chemical solutions and materials of construction is necessary in proper pump selection.

Glass, porcelain, enamel-lined, and stoneware pumps have but limited application owing to the inability to withstand mechanical and severe thermal shock. The high silicon-iron alloys such as duriron, tantiron, and corrosiron are more applicable to severe mechanical shock than the materials mentioned above, but nonmachinability and the high cost of grinding these very hard and brittle materials restrict their application. Lead and lead alloys are applicable for pump materials but are limited to uses wherein lead does not enter into the reaction. Special hard lead, firmly adherent to a supporting outer shell of cast iron or other metal, has been adopted as practicable. Hard-rubber-lined, Pyrex, and plastic pumps are also available and highly desirable for pumping hydrochloric acid.

Brasses and bronzes, iron-base alloys, nickel, monel, magnesium alloys, hard rubber, plastics, tin, aluminum, and like metals must be added to ordinary gray and white cast iron as materials for pump construction. Practically any alloy or modern metal can be fabricated into pumps, and it remains only for the chemical engineer to stipulate the kind of solution he wishes to handle, or the kind of metal, and the pump manufacturer will attempt to construct a pump for the service demanded. (See Perry's "Chemical Engineers' Handbook," 3d ed., p. 1424.)

Summary of Information Required for Selecting Pumps[1]

1. Capacity and head:
    a. Maximum for each.
    b. Permissible or desirable range for each.
    c. Possible future change in requirements.
2. Desirable operating characteristics:
    a. Constant head and capacity, or
    b. Variable capacity and nearly constant head, or
    c. Variable head with some variation in capacity, or
    d. Constant capacity against variable heads, etc.
    (Some of these may call for variable speed, throttling, or bypassing. Manual or automatic? These factors influence selection of drive. See item 8 below.)

[1] R. E. Dolman, Pumps, *Chem. Eng.*, **59**(3): 155 (1952).

3. Nature of the liquid—Is it volatile? Lubricating? Corrosive? Unstable? Effect of temperature, pressure, shear rates, time, etc., on its properties. (The name of the liquid may be helpful in pump selection, but is not essential in itself if complete data are furnished.)
4. Nature and size of solids in suspension.
5. Corrosion data—suitable materials of construction.
6. Range of operating temperature—Possible crystallization, solidification? Expansion problems? (Temperature may also influence materials of construction.)

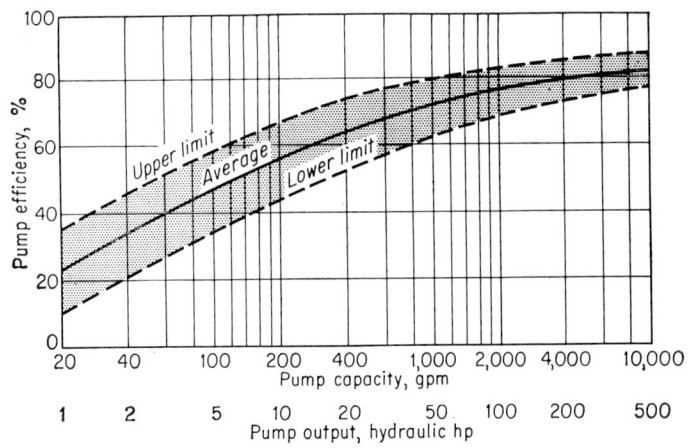

Fig. 4-7. Efficiencies of a centrifugal pump. [*Courtesy of R. M. Braca and J. Happel, Chem. Eng.*, **60**(1): 180 (1953).]

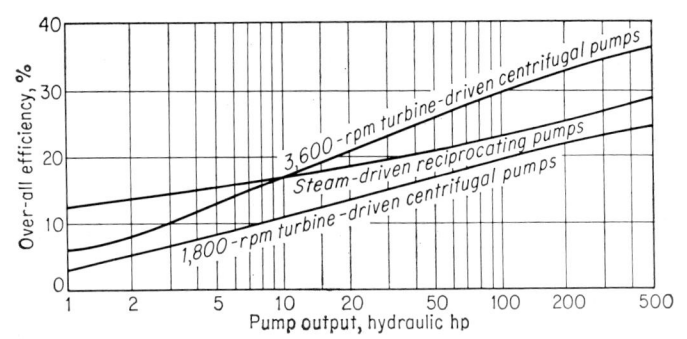

Fig. 4-8. Over-all efficiencies of pumps and drivers. [*Courtesy of R. M. Braca and J. Happel, Chem. Eng.*, **60**(1): 181 (1953).]

7. Viscosity range—affects suction flow, slip, horsepower, etc.
8. Type of power or prime mover available or allowable (see item 3 above). Plant steam balance should be considered. Is there need for emergency standby equipment of special type, with special drives in case of electric-power failure?
9. Load factor.
10. Efficiencies. Efficiencies for centrifugal pumps and over-all efficiencies of motor-driven centrifugal pumps are presented in Figs. 4-7 to 4-11 in terms of hydraulic horsepower.
11. Costs. Costs of pump equipment for 1958 are presented in Fig. 6-24.

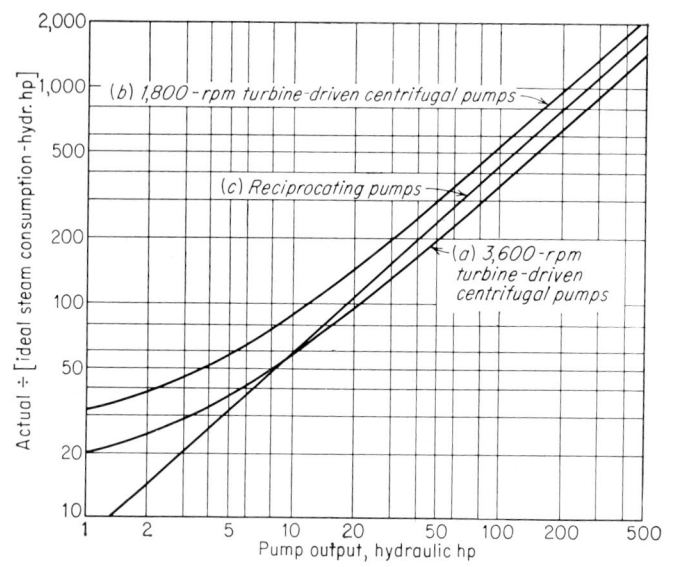

FIG. 4-9. Steam consumption of steam-driven pumps. [*Courtesy of R. M. Braca and J. Happel, Chem. Eng.*, **60**(1): 181 (1953).]

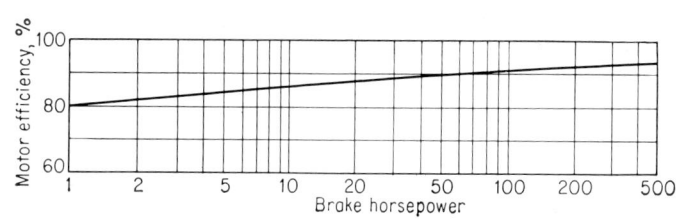

FIG. 4-10. Efficiencies of three-phase induction motors. [*Courtesy of R. M. Braca and J. Happel, Chem. Eng.*, **60**(1): 181 (1953).]

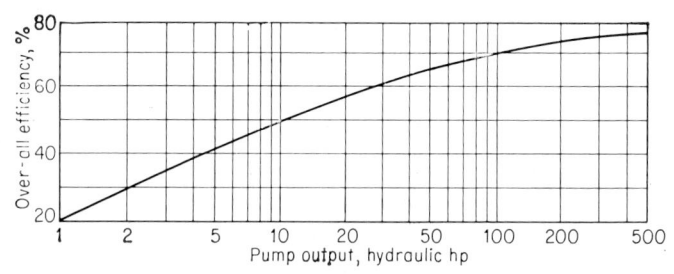

FIG. 4-11. Over-all efficiencies of motor-driven centrifugal pumps. [*Courtesy of R. M. Braca and J. Happel, Chem. Eng.*, **60**(1): 181 (1953).]

Process Auxiliary Systems

Design and cost information for process piping, insulation, instrumentation, and power systems is discussed in Chap. 9. Methods of handling preconstruction cost estimation of these auxiliary systems are given in Chap. 6.

MECHANICAL DESIGN AND FABRICATION OF EQUIPMENT

Although chemical engineers in plant design work are not usually responsible for detailed mechanical design and fabrication of chemical engineering equipment, they should have some knowledge of these methods.

Development of Equipment Design

This is the province of mechanical engineers who work from the design conditions as set up by the process engineering group and transmitted via specification sheets. These sheets give pressure, temperature, power, and approximate size requirements in addition to suggesting materials of construction. Standard equipment and construction items are utilized where possible, and designs are monitored by code regulations, such as the ASME Code for Unfired Pressure Vessels[1] and others.[2] Designs take into consideration such items as allowable stress under maximum working pressure and temperature, economic design, using such items as standard dished heads, wind loading on outside equipment, vibrational analysis, power and torque requirements for agitation, and many similar problems. For further background on this subject, the reader is referred to Hesse and Rushton[3] or Rase and Barrow.[4]

The results of these calculations are conveyed by sketches to engineering draftsmen who prepare detailed drawings for shop fabrication.

Equipment Fabrication

The type of fabrication depends on the specific equipment being manufactured. The most common fabrication methods for chemical engineering equipment are those related to the preparation of tanks and pressure vessels.

---

[1] American Society of Mechanical Engineers, 29 W. 39th St., New York, 1956.

[2] Synopsis of Boiler Laws, Rules and Regulations by States, Provinces, and Cities (USA and Canada), National Bureau of Casualty and Surety Underwriters, Boiler and Machine Division, 60 John St., New York.

[3] H. C. Hesse and J. H. Rushton, "Process Equipment Design," D. Van Nostrand Company, Inc., Princeton, N.J., 1945.

[4] H. F. Rase and M. H. Barrow, "Project Engineering of Process Plants," John Wiley & Sons, Inc., New York, 1957.

Fabrication of Vessels. Shop construction men, working from shop prints, make the basic *layout* of the equipment by marking on flat sheets and other required metal forms. Excess material is removed by *shearing* of straight-line, thin sheet sections or by *flame cutting* of sections up to 16 in. in thickness. Ease of oxidation of the metal is a requirement for the oxygen-gas torch, and carbon steel is easily cut. Alloy metals require an inert gas plasma jet flame. If the material is heat labile, expensive sawing methods may have to be substituted. Holes are also cut out for nozzles and manholes, and then the edges are ground in preparation for butt welding.

Shaping of the cast pieces to design dimensions is done by such mechanical operations as bending, pressing, rolling, stamping, or die spinning, with or without heat addition and subsequent annealing. Jigs, straps, jacks, hoists, and wedges are used to assemble the pieces for seam-joining procedures. Final hand grinding and edge alignment are followed by tack welding. Butt-welding joints and other parts are then fused together by electric-arc welding. The shielded-arc method is used almost entirely for this type of joining. The compatible metal-wire or rod electrode used is covered with a chemical flux coating which partly vaporizes in the heat of the arc to form a protective inert gas blanket over the molten pool of base and weld metal during cooling. Automatic, continuous-feed welding machines are generally used for routine welding of long seams with hand operation reserved for small or more complex sections. Heliarc welding, in which a gas blanket is maintained by a jet of helium or argon, is required for highly oxygen-sensitive alloys, such as Hastelloy. Brazing and silver soldering are common for joining small sections of cast-iron, steel, or copper-base alloys.

Preliminary testing is done next to uncover flaws in joining, and sometimes in the metal itself. Nondestructive test methods, such as ultrasonics and X- or gamma-ray inspection, are used on main welding seams. Dye penetration, hydrostatic, or air-soap tests are also used to locate leaks and cracks. Helium or Freon leak detectors are necessary for more critical leak testing, particularly on high-vacuum equipment.

Fabrication methods develop working stresses in the equipment which may have to be relieved by *heat treating* to impart maximum strength and prevent stress corrosion. A final hydraulic test at two to three times the rated operating pressure, as specified by code or local regulations, must be made before the vessel is delivered to the customer.

Fabrication of Other Equipment. Equipment, such as pumps, compressors, and centrifuges, is made from castings, forgings, and stampings. Many of the required parts are machined to close tolerances by conventional machine-shop operations, such as milling, boring, drilling, tapping, honing, and polishing. Parts are assembled by means of machine-screw fittings or by brazing or welding, as described above.

EXAMPLE OF AN EQUIPMENT SPECIFICATION JOB[1]

For this example, representative equipment suppliers have been specified by use of the "Chemical Engineering Catalog."[2] This gives the student practice in using this excellent reference. Additional specifications were obtained from the equipment suppliers' pamphlets. Detailed information of this type is not required for preconstruction cost estimating as developed in Chap. 6. Only materials of construction and the data required to use the abscissas of the cost estimating curves are generally sufficient.

Sample Calculations for Equipment Design

A-1. Benzol transfer pump

   Duty: Explosion-proof pump capable of delivering tank carlot of technical benzol to storage in 3 hr

   Benzol rate: $\dfrac{8,000}{3 \times 60} = 44.5$ gpm

   Maximum head: 15 ft

   Selection: 50-gpm pump with 1-hp motor at 1,200 rpm recommended

A-2. Benzol storage tank

   Duty: Storage capacity for 2 weeks' processing required plus allowance for tank carlot

   Daily consumption: $\dfrac{2,029}{0.879 \times 8.33} = 278$ gal

   Two weeks' supply: $10 \times 278 = 2,780$ gal

   Tank carlot: 8,000 gal

   Minimum tank capacity: $2,780 + 8,000 = 10,780$ gal

   Safety allowance: 10 per cent or 1,078 gal

   Selection: Two tanks of 6,000 gal (nominal) capacity each. Dimensions and specifications as recommended by Henderson.[3]

A-3. Benzol feed pump

   Duty: Explosion-proof pump to feed technical benzol continuously into chlorinator

   Benzol rate: $\dfrac{10,483}{0.879 \times 8.33 \times 24 \times 60} = 1.0$ gpm

   Maximum head: 20 ft

   Selection: 7-gpm centrifugal pump with ⅛-hp motor

A-4. Water demineralizer

   Water requirements: $\dfrac{589}{8.33 \times 24} = 2.94$ gph

   Selection: 10-gph Barnstead cation-anion exchange unit with full automatic controls

[1] *Reference:* Manufacture of Benzene Hexachloride, Chap. 3. (Note that the letter-number references, for example, A-1, of equipment in the sections which follow are based on the equipment flow sheet in Fig. 3-2.)

[2] Reinhold Publishing Corporation, New York, 1959.

[3] J. G. Henderson, *Chem. Eng.*, **54**(3): 106 (1947).

CHAP. 4]   SELECTION OF EQUIPMENT AND MATERIALS   153

A-5. Chlorine evaporator

Duty: A device capable of evaporating 4,228 lb of liquid chlorine per day using hot water at 150°F as the heating medium

Heat requirements:
Latent heat of chlorine:[1] 121.0 Btu/lb at 30.1°F
Max. heater temperature:[1] 150°F as the heating medium

Net heat of evaporation: $\dfrac{4{,}228}{24} \times 121 = 21{,}300$ Btu/hr

Sensible heat: $\dfrac{4{,}228}{24} (150 - 30.1) \times 0.223 = 4{,}700$ Btu/hr

Total heat input: $21{,}300 + 4{,}700 = 26{,}000$ Btu/hr
Log mean temp. difference: 148°F
Assume over-all heat-transfer coefficient: $U = 75$ Btu/(hr) (ft²) (°F)

Heating area required: $\dfrac{Q}{U \times \Delta T} = \dfrac{26{,}000}{75 \times 148} = 2.34$ ft²

Selection: Hairpin-type heat exchanger, mounted vertically, with 5 ft² of heating surface and ample free space on shell side

B-1. Chlorinator

Basis of design: Extrapolation of subpilot reactor conditions and results: Reactor charge consisting of 2,038 g of organic phase and 285 g of the aqueous phase permitted the absorption of 20 liters/hr (59 g/hr) of chlorine at a reaction temperature of 70°C with 99 per cent conversion of the chlorine

Deductions: To absorb $4{,}228 \div 24 = 176$ lb of chlorine per hour will require a minimum of

$$176\tfrac{5}{59} \times 2{,}038 = 6{,}080 \text{ lb of product phase}$$

Corresponding weight of aqueous phase:

$$\dfrac{3{,}127}{14{,}519} \times 6{,}080 = 1{,}309 \text{ lb}$$

Calculation of minimum volume of reactor:

Volume of product phase: $\dfrac{6{,}080}{1.0340 \times 8.33}$ .... 702 gal

Volume of aqueous phase: $\dfrac{1{,}342}{1.0975 \times 8.33}$ 148

Add 50 per cent extra capacity for safety factor.................... 430
Add 20 per cent for free board.. 170
Total capacity required... 1,450 gal

Selection: Jacketed-type XXL Pfaudler glass-lined reaction kettle with glass-lined agitator, nominal 1,500 gal capacity

B-2. Reflux condenser

The basic heat balance equation is

$$\text{Heat input} + \text{heat of reaction at 25°C} = \text{heat output}$$

---

[1] W. T. Anderson, *Ind. Eng. Chem.*, **39**: 845 (1947).

154   CHEMICAL ENGINEERING PLANT DESIGN   [CHAP. 4

Heat and material balance data as developed in Chap. 3 were used to prepare Fig. 3-6. The reflux heat duty shown on this figure is 3,718,568 Btu/24 hr, or 155,000 Btu/hr.

Heat-transfer requirements:
  Assume: Cooling water enters at 25°C and leaves at 35°C
  Vapors are condensed at 70°C and condensate is cooled to 30°C
  Over-all heat-transfer coefficient $U = 100$
  Log mean temperature difference: 18°C (32.4°F)
  Minimum condensing surface: $\dfrac{Q}{U \times \Delta T} = \dfrac{155,000}{100 \times 32.4} = 46.2 \text{ ft}^2$

Selection: Pfaudler glass-lined double-jacketed condenser with 62 ft² of condensing surface. (*Note:* A further factor of safety is contained in the cooling afforded by 163 ft² of cooling area of the reaction kettle jacket.)

B-3. Vent gas separator

Duty: An expansion chamber in which the condensate from B-2 may be separated from the vent gases without entrainment
Selection: Pfaudler glass-lined vacuum receiver of 50 gal capacity

D-1. Dilute acid receiver

Duty: Storage for 1 week's supply of dilute acid
Daily storage requirements: $\dfrac{3,127}{1.0175} \times 8.33 = 286$ gal
Weekly storage requirements: $5 \times 286 = 1,430$ gal
Selection: 1,500-gal horizontal glass-lined storage tank

D-2. Acid still feed pump

Acid flow rate: $\dfrac{3,127}{1.0975 \times 8.33 \times 24} = 14.28$ gph
Head: 20 ft
Selection: proportioning pump

D-3. Acid stripping still

Duty: Required to strip a 36 per cent HCl product from a 25 per cent feed; bottoms to contain 20 per cent HCl

Still conditions:

	Temp., °C	HCl, wt %	HCl, mol %	Lb/hr
Feed................	65	25.0	14.1	130.4
Distillate............	75	26.0	21.7	38.8
Bottoms.............	110	20.4	11.2	91.6

Mols vapor rising (mole distillate): 1.765 mols/hr
Mols liquid descending: 4.562 mols/hr
Reflux ratio (still used as stripper only): $\dfrac{4.562}{1.765} = 2.59$

CHAP. 4]   SELECTION OF EQUIPMENT AND MATERIALS   155

Determination of theoretical plates required: The calculation of the plate requirements was made by the McCabe-Thiele method.[1] The equilibrium curve for HCl-water was calculated from data given in the "Chemical Engineers' Handbook."[2]
Number of theoretical plates required: 2 + reboiler
  Assume plate efficiency of 50%
Actual plates required: 2 ÷ 0.50 = 4 + reboiler
  Assume maximum free vapor velocity equals 1.0 fps

Max. vapor rate: $\dfrac{1.765 \times 359 \times 383}{60 \times 60 \times 275} = 0.246$ cfs

Cross-sectional area of column: 0.246 ÷ 1.0 = 0.246 ft²
Required diameter of column: 6.72 in.
Selection: A Karbate column of 8 in. nominal diameter will be used. Four bubble cap plates and a reboiler will be furnished. Using a plate spacing of 18 in. requires a column height of 12 ft.

D-4. Acid still condenser

Heat of condensation of water: 589 × 540 × 1.8 .... 572,500 Btu/day
Heat of condensation of benzene: 7 × 95 × 1.8 1,200
Heat of solution of HCl:[3] $\dfrac{331}{36.5}$ × 17,400 × 1.8 505,400
Sensible heat: 931 × 1.0 × (75−30) × 1.8 75,400
  Total heat transferred .... 856,200 Btu/day

Assume: Over-all heat-transfer coefficient, $U = 100$
Log mean temperature difference: 25°C (41°F)

Heat-transfer area required: $\dfrac{Q}{U \times \Delta T} = \dfrac{856,200}{24 \times 100 \times 41} = 8.7$ ft²

Selection: Heat exchanger using 10 BWG 18 tubes, ¾ in. diam., 4 ft long, in nominal 6-in. shell

D-5. Conc. HCl receiver

Duty: Storage of one day's supply of 36 per cent HCl acid

Daily storage requirements: $\dfrac{931}{1.1789 \times 8.33} = 95.9$ gpd

Selection: 100-gal glass-lined receiver

D-6. Conc. HCl pump

Duty: Transfer of 36°Bé acid from receiver to storage and from storage to tank car
Selection: 10-gpm pump

D-7. Conc. HCl storage

Selection: A 10,000-gal rubber-lined steel tank recommended to allow sufficient capacity to store a tank carlot of muriatic acid in addition to 2 weeks' production

---

[1] J. H. Perry (ed.), "Chemical Engineers' Handbook," 3d ed., p. 591, McGraw-Hill Book Company, Inc., New York.
[2] *Ibid.*, p. 166.
[3] N. A. Lange, "Handbook of Chemistry," p. 1504, Handbook Publishing Co., Sandusky, Ohio, 1956.

156   CHEMICAL ENGINEERING PLANT DESIGN   [CHAP. 4

D-8. Acid recycle pump

$$\text{Liquid rate: } \frac{2{,}196}{1.057 \times 8.33 \times 24} = 10.4 \text{ gph}$$

Selection: 20-gph volumetric pump

E. Crude product pump

$$\text{Flow rate: } \frac{14{,}319}{1.034 \times 8.33 \times 24 \times 60} = 1.152 \text{ gpm}$$

Selection: 10-gpm pump

F-1. Caustic make-up tank

$$\text{Daily caustic make-up: } \frac{3{,}101}{1.006 \times 8.33} = 371 \text{ gal}$$

Weekly make-up: $371 \times 5 = 1{,}855$ gal
Selection: 2,000-gal steel tank equipped with Lightnin angular off-center agitator

F-2. Caustic transfer pump

Duty: Transfer 1,855 gal of 1 per cent caustic solution to storage in 3 hr
Flow rate: $1{,}855 \div 3(60) = 10.3$ gpm
Selection: 10-gpm cast-iron centrifugal pump

F-3. Caustic storage

Selection: Horizontal steel tank to hold 2 weeks' supply of 1 per cent caustic, i.e., 4,000 gal capacity

F-4. Caustic feeder

$$\text{Flow rate: } \frac{3{,}101}{1.006 \times 8.33 \times 24} = 15.4 \text{ gph}$$

Selection: Proportioning pump with automatic pH control

F-5. Neutralizer

Duty: A resistant vessel of sufficient capacity to permit 1-hr retention of materials

$$\text{Daily flow rate: } \frac{14{,}319}{1.0340 \times 8.33} + \frac{3{,}101}{1.006 \times 8.33} = 2{,}034 \text{ gpd}$$

Hourly flow rate: $2{,}034 \div 24 = 85$ gph
Selection: 100-gal glass-lined kettle with agitator

F-6. Spent caustic separator

Duty: One hour's retention of process materials required to effect satisfactory separation of the two phases involved
Selection: 200-gal steel tank with conical bottom

G-1. Flash still feed pump

$$\text{Flow rate: } \frac{14{,}319}{1.0340 \times 8.33 \times 24 \times 60} = 1.15 \text{ gpm}$$

Selection: Same as item A-3

CHAP. 4]   SELECTION OF EQUIPMENT AND MATERIALS   157

G-2. Flash still

Requirements: A continuous evaporator of sufficient size to hold 8-hr charge with enough heat-transfer area to evaporate 391 lb of solvents per hour

Heat-transfer calculations:

Heat of evaporation (Btu/day):
For benzene, $8{,}337 \times 90.6 \times 1.8$ 1,359,600
For monochlor, $969 \times 77.6 \times 1.8$ 135,350
For BHC, $24 \times 70.0 \times 1.8$ 3,000
For water, $45 \times 540 \times 1.8$ 43,750

Sensible heat:
Feed stream, $14{,}299 \times (100 - 60) \times 1.8$ 1,029,500
Recycle stream, $7{,}230 \times (100 - 30) \times 1.8$ 912,000

Total heat transfer, Btu/day.............................. 3,483,200

Hourly heat transfer: $\dfrac{3{,}483{,}200}{24} = 145{,}100$ Btu/hr

Assume steam is available at 15 psig (122°C) and the over-all heat-transfer coefficient $U = 200$

Mean temperature difference: $122 - 100 = 22°C$, or $39.6°F$

Heating surface required: $\dfrac{Q}{U \times \Delta T} = \dfrac{145{,}100}{200 \times 39.6} = 18.21 \text{ ft}^2$

Required capacity of vessel: $\dfrac{12{,}153 \times 8}{1.025 \times 8.33 \times 24} = 474.2$ gal

Selection: Continuous evaporator and stripper of 500 gal capacity with heating surface of 20 ft²

G-3. Flash still condenser

Heat-transfer requirements (Btu/day):
Heat of evaporation (see G-2)................................ 1,541,700
Sensible heat: $9{,}376 \times 0.45 \times (100 - 30) \times 1.8$ 531,100

Total heat transfer.. 2,072,800

Hourly heat transfer: $\dfrac{2{,}072{,}800}{24} = 86{,}366$ Btu/hr

Assume over-all heat-transfer coefficient $U = 50$

Log mean temperature difference: $23°C$, or $41°F$

Condensing surface required: $\dfrac{Q}{U \times \Delta T} = \dfrac{86{,}366}{50 \times 41} = 42.1 \text{ ft}^2$

Selection: Single-pass stainless-steel condenser of 50 ft² condensing surface

G-4. Solvent receiver

Duty: Retention of 1 week's production of solvents

Daily capacity requirements: $\dfrac{9{,}499}{1.011 \times 8.33} = 1{,}128$ gal

Weekly capacity requirements: $5 \times 1{,}128 = 5{,}640$ gal
Selection: 6,000-gal, mild-steel tank, lead-lined

H-1. Crystallizer feed pump

Flow rate: $\dfrac{12{,}153}{1.034 \times 8.33 \times 24 \times 60} = 9.79$ gpm

Selection: 10-gpm centrifugal pump

H-2. Crystallizer

Duty: To crystallize out major fraction of dissolved BHC from still bottoms by cooling to 30°C

Heat transferred (per hour):

Sensible heat:

For benzene, $\dfrac{3{,}415}{24} \times 0.444 \times (100 - 30) \times 1.8$ 7,920

For monochlor, $\dfrac{1{,}847}{24} \times 0.309 \times (100 - 30) \times 1.8$ 3,000

For BHC, $\dfrac{6{,}806}{24} \times 0.211 \times (100 - 30) \times 1.8$ 7,540

For water, $24\frac{2}{24} \times 1.000 \times (100 - 30) \times 1.8$ 126

For trichlor, $6\frac{1}{24} \times 0.350 \times (100 - 30) \times 1.8$ 116

    Total sensible heat, Btu/hr 18,702

Heat of crystallization: $\dfrac{4{,}776}{24} \times 5.87$ 1,166

    Total heat transferred, Btu/hr 19,868

Over-all heat-transfer coefficient U: In a similar crystallization (66 per cent hexachloroethane in perchloroethylene) using a 6-in. Vogt crystallizer, the overall coefficient was found to be 12 Pcu/(hr)(ft²)(°C). It is assumed that this value of U is valid for the present requirements and that cooling water enters at 25°C and leaves at 35°C.

Log mean temp. difference: 23°C, or 41°F

Cooling surface required: $\dfrac{Q}{U \times \Delta T} = \dfrac{19{,}868}{12 \times 41} = 40.4$ ft²

Selection: Double-pipe crystallizer, nominal 8 in. diameter. To consist of 4 sections each 20 ft long; each section shall be jacketed for 18 ft of its length. Effective heat-transfer area will be 153 ft²; total volume 24.9 ft³; process material retention time 3.33 hr.

I-1. Continuous centrifuge

Solids separated: $4{,}923 \div 24 = 205$ lb/hr

Clarified liquor: $\dfrac{7{,}230}{1.100 \times 8.33 \times 24} = 32.9$ gph

Selection: Continuous centrifugal having a capacity of 550 lb/hr

I-2. Recycle liquor receiver

Duty: Storage of one day's supply of recycle liquor

Daily storage requirements: $\dfrac{7{,}230}{1.100 \times 8.33} = 789$ gal

Selection: 1,000-gal horizontal storage tank of steel construction with lead lining

I-3. Wet crystal hopper

Duty: Retention of one day's supply of wet crystals

Capacity requirements: (apparent density assumed to be 1.5)

$$\dfrac{4{,}800}{1.5 \times 62.5} = 51.2 \text{ ft}^3$$

Selection: 4 ft. diam $\times$ 6 ft stainless-steel ($\frac{3}{16}$ in.) bin with tapered bottom to fit on dryer screw feeder. Capacity equals 80 ft³.

CHAP. 4]   SELECTION OF EQUIPMENT AND MATERIALS   159

J-1. Rotary vacuum dryer

Duty: 200 lb/hr of BHC crystals containing 3 per cent volatiles (largely benzene) is to be dried to ½ per cent volatiles. The solvent is to be recovered.

Type of equipment: A continuous vacuum rotary dryer is recommended for case of solvent recovery, rapid evaporation, reduced drying time, and a minimum of loss and injury to the BHC product.

Heat balance assumptions: Operating temperature, 60°C; operating pressure, 20 in. vacuum (*Note:* Boiling point of benzene at 20 in. of vacuum is 45.5°C); drying time required, 1 hr minimum; over-all heat-transfer coefficient, 10 Btu/hr; water (in jacket) in at 80°C, out at 60°C.

Heat balance calculations (Btu/hr):

Heat of evaporation:

For benzene, $\dfrac{83 \times 97.5 \times 1.8}{24}$ 607

For monochlor, $\dfrac{39 \times 77.6 \times 1.8}{24}$ 227

For water, $\dfrac{1 \times 563 \times 1.8}{24}$ 42

Sensible heat:

Of BHC, $\dfrac{4{,}776 \times 0.211 \times (60 - 30) \times 1.8}{24}$ 2,267

Others, $\dfrac{24 \times 0.30 \times (60 - 30) \times 1.8}{24}$ 16

Radiation loss: 20 per cent... 630

Total heat requirements... 3,789

Log mean temp. difference: 25°C, or 45°F

Heat-transfer area required: $\dfrac{Q}{U \times \Delta T} = \dfrac{3{,}789}{10 \times 45} = 8.42 \text{ ft}^2$

Assuming a safety factor of 4, the actual design area will be 33.68 ft².

Selection: Solvent recovery dryer. Nominal diameter, 1 ft 6 in.; length, 7 ft 0 in.

Actual drying time: Assuming 20 per cent of the dryer volume is occupied by the solids, the retention time will be 1.74 hr.

K-1. Dry BHC hopper

Requirements: Two hoppers to hold one day's supply of BHC product each

Daily capacity requirement: $\dfrac{4{,}800}{1.5 \times 62.5} = 51.2 \text{ ft}^3$

Selection: 4 ft diam. × 6 ft stainless-steel hopper with conical bottom terminating in 12 in. diam. outlet fitted with sliding-vane valve

K-2. BHC packaging and storage

Duty: Packing area and storage for 2 weeks' production of dry benzene hexachloride
Containers: 100 lb fiberboard drums, 16 in. diam. × 25 in.
Drums filled per day: 4,800 ÷ 100 = 48
Drums per two weeks: 48 × 20 = 960
Height of stack: 3 drums
Floor area required: 960 × 1.78/3 = 570 ft²

L-1. Solvent still feed pump

Flow rate: $\dfrac{9{,}499}{0.908 \times 8.33 \times 60 \times 24} = 0.872 \text{ gpm}$

Selection: Same as item H-1

L-2. Solvent still

Calculation of number of theoretical plates: The McCabe-Thiele method will again be used with benzene and chlorobenzene as the key components. The equilibrium data will be calculated from vapor-pressure data given by Perry.[1]

Equilibrium data:

Temp., °C	Vapor pressure, mm Hg		Mol fr of benzene	
	C_6H_6	C_6H_6Cl	In liq.	In vapor
80	760	.....	1.000	1.000
90	1,016	208.4	0.683	0.914
100	1,344	293	0.444	0.786
110	1,748	405	0.256	0.589
120	2,238	543	0.128	0.378
130	2,825	719	0.0195	0.0726
131	.....	760	0.000	0.000

Still conditions:

Feed: (Temp., 40°C; bp, 84°C)

Component	Lb	Mols	Mol fr
Benzene.....	8,420	108.10	0.904
Monochlor...	1,008	8.96	0.075
BHC.........	24	0.08	0.001
Trichlor....	3		
Water.......	44	2.44	0.020
Total.......	9,499	119.58	1.000

Bottoms:

Component	Lb	Mols	Mol fr
Benzene.....	10	0.128	0.014
Monochlor...	1,008	8.960	0.975
BHC.........	24	0.084	0.010
Trichlor....	3		
Total.......	1,045	9.172	1.000

Distillate:

Component	Lb	Mols	Mol fr
Benzene.....	8,410	107.8	0.9777
Water.......	44	2.44	0.0223
Total.......	8,454	110.24	1.0000

[1] *Op. cit.*, pp. 153, 155, 165.

CHAP. 4]   SELECTION OF EQUIPMENT AND MATERIALS   161

q-line calculations:
  Feed enters at 40°C; average mol wt, 79.3
  Average specific heat, 37.1 cal/(g mol)(°C)
  Average latent heat, 7,330 cal/g mol
  To heat 1 g mol of feed at 40°C to saturated vapor at 84°C requires

$$37.1 \times (84 - 40) + 7,330 = 8,960 \text{ cal/g mol}$$

  Therefore, $q = \dfrac{8,960}{7,330} = 1.22$

  Slope of q line, $\dfrac{1.22}{1.22 - 1} = 5.45$

  Assuming reflux ratio equals 2, then the y intercept $= \dfrac{0.99}{2 + 1} = 0.33$

  Theoretical plates required, 7
  As a safety factor, assume plate efficiency is 33 per cent
  Actual number of plates required, 21

Determination of column diameter:
  Superficial vapor velocity in column (fps):

$$u = Kv \sqrt{\dfrac{P_1 - P_2}{P_2}} = 0.56 \text{ fps}$$

  where $Kv = 0.14$
     $P_1 = 969$
     $P_2 = 56.2$

  Free vapor flow, $\dfrac{110.24 \times 359}{24 \times 60 \times 60} = 0.458$ fps

  Column area required, $0.458 \div 0.56 = 0.81$ ft^2
  Minimum column diameter, 6.5 in.
Selection: An 8-in.-diam. column with 21 bubble cap plates spaced 18 in. apart. Each plate is to contain six 2-in. standard bubble caps. Feed connections will be provided on the fourth and adjacent plates.
Reboiler:
  Nominal capacity: Sufficient to hold one day's supply of feed liquor, i.e., 114 gal
  Heat-transfer requirements:
    Product rate, $110.24 \div 24 = 4.59$ lb mols/hr
    Hourly vaporization, $3 \times 4.59 = 13.77$ mols/hr
    Heat of vaporization, $13.77 \times 7,330 \times 1.8 = 181,700$ Btu/hr

  To heat feed to boiling point, $\dfrac{9,499}{24} \times 0.467 \times (84 - 40) \times 1.8$

$$= 14,500 \text{ Btu/hr}$$

  Total heat requirements: 196,200 Btu/hr
  Assume that 75 psig steam is to be used as the heating medium and the over-all heat-transfer coefficient $U = 150$
Log mean temp. difference: 39°C, or 70°F

  Heating surface required: $\dfrac{Q}{U \times \Delta T} = \dfrac{196,200}{150 \times 70} = 18.7$ ft^2

Selections: A 100-gal still pot containing 71 ft of 1½-in. BWG 16 stainless-steel tubing will be used. The designed heat-transfer area is 37.2 ft^2.

L-3. Solvent still condenser

Condensing surface required:
Assume over-all heat-transfer coefficient $U = 100$
Cooling water enters at 25°C and leaves at 35°C
Heat transferred, 181,700 Btu/hr
Log mean temp. difference, 50°C, or 90°F

Condensing surface, $\dfrac{Q}{U \times \Delta T} = \dfrac{181,700}{100 \times 90} = 20.1 \text{ ft}^2$ min.

Selection: Heat exchanger with 41 ft² of condensing surface. (This is a duplicate of item G-3.)

L-3a. Condensate cooler

Duty: To cool condensate from 80°C to 30°C
Assume: Over-all heat-transfer coefficient $U = 75$
Cooling water enters at 25°C and leaves at 35°C
Log mean temp. difference: 18°C, or 32.4°F

Heat transferred: $\dfrac{8,454 \times 0.482 \times (80 - 30) \times 1.8}{24} = 15,273$ Btu/hr

Min. cooling surface required: $\dfrac{Q}{U \times \Delta T} = \dfrac{15,273}{75 \times 32.4} = 6.28 \text{ ft}^2$

Selection: Double-pipe heat exchanger with 9.5 ft² of surface

L-4. Wet benzol receiver

Duty: Storage capacity for one day's receipts

Required capacity: $\dfrac{8,454}{0.880 \times 8.33} = 1,127$ gal

Selection: Nominal 1,200-gal welded horizontal steel tank

L-5. Wet benzol pump

Flow rate: $\dfrac{8,454}{0.880 \times 8.33 \times 24 \times 60} = 0.80$ gpm

Selection: Same as item F-2 except bronze

M-1. Monochlor pump

Flow rate: $\dfrac{1,045}{1.107 \times 8.33 \times 24} = 4.73$ gph

Selection: Same as item H-1

M-1a. Monochlor cooler

Duty: Cool monochlor from 131 to 30°C
Assume: Over-all heat-transfer coefficient $U = 75$
Log mean temp. difference: 30°C, or 54°F

Heat transferred: $\dfrac{1,045 \times 0.320 \times (131 - 30) \times 1.8}{24} = 2,533$ Btu/hr

Required cooling surface: $\dfrac{2,520}{75 \times 54} = 0.625 \text{ ft}^2$

Selection: Double-pipe heat exchanger of 1-in. iron pipe with 1½-in. jacket; length, 6 ft; cooling surface, 2.06 ft²

CHAP. 4] SELECTION OF EQUIPMENT AND MATERIALS 163

M-2. Monochlor storage

Duty: Storage capacity for a tank carlot plus 2 weeks' production

Daily production: $\dfrac{1{,}045}{1.107 \times 8.33} = 113.5$ gal

Two weeks' production: 10 × 113.5.............................. 1,135 gal
Add tank carlot.. 8,000
  Required tank capacity.. 9,135 gal

Selection: 10,000-gal horizontal all-welded steel tank

Equipment Specifications

A-1. Benzol transfer pump

No. required: Two
Material handled: Technical benzol at 30°C
Capacity rating: 50 gpm
Type: 1CNF-52 open impeller centrifugal
Maximum head: 15 ft
Material of construction: Bronze
Size inlet: 1½ in.
Size outlet: 1 in.
Mounting: Rigid steel base, flexible motor coupled
Motor: 1-hp, 220–400-volt, 3-phase, 60-cycle, a-c, squirrel-cage induction, 1,750 rpm
Available from: Worthington Corp., East Orange, N.J.

A-2. Benzol storage tank

No. required: Two
Type: Horizontal, all-welded
Nominal capacity: 6,000 gal
Dimensions: OD 8 ft 0 in., length 18 ft 0 in.
Construction: In accordance with ASME Code U 68 for Unfired Pressure Vessels; all-welded with flanged and dished heads; shell thickness ½ in., head thickness ⁵⁄₁₆ in.
Connections: Manhole 18 in., skimmer 6 ips, drains 3 in. (2 required), gauge connections 1½ in.; 2-in. vent
Material of construction: Mild steel
Available from: Lancaster Iron Works, Inc., Lancaster, Pa.

A-3. Benzol feed pump

No. required: Two
Capacity rating: 7 gpm (max)
Materials handled: Wet technical benzol at 30°C
Type: D-11 centrifugal, single-stage
Total head: 20 ft
Material of construction: Type 316 stainless steel
Size outlet: ¼ in.
Size inlet: ¼ in.
Mounting: Integral with motor
Motor: ⅛-hp, split-phase induction, 60-cycle, 110-volt, a-c, General Electric explosion-proof motor
Available from: Eastern Industries, Hamden, Conn.

A-4. Water demineralizer

No. required: One
Capacity rating: 10 gph
Type: DM-10 Standard
Service: 24-hr operation per day
Controls: Full automatic control system
Dimensions (space requirements): Length 34 in., width 17 in., height 46 in.
Available from: Barnstead Still and Demineralizer Co., Inc., Boston, Mass.

A-5. Chlorine vaporizer

No. required: One
Type: Pfaudler type UB, hairpin tube bundle heat exchanger
Heat transfer area: 5 ft^2
Service: Vaporize 4,228 lb of liquid chlorine per day
Tube size: ¾-in. No. 14 BWG copper
Average length of hairpin: 4 ft
No. of hairpins: 4
Shell dimensions: Nominal 6-in. diameter, 60-in. over-all length
Connections: Liquid chlorine inlet 1 in., chlorine gas outlet 2 in., hot-water connections 1½ in.
Maximum pressure: 160 psi
Gaskets: Garlock 900, liquid chlorine service
Available from: The Pfaudler Co., Rochester 4, N.Y.

B-1. Chlorinator

No. required: Two
Type: XXL series Pfaudler glass-lined, jacketed, reaction kettle with agitator
Model No.: R-78-1500-1000
Nominal capacity: 1,500 gal
Jacket heating surface: 158 ft^2
Kettle dimensions: Outside diameter 84 in., over-all height 110 in.
Materials of construction: ⅝-in. open-hearth, steel-lined with acid-resistant glass
Kettle connections: Outlet 3 in. flanged (supported stuffing-box type), manhole 18 in. with full-vision peephole, vapor outlet 8 in. vertical flanged, agitator opening 6 in. ID, inlets 4 in. vertical flanged (3 required)
Jacket connections: Overflow 1½ in. (2 required), steam or water inlet 2 in., bottom drain 1½ in.
Mounting: Six 3-in. pipe legs, setting kettle 18 in. off floor
Agitator: Type: 54-in. three-blade impeller, glass-covered, hollow shaft for thermometer
   R.p.m.: 90
   Drive: Type R-LBH with Pfaudler special gears
   Motor: 20-hp, 220–440-volt, 3-phase, 60-cycle a-c, squirrel-cage induction, 1,750 rpm
   Stuffing box: Pfaudler high-duty type with bronze sleeve
Chlorine diffuser: 1-in. silica tube with twenty-five ¼-in. tangential jets
Gaskets: Corning Type R-3
Available from: The Pfaudler Co., Rochester, N.Y.

CHAP. 4] SELECTION OF EQUIPMENT AND MATERIALS 165

B-2. Reflux condenser

No. required: Two
Type: Pfaudler glass-lined, double-jacketed condenser
Condensing surface: 62 ft^2
Dimensions: OD 30½ in., ID 22 in., over-all length 70 in.
Material of construction: Open-hearth steel; annular condensing chamber covered with acid-resistant glass
Connections: Vapor inlet 6 in., condensate outlet 2 in., cooling-water connections 2 in.
Gaskets: Corning Type R-1
Available from: The Pfaudler Co., Rochester, N.Y.

B-3. Vent gas separator

No. required: Two
Type: Glass-lined vacuum receiver, Type VR-50
Rated capacity: 50 gal
Duty: Separate benzene-chlorobenzene–muriatic acid condensate from chlorine-containing gases
Maximum temperature: 30°C
Dimensions: Diameter 24 in., over-all length 41 in.
Connections: Handhole 5 in., openings 2 in. (2 required), outlet 2 in.
Gaskets: Corning Type R-1
Available from: The Pfaudler Co., Rochester, N.Y.

C. Decanter

No. required: One
Type: Glass-lined receiver
Rated capacity: 100 gal
Materials handled: Chlorobenzene layer (sp gr 1.0340), muriatic acid layer (sp gr 1.0975)
Maximum temperature: 70°C
Material of construction: 7⁄16-in. open-hearth steel covered with acid-resistant glass on inside
Dimensions: Diameter 30 in., over-all height 43 in.
Connections: Handhole 5 in. with peephole, openings 2 in. (2 required), bottom outlet 2 in., side outlet 2 in.
Gaskets: Corning Type R-3
Available from: The Pfaudler Co., Rochester, N.Y.

D-1. Dilute acid receiver

No. required: One
Type: Horizontal glass-lined tank
Rated capacity: 1,500 gal
Actual capacity: 1,841 gal
Materials handled: Muriatic acid at 60°C
Over-all dimensions: Diameter 78 in., length 100 in.
Connections: Manhole 18 in., outlet 2 in., opening 2 in.
Materials of construction: 7⁄16-in. open-hearth steel, glass-lined interior
Model No.: HW-508
Available from: The Pfaudler Co., Rochester, N.Y.

D-2. Acid still feed pump

No. required: Two
Capacity rating: 20 gph
Type: Milton Roy Simplex controlled volume model MD1-25-56 HB
Material handled: 25 per cent muriatic acid at 60°C
Maximum head: 20 ft
Materials of construction: Hastelloy B stainless steel with Hastelloy B catchall gland cup and Teflon packing
Motor: ¼-hp, 110-volt, single-phase, 60-cycle, a-c, General Electric right-angle gearhead open motor, 56 rpm
Available from: Milton Roy Company, Philadelphia, Pa.

D-3. Acid stripping still

No. required: One
Type: Karbate bubble cap column
Duty: Strip product containing 36 per cent HCl from a 25 per cent feed with 20 per cent HCl in bottoms
Maximum pressure: 5 psi
Maximum temperature: 110°C
Material of construction: No. 28 Karbate column and bubble caps carbon-steel supports
Dimensions: Nominal diameter 8 in., length 8 ft
No. of plates: 6 standard 8 in.
Reboiler: Lower 2 ft for column to be jacketed for 40 psi steam
Registry: U.S. Bureau of Internal Revenue
Available from: National Carbon Co., Inc., New York

D-4. Acid still condenser

No. required: One
Type: Fixed-tube sheet condenser
Service: Condense 36 per cent, muriatic acid vapor at 100°C
Tube length: 4 ft
Tube size: ¾-in. BWG 18
No. of tubes: 10
Shell diameter: 6 in.
Over-all dimensions: 12 in. diameter $\times$ 5 ft length
Materials of construction: Everdur tubes; tube sheets and bonnets, mild steel shell
Available from: The Pfaudler Co., Rochester, N.Y.

D-5. Conc. HCl receiver

No. required: One
Capacity rating: 100 gal
Type: Pfaudler glass-lined receiver, model VR 100
Character of contents: 22°Bé muriatic acid at 30°C
Dimensions: OD 30 in., height 43 in.
Connections: Handhole 5 in., openings 2 in. (2 required), outlet 2 in.
Material of construction: 7/16-in. open-hearth steel covered with acid-resistant glass
Available from: The Pfaudler Co., Rochester, N.Y.

CHAP. 4]   SELECTION OF EQUIPMENT AND MATERIALS   167

D-6. Conc. HCl pump

No. required: One
Rated capacity: 7 gpm
Type: D-11 centrifugal, single-stage, open impeller
Service: Intermittent transfer of 36 per cent muriatic acid from receiver to storage, and from storage to tank car
Maximum head: 25 ft
Material of construction: Hastelloy C
Size inlet: 1/4 in.
Size outlet: 1/4 in.
Motor: 1/8-hp, 110-volt, split-phase induction, 60-cycle, a-c, General Electric totally enclosed motor
Available from: Eastern Industries, Inc., Hamden, Conn.

D-7. Conc. HCl storage

No. required: One
Nominal capacity: 10,000 gal
Type: Horizontal all-welded steel tank, rubber-lined, vented
Character of contents: 22°Bé muriatic acid
Dimensions: OD 9 ft 0 in., length 23 ft 6 in.
Construction: All welded mild steel with flanged and dished heads. Head thickness 5/16 in., shell thickness 1/4 in., internal lining 3/16 in., acid-resistant rubber
Connections: Manhole 18 in., skimmer 6 in., drains 3 in. (2 required), gauge connections 1 1/2 in.
Supports: 4 concrete saddles
Available from: Lancaster Iron Works, Inc., Lancaster, Pa.

D-8. Acid recycle pump

No. required: Two
Capacity rating: 20 gph
Type: Milton Roy Simplex controlled volume pump model No. MD1-25-56-HB
Liquid handled: 20 per cent muriatic acid at 110°C
Maximum head: 30 ft
Material of construction: Hastelloy B stainless steel with Hastelloy B catchall gland cap and Teflon packing
Motor: 1/4-hp, 110-volt, single-phase, 60-cycle, a-c, General Electric right-angle gearhead open motor, 56 rpm
Available from: Milton Roy Company, Philadelphia, Pa.

E. Crude product pump

No. required: Two
Rated capacity: 7 gph
Type: D-11 centrifugal, single-phase, open impeller
Materials handled: Chlorobenzenes containing muriatic acid at 70°C
Maximum head: 20 ft
Material of construction: Hastelloy C
Size inlet: 1/4 in.
Size outlet: 1/4 in.
Motor: 1/8-hp, 110-volt, split-phase induction, 60-cycle, a-c, General Electric totally enclosed motor
Available from: Eastern Industries, Inc., Hamden, Conn.

F-1. Caustic make-up tank

No. required: One
Nominal capacity: 2,000 gal
Type: All-welded vertical tank
Character of contents: 1 per cent caustic at 30°C
Over-all dimensions: Diameter 7 ft 0 in., height 7 ft 0 in.
Construction: 1/4-in. mild steel, all-welded open tank
Connections: Skimmer drain 1 in., final drain 1 in., agitator opening 1 in., hole 6 in. from bottom of tank, 2-in. vent
Supports: Four 3-in. pipe legs, 6 in. off floor
Available from: Lancaster Iron Works, Inc., Lancaster, Pa.

F-1a. Caustic make-up tank agitator

No. required: One
Type: Lightnin 1-hp angular off-center agitator
Dimensions: Propeller diameter 6 in., shaft diameter 7/8 in., effective shaft length 8 in.
Material of construction: Steel propeller and shaft, bronze bearings
Motor: 1-hp, 220–440-volt, 3-phase, 60-cycle, a-c, squirrel-cage induction, 1,750 rpm
Available from: Mixing Equipment Co., Inc., Rochester, N.Y.

F-2. Caustic transfer pump

No. required: One
Nominal capacity: 10 gpm
Type: 3/4 CNF-42
Materials handled: 1 per cent caustic at 30°C
Maximum head: 20 ft
Size inlet: 1 in.
Size outlet: 3/4 in.
Material of construction: Cast iron
Motor: 1/3-hp, 110-volt, single-phase, 60-cycle, a-c, squirrel-cage, induction, 1,200 rpm
Mounting: Rigid steel base, flexible motor coupling
Available from: Worthington Corp., East Orange, N.J.

F-3. Caustic storage

No. required: One
Nominal capacity: 4,000 gal
Type: Horizontal all-welded steel
Character of contents: 1 per cent caustic at 30°C
Over-all dimensions: Diameter 7 ft 0 in., length 15 ft 6 in.
Construction: All-welded mild steel with flanged and dished head; head thickness 1/4 in., shell thickness 5/16 in.
Connections: Manhole 18 in., skimmer 1 in., drain 2 in., openings 1 in. and 2 in., gauge connections 1 1/2 in., vent 2 in.
Available from: Lancaster Iron Works, Inc., Lancaster, Pa.

F-4. Caustic feeder

No. required: Two
Capacity rating: 23 gph

Type: Milton Roy Simplex volumetric feeder pump MD1-25-56
Material handled: 1 per cent caustic at 30°C
Maximum head: 95 ft
Material of construction: Cast-iron liquid end with 316 stainless-steel valve seats and plunger, 440 stainless-steel ball valves and neoprene packing
Size inlet: ½ in.
Size outlet: ⅜ in.
Motor: ¼-hp, 110-volt, split-phase induction, 60-cycle, a-c, General Electric, series right-angle gearhead open motor, 56 rpm
Mounting: Integral cast-iron base
Available from: Milton Roy Company, Philadelphia, Pa.

F-5. Neutralizer

No. required: One
Nominal capacity: 100 gal
Type: Pfaudler type P glass-lined reaction kettle, unjacketed, with agitator
Kettle dimensions over-all: Diameter 34 in., height 61½ in.
Connections: Openings 2 in. (2 required) and 3 in. (2 required), bottom outlet 2 in., handhole 5 × 8 in. with peephole, agitator opening 2 in.
Materials of construction: ⁵⁄₁₆-in. open-hearth steel-lined with acid-resistant glass
Agitator: Type: 36-in. impeller with baffle (90 rpm)
  Drive: ⅓-hp, 110-volt, single-phase, 60-cycle, a-c, squirrel-cage, induction with type PB reduction from 1,200 rpm
Available from: The Pfaudler Co., Rochester, N.Y.

F-6. Spent caustic separator

No. required: One
Nominal capacity: 100 gal
Type: Vertical tank with conical bottom
Character of contents: Chlorobenzene layer (sp gr 1.0340), ½ per cent caustic layer (sp gr 1.006)
Dimensions: Diameter 24 in., height 66 in., cone angle 45°
Connections: Handhole 5 in., openings 2 in. and 1 in., bottom drain 1 in., side drain 1 in.
Construction: All-welded of ¼-in. mild steel
Available from: Lancaster Iron Works, Inc., Lancaster, Pa.

G-1. Flash still feed pump

No. required: One
Capacity rating: 7 gpm
Type: D-11 centrifugal, single-stage, open-impeller
Character of liquid: Chlorobenzenes
Total head: 20 ft
Material of construction: Type 316 stainless steel
Size inlet: ¼ in.
Size outlet: ¼ in.
Motor: ⅛-hp, split-phase induction, 60-cycle, 110-volt, a-c, General Electric explosion-proof motor
Mounting: Integral pump and motor
Available from: Eastern Engineering Co., New Haven, Conn.

G-2. Flash still

No. required: One
Nominal capacity: 500 gal
Type: Artisan combined continuous evaporator and stripper
Character of contents: Chlorobenzenes at 105°C
Services: Continuous
Over-all dimensions: Diameter 5 ft 0 in., height 7 ft 2 in.
Connections: Handhole 5 in., openings 2 in. (2 required), vapor outlet 4 in., bottom drain 3 in.
Heating coil: Surface area 20 ft^2, dimensions 76 ft of 1-in. BWG 18 stainless-steel tubing, type 304
Accessories: Knockback coil, spiral baffle, entrainment plate, safety valve
Materials of construction: $7/16$ stainless steel, type 304
Gaskets: Corning type R-3
Available from: Artisan Metal Products, Inc., Boston, Mass.

G-3. Flash still condenser

No. required: One
Type: Fixed-tube sheet heat exchanger
Condensing surface: 41.3 ft^2
Over-all dimensions: Nominal diameter 6 in., length 60 in.
Tube size: $3/4$-in. BWG 18
Tube length: 4 ft
No. of tubes required: 12
Operating pressure: 5 psi
Materials of construction: Tubes, tube sheets, and bonnets of type 304 stainless steel, shell of carbon steel
Available from: The Pfaudler Co., Rochester, N.Y.

G-4. Solvent receiver

No. required: One
Nominal capacity: 5,000 gal
Type: Horizontal welded steel, lead-lined
Character of contents: Chlorobenzenes at 30°C
Dimensions: Diameter 8 ft 0 in., length 18 ft 0 in.
Construction: In accordance with ASME Code for Unfired Pressure Vessels; all-welded with flanged and dished heads; shell thickness $1/2$ in., head thickness $5/16$ in.
Material of construction: Mild steel lined with $3/16$-in. homogeneously bonded lead
Supports: 3 concrete saddles
Available from: Lancaster Iron Works, Inc., Lancaster, Pa.

H-1. Crystallizer feed pump

No. required: Two
Rated capacity: 10 gpm
Type: Model E-7 centrifugal, single-stage, open impeller
Character of liquid: Chlorobenzenes at 100°C
Total head: 35 ft
Size inlet: $1/4$ in.
Size outlet: $1/4$ in.

CHAP. 4]   SELECTION OF EQUIPMENT AND MATERIALS   171

Material of construction: Type 316 stainless-steel shaft, casing, and impeller; cast-iron base
Seal: Mechanical rotary type with Teflon sealing rings
Motor: ¼-hp, split-phase induction, 110-volt, 60-cycle, a-c, explosion-proof frame
Available from: Eastern Industries, Inc., Hamden, Conn.

H-2. Crystallizer

No. required: One unit
Type: Double-pipe crystallizer
Materials handled: Benzene hexachloride in chlorobenzenes
Terminal process temperatures: In, 100°C; out, 30°C
Effective heat-transfer area: 153 ft^2
Over-all dimensions: Height 5 ft, length 22 ft, width 2 ft
No. of sections: 4
Length per section: 20 ft
Inner pipe: Nominal 8 in.
Scrapers: Vogt spiral spring type (36 rpm)
Jacket: 10-in. pipe
Materials of construction: Inner pipe, 304 stainless steel; scrapers, 304 stainless steel; jackets, carbon steel
Connections: For process material, 4 in.; for cooling water, 2 in.
Motor: 5-hp, 220–440-volt, 60-cycle, 3-phase, a-c, squirrel-cage induction, 1,750 rpm
Speed reducer: Type E Westinghouse, 5-hp double reduction unit, ratio 47.3:1
Drive: Sprocket chain coupled
Mounting: Upright integral unit on four 6-in. channel iron supports
Available from: Henry Vogt Machine Co., Louisville, Ky.

I-1. Continuous centrifuge

No. required: One
Nominal capacity: 550 lb/hr
Type: ter Meer continuous centrifugal
Size: 5–12
Over-all dimensions (approx.): Length 40 in., width 24 in., height 32 in.
Connections: Solids discharge 4 in. × 24 in. approx.; filtrate discharge 3 in. × 6 in. approx.; slurry entrance 4 in.
Drive motor: 2-hp, 220–440-volt, 3-phase, 60-cycle, a-c, squirrel-cage induction, 1,750 rpm
Materials of construction: Basket, screening, and piping of 304 stainless steel; housing and base of cast iron
Mounting: Cast-iron base with four foundation boltholes
Available from: Baker, Perkins, Inc., Saginaw, Mich.

I-2. Recycle liquor receiver

No. required: One
Type: Horizontal, welded steel, lead-lined
Nominal capacity: 1,000 gal
Character of contents: Chlorobenzenes at 30°C
Dimensions: Diameter 4 ft 0 in., length 11 ft 6 in.
Construction: In accordance with ASME Code for Unfired Pressure Vessels; all-welded with flanged and dished heads; shell thickness $3/16$ in., head thickness $3/16$ in.

Connections: Manhole 18 in., skimmer 2 in., drain 4 in., nozzles 2 in. (2 required), gauge connections 1½ in.
Materials of construction: Mild steel, lined with ³⁄₃₂-in. homogeneously bonded lead
Supports: 3 concrete saddles
Available from: Lancaster Iron Works, Inc., Lancaster, Pa.

I-3. Wet crystal hopper

No. required: One
Type: Stainless-steel bin
Nominal capacity: 80 ft³
Character of contents: Wet BHC crystals
Dimensions: Cross section 4 ft diam., over-all height 6 ft
Connections: Top opening 3 in. × 6 in., discharge 12 ft 8 in. diam., peephole centered in one side 5 in.
Construction: All-welded of ³⁄₁₆-in. stainless steel, 45° taper at bottom
Available from: Acme Coppersmithing and Machine Co., Oreland, Pa.

J-1. Rotary vacuum dryer

No. required: One
Type: Acme continuous jacketed solvent recovery dryer
Material handled: 200 lb/hr of BHC containing 3 per cent volatiles (largely benzene), to be dried to 0.5 per cent moisture
Maximum temperature: 60°C
Dryer dimensions: Nominal diameter 1 ft 6 in., length 7 ft 0 in.
Heating surface: 34 ft²
Working capacity: 6 ft³
Over-all dimensions: Height 4 ft 4 in., width 2 ft 0 in., length 8 ft 6 in.
Accessories: Screw conveyor with lifting flights, variable screw feeder (0 to 5 rpm), automatic discharging device
Material of construction: 304 stainless steel for process contact sections; carbon steel for jacket, supports, etc.
Drive: 2-hp, 220–440-volt, 3-phase, 60-cycle, a-c, squirrel-cage induction motor, coupled to appropriate gear reducer driving scraper at 36 rpm (variable-speed type)
Available from: Acme Coppersmithing and Machine Co., Oreland, Pa.

J-2. Vacuum condenser and receiver

No. required: One (combined unit)
Type: Stokes No. 85-0 surface condenser
Size: 0
Cooling surface: 10 ft²
Receiver capacity: 8.5 gal
Dimensions: Over-all height 6 ft 8 in., receiver diameter 17 in.
Connections: Vapor inlet 2 in., vacuum pump connections 2 in., water connections 1 in., condensate outlet 1 in.
Material of construction: Brass condensing tubes, stainless-steel shell receiver
Available from: F. J. Stokes Machine Co., Philadelphia, Pa.

J-3. Vacuum pump

No. required: one
Type: Type W reciprocating vacuum pump

CHAP. 4] SELECTION OF EQUIPMENT AND MATERIALS 173

Vacuum required: 28 ft
Model: 3275-W
Displacement: 10 cfm
Pump speed: 810 rpm
Material of construction: Bronze
Motor drive: 1¾-hp, 220–440-volt, 3-phase, 60-cycle, a-c, squirrel-cage induction motor, 1,750 rpm with V-belt drive
Mounting: On 10-in. channel iron base
Available from: F. J. Stokes Machine Co., Philadelphia, Pa.

J-4. Recovered solvent pump

No. required: One
Capacity rating: 7 gpm
Type: D-11 centrifugal, single-stage
Materials handled: Benzene and chlorobenzene at 30°C
Total head: 20 ft
Material of construction: Type 316 stainless steel
Size inlet: ¼ in.
Size outlet: ¼ in.
Motor: ⅛-hp, split-phase induction, 60-cycle, 110-volt, a-c, General Electric explosion-proof motor
Mounting: Integral with motor
Available from: Eastern Industries, Inc., Hamden, Conn.

K-1. Dry BHC hopper

No. required: Two
Nominal capacity: 80 ft^3
Type: Stainless-steel bin
Character of contents: Dry BHC crystals
Dimensions: Cross section 4 ft diam., over-all height 6 ft
Construction: All-welded 3/16-in. stainless steel with conical bottom terminating in 12-in.-diam. outlet
Connections: Top, 8 in. × 12 in. flanged opening, outlet, 12 in. diam., fitted with sliding vane valve; 5-in. peephole centered in one side
Available from: Acme Coppersmithing and Machine Company, Oreland, Pa.

K-2. BHC packaging

No. required: One
Type: No. 41 Vibrox drum packer
Capacity rating: 100- to 750-lb drums
Over-all dimensions: Length 40⅜ in., width 23 in., height 25½ in.
Motor: 1-hp, 220–440-volt, 3-phase, 60-cycle, a-c, squirrel-cage induction, 1,750 rpm
Accessories: Jack-type lift and casters for portable use
Available from: B. F. Gump Co., Chicago, Ill.

K-2a. BHC storage

Capacity of area: To hold 4 weeks' supply of BHC
Containers: 110-lb drums
Max. drums stored: 960
Height of drum storage pile: 3 drums
Dimensions of drum: 16 in. diam. × 25 in. height

Floor area required: 570 ft²
Height of ceiling: 9 ft
Type of floor: Monolithic concrete

L-1. Solvent still feed pump

No. required: Two
Rated capacity: 10 gpm
Type: Model D-11 single-stage centrifugal
Character of liquid: Chlorobenzene at 30°C
Total head: 20 ft
Size inlet: ¼ in.
Size outlet: ¼ in.
Material of construction: Type 316 stainless-steel shaft, casing and impeller; cast-iron base
Seal: Mechanical rotary type, Teflon sealing rings
Motor: ¼-hp, split-phase induction, 110-volt, a-c, 60-cycle, General Electric explosion-proof motor
Available from: Eastern Industries, Inc., Hamden, Conn.

L-2. Solvent still

Column:

No. required: One
Type: Bubble cap tower
Nominal diameter: 8 in.
No. of plates: 21
Plate specifications: 6 bubble caps 2 in. diam., downspout 1 in. diam.
Plate spacing: 18 in.
Working pressure: Atmospheric
Max. working temperature: 131°C
Connections: Vapor outlet 4 in., feed connections ¾ in. (on 3d, 4th, and 5th plates), reflux inlet 1½ in.
Over-all height: 33 ft
Material of construction: Stainless steel

Still pot:

No. required: One
Type: Unjacketed kettle with heating coil
Nominal capacity: 100 gal
Heating surface: 37.2 ft²
Over-all kettle dimensions: Diameter 32 in., height 51 in.
Connections: Vapor outlet 8 in. flanged, openings 2 in. (2 required), handhole 5 in., bottom drain 1 in., level gauge connections ¾ in., relief valve 1 in.
Mounting: Four 2-in. pipe legs, 18 in. high
Material construction: ¼-in. stainless steel
Available from: The Pfaudler Co., Rochester, N.Y.

L-3. Solvent still condenser

No. required: One
Type: FTS heat exchanger
Condensing surface: 30 ft²

Over-all dimensions: Nominal diameter 6 in., length 6 in.
Tube size: ¾-in. BWG 18
Tube length: 48 in.
No. of tubes required: 12
Operating pressure: 5 psi
Materials of construction: Tubes, tube sheets, and bonnets of 304 stainless steel, shell of carbon steel
Available from: The Pfaudler Co., Rochester, N.Y.

L-3a. Condensate cooler

No. required: One
Type: Double-pipe heat exchanger
Cooling surface: 9.5 ft^2
Construction: Two sections, 14 ft each, of 1-in. stainless-steel pipe jacketed with 1½-in. wrought-iron pipe; to be fabricated on site

L-4. Wet benzol receiver

No. required: One
Type: Enclosed kettle
Nominal capacity: 100 gal
Character of contents: Wet benzene
Over-all dimensions: Diameter 30 in., height 48 in.
Connections: Handhole 5 in., openings 1 in. and 2 in., bottom drain 2 in.
Material of construction: ¼-in. mild steel
Supports: Three 2-in. pipe legs, 12 in. high
Available from: Lancaster Iron Works, Inc., Lancaster, Pa.

L-5. Wet benzol pump

No. required: One
Rated capacity: 10 gpm
Type: 1 CNF-52 open-impeller centrifugal pump
Materials handled: Wet benzene at 30°C
Maximum head: 30 ft
Size inlet: 1½ in.
Size outlet: 1 in.
Material of construction: Bronze
Motor: ⅓-hp, 110-volt, single-phase, 60-cycle, a-c, squirrel-cage induction, 1,750 rpm
Mounting: Rigid steel base, flexible motor coupling
Available from: Worthington Corp., East Orange, N.J.

M-1. Monochlor pump

No. required: One
Rated capacity: 10 gpm
Type: Model D-11 single-state centrifugal pump
Character of liquid: Chlorobenzene
Total head: 20 ft
Size inlet: ¼ in.
Size outlet: ¼ in.
Material of construction: Type 316 stainless-steel shaft, casing, and impeller, cast-iron base

Seal: Mechanical rotary-type
Motor: $\frac{1}{4}$-hp, split-phase induction, 60-cycle, 110-volt, a-c, General Electric explosion-proof motor
Available from: Eastern Industries, Inc., Hamden, Conn.

M-1a. Monochlor cooler

No. required: One
Type: Double-pipe heat exchanger
Cooling surface: 2.1 ft^2
Construction: 6 ft of 1-in. stainless-steel pipe jacketed with 1$\frac{1}{2}$-in. wrought-iron pipe; to be fabricated on site

M-2. Monochlor storage

No. required: One
Type: Horizontal, all-welded steel
Nominal capacity: 10,000 gal
Character of contents: Monochlorobenzene at 30°C
Over-all dimensions: Diameter 9 ft 0 in., length 23 ft 6 in.
Construction: In accordance with ASME Code for Unfired Pressure Vessels, all-welded with flanged and dished heads; shell thickness $\frac{1}{2}$ in., head thickness, $\frac{5}{16}$ in.
Connections: Manhole 18 in., skimmer 6 in., nozzles 3 in. (2 required), drain 6 in. gauge connections 1$\frac{1}{2}$ in.
Material of construction: Mild steel
Available from: Lancaster Iron Works, Inc., Lancaster, Pa.

CHAPTER 5

Plant Layout[1]

The arrangement of equipment and facilities specified from process flow-sheet considerations is a necessary requirement for accurate preconstruction cost estimating or for future detailed design involving piping, structural, and electrical facilities. Careful attention to the development of plot and elevation plans will point out unusual plant requirements and, therefore, give reliable information on building and site costs required for precise preconstruction cost accounting.

PLANNING LAYOUTS

Factors in Planning Layouts

Rational design must include arrangement of processing areas, storage areas, and handling areas in efficient coordination and with regard to such factors as:
1. New site development or addition to a previously developed site
2. Future expansion
3. Economic distribution of services—water, process steam, power, and gas
4. Weather conditions—are they amenable to outdoor construction?
5. Safety considerations—possible hazards of fire, explosion, and fumes
6. Building code requirements
7. Waste-disposal problems
8. Sensible use of floor and elevation space

[1] Plant layout study is not an entirely necessary requirement for preliminary cost estimation work, as shown by several methods in Chap. 6 on economic analysis. However, where more accurate cost analysis work is required, the information given in this chapter is very important.

Methods of Layout Planning

To start a detailed planning study, space requirements must be known for various products, by-products, and raw materials, as well as for process equipment. A starting or reference point, together with a directional schematic flow pattern, will enable the design engineers to make a trial plot plan, as explained below. A number of such studies will be required before a suitable plot and elevation plan is chosen.

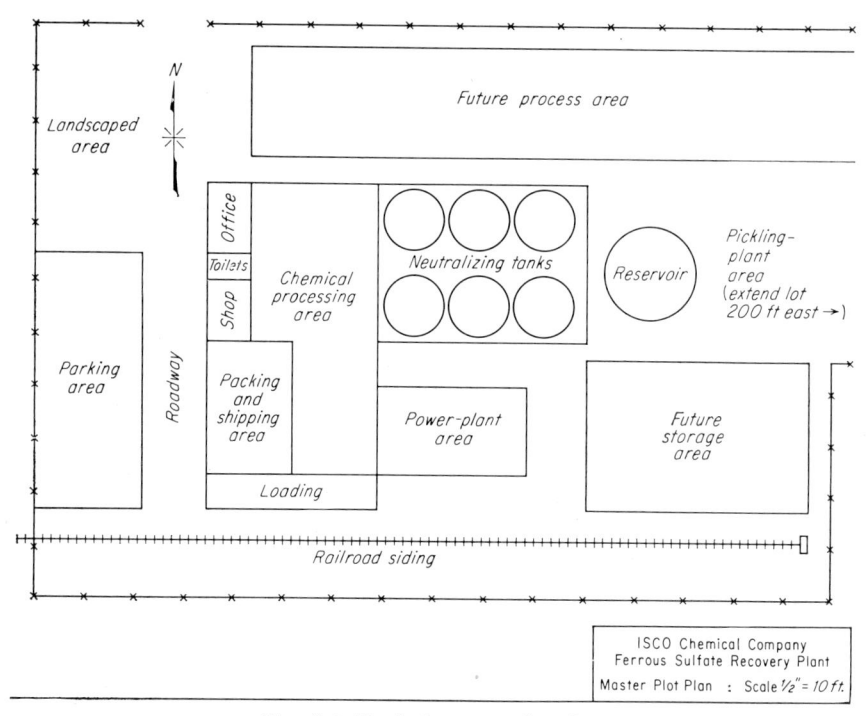

Fig. 5-1. Typical master plot plan.

Unit Areas Concept. The basic blocks with which to build an arrangement for plot plans are often used in the *unit area* concept. This method of planning is particularly well adapted to large plant layouts. Unit areas are often delineated by means of distinct process phases and operational procedures, by reason of contamination, and by safety requirements. Thus, the delineation of the shape and extent of a unit area and the interrelationships of each area in a *master plot plan* is one of the first tasks of layout planning. Figure 5-1 is an example of this type of planning.

Two-dimensional Layouts. To visualize the layout problem, two-dimensional scaled templates or small cutouts of unit areas and equipment

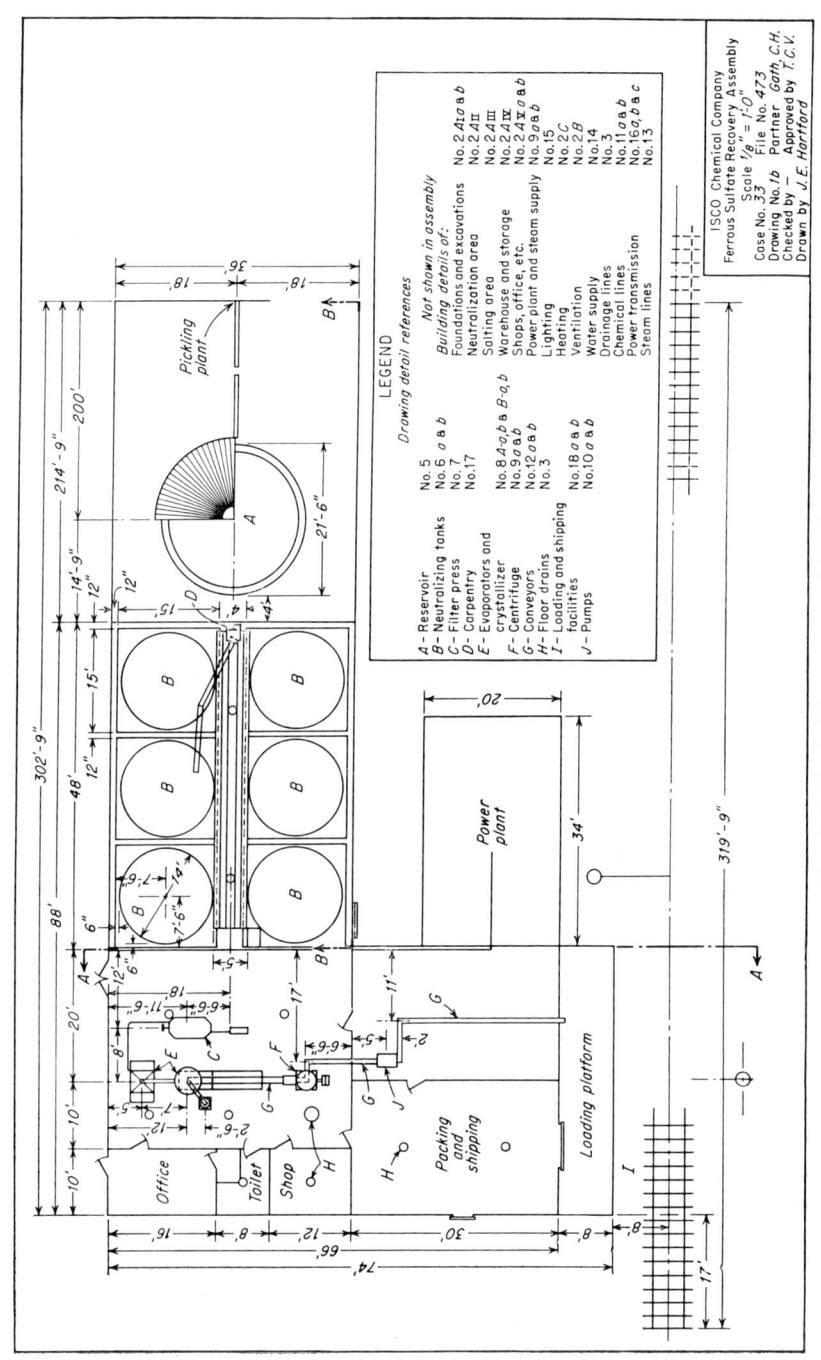

FIG. 5-2. Typical layout assembly plan.

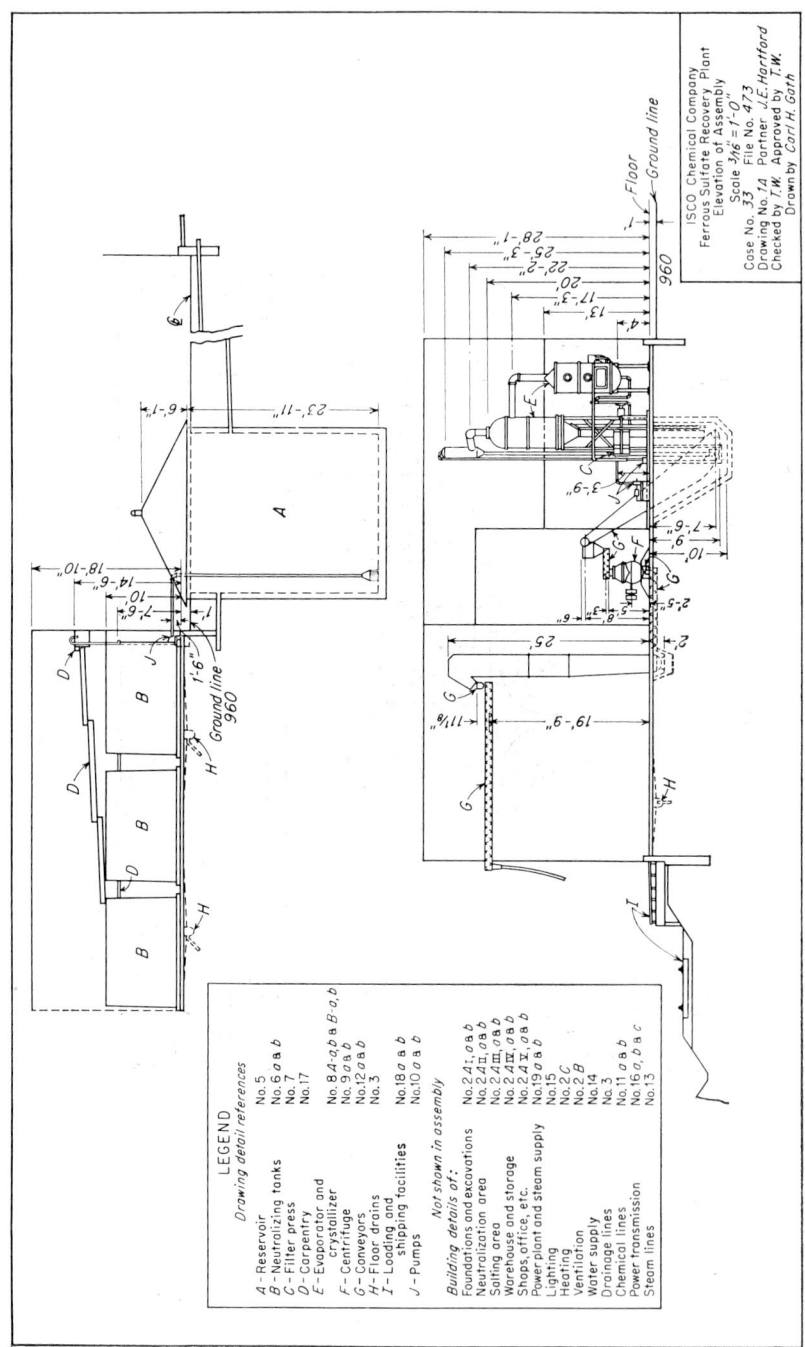

FIG. 5-3. Typical layout assembly elevation.

within each area are shifted about on crosshatched scale paper. A group of experienced engineers will work together with this method to provide a basic plot plan from which can be prepared detailed two-dimensional diagrams, as shown in a related series of drawings (Figs. 5-1 to 5-3). The preparation of a *scale model* of the plant can be started simultaneously in the layout planning procedures.

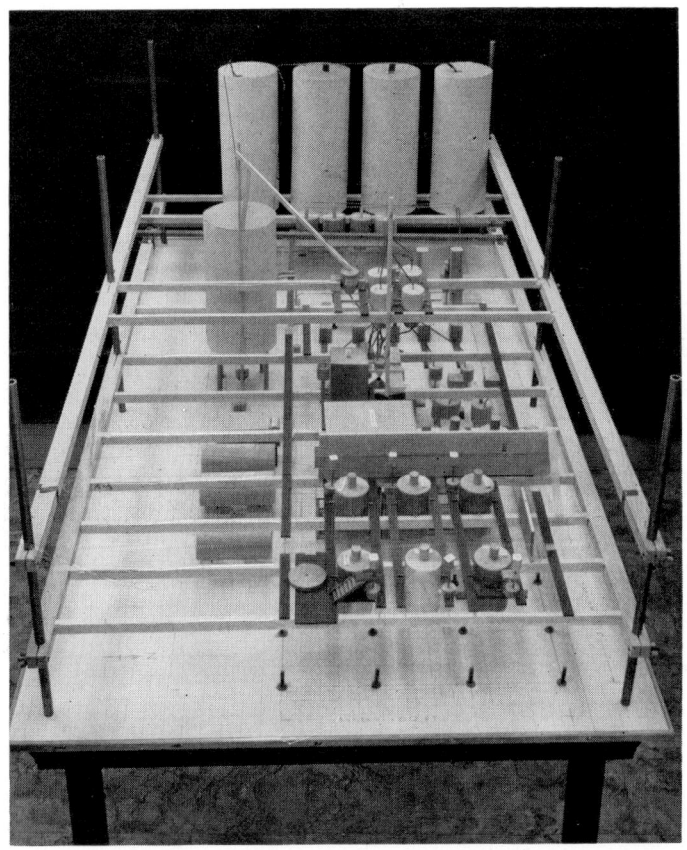

Fig. 5-4. Inexpensive scale model. (*Courtesy of Blaw-Knox Co., Pittsburgh, Pa.*)

Scale Models. Recent developments in the use of scale models have shown the advantages of this method over the detailed two-dimensional method. Figure 5-4 is a view of a study model costing less than $1,000. It was made from blocks of wood and cardboard set on $\frac{1}{4}$ in. to the foot scaled paper. This low-cost model is used chiefly to develop plot and elevation plans and cannot be used for piping and utilities layout. Figure 5-5 is a more complete model costing $5,000 to $10,000. It is

Fig. 5-5. Elaborate scale model. (*Courtesy of Blaw-Knox Co., Pittsburgh, Pa.*)

Fig. 5-6. Commercial chemical plant designed with aid of scale model in Fig. 5-5. (*Courtesy of Blaw-Knox Co., Pittsburgh, Pa.*)

used for detailed layout of process piping, utilities, and control facilities.[1] Dimensional accuracy on models is about $\pm \frac{1}{32}$ in. so that most scale-ups are accurate to within an inch. Figure 5-6 shows the actual plant built from this model. Cost estimation of these facilities and isometric layout diagrams to scale can be obtained readily from the model with as much as 25 per cent savings over two-dimensional methods. This more expensive model is also used during the construction and operator training periods to great advantage. Often these models find a permanent place in the control room, where they permit operators to trace lines quickly instead of walking and climbing over an extensive part of the plant. The advantage of three-dimensional models can be summarized as:

1. Optimum design selection
2. Effective construction planning
3. Savings in engineering design, construction, operating, and maintenance costs
4. More rapid and safer training of personnel

Principles of Plant Layout

Some of the guiding principles for detailed plant layout will be discussed next for the benefit of students making layout decisions for the first time.

Storage Layout. Storage facilities for raw materials and intermediate and finished products may be located in isolated areas or in adjoining areas. Hazardous materials become a decided menace to life and property when stored in large quantities and should consequently be isolated. Storage in adjoining areas to reduce materials handling may introduce an obstacle toward future expansion of the plant. Arranging storage of materials so as to facilitate or simplify handling is also a point to be considered in design. Where it is possible to pump a single material to an elevation so that subsequent handling can be accomplished by gravity into intermediate reaction and storage units, costs may be reduced. Liquids can be stored in small containers, barrels, horizontal or vertical tanks and vats, either indoors or out of doors.

Equipment Layout. In making a layout, ample space should be assigned to each piece of equipment; accessibility is an important factor

---

[1] Materials available for this type of model are:

Equipment and structures: balsa or pine blocks, hardwood dowels and molding, plywood, Lucite or plexiglas, polyurethane or polystyrene foam.

Piping and wiring: hard brass or plastic extruded rod, colored plastic-coated electric lead wire.

Molded models of common engineering equipment and piping hardware in scales of $\frac{1}{4}$, $\frac{3}{8}$, and $\frac{1}{2}$ in. to the foot can be obtained from such suppliers as F. W. Harman Associates, Halecite, N.Y.; Engineering Model Associates, Los Angeles; Industrial Models, Inc., Wilmington, Del.; United Scale Models, Inc., Chester, Pa. Complete plant models can be custom-fabricated by these companies.

for maintenance (Fig. 5-2). Unless a process is well seasoned, it is not always possible to predict just how its various units may have to be changed in order to be in harmony with each other. This is especially true if the process is being developed from pilot plant scale operation, as usually carried out, with equipment not specially designed for the process. It is well known that in chemical manufacturing, processes may be adopted which appear to be sound after a reasonable amount of investigation in the pilot plant stage, yet frequently require minor or even major changes before all parts are properly operating together.

It is extremely poor economy to fit the equipment layout too closely into a building. A slightly larger building than appears necessary will cost little more than one that is crowded. The extra cost will indeed be small in comparison with the penalties that will be extracted if, in order to iron out the kinks, the building must be expanded.

The operations that constitute a process are essentially a series of unit operations that may be carried on simultaneously. These include filtration, evaporation, crystallization, separation, and drying. Since these operations are repeated several times in the flow of materials, it should be possible to arrange the necessary equipment into groups of the same kinds. This sort of layout will make possible a division of operating labor so that one or two operators can be detailed to tend all equipment of a like nature.

The relative levels of the several pieces of equipment and their accessories determine their placement. Although gravity flow is usually preferable, it is not altogether necessary because liquids can be transported by blowing or by pumping, and solids can be moved by mechanical means. Gravity flow may be said to cost nothing to operate, whereas the various mechanical means of transportation involve the first cost of the necessary equipment and the cost of operation and maintenance. But material must be elevated to a level where gravity flow must start. However, gravity flow usually means a multistory layout, whereas the factors favoring a single-story plant may largely, if not entirely, compensate for the cost of mechanical transportation.

Access for initial construction and maintenance is a necessary part of planning. For example, overhead equipment must have space for lowering into place, and heat-exchange equipment should be located near access areas where trucks or hoists can be placed for pulling and replacing tube bundles. Thus, space should be provided for repair and replacement equipment, such as cranes and forked trucks, and for snow-removal equipment, as well as access way around doors and underground hatches.

Safety. A great deal of planning is governed by local and national safety and fire code requirements. Fire protection consisting of reservoirs, mains, hydrants, hose houses, fire pumps, reservoirs, sprinklers in buildings, explosion barriers and directional routing of explosion forces to clear areas, and dikes for combustible-product storage tanks must be incorporated to protect costly plant investment and reduce insurance rates. (See Chap. 8 and Perry's "Chemical Engineers' Handbook," 3d ed., sec. 30, Safety and Fire Protection, pp. 1847–1874.)

Plant Expansion. Expansion must always be kept in mind. The question of multiplying the number of units or increasing the size of the prevailing unit or units merits more study than it can be given here. Suffice it to say that one must exercise engineering judgment; that as a penalty for bad judgment, scrapping of present serviceable equipment constitutes but one phase, for shutdown due to remodeling may involve a greater loss of money than that due to rejected equipment. Nevertheless, the cost of change must sometimes be borne, for the economies of larger units may, in the end, make replacement imperative.

Floor Space. Floor space may or may not be a major factor in the design of a particular plant. The value of land may be a considerable item. The engineer should, however, follow the rule of practicing economy of floor space, consistent with good housekeeping in the plant and with proper consideration given to line flow of materials, access to equipment, space to permit working on parts of equipment that need frequent servicing, and safety and comfort of the operators.

Utilities Servicing. The distribution of gas, air, water, steam, power, and electricity is not always a major item, inasmuch as the flexibility of distribution of these services permits designing to meet almost any condition. But a little regard for the proper placement of each of these services, practicing good design, aids in ease of operation, orderliness, and reduction in costs of maintenance. No pipes should be laid on the floor or between the floor and the 7-ft level, where the operator must pass or work. Chaotic arrangement of piping invites chaotic operation of the plant. The flexibility of standard pipe fittings and power-transmission mechanisms renders this problem one of minor difficulty.

Building. After a complete study of quantitative factors, the selection of the building or buildings must be considered. Standard factory buildings are to be desired, but, if none can be found satisfactory to handle the space and process requirements of the chemical engineer, then a competent architect should be consulted to design a building around the process—not a beautiful structure into which a process *must* fit. It is fundamental in chemical engineering industries that the buildings should be built around the process, instead of the process being made to fit buildings of conventional design.

In many cases only the control area requires housing, with the process equipment erected outdoors. This is known as *outdoor construction*, and such layouts should be considered for many types of plants (see Chap. 8).

What consideration must be given to buildings depends upon conditions. If the designer must adapt his design to fit an old building or building space already erected, his problem is cut out for him and he has limiting conditions. However, the selection of the design of a new building to meet the requirements of the process is more scientific. In this case, one finds before him practically all types of standard building, built in units, interlocking or otherwise, ready for shipment and erection (see Chap. 8).

Throughout chemical industry, much thought must be given to the disposal of waste liquors, fumes, dusts, and gases. Ventilation, fume elimination, and drainage may require the installation of extra equipment. This may involve the design of the individual pieces of operating equipment, or it may require the installation of isolated equipment. If the latter be the case, the location of such equipment where it will not interfere with the flow of materials in process should be practiced. The selection of the proper piece of equipment for doing this service is also an important point; the less attention the ventilating, fume, or waste-elimination systems require, the better service they may render. Sometimes air conditioning of the plant is called for and may require an elaborate setup. But the installation of such equipment, when needed, pays in better service from operators, less discomfort, greater production, and a better morale than when such conditions are left to nature.

It must be recognized that there is not only one solution to the problem of layout of the equipment. There are many rational designs. Which plan to adopt must be decided upon after exercise of engineering judgment and after striking a balance of the advantages and disadvantages of each possible choice.

Materials-handling Equipment. Consideration of equipment for materials handling is only a minor factor in most cases of arrangement, owing to the multiplicity of available materials-handling devices. But where this operation is paramount in a process, serious thought must be given to it. Again it should be said that engineering judgment must be exercised. Whenever possible, one should take advantage of the topography of the site location, if such will serve to advantage in the process.

Railroads and Roads. Existing or possible future railroads and highways adjacent to the plant must be known in order to plan rail sidings and access roads within the plant. Railroad spurs and roadways of the correct capacity and at the right location should be provided for in a traffic study and over-all master track and road plan of the plant area. Some of the factors in rail-track planning are:

1. Existing and future off-site main rail facilities
2. Permissible radius of curvature for spurs—consult local rail authorities
3. Provision for traffic handling—arrangement of spurs and ladder track and switching

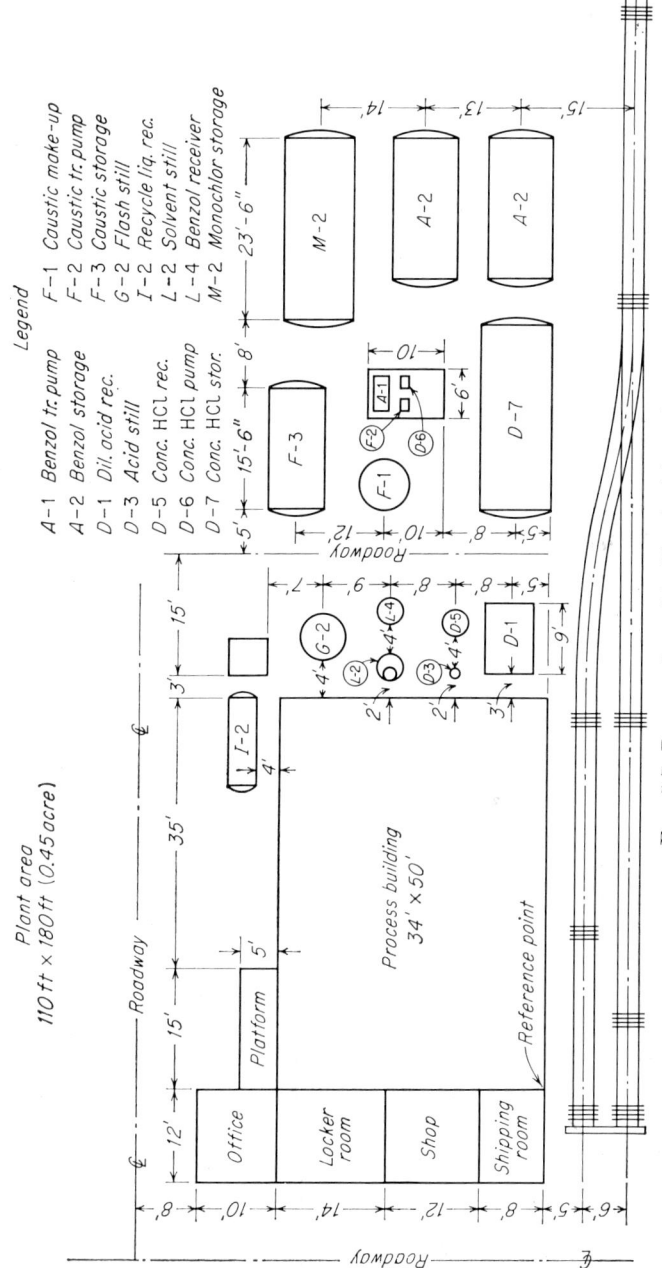

FIG. 5-7. Benzene hexachloride plant layout.

4. Adequate spur facilities
   a. Loading and unloading facilities for initial plant construction and subsequent operations
   b. Rack stations for liquid handling
   c. Storage space for full and empty cars
   d. Space for cleaning and car repairs

Major provisions in road planning for multipurpose service are:
1. A means of interplant movement for road traffic, both pedestrian and vehicular
2. Heavier and wider roads for large-scale traffic
3. Routing of heavy traffic outside the operational areas
4. Roadways for access to initial construction, maintenance, and repair points
5. Roadways to isolated points, storage tanks, and safety equipment, such as fire hydrants

SAMPLE PROBLEM

Benzene Hexachloride Plant Layout

As a preliminary step to the layout of process equipment, storage facilities, and materials-handling equipment, scale models of the pertinent items were constructed and manipulated until a rational arrangement appeared. It was desired to effect efficient coordination of the units with due regard for economic distribution of water, steam, and power, economic use of floor space, and elimination of predictable hazards. Figure 5-7 depicts the over-all layout of the proposed plant area as finally modified. From it the complete equipment layouts and elevations in the process building should be sketched.

As a result of the layout, it is determined that 1.0 acre of land is sufficient area for containing designed facilities plus those necessary for an estimated 100 per cent future expansion of production. The real estate and facilities requirements as developed from the layout design may thus be taken into account in a preconstruction estimation.

CHAPTER 6

Economic Evaluation of the Project

> For which of you, intending to build a tower, sitteth not down first, and counteth the cost, whether he have sufficient to finish it? Lest haply, after he hath laid the foundation, and is not able to finish it, all that behold it begin to mock him, saying, this man began to build, and was not able to finish.   *St. Luke* 14:28–30

The decision by company management to commercialize a process is based almost entirely on the economic evaluation made with the cooperative efforts of many groups within the company. These groups include accounting and finance, marketing and sales, and engineering design and operation.

The methods of preparing an economic evaluation are numerous and vary from one company to the next. In some cases short-cut or "quickie" evaluation procedures may be required as a basis for justifying further expenditures on a project. Much more exacting information is required when a final decision to commercialize a process is to be made. There are all degrees of economic analysis between these two extremes. It is beyond the scope of this text to delve into every ramification of economic analysis. Rather it is the aim of this chapter to present several general-purpose methods with suitable reference tables and discussion so that the student has a sound basis for making reliable estimates and for proceeding further on his own if he does this type of work in his industrial position.

Every method of economic evaluation requires cost analysis of the following principal subjects. Typical items which might appear under each subject in a detailed analysis are included for orientation purposes.

I. Capital investment
   A. Fixed capital for plant facilities
      1. Site
      2. Building
      3. Utilities plants
      4. Process equipment
      5. Storage facilities
      6. Auxiliary utilities and emergency facilities
   B. Working capital
      1. Raw materials inventory
      2. In-process inventory
      3. Product inventory
      4. Maintenance and repair inventory
      5. Accounts receivables credit carry-over
      6. Minimum cash reserve
II. Total product costs
   A. Manufacturing costs
      1. Raw material
      2. Shipping containers
      3. Operating costs
         a. Operating labor
         b. Operating supervision
         c. Maintenance and repair
         d. Operating supplies
         e. Utilities
            (1) Electricity
            (2) Steam
            (3) Water
            (4) Fuel
         f. Control laboratory
         g. Miscellaneous
      4. Overhead costs
         a. Employee benefits
         b. Medical service
         c. Cafeteria
         d. Purchasing
         e. Shops
         f. Property protection
         g. General plant supervision
      5. Depreciation
      6. Property taxes and insurance
   B. General expenses
      1. Freight and delivery
      2. Administration expense
      3. Sales expense
      4. Research expense
III. Economic analysis
   A. Selling price
      1. Market analysis
         a. Price-volume relationship: present and anticipated
         b. Application of products
         c. Competition

      2. Federal income tax
      3. Net or new earnings
  B. Profitability
      1. Return on investment
      2. Cost and profit charts
      3. Other methods of economic analysis
          a. Pay-out time
          b. Accounting method
          c. Investor's method
          d. Project present worth

The above outline serves to orient the student on the subject matter which will be covered in more detail in the sections which follow. A thorough background of process design based on material in the preceding chapters will be required to make a satisfactory economic evaluation.

CAPITAL INVESTMENT

Capital investment is divided into two parts: (1) fixed capital; (2) working capital. A fixed capital cost estimate is useful in determining maintenance and depreciation charges under operating costs and it is a necessity in making economic analyses such as return on investment. The working capital (10 to 15 per cent of fixed capital) is the less critical, economically speaking.

Fixed Capital Cost Estimates

Fixed capital includes the capital requested for (1) all the process and manufacturing machinery and equipment, installed and ready for operation, and (2) nonmanufacturing machinery and equipment items which include land and buildings; installations for generation or distribution of utilities (steam, electricity, water, air, gas); shops, warehouses, and transportation facilities; employee and office facilities; research and control laboratories; and miscellaneous items such as fences, railroads, roads, yard lighting, and telephones. If a plant is to be built in a new location where none of these items of service is available, the nonmanufacturing capital required must be estimated in the same manner as the items of process equipment.

The degree of accuracy in making cost estimates should be considered in making the study. Actual plant costs should be anticipated on the basis that it takes 2 to 3 years to design and construct a plant. Time and money available must be balanced against the reliability desired. Preliminary project estimates, for example, are made to within 30 per cent. Estimates for budget requests are within 10 per cent, whereas a final construction estimate, including all bids, should be within 5 per cent. A guide to the accuracy of cost estimation methods has been prepared by

W. T. Nichols[1] and this will be used as the basis for illustrating various capital cost estimating methods. Figure 6-1 graphs the type of estimate versus cost of preparing the estimate, probable capitalization cost with percentages, and per cent contingency which represents additional money

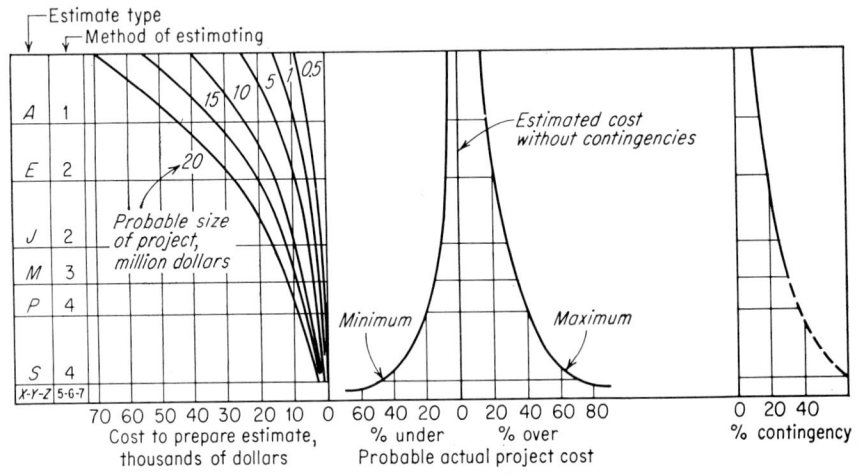

Fig. 6-1. Graph of accuracy of cost estimations (reference—Table 6-1). [*Courtesy of W. T. Nichols, Chem. Eng.*, **58**(6): 248 (1951).]

required to meet unforeseen expenses and price increases. Table 6-1 is an explanation of information required for each estimate type A to Z. The assignment of dollar values is covered under Methods of Estimating (1 to 7) which follow.

Method 1 (Estimate Type A)

1. Quoted equipment cost (delivered).................................. $ 00,000
2. Estimated installation cost for each item of equipment............. 00,000
3. Material cost take-off for pipe, electrical, architectural, structural, instruments, etc.. 00,000
4. Labor estimated from practice factors for items above............... 00,000
5. Subtotal... $000,000
6. Engineering and contingency (% of subtotal)........................ 00,000
7. Total.. $000,000

Method 2 (Estimate Type E, J)

1. Installed equipment cost = delivered equipment cost (quoted and estimated) × 1.43... $ 00,000
2. Buildings, pipe, sewers, electrical, instruments, structural, foundations, insulation, painting, etc., based on unit cost per item, length, area, or volume from preliminary design and take-off......................... 00,000
3. Subtotal... $000,000
4. Engineering and contingency (% subtotal)............................ 00,000
5. Total.. $000,000

[1] W. T. Nichols, *Chem. Eng.*, **58**(6): 248 (1951).

CHAP. 6] ECONOMIC EVALUATION OF THE PROJECT 193

TABLE 6-1. ACCURACY OF COST ESTIMATIONS*
Reference: Fig. 6-1

Estimate types	A	E	J	M	P	S	X	Y	Z
Information available:									
General design basis[1]	x†	x	x	x	x	x	x	x	x
Flow-sheet and material balance	x	x	x	x	x	x	x	x	
Heat and energy balance	x	x	x	x	x	x	x		
Equipment and instrument list[2]	x	x	x	x	x	x			
Performance data sheets	x	x	x	x	x	x			
Survey of plant area[3]	x	x							
Availability of utilities and transportation[4]	x	x							
Information developed by engineers and designers:[10]									
Design sketches[5]	x	x	x	x	x	x			
Layout of mfg. facilities[6]	x	x	x	x	x				
Layout of non-mfg. facilities[7]	x	x	x	x					
General plant layout and land development	x	x	x	x					
Type of construction	x	x	x						
Schedule of pipelines[8]	x	x	x						
Piping layout[9]	x	x	x						
Elec. layout and one-line diagram	x	x	x						
Detailed piping and instrument flow sheets	x	x							
Instrument specifications	x								
Electrical control and interlocks	x								
Soil-bearing values	x								
Architectural and structural design (approx.)	x								

* W. T. Nichols, *Chem. Eng.*, **58**(6): 248 (1951).
† x = completed.
[1] (a) Product to be made, specifications, and storage provisions; (b) plant capacity; (c) operating stream time, (d) provisions for expansion, (e) raw materials, storage, and source.
[2] Showing number of each item required and capacity and materials of construction.
[3] Land values and land developments.
[4] Description of laboratory and service facilities.
[5] Unusual items of equipment.
[6] Showing equipment in plan and elevation.
[7] Utilities and service facilities, roads, sewers, etc.
[8] Sizes and materials of construction.
[9] Superimposed on equipment and plant layout.
[10] Cost for preparing estimates contemplates no alternative studies and will not result in any construction drawings and specifications. Does not include cost of preparing available information on subsoil investigations to determine soil-bearing values.

Method 3 (Estimate Type M).[1,2] This method, based on published equipment costs, is the most accurate method available to the student. Its accuracy averages + 15 to −30 per cent (e.g., a $1,000,000 estimated cost would run $870,000 minimum or $1,430,000 maximum).

[1] C. H. Chilton, *Chem. Eng.*, **56**(6): 97 (1949).
[2] H. J. Lang, *Chem. Eng.*, **54**(10): 118 (1947).

1. Delivered equipment costs from references (e.g., Table 4-6) and on current index basis... $ 00,000
2. Installed equipment costs
    a. From references and on current index basis, or
    b. Item 1 × 1.43... 00,000
3. Process piping... 00,000

    Type plant   Per cent of item 2

    Solid... 7–10 ⎫ See also
    Solid-fluid................................... 10–30 ⎬ Table 6-4,
    Fluid... 30–60 ⎭ p. 215

4. Instrumentation... 00,000

    Amount of automatic control   Per cent of item 2

    None.. 3–5
    Some.. 5–12
    Extensive..................................... 12–20

5. Buildings and site development............................ 00,000

    Type of plant   Per cent of item 2

    Outdoor....................................... 10–30 ⎫ See also
    Outdoor-indoor................................ 20–60 ⎬ Table 6-5,
    Indoor.. 60–100 ⎭ p. 218

6. Auxiliaries (e.g., electric and steam power)................. 00,000

    Extent   Per cent of item 2

    Existing...................................... 0
    Minor additions............................... 0–5
    Major additions............................... 5–25
    New facilities................................ 25–100

7. Outside lines... 00,000

    Average length   Per cent of item 2

    Short... 0–5
    Intermediate.................................. 5–15
    Long.. 15–25

8. Total physical plant costs:
    Sum of items 2 + 3 + 4 + 5 + 6 + 7 = subtotal.............. $000,000
9. Engineering and construction............................... 00,000

    Complexity   Per cent of item 8

    Simple.. 20–35 ⎰ See also Table
    Difficult..................................... 35–60 ⎱ 6-6, p. 220

10. Contingencies... 00,000

    Type process   Per cent of item 8

    Firm.. 10–20
    Subject to change............................. 20–30
    Speculative................................... 30–50
    Average from Fig. 6-1......................... 30

11. Size factor... 00,000

    Size plant   Per cent of item 8

    Large commercial plant (>$2,000,000)........... 0–5
    Small commercial plant ($500,000–$2,000,000)... 5–15
    Experimental unit (<$500,000).................. 15–35

12. Total plant or fixed capital costs:
    Sum of items 8 + 9 + 10 + 11 = total......................... $000,000

A comparison of this estimated distribution of expenditures with actual chemical plant cost data can be made by studying Tables II and III in *Chem. Eng.*, **60**(11): 193 (1953).

Method 4 (Estimate Type P, S)

1. Capital cost = *factor* × total cost of equipment determined by current quotations on unusual items, experience factors, and established or published data.
2. *Factor* to be used above:[1]

Type plant	Delivered equipment cost factor	Installed equipment cost factor
Solids processing..................	3.10	2.16
Solids-fluids processing.............	3.63	2.56
Fluids processing..................	4.74	3.30

Examples of type of plants:
    Solids processing: Coal briquetting; agricultural limestone
    Solids-fluids processing: Solvent extraction plant for minerals with ore-preparation facilities
    Fluids processing: Petroleum distillation; continuous saponification of fats

Method 5 (Estimate Type X)

1. Capital cost = *factor* × total cost of equipment from published data × cost index factor to up-date (Fig. 6-31). *Factor* is same as given in **method 4**, item 2.

Method 6 (Estimate Type X, Y)

1. Capital cost = annual capacity in tons × $ per annual ton of capacity from published figures (Fig. 6-2) × cost index factor to bring up to date (Fig. 6-31).

Method 7 (Estimate Type Z)

1. Capital cost = annual capacity in tons × product sales value in $/ton ÷ *ratio*

$$Ratio = \text{turnover ratio} = \frac{\text{product sales value, \$/ton}}{\text{current investment, \$ per annual ton}}$$

The turnover ratio varies from 0.2 to 8.0, depending on the product. An average value of 1.0 is an educated guess. Kiddoo[2] lists 82 typical processes.

Plant Cost Estimating

Equipment Costs. Methods 3, 4, and 5 under fixed capital cost estimating methods require equipment cost estimates based on published references. Table 4-6 is an index of cost sources for process equipment. In the next few pages of this chapter there are presented Figures 6-3 to 6-30, which are installed cost data derived largely from Chilton[3] or Aries and Newton.[4] Arrangement is by alphabetical order of equipment titles.

[1] H. J. Lang, *Chem. Eng.*, **55**(6): 112 (1948).
[2] G. Kiddoo, *Chem. Eng.*, **58**(10): 145 (1951).
[3] C. H. Chilton, *Chem. Eng.*, **56**(6): 97 (1949).
[4] R. S. Aries and R. D. Newton, "Chemical Engineering Cost Estimation," McGraw-Hill Book Company, Inc., New York, 1955.

196 CHEMICAL ENGINEERING PLANT DESIGN [CHAP. 6

Dollars per annual ton of capacity:

- 30 —
- Catalytic desulf. of gasoline
- Contact sulfuric acid ex smelter gas
- Taconite beneficiation
- TCC gasoline
- Low-purity oxygen
- 20 —
- Portland cement
- Sulfur ex low-grade ores
- Cat. poly. of refy. gas (liq. prod.)
- 37% formaldehyde ex methanol
- Contact sulfuric acid ex sulfur
- 15 — Alkylation via sulfuric acid process
- Superphosphate
- Soybean extraction (bean basis)
- Ammonium nitrate
- Sulfur ex hydrogen sulfide
- Aluminum sulfate
- Solvent dewaxing of lube oil
- 10 — Refined NaCl ex brine
- NaOH purification via ammonia proc.
- Natural gasoline (on gas throughput)
- Sodium silicate
- 8 — Catalytic cracking (charge basis)
- Platforming (charge basis)
- Ammonium sulfate
- Solvent extraction of lube oil
- Tung nut extraction (nut basis)
- 6 — Delayed coking (charge basis)
- Lime
- 5 —
- Thermal cracking (charge basis)
- 4 —
- Hypersorption (on gas throughput)
- 3 —
- Vacuum distillation of lube oil
- 2 —
- H$_2$S ex nat. gas (on gas throughput)
- 1.5 —
- 1.0 — Crude oil topping
- 0.8 —
- 0.6 —
- 0.5 — Vacuum flashing of crude oil
- 0.4 —
- 0.3 —

Dollars per annual ton of capacity:

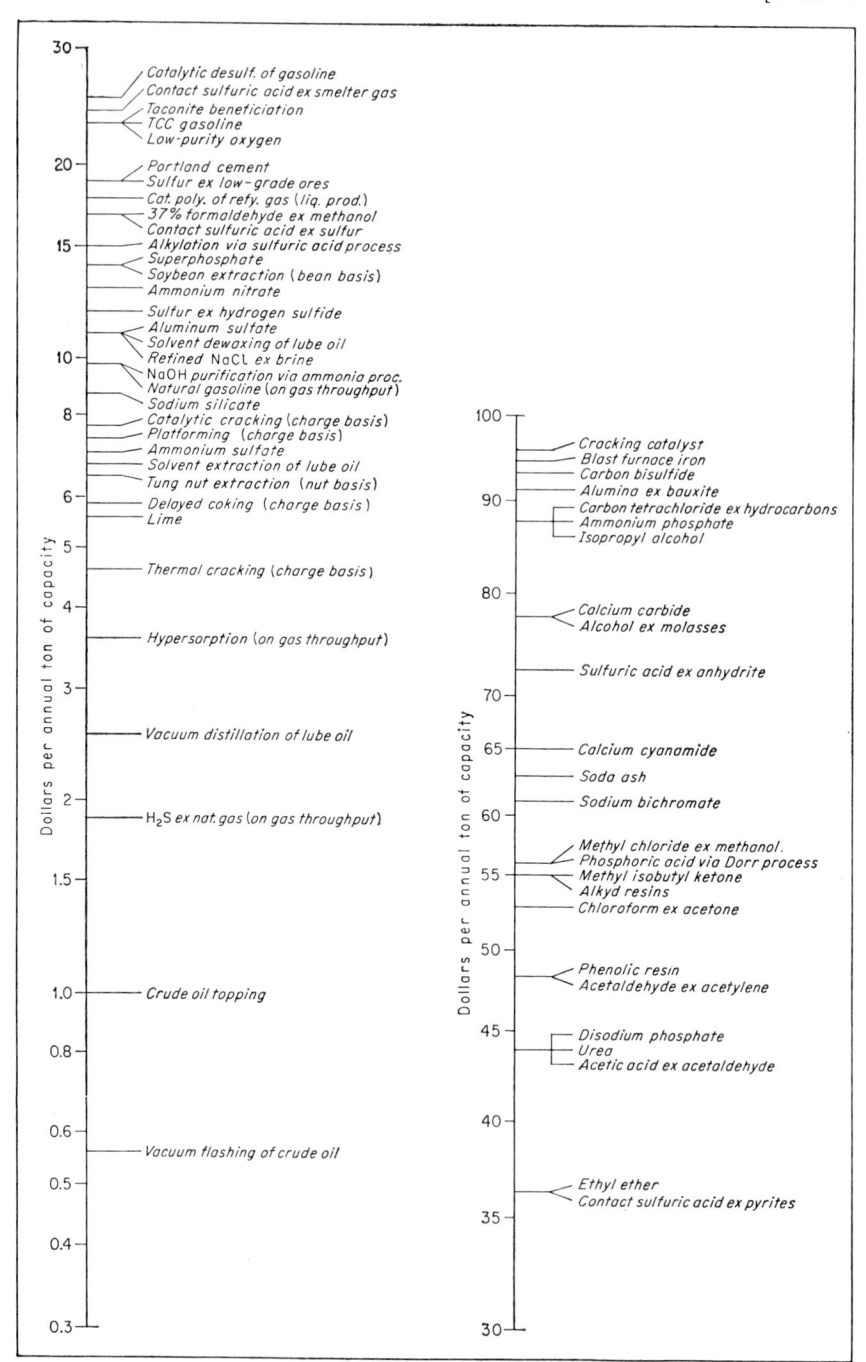

- 100 —
- Cracking catalyst
- Blast furnace iron
- Carbon bisulfide
- Alumina ex bauxite
- 90 — Carbon tetrachloride ex hydrocarbons
- Ammonium phosphate
- Isopropyl alcohol
- 80 —
- Calcium carbide
- Alcohol ex molasses
- Sulfuric acid ex anhydrite
- 70 —
- 65 — Calcium cyanamide
- Soda ash
- Sodium bichromate
- 60 —
- Methyl chloride ex methanol
- Phosphoric acid via Dorr process
- 55 — Methyl isobutyl ketone
- Alkyd resins
- Chloroform ex acetone
- 50 —
- Phenolic resin
- Acetaldehyde ex acetylene
- 45 — Disodium phosphate
- Urea
- Acetic acid ex acetaldehyde
- 40 —
- Ethyl ether
- Contact sulfuric acid ex pyrites
- 35 —
- 30 —

Fig. 6-2. Capital investment per annual ton of designed plant capacity—

ECONOMIC EVALUATION OF THE PROJECT

Left scale — Dollars per annual ton of capacity:

- 300 — Aniline ex nitrobenzene; Diphenylamine; Furfural
- 280 — Ethylene glycol; Toluene via hydroforming; GR-S copolymer; Alcohol ex wood waste
- 260 — Alcohol ex sulfite liquor; Electrolytic chlorine
- 240 — Synthetic methanol; Acetic acid via alcohol oxidation
- 220 — Synthetic ammonia
- — Acetylene via calcium carbide
- 200 — Sulfate pulp
- — Hydrofluoric acid
- 180 — Acetic anhydride ex acetic acid; Oxalic acid via oxidation
- 160 — Phosphoric acid via blast furnace; Acrylonitrile ex cyanohydrin; Silicon carbide; 37% formaldehyde ex hydrocarbons
- — Synthetic nitric acid; HCl ex salt
- 140 — Alcohol ex grain
- — Kraft paper or newsprint ex pulp; Aviation gasoline
- 120 — Ethylene ex refinery gas; Chlorine via nitrosyl chloride
- — Dialkyl phthalates; Carbon black; Trichlorethylene ex acetylene; High-purity oxygen
- 100

Right scale — Dollars per annual ton of capacity:

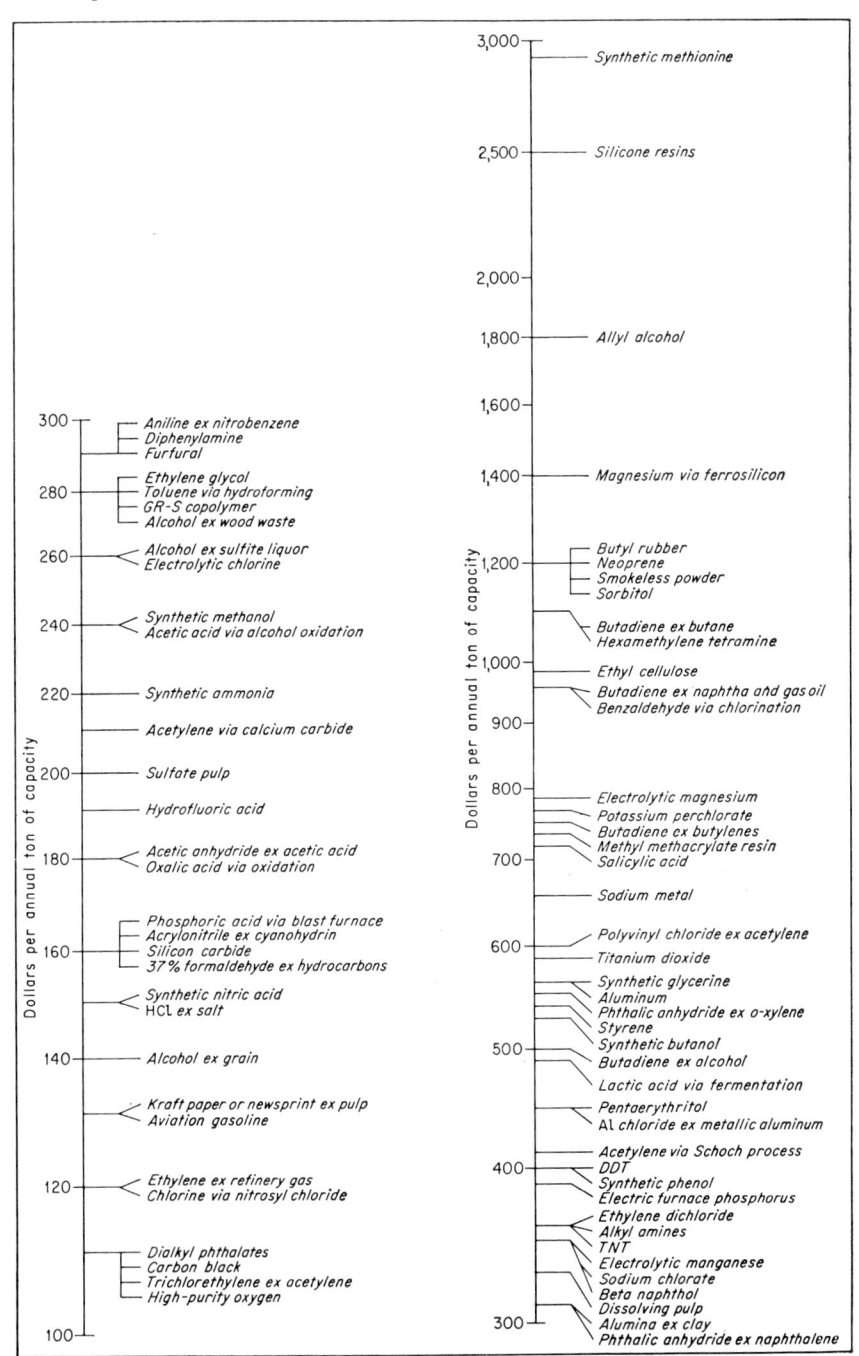

- 3,000 — Synthetic methionine
- 2,500 — Silicone resins
- 2,000
- 1,800 — Allyl alcohol
- 1,600
- 1,400 — Magnesium via ferrosilicon
- 1,200 — Butyl rubber; Neoprene; Smokeless powder; Sorbitol
- — Butadiene ex butane; Hexamethylene tetramine
- 1,000 — Ethyl cellulose
- — Butadiene ex naphtha and gas oil; Benzaldehyde via chlorination
- 900
- 800 — Electrolytic magnesium; Potassium perchlorate; Butadiene ex butylenes; Methyl methacrylate resin
- 700 — Salicylic acid
- — Sodium metal
- 600 — Polyvinyl chloride ex acetylene; Titanium dioxide
- — Synthetic glycerine; Aluminum; Phthalic anhydride ex o-xylene; Styrene
- 500 — Synthetic butanol; Butadiene ex alcohol
- — Lactic acid via fermentation
- — Pentaerythritol; Al chloride ex metallic aluminum
- — Acetylene via Schoch process
- 400 — DDT; Synthetic phenol; Electric furnace phosphorus
- — Ethylene dichloride; Alkyl amines; TNT; Electrolytic manganese; Sodium chlorate; Beta naphthol; Dissolving pulp
- 300 — Alumina ex clay; Phthalic anhydride ex naphthalene

1950 prices. [*Courtesy of C. H. Chilton, Chem. Eng.*, **58**(5): 164 (1951).]

Three problems must be considered in compiling an up-to-date summary of equipment costs: (1) bringing published reference costs up to date; (2) estimating equipment costs for equipment or plants of various capacities when the cost data are available for only one capacity; (3) estimating installation charges when delivered equipment prices are listed.

Cost Estimates and Price Indices. One should note that the date prices are quoted is a very important piece of data since basic costs change continually. To approximate current costs, the use of *cost indices*

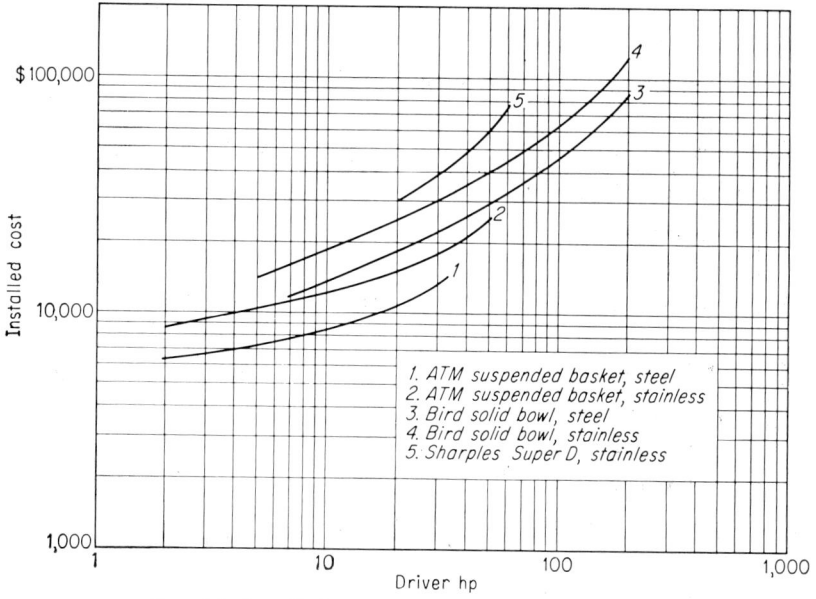

Fig. 6-3. Installed cost of centrifuges (ENR = 750).

1. ATM suspended basket, steel
2. ATM suspended basket, stainless
3. Bird solid bowl, steel
4. Bird solid bowl, stainless
5. Sharples Super D, stainless

is satisfactory if the elapsed time span is not too great so that error arises from technological advances.

While various indices are available for estimating purposes, only the four most applicable to chemical engineering cost estimating will be discussed.

1. *Engineering News-Record Construction Cost Index* (*ENR*). This is a construction index based on the economics of structural steel, portland cement, lumber, and common labor on a 20-city average with the value for the year 1913 set at 100. The base year of the index is sometimes changed to 1926 and 1949 so that care should be taken to see that a consistent base year is used throughout the cost evaluation procedure. Appearing weekly in *Engineering News-Record*, the index reflects wage rates and material price trends. Figure 6-31 (p. 220) is a plot of ENR

CHAP. 6] ECONOMIC EVALUATION OF THE PROJECT 199

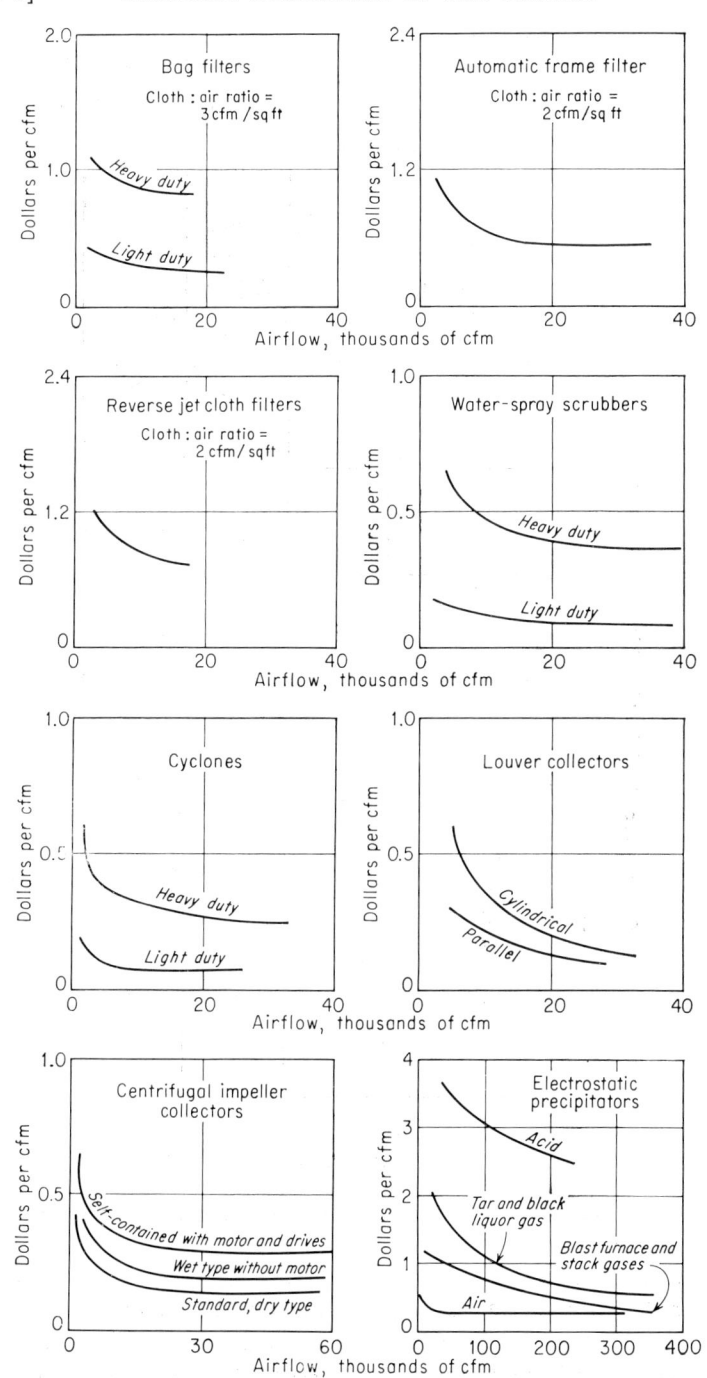

FIG. 6-4. Installed cost of dust collectors (ENR = 750).

values from 1940 to 1958 with space available up to 1970 for bringing the chart up to date or for forecasting. This index differs from the Building Cost Index in that the latter is based on skilled labor instead of common

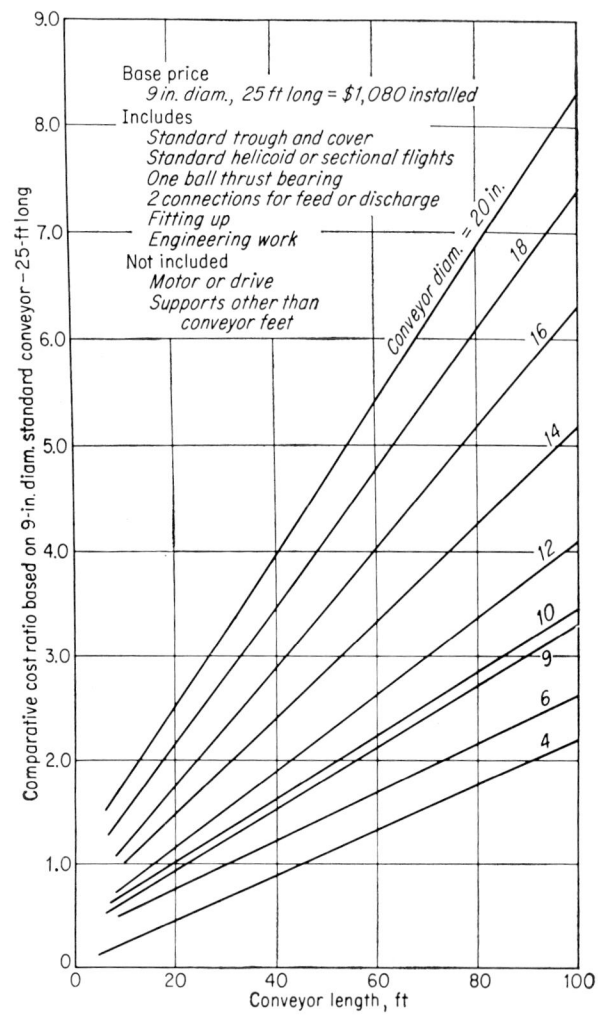

FIG. 6-5. Installed cost of standard conveyors (ENR = 750).

labor. Both indices are explained in more detail in the *Engineering News-Record*.[1]

2. *Marshall and Stevens Equipment Index (MS)*. An index with the value for the year 1926 equal to 100 is published quarterly in the Process Equipment Section of *Chemical Engineering* as the Marshall-Stevens

[1] *Eng. News-Record*, **143**(9): 398 (Sept. 1, 1949); **157**(22): 81 (Dec. 6, 1956).

process equipment index. It is based on detailed equipment cost appraisals in eight typical process industries with a weighting based on the total product value of each industry. Values of this index as a

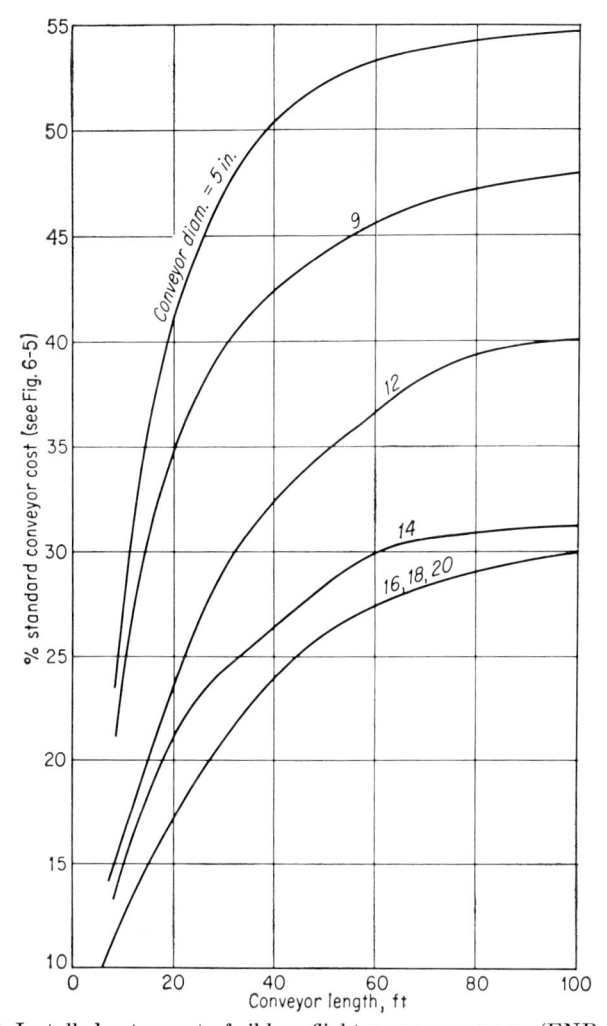

FIG. 6-6. Installed extra cost of ribbon flight screw conveyors (ENR = 750).

function of time are plotted in Fig. 6-31. More detailed information on this index is given in *Chemical Engineering*.[1]

3. *Nelson Refinery Construction Index.*[2] This index is a refinement of the ENR index with the following weighted components: materials—iron

[1] R. W. Stevens, *Chem. Eng.*, **54**(11): 124 (1947); **62**(3): 178 (1955).
[2] *Oil Gas J.*, **54**(74): 110 (Oct. 1, 1956).

202   CHEMICAL ENGINEERING PLANT DESIGN   [CHAP. 6

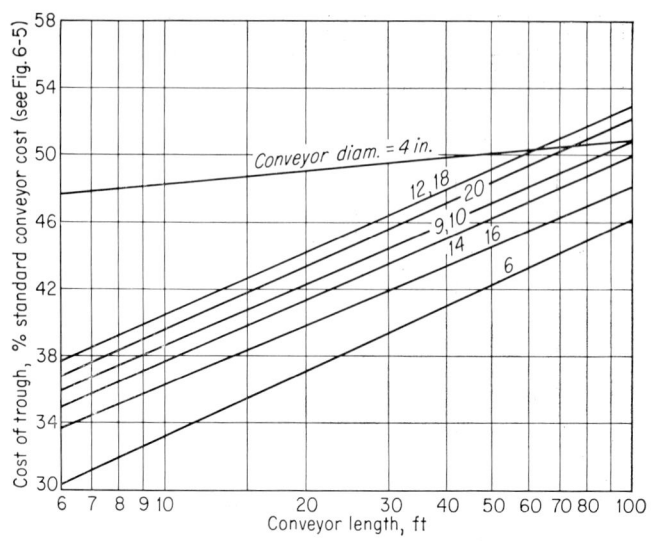

FIG. 6-7. Installed extra cost for standard thickness conveyor troughs of varying diameter (ENR = 750).

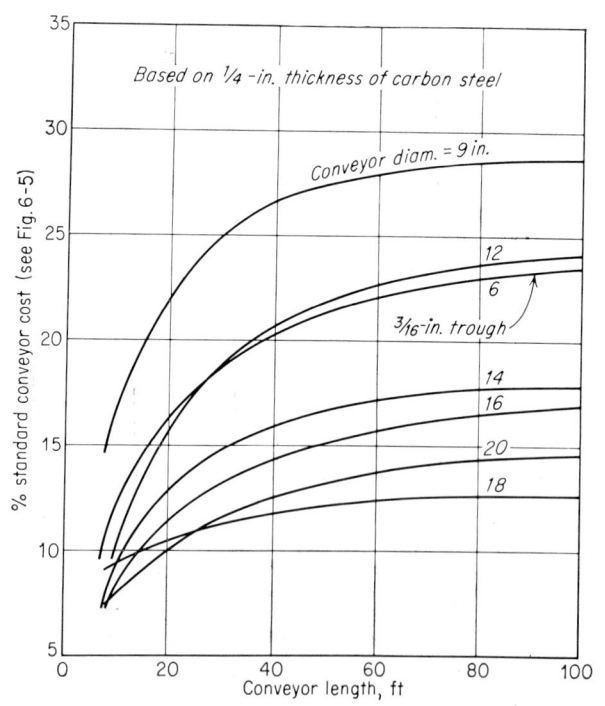

FIG. 6-8. Installed extra cost for heavy gauge thickness conveyor troughs based on ¼-in. thickness of carbon steel (ENR = 750).

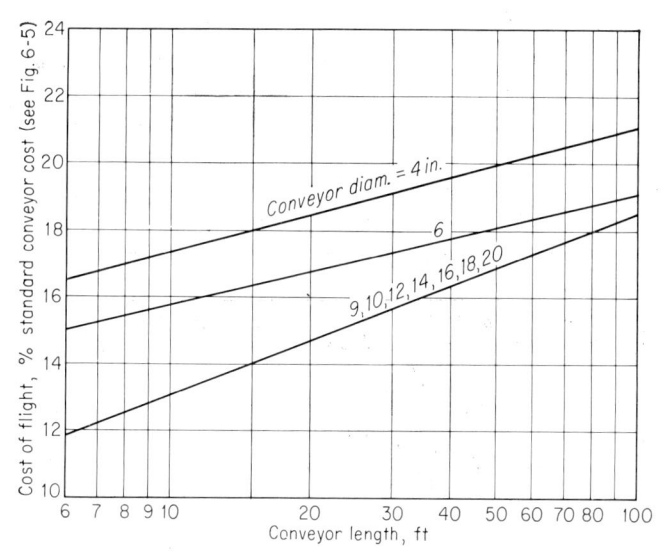

FIG. 6-9. Installed extra cost for screw conveyor flights (ENR = 750).

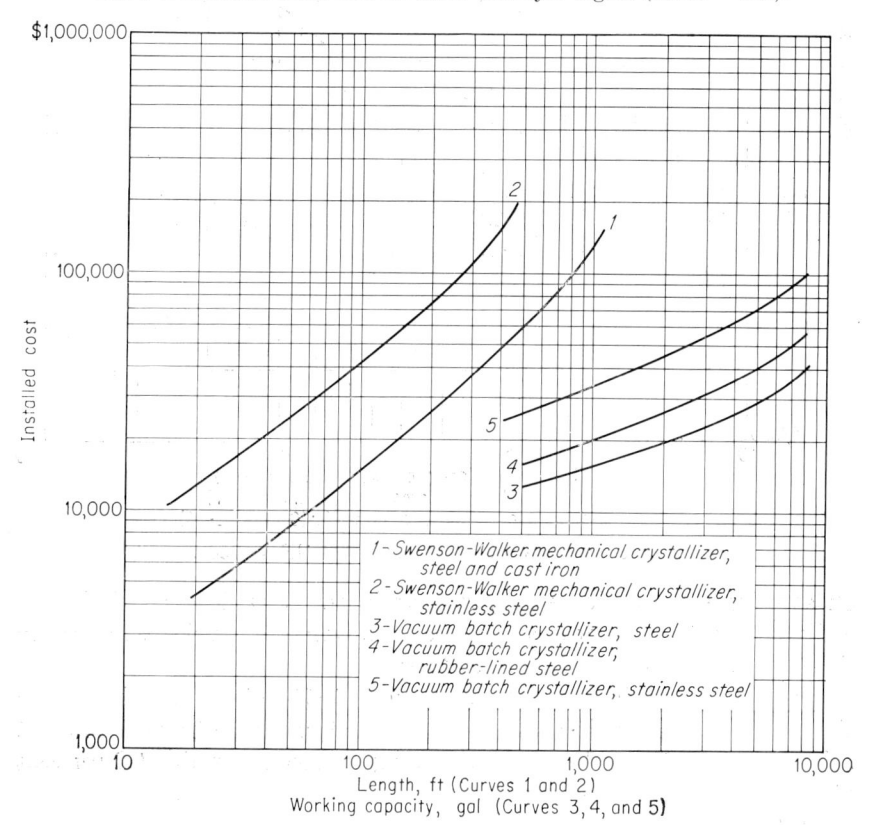

FIG. 6-10. Installed cost of crystallizers (ENR = 750).

1 - Swenson-Walker mechanical crystallizer, steel and cast iron
2 - Swenson-Walker mechanical crystallizer, stainless steel
3 - Vacuum batch crystallizer, steel
4 - Vacuum batch crystallizer, rubber-lined steel
5 - Vacuum batch crystallizer, stainless steel

and steel, 24 per cent; building materials, 8 per cent; miscellaneous equipment, 8 per cent; labor—skilled, 30 per cent; common, 30 per cent. Values of the index appear in the Engineering Section of the *Oil and Gas Journal* the first week of each month.

4. *Bureau of Labor Wages and Material Price Index.* The Bureau of Labor Statistics, U.S. Department of Labor, issues reports on a labor and materials index in the *Monthly Labor Review.* The labor index is based

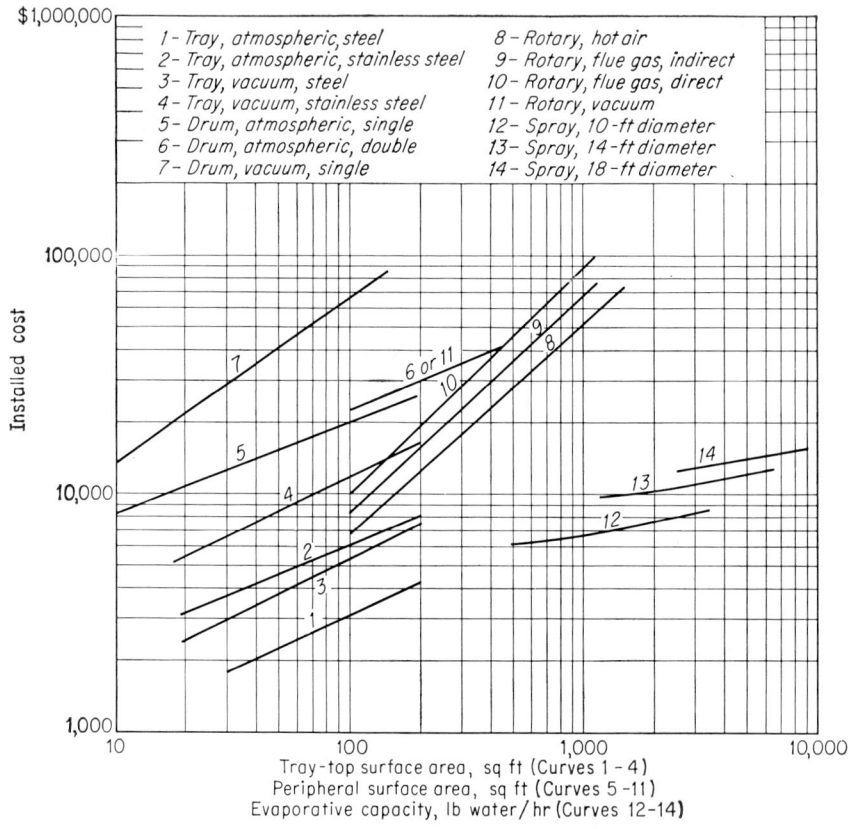

FIG. 6-11. Installed cost of dryers (ENR = 750).

on average earnings in cents per hour in the durable goods industry, while the material index is derived from whosesale price averages in the metal and metal-product industries.

A comparison of these indices for the years 1951 to 1957 made by Smith[1] showed that all the indices followed the same trend in cost rise, with ENR being the highest and the MS index being the lowest by about 8 per cent at the end of the 6-year period.

[1] C. A. Smith, *Cost Eng.*, **2**(4); 110 (1957).

CHAP. 6] ECONOMIC EVALUATION OF THE PROJECT 205

The use of the index method is very simple.
Let I_k = known index value for the date k
I_x = known index value for the date x
C_k = known equipment cost for the date k
C_x = unknown equipment cost for the date x

Then
$$C_x = C_k \frac{I_x}{I_k} \qquad (6\text{-}1)$$

Most of the cost data presented in this book are based on an installed 1958

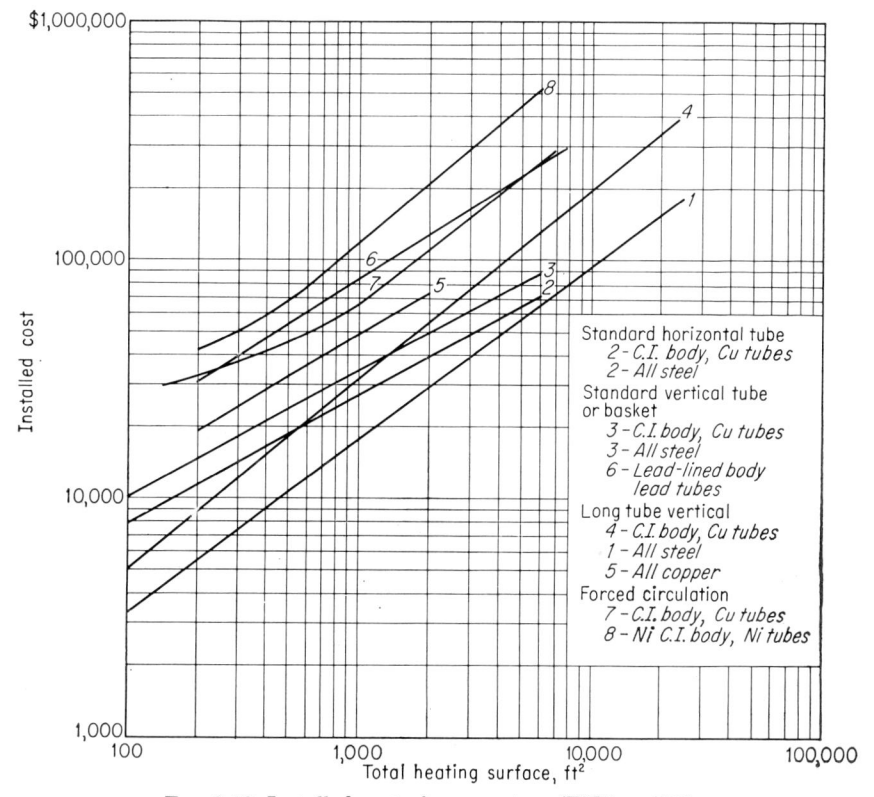

FIG. 6-12. Installed cost of evaporators (ENR = 750).

ENR value of 750. Other estimates can be made current by use of the above formula and the data plotted in Fig. 6-31.

Equipment Cost Comparisons. Several methods have been proposed for approximating costs of equipment of different size or capacity when the cost of a given unit is known. Williams[1] offers the general "six-tenths factor." If the cost of a given unit is known at one capacity and

[1] R. Williams, Jr., *Chem. Eng.*, **54**(12): 124 (1947).

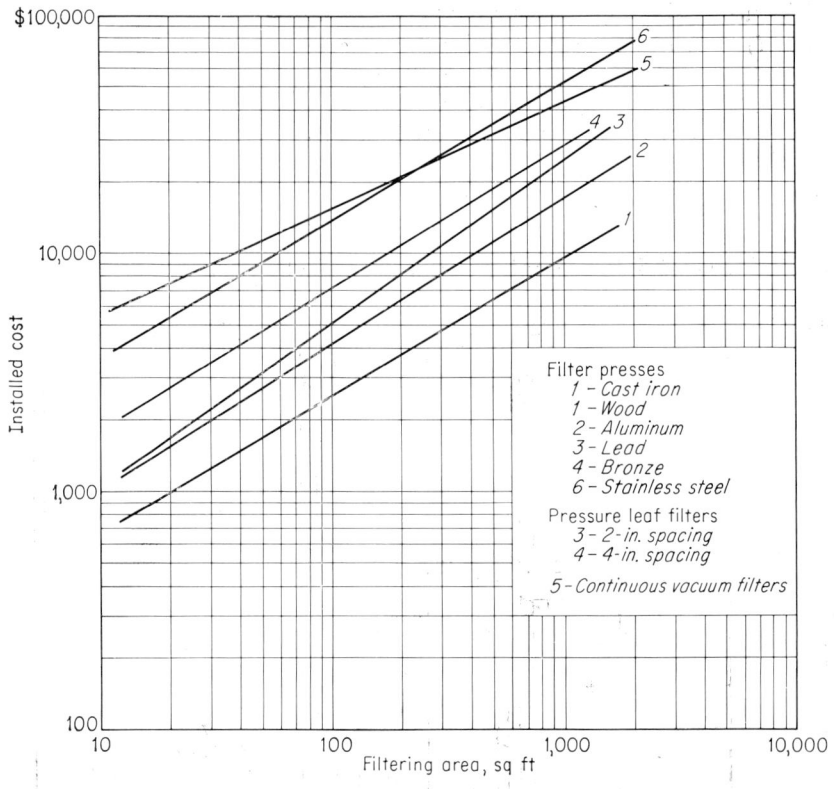

FIG. 6-13. Installed cost of filters (ENR = 750).

the cost is desired at a second capacity x times the first, multiply the known cost by $x^{0.6}$ to obtain the cost at the second capacity.

Equipment costs may be correlated by equations of form

$$y = ax^n \qquad (6\text{-}2)$$

where y = cost
 a = constant for type of equipment involved
 x = capacity of unit
 n = constant

When x_1 = capacity of first unit
 x_2 = capacity of second unit
 y_1 = cost of first unit

then

$$\frac{y_2}{y_1} = \left(\frac{x_2}{x_1}\right)^n \qquad \text{or} \qquad y_2 = y_1 \left(\frac{x_2}{x_1}\right)^n \qquad (6\text{-}3)$$

The latter equation becomes the six-tenths factor expressed in mathematical form where $n = 0.6$. This is a general constant and not specific for a given type of equipment.

This 0.6 factor also applies to total plant cost estimations, provided the scale-up is by size of similaru nits not greater than tenfold and not by multiplying the number of units. See Eq. (6-4) (p. 222) for possible ranges of the exponent n.

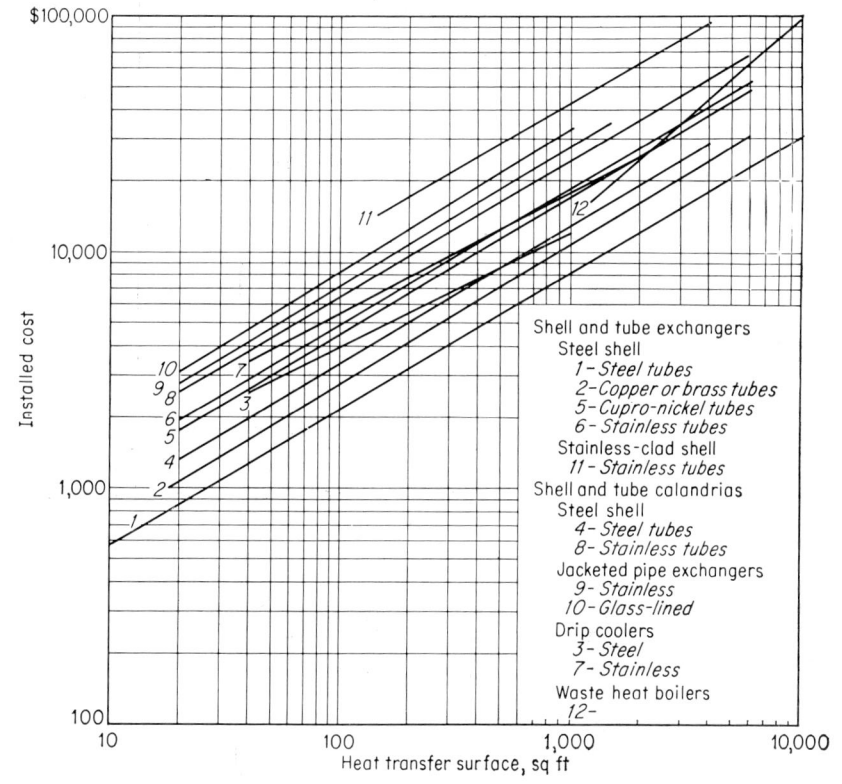

FIG. 6-14. Installed cost of heat exchangers (ENR = 750).

Computing Installation Costs. When delivered equipment costs are the only data available, installation costs must be estimated. These costs include labor, foundations, supports, platforms, site preparation, normal temperature-level insulation, and other factors directly related to the erection of equipment. Table 6-3 (p. 214) gives installation costs as a percentage range of delivered equipment costs for various types of equipment. As a general rule, costs for equipment installation are often estimated as 43 per cent of the delivered equipment costs.[1]

[1] H. J. Lang, Simplified Approach to Preliminary Cost Estimates, *Chem. Eng.*, **54**(10): 130 (1947).

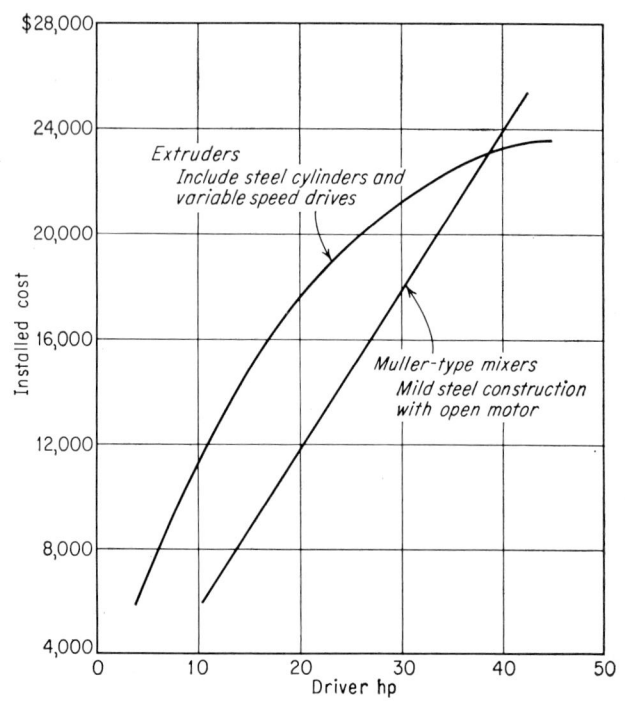

FIG. 6-15. Installed cost of extruders and Muller-type mixers (ENR = 750).

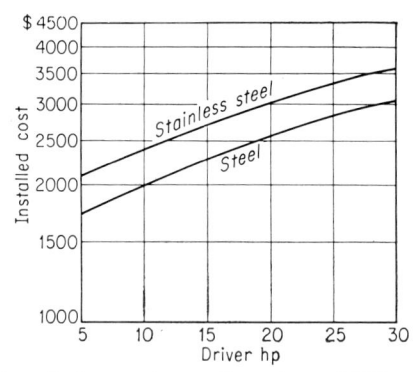

FIG. 6-16. Installed cost of propeller mixers (ENR = 750).

Piping Costs. The cost of piping involves pipe and fittings, supports, and labor for all service and process lines. Making this cost estimate requires detailed piping layout drawings or plant models to determine piping lengths, fittings, etc. (see Chap. 9). Piping cost estimates can be made on the basis of installed equipment costs as shown in Table 6-4 (p. 215). A more accurate method is discussed in Chap. 9.

CHAP. 6] ECONOMIC EVALUATION OF THE PROJECT 209

Insulation Costs. Where very high or very low temperatures exist for precise conditions, insulation costs may become a significant factor. For normal temperature levels the insulation expense is included under equipment installation costs, amounting to 6 to 10 per cent of delivered equipment cost with 40 per cent of the cost charged to materials, the

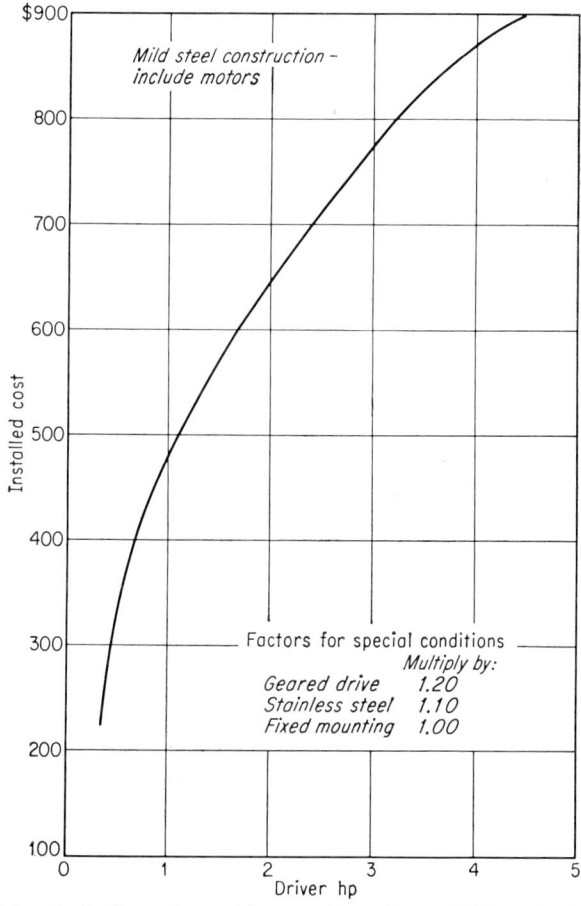

FIG. 6-17. Cost of portable propeller mixers (ENR = 750).

balance to labor. For more detailed information on insulation, see Chap. 9.

Instrumentation. Extensive automatic control of chemical plants is required when tolerances on variables are small and savings in labor and utilities are necessary. Unit operations such as distillation, extraction, and heat transfer as well as most of the unit processes such as sulfonation and nitration require expensive instrumentation. On the other hand,

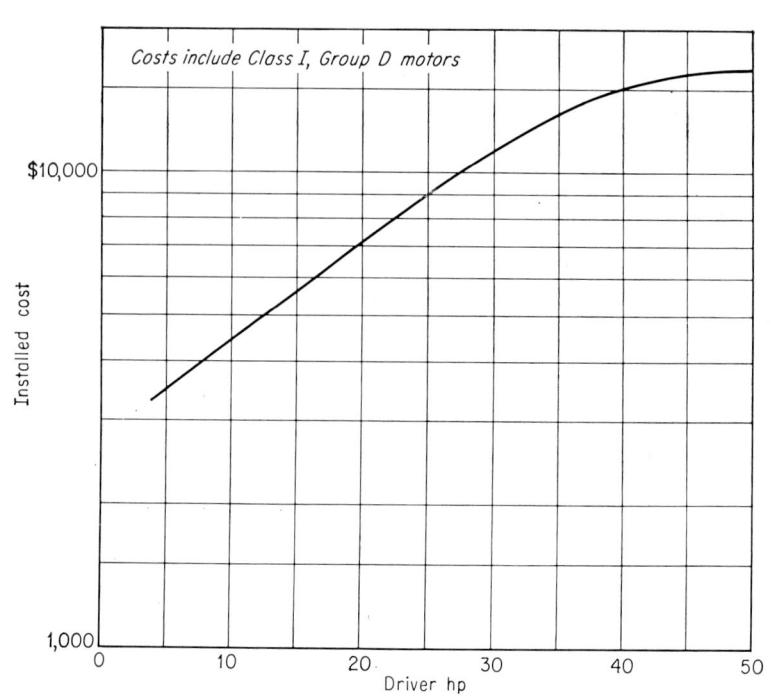

FIG. 6-18. Installed cost of spiral ribbon mixers (ENR = 750).

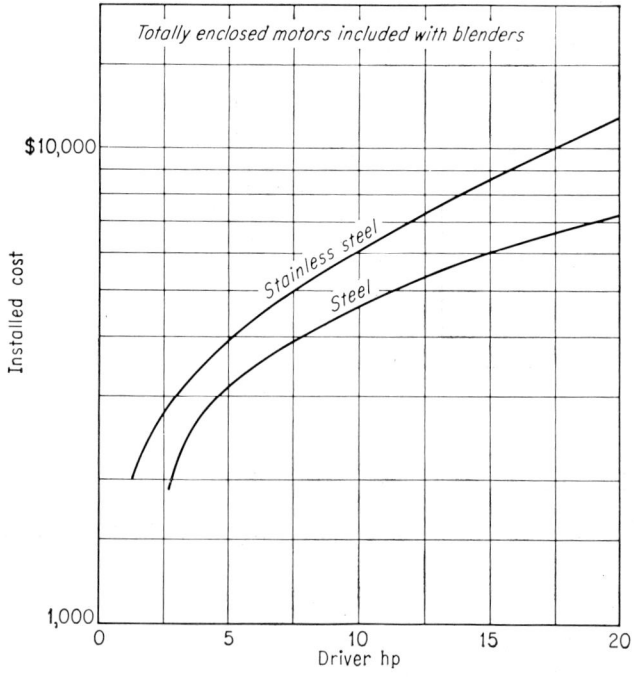

FIG. 6-19. Installed cost of rotary double-cone blenders (ENR = 750).

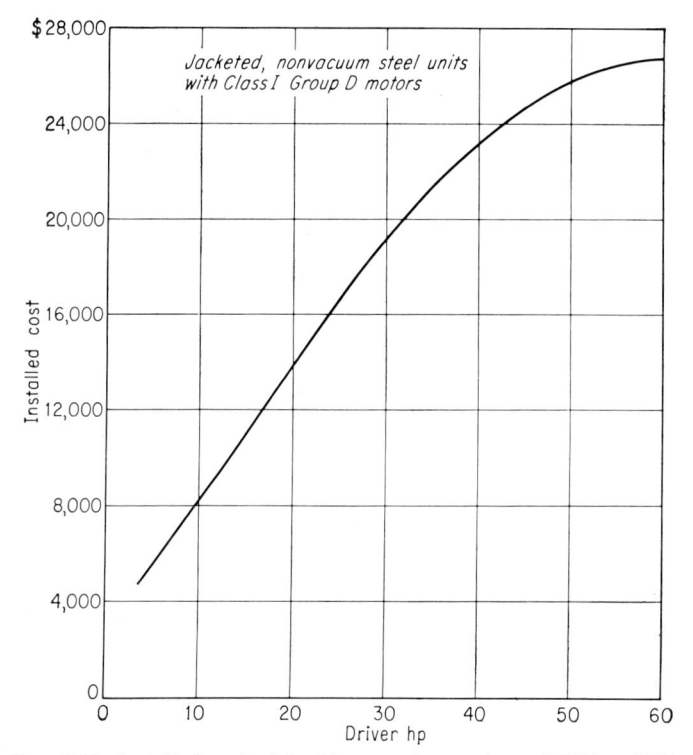

FIG. 6-20. Installed cost of double-arm sigma mixers (ENR = 750).

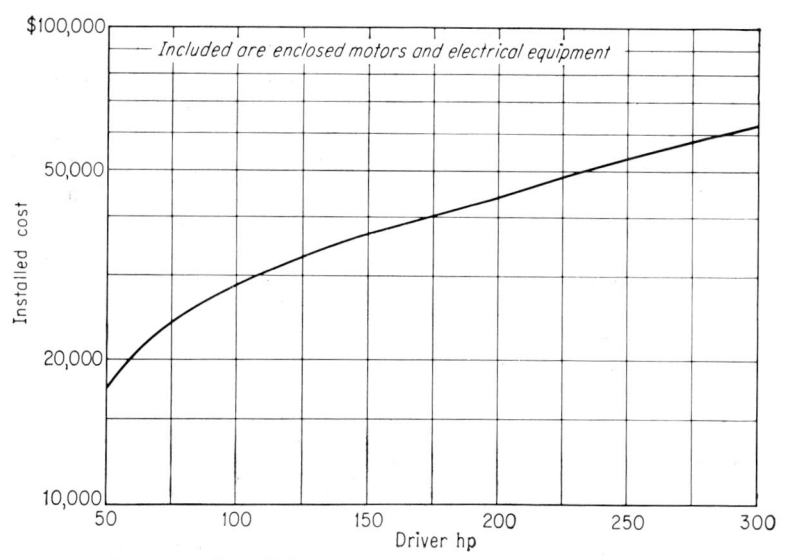

FIG. 6-21. Installed cost of two-roll mixers (ENR = 750).

212   CHEMICAL ENGINEERING PLANT DESIGN   [CHAP. 6

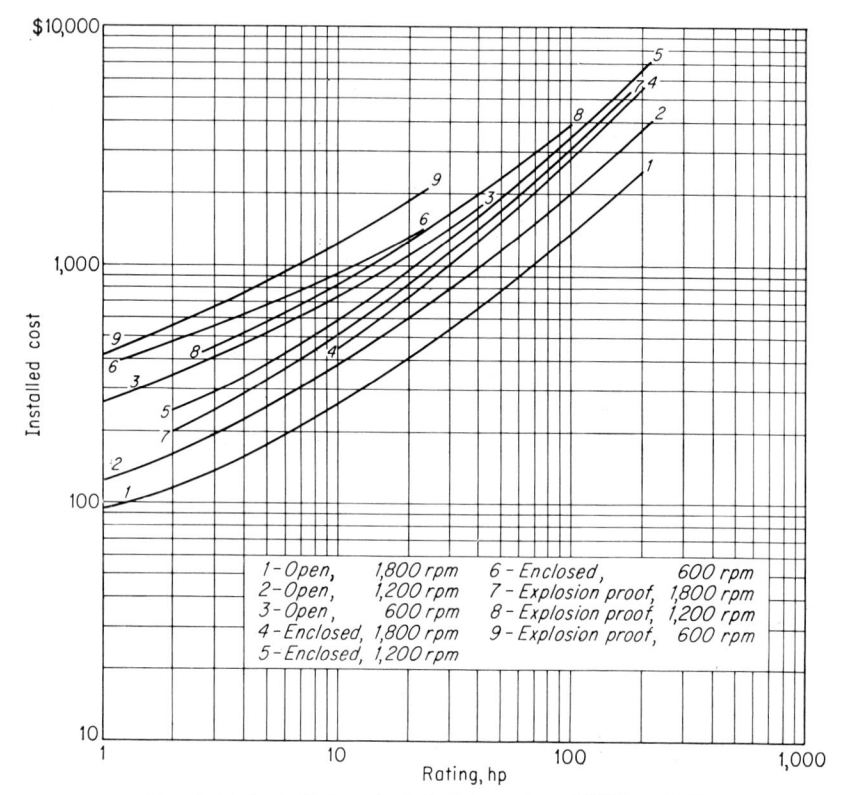

FIG. 6-22. Installed cost of electric motors (ENR = 750).

1 - Open, 1,800 rpm
2 - Open, 1,200 rpm
3 - Open, 600 rpm
4 - Enclosed, 1,800 rpm
5 - Enclosed, 1,200 rpm
6 - Enclosed, 600 rpm
7 - Explosion proof, 1,800 rpm
8 - Explosion proof, 1,200 rpm
9 - Explosion proof, 600 rpm

TABLE 6-2. RELATIVE COST OF MOTORS*

| | Approximate relative cost |||
Motor design	Motor only, %	Control only, %	Motor and control, %
Normal-torque squirrel-cage	100	100	100
High-torque squirrel-cage	115	100	113
High-slip (5–8%) squirrel-cage	134	100	130
High-slip (8–13%) squirrel-cage	155	100	150
Wound-rotor induction	130	665	190
Synchronous (100% PF)	130	860	205
Synchronous (80% PF)	155	860	225
Shunt-wound d-c	175	640	220
Series or compound d-c	180	640	225

* Performance characteristics usually determine the particular type of motor to be selected for a given drive. There are applications, however, where several designs of motors have suitable characteristics and the decision may be strongly influenced by

CHAP. 6] ECONOMIC EVALUATION OF THE PROJECT 213

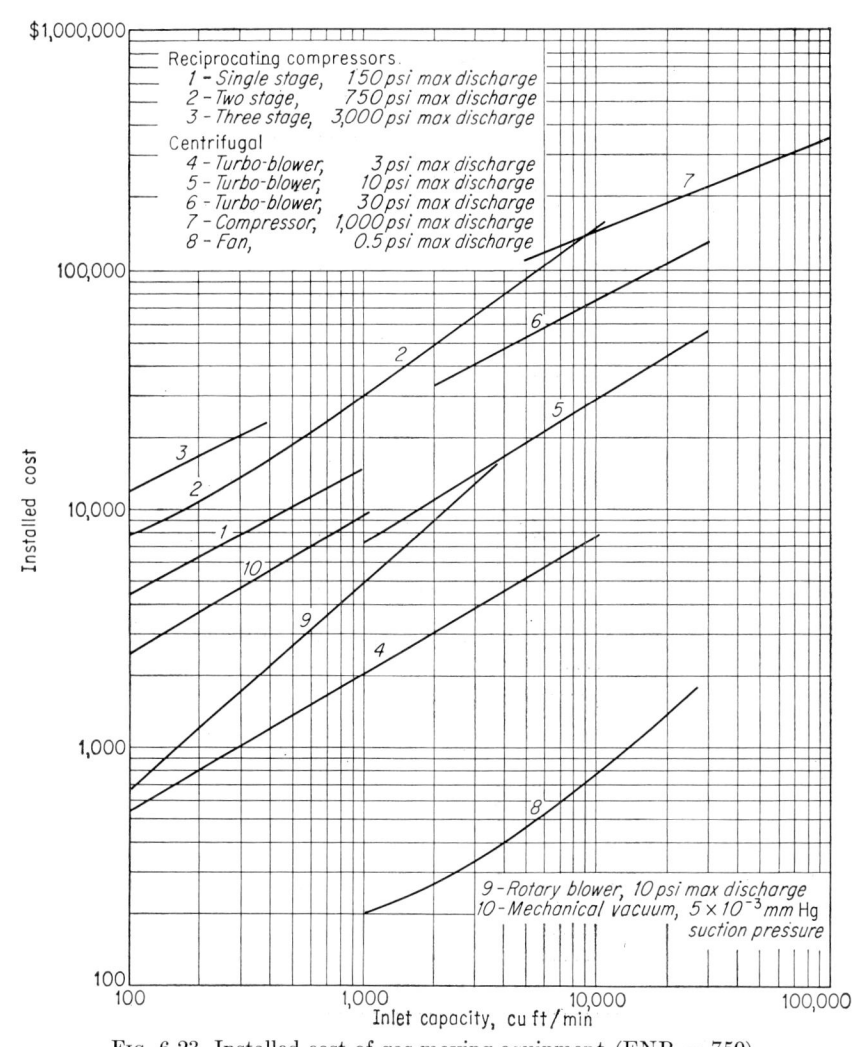

FIG. 6-23. Installed cost of gas-moving equipment (ENR = 750).

mixing, filtration, and size-reduction operations need few controls. Purchased costs of instruments can be obtained from Table 9-18 with

the relative cost of the particular motors involved and their associated controls. It is impossible to establish a fixed relationship between costs of various types of motors and controls for all ratings since the relative costs vary widely with horsepower, voltage, and type of enclosure. However, Table 6-2 shows an approximate cost comparison between the various types of motors and controls commonly used in the process industries. It is based on 440-volt three-phase 60-cycle open motors having an average rating of 100 hp at 1,200 rpm with full-voltage magnetic control in general-purpose type of enclosures. Cost approximations for motors are given in Fig. 6-22.

TABLE 6-3. INSTALLATION COSTS OF EQUIPMENT

Equipment	Per cent of delivered equipment cost
Beaters	25
Blowers	5–10
Boilers	40–50
Classifiers	3–5
Compressors	10–20
Conveyors	20–25
Dryers	20–30
Elevators (bucket)	25–40
Evaporators	25
Feeders	5–10
Filters:	
Plate and frame	30
Pressure leaf	30
Vacuum, continuous	20
Flotation machines	5–8
Gyratory crushers	5–13
Jaw crushers	8–10
Jigs	12–15
Mills:	
Ball, rod, pebble	4–10
Roller	15–20
Hammer	20
Motors (with line starter)	6–20
Pumps:	
Centrifugal	8–15
Duplex	3–5
Triplex	6
Vacuum	3–7
Stainless-steel	20–30
Glass-lined	50–100
Roaster	12
Rolls	4–12
Scales (platform)	40
Screens	5–15
Stills, small:	
Cast iron with agitator	50–70
Glass-lined	80–100
Tanks:	
Wood, large	40–60
Wood, small	100
Steel	20–25
Thickeners	12–15
Towers:	
Large	25–50
Small	100–150
Vats, circular, redwood	40–100

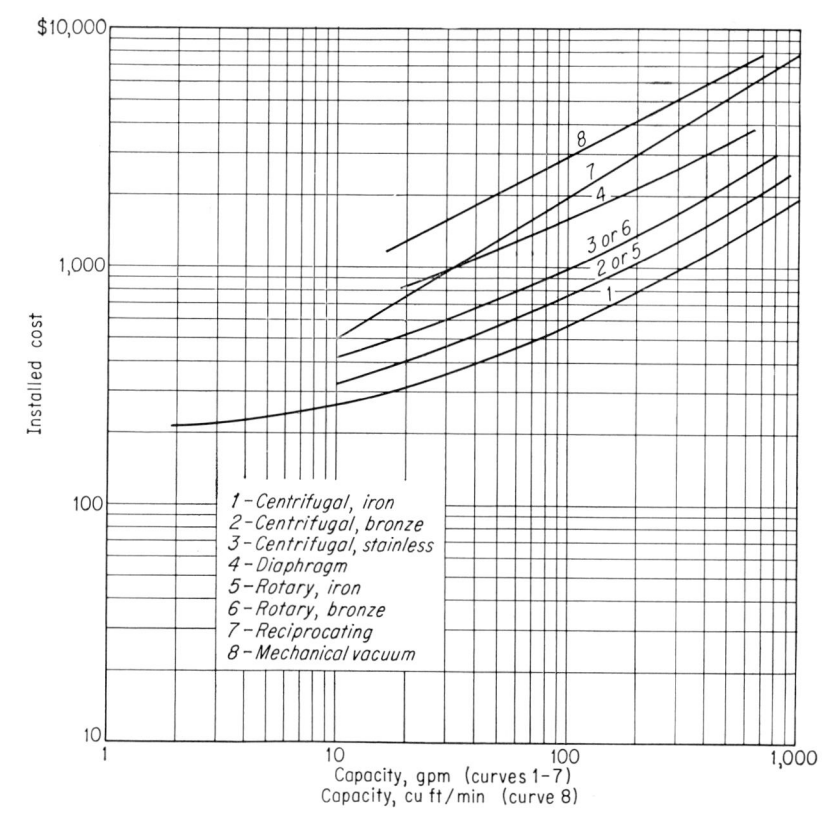

FIG. 6-24. Installed cost of pumps (ENR = 750).

Curves:
1 - Centrifugal, iron
2 - Centrifugal, bronze
3 - Centrifugal, stainless
4 - Diaphragm
5 - Rotary, iron
6 - Rotary, bronze
7 - Reciprocating
8 - Mechanical vacuum

Capacity, gpm (curves 1-7)
Capacity, cu ft/min (curve 8)

TABLE 6-4. PIPING COST ESTIMATES

Type process	Per cent of installed equipment cost		
	Material	Labor	Total
Solid...............	5	4	9
Solid-fluid...........	15	10	25
Fluid...............	35	26	61

35 per cent added for installation. Rough estimates can be made as follows as a percentage of installed equipment costs: minor instrumentation, 3 to 5 per cent; moderate, 10 per cent; extensive, 20 per cent.

Electrical Installation. In most estimates this cost is absorbed in the building and installed equipment costs. Electrical installation costs generally amount to 10 to 15 per cent of the delivered equipment costs.

216　CHEMICAL ENGINEERING PLANT DESIGN　[CHAP. 6

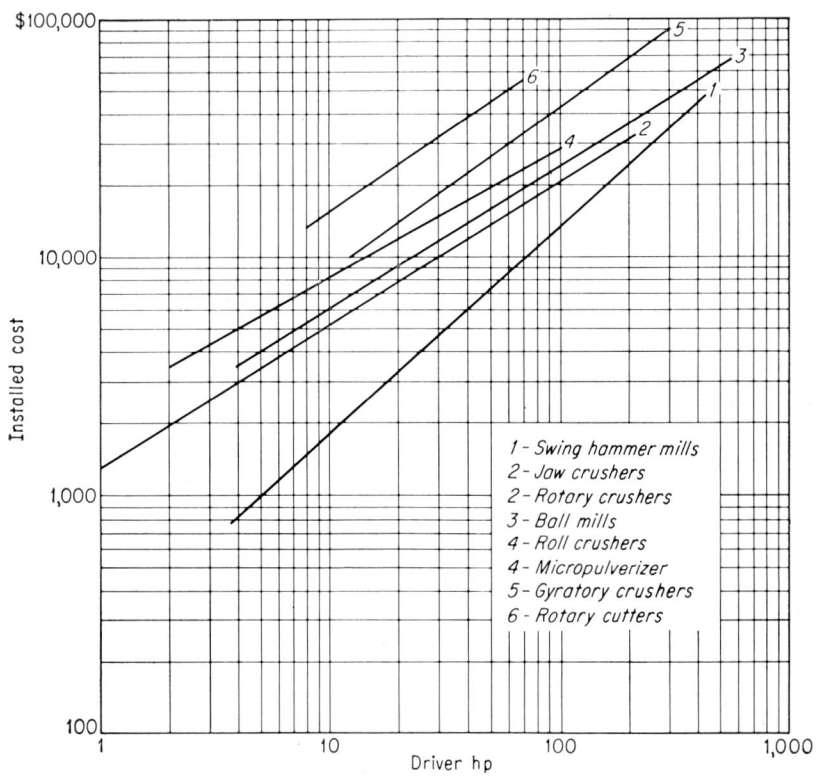

FIG. 6-25. Installed cost of size-reduction equipment (ENR = 750).

1 - Swing hammer mills
2 - Jaw crushers
2 - Rotary crushers
3 - Ball mills
4 - Roll crushers
4 - Micropulverizer
5 - Gyratory crushers
6 - Rotary cutters

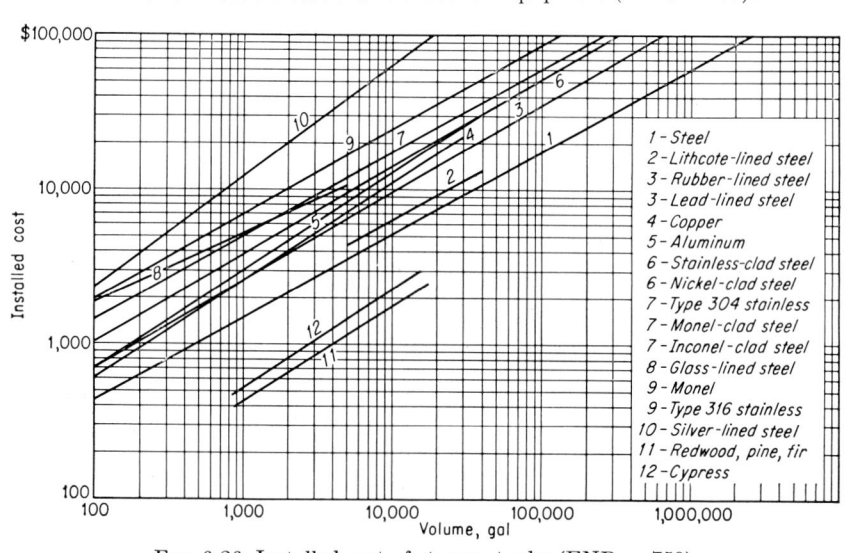

FIG. 6-26. Installed cost of storage tanks (ENR = 750).

1 - Steel
2 - Lithcote-lined steel
3 - Rubber-lined steel
3 - Lead-lined steel
4 - Copper
5 - Aluminum
6 - Stainless-clad steel
6 - Nickel-clad steel
7 - Type 304 stainless
7 - Monel-clad steel
7 - Inconel-clad steel
8 - Glass-lined steel
9 - Monel
9 - Type 316 stainless
10 - Silver-lined steel
11 - Redwood, pine, fir
12 - Cypress

Site and Building Costs. These costs vary considerably, depending on site location and type of construction. Table 6-5 gives preconstruction cost estimating figures. Further detailed information on site development and structure costs requires plant layout diagrams, as explained in Chaps. 5 and 8, with cost figures as given at the end of Chap. 8.

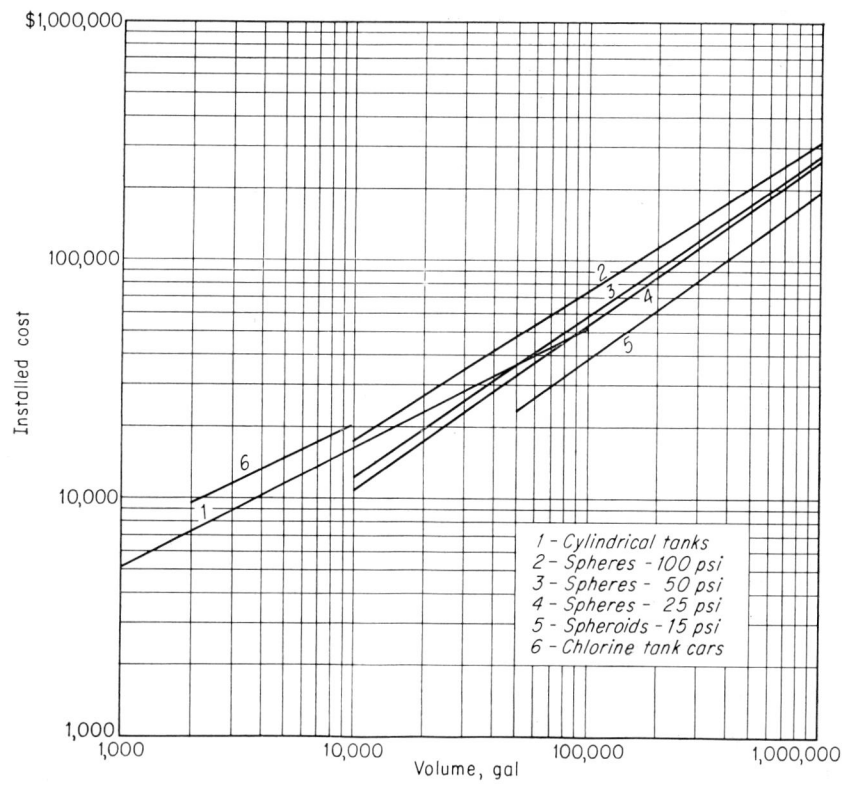

FIG. 6-27. Installed cost of steel pressure vessels (ENR = 750).

Engineering, Construction, and Contractor Costs. These costs include design, engineering, field supervision of temporary and permanent construction, and inspection. Table 6-6 shows cost estimating factors based on physical plant costs and degree of complexity.

The cost of engineering invariably depends on the type and complexity of the job and on the amount of overhead applied by the manufacturing corporation, the architect-engineer, or both. Engineering cost is defined to include the salaries of engineers, draftsmen, and supporting personnel; space, light, and heat; engineering supplies; depreciation of engineering equipment; travel; field liaison; and overhead. Depending on the accounting system of a corporation, the overhead has been at least 100

218　CHEMICAL ENGINEERING PLANT DESIGN　[CHAP. 6

TABLE 6-5. SITE AND BUILDING COSTS

Range of installed equipment costs	Type of construction (as per cent of installed equipment costs)		
	Outdoor*	Outdoor-indoor†	Indoor‡
Less than $350,000............	30	60	100
$350,000–$1,500,000..........	20	40	60
More than $1,500,000.........	10	20	35

* Outdoor. All major equipment unhoused; prefabricated building construction for control and administration structures.

† Outdoor-indoor. Some major equipment housed in permanent buildings; storage facilities and remaining major equipment unhoused; all permanent buildings custom-designed and built.

‡ Indoors. All major equipment housed in permanent buildings; all permanent buildings custom-designed and built.

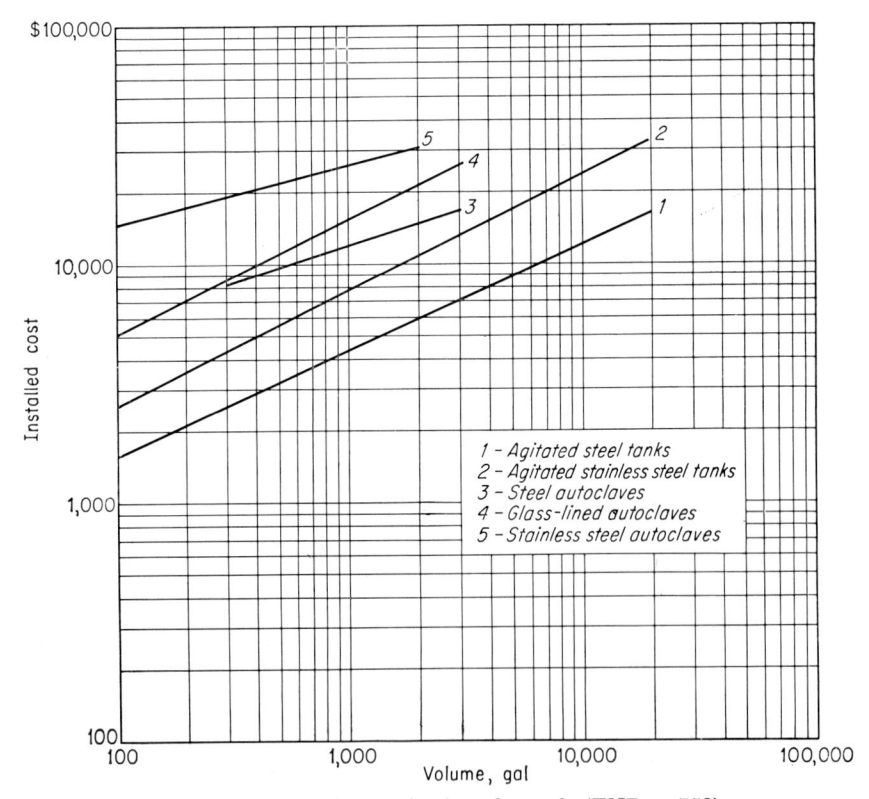

FIG. 6-28. Installed cost of agitated vessels (ENR = 750).

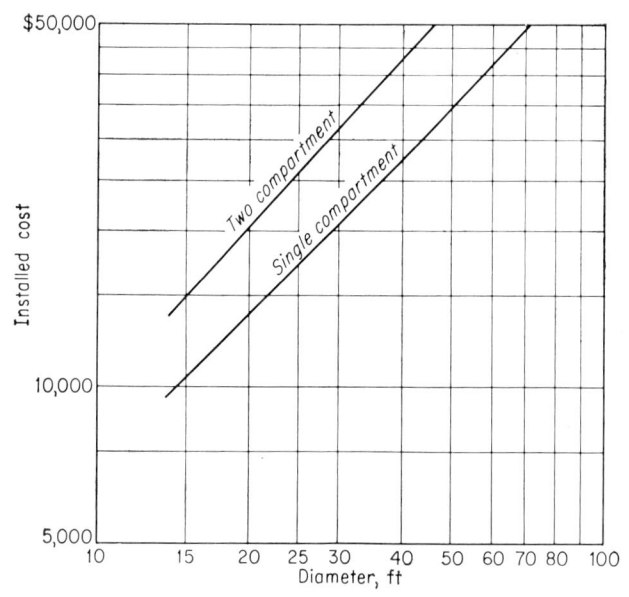

Fig. 6-29. Installed cost of continuous thickeners (ENR = 750).

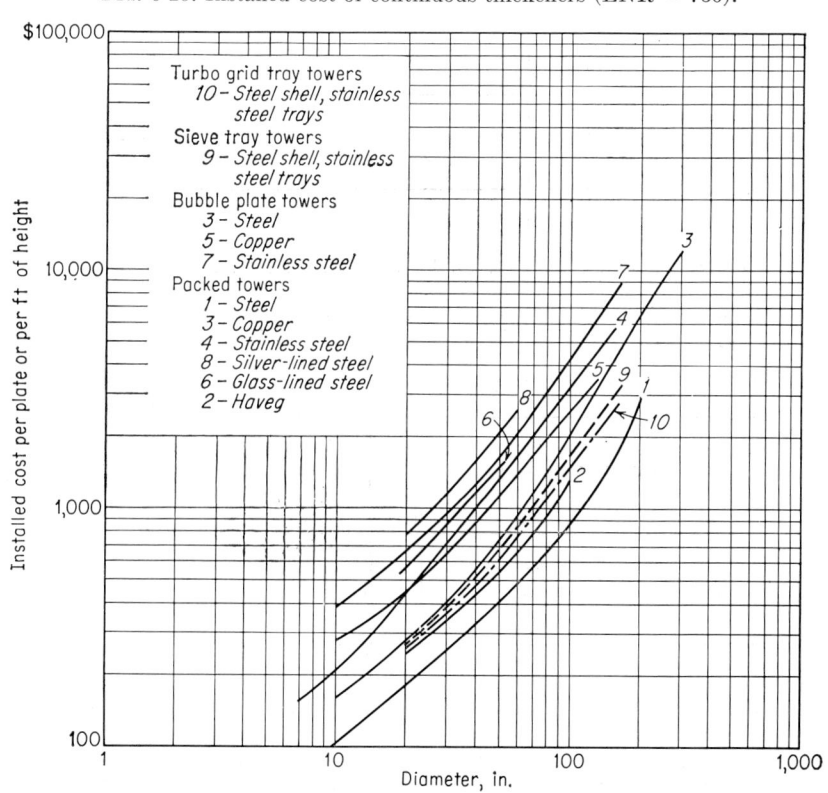

Fig. 6-30. Installed cost of packed and bubble plate towers (ENR = 750).

219

FIG. 6-31. Cost estimating indices.

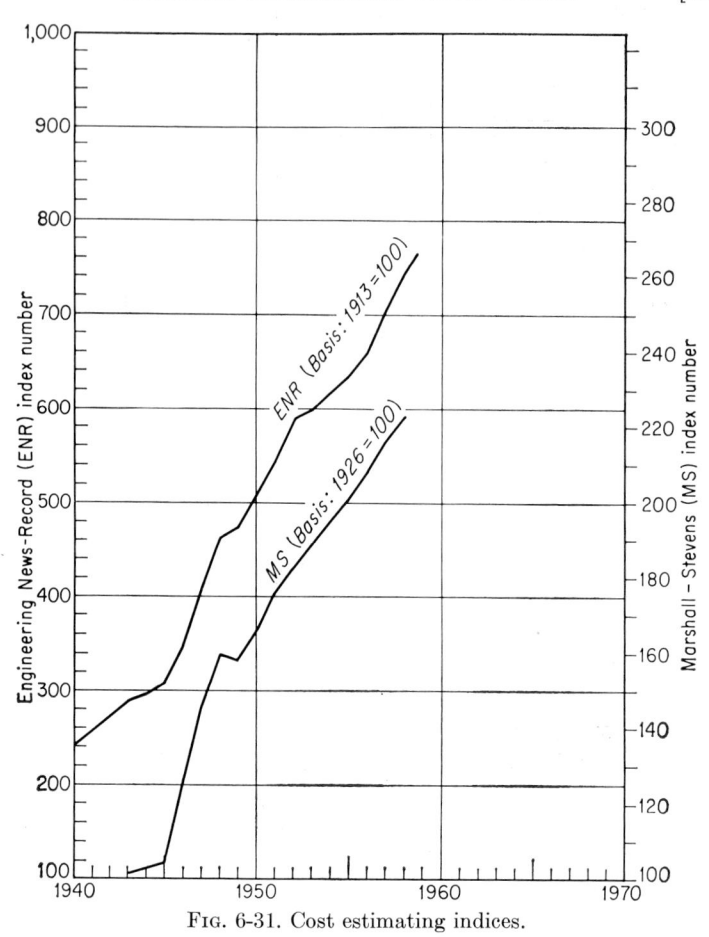

TABLE 6-6. COST OF ENGINEERING, CONSTRUCTION, AND CONTRACTOR FEES

Physical plant cost	Per cent of physical plant cost	
	Average	Complex engineering
Less than $1,000,000	40	60
$1,000,000–$5,000,000	33	50
More than $5,000,000	25	37

per cent on engineering personnel salaries. Engineering design costs, exclusive of construction supervision, range from 5 per cent minimum to 15 per cent maximum of the total cost of the engineering job, where existing chemical plant facilities were duplicated in like kind with a

5 per cent design engineering cost. A pioneering chemical plant was built based on incomplete process data with a 15 per cent engineering design change. For an approximation, the manpower required, with about 120 per cent overhead, is $100 to $125 per man-day worked.

When the job is adequately defined, a drawing list can be prepared and related to drafting and design engineering man-days. A rough average figure for preparing and checking a drawing on sizable chemical plant construction jobs is one hundred man-hours. Another approximation is that one hundred drawings are required for each 1 million dollars of investment on large chemical jobs. For small, or semiworks, plants, the number of drawings per unit of plant cost may be higher. In the majority of cases the mechanical engineering cost will be high, usually about 40 per cent of the engineering cost because of the large amount of piping and pressure-vessel work, with the rest divided as follows: chemical, 25; structural, 20; electrical, 10; and instrument, 5.

Working Capital

The working capital requirements have been itemized by Wessel:[1]
1. Raw materials inventory—one month's supply at cost
2. Materials in-process inventory—one week at manufactured cost
3. Product inventory—one month at manufactured cost
4. Accounts receivable—one month at selling price
5. Available cash to meet current expenses of wages, raw materials, utilities, and supplies—one month at manufacturing cost

The above quantities are standard for estimation purposes. The quantity of raw material that needs to be held in inventory will vary with each raw material. Capital is required to cover credit (or accounts receivable) extended to customers according to the terms of the rate, generally 30 days. Additional cash is required to pay wages and salaries and to purchase raw materials and pay for other operating expenses.

To simplify preconstruction cost estimation, use 15 to 20 per cent of the fixed capital investment for working capital investment. Another method of estimation is based on fixed percentage of the annual sales dollar, averaging 25 per cent for 100 chemical process companies.

Total Capital Investment

The sum of fixed and working capital investments as determined by one or more of the methods outlined above constitutes the total capital investment. Since the fixed capital costs are generally greater than 85 per cent of the total capital costs, the errors inherent in using **methods 1 to 7** for determining fixed capital costs also apply to total capital costs (Fig. 6-1). These derived figures are thus available for making manu-

[1] H. E. Wessel, *Chem. Eng.*, **60**(1): 171 (1953).

facturing cost estimates, setting the selling price of products, and evaluating the over-all process economics.

As there is a direct relationship between fixed and working capital costs, **method 4** can be expanded to get a "quick" estimate of total capital investment within an error of 20 to 40 per cent overestimated to 40 to 60 per cent underestimated. Table 6-7 summarizes this method.

TABLE 6-7. QUICK ESTIMATION METHOD FOR CAPITAL COSTS*

Type of plant	Delivered equipment cost		Installed equipment cost	
	Fixed capital	Total capital	Fixed capital	Total capital
Solids processing................	3.10	3.60	2.16	2.50
Solids-fluids processing............	3.63	4.20	2.56	3.00
Fluids processing................	4.74	5.50	3.30	3.80

* Reference: Fixed Capital Cost Estimation, **method 4** (p. 195).

In many cases there is a requirement to estimate the fixed or total capital investment for another plant using the same process but at a different production capacity. The capacity factor used for separate equipment can also be applied to the complete plant investment:[1]

$$\text{Investment B} = \text{investment A} \left(\frac{\text{prod. capacity B}}{\text{prod. capacity A}}\right)^n \quad (6\text{-}4)$$

where n = 0.6–0.7 for scale-up accomplished by increasing size of units
  = 0.8–1.0 for scale-up accomplished by multiple units of identical size
  = 0.3–0.5 for pilot plants, high-pressure and/or high-temperature commerical plants

These factors should not be applied for scale-up ratios greater than 50.

TOTAL PRODUCT COSTS

Manufacturing Costs

An estimate of manufacturing costs is the next step required for making an economic evaluation of the project. Only costs entering actual production are taken into account. Other important costs such as selling, research, and administrative expenses, income taxes, and return on investment will not be considered until later in this chapter. A

[1] C. H. Chilton, *Chem. Eng.*, **57**(4): 112 (1950).

CHAP. 6] ECONOMIC EVALUATION OF THE PROJECT 223

TABLE 6-8. DETAILED MANUFACTURING COST ESTIMATE

1. Product ____X____ in _____ (drums, bags, tank cars, etc.)
2. Production Rate, __0,000,000__ units (lb, tons, gallons, etc.) per year of __0,000__ hr, __000__ days
3. Plant location _____
4. Plant investment:
   Machinery and equipment (M and E).. $000,000
   Building... 00,000

5. Raw materials:	Unit	Quantity per year	Unit cost, $	Allocated costs, $/yr	$ per unit of product
Y...............................	Lb	0,000,000	0.00	000,000	0.0000
Z...............................	Ton	00,000	0.00	000,000	0.0000
Subtotal........................	.....	....	.....	000,000	0.0000
Credit for by-products:					
B...............................	Lb	000,000	0.00	−000,000	−0.0000
N...............................	Lb	00,000	0.00	− 00,000	−0.0000
Credit subtotal.................	.....	....	.....	−000,000	−0.0000
Net raw material cost...........	.....	....	.....	000,000	0.0000
6. Direct conversion expense (DCE):					
Labor (D.L.)....................	Man-hr	00,000	0.00	00,000	0.0000
Supervision.....................	.....	....	.....	0,000	0.0000
Payroll charges (12–20% D.L.-Sup.).....	.....	....	.....	0,000	0.0000
Utilities:					
Steam...........................	M lb	00,000	0.00	00,000	0.0000
Electricity.....................	Kwhr	0,000,000	0.0000	00,000	0.0000
Fuel............................	Therms	00,000	0.0000	0,000	0.0000
Compressed air..................	Mcf				
Refrigeration...................	Tons				
Inert gas.......................	MCF				
Water cooling...................	M gal	000,000	0.000	0,000	0.0000
Water process...................	M gal	0,000	0.0000	000	0.0000
Repairs (2–10% of M and E)......	.....	....	.....	00,000	0.0000
Factory supplies (0.5–1% of M and E)...	.....	....	.....	0,000	0.0000
Clothing and laundry ($1/wk for ea. worker).......................	.....	....	.....	000	0.0000
Laboratory (salaries and 100% overhead).......................	.....	....	.....	0,000	0.0000
Royalty (if applicable).........	.....	....	.....	0,000	0.0000
Total DCE.......................	.....	....	.....	000,000	0.0000
7. Indirect conversion expense (IDCE):					
Depreciation (10% of M and E)...	.....	....	.....	00,000	0.0000
Depreciation (5% of bldg.)......	.....	....	.....	0,000	0.0000
Taxes (0.2% of M and E + bldg.).	.....	....	.....	0,000	0.0000
Insurance (2% of M and E + bldg.).....	.....	....	.....	0,000	0.0000
Controllable IDCE (40–60% of D.L. or 15–30% of DCE).................	.....	....	.....	00,000	0.0000
Rent (if any rented buildings 8–10% of value or land are included)...........	.....	....	.....	0,000	0.0000
Total IDCE......................	.....	....	.....	00,000	0.0000
8. Bulk conversion cost (items 6 + 7)....	.....	....	.....	000,000	0.0000
9. Bulk manufacturing cost (items 5 + 8)	.....	....	.....	0,000,000	0.0000
10. Packaging and shipping:					
Containers......................	.....	....	.....	00,000	0.0000
Packing labor...................	.....	....	.....	0,000	0.0000
Shipping supplies and labor.....	.....	....	.....	0,000	0.0000
Freight allowances..............	.....	....	.....	−0,000	−0.0000
Total...........................	.....	....	.....	00,000	0.0000
11. Total manufacturing cost, f.o.b. (items 9 + 10)...................	.....	....	.....	0,000,000	0.0000

TABLE 6-9. SHORT-CUT MANUFACTURING COST ESTIMATE

1. Product ___X___ in _____ (drums, bags, tank cars, etc.)
2. Production rate, ___0,000,000___ units (lb, tons, gallons, etc.) per year of ___0,000___ hr, ___000___ days
3. Plant location _____
4. Plant investment:*
   Machinery and equipment (M and E)... $000,000
   Building.. 00,000

5. Raw materials:	Unit	Quantity per year	Unit cost, $	Allocated costs, $/yr	$ per unit of product
Y...................................	Lb	0,000,000	0.00	000,000	0.0000
Z...................................	Ton	00,000	0.00	000,000	0.0000
Subtotal.............................	......	....	......	000,000	0.0000
Credit for by-products:					
B...................................	Lb	000,000	0.00	−000,000	−0.0000
N...................................	Lb	00,000	0.00	− 00,000	−0.0000
Credit subtotal......................	......	....	......	−000,000	−0.0000
Net raw material cost................	......	....	......	000,000	0.0000
6. Direct conversion expense (DCE):					
a. Utilities:					
Steam............................	M lb	00,000	0.00	00,000	0.0000
Electricity.......................	Kwhr	0,000,000	0.0000	00,000	0.0000
Fuel.............................	Therms	00,000	0.0000	0,000	0.0000
Compressed air...................	Mcf				
Refrigeration....................	Tons				
Inert gas........................	Mcf				
Water cooling....................	M gal	000,000	0.0000	0,000	0.0000
Water process...................	M gal	0,000	0.0000	000	0.0000
b. Labor............................	Man-hr	00,000	0.00	00,000	0.0000
c. Supervision......................	......	....	......	0,000	0.0000
d. Total direct-labor charge [(6b + 6c) × 1.79].............................	......	....	......	00,000	0.0000
e. Laboratory expenses (salaries + 100% overhead)......................	......	....	......	0,000	0.0000
f. Total DCE (items 6a + 6d + 6e)....	......	....	......	00,000	0.0000
7. Indirect conversion expense (IDCE):					
a. Fixed charges and repairs—0.20 of M and E...........................	......	....	......	00,000	0.0000
b. Building charges—0.05 of bldg.....	......	....	......	0,000	0.0000
c. Total IDCE (items 7a + 7b)........	......	....	......	00,000	0.0000
8. Bulk conversion cost (items 6f + 7c)....	......	....	......	000,000	0.0000
9. Bulk manufacturing cost (items 5 + 8)..	......	....	......	0,000,000	0.0000

* When building costs are not estimated separately, calculate item 7 (IDCE) as 18% of the total fixed capital investment of the plant.

detailed and a short-cut manufacturing cost estimate based on the work of Dybdal[1] are shown in Tables 6-8 and 6-9. Average percentage factors based on fixed capital and labor charges are used. A discussion of the possible variation in many of these items will follow. Table 6-10 shows cost distribution of manufacturing costs for 14 different chemical industries.

[1] E. C. Dybdal, *Chem. Eng. Progr.*, **46**(20): 57 (1950).

TABLE 6-10. COST DISTRIBUTION AS PERCENTAGE OF MANUFACTURING COST

	Formaldehyde ex methane[1]	Monsanto "X" (liquid-solid)[2]	Ammonia (as part of nitrate or urea, steam re-forming)	$HNO_3 + NH_4NO_3$	Urea	Pharmaceutical	Tonnage oxygen[3]	Alcohol ex molasses	Electrolytic chlorine + caustic[4]	Phenol via sulphonation[5]	Aniline[5]	Soda ash, ammonia-soda process	Platforming + Udex[6]	Gasoline refining (complete)[5]
Raw materials	28	75	12	68[7]	60[7]	83		75	9	76	82	36	86	80
Steam, water, fuel	3	}5	21	3	7	0.5	13	6	18	2.5	6	18		}3
Power	16						40			1.5		1		
Subtotal	47	80	33	71	67	83.5	53	81	27	80	88	55	86	83
Labor, salaries	23	5	13	6	10	10	2	13	11	8	8	13	6	5
Maintenance		3	4	2	2.5	2	8	2	18	6		3	2	4
Miscellaneous	30[8]	5	6	5	3.5	2.5								
Taxes, depreciation		7	44	16	17	2	37	4	44	6	4		4	8
Total	100	100	100	100	100	100	100	100	100	100	100	100	100	100
Production rate, tons/day		7.5	200	300	100	1	1,000	100	50 + 56.4	50	50	300	1,500[9]	7,500[10]
Selling price, ¢/lb			4.0	3.5	5.5	200			2.7 (ea.)	20	18	1.3	1.8 (avg)	1.4 (avg)
Manufacturing cost, ¢/lb			2.5	2.0	3.0	120	1.8		2.0 (ea.)	11	12			1.1 (avg)

[1] FIAT Report No. 1035.
[2] See Ref. 6.
[3] C. R. Downs, *Chem. Eng.*, August, 1948, p. 115.
[4] A. J. P. Wilson, *Chem. Eng.*, July, 1951, p. 287.
[5] "Chemical Engineering Flow Sheets," McGraw-Hill Book Company, Inc., New York, 1944.
[6] *Petr. Proc.*, **7**: 839 (1952).
[7] Ammonia at $80 per ton.
[8] Includes amortization.
[9] 10,000 bbl/day.
[10] 50,000 bbl/day.

SOURCE: H. E. Wessel, *Chem. Eng.*, **60**: 1–171 (1953).

Raw Material Costs

The costs of raw materials and the value of by-products and products must be determined from statistics in the company office or from a study of long-time price curves from such sources as *U.S. Census of Manufactures, Chemical Engineering, Chemical and Engineering News, Industrial and Engineering Chemistry, Chemical Week, Oil, Paint and Drug Reporter,* and the daily papers.

For firm production cost estimates, the prices listed in these journals should be verified by consulting the suppliers to make sure that they apply to the quality and quantity of the individual chemicals required in the projected manufacturing operations. Prices for chemicals not listed in the journals should be obtained from the suppliers. Current market prices of chemicals the company purchases for its manufacturing operations as well as interdepartmental or divisional transfer prices of chemicals produced and consumed within the company may be obtained from company files. Such price information is not available to individuals outside the company. Credit for by-products varies with each particular case. Generally, by-products are credited at their sales price minus additional selling, shipping, and purification costs, if any.

Utilities

The auxiliary process requirements are listed as utilities under item 6 in Tables 6-8 and 6-9. These may be supplied in several ways, with charges made according to various accounting procedures:

1. Purchased service from noncompany facilities at a fixed rate, e.g., a public utility or adjacent plant with central station facilities.
2. Utility supplied from company-owned central station facilities.
3. Service may be generated at the site and used for only the one process. This must be included as a part of capital cost and operating expenses accounted accordingly.

The cost of utilities will show considerable variation so that a check in established rates at the plant location site will be necessary for accurate cost accounting. For preconstruction cost estimating, the rates shown in Tables 6-11 and 6-13 will suffice. A brief discussion of the most important utilities follows with a more adequate coverage included in Chap. 7.

Water. Process water is used in chemical reactions and in washing, extracting, dissolving, and similar processing operations, for drinking, for sanitary facilities, and for general cleanup and washing. Estimates should be generous for the latter requirements; requirements are given for 100 industries in Fig. 6-32. Fresh water, treated or untreated well or city water, and in some cases distilled or deionized water may be

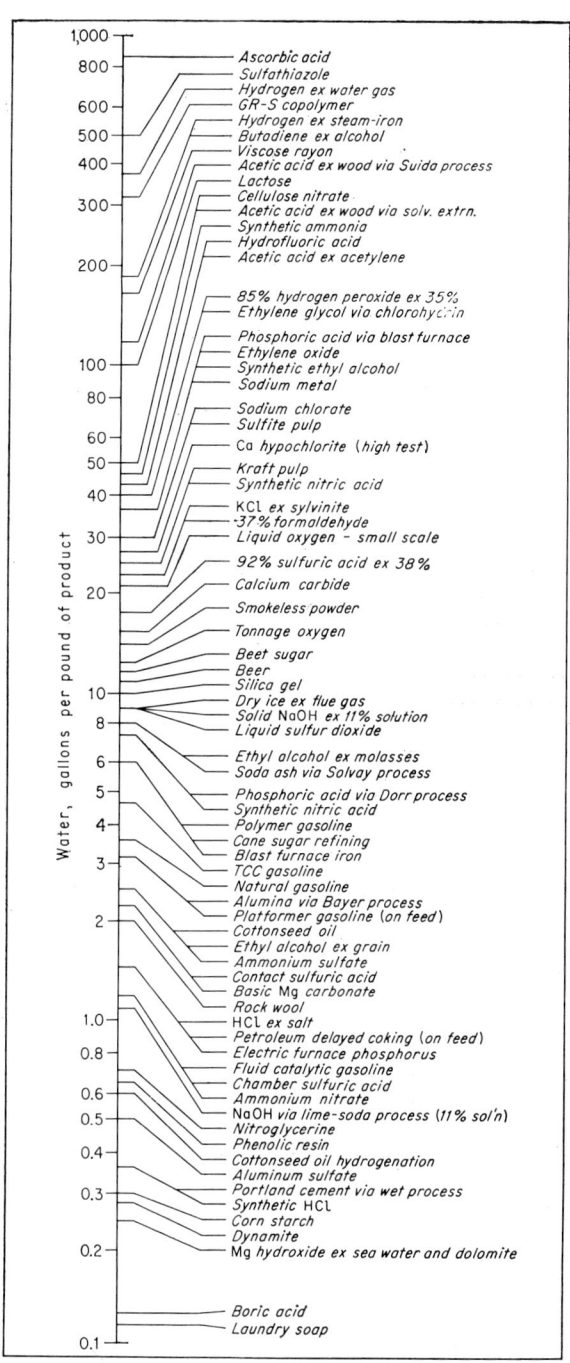

FIG. 6-32. Process water requirements for chemical industries. [*Courtesy of C. H. Chilton, Chem. Eng.*, **58**(4): 111 (1951).]

required. The cost of process water is generally within the range of $0.02 to $0.25 per 1,000 gal. Cost of deionizing water is comparable to the cost of distilling, which approximately doubles the cost of normal process water.

The total annual quantity of *cooling water* is summarized and this is multiplied by a factor of 1.10 to 1.50 to allow for losses and contingencies.

TABLE 6-11. ESTIMATION RATES FOR UTILITIES*

Type of utility	Rate
Steam:	
400 psi	$0.50–$0.90/1,000 lb
100 psi	0.25–0.70/1,000 lb
Exhaust	0.15–0.30/1,000 lb
Electric power:	
Purchased	0.01–0.02/kwhr
Self-generated	0.005–0.01/kwhr
Cooling water:	
Well	0.02–0.10/1,000 gal
River or salt	0.01–0.04/1,000 gal
Tower	0.01–0.05/1,000 gal
Process water:	
City	0.07–0.25/1,000 gal
Well	0.02–0.10/1,000 gal
Filtered and softened	0.10–0.20/1,000 gal
Distilled	0.60–1.00/1,000 gal
Compressed air:	
Process air	0.015–0.03/1,000 ft^3
Filtered and dried for instruments	0.04–0.10/1,000 ft^3
Coal	6.00–10.00/ton
Fuel oil, No. 6	0.04–0.08/gal
Gas:	
Natural	0.30–0.80/1,000 ft^3
Manufactured	0.60–1.30/1,000 ft^3
Refrigeration:	
Steam-jet, 50°F	0.55/ton-day
Ammonia, 34°F	0.50/ton-day
Ammonia, 0°F	0.90/ton-day
Ammonia, −17°F	1.20 ton-day

* R. S. Aries and R. D. Newton, "Chemical Engineering Cost Estimation," McGraw-Hill Book Company, Inc., New York, 1955.

Cooling-water costs vary greatly, depending on the geographical location, water-treating problems, pumping costs, etc. (see Chap. 7).

Steam. Process steam requirements may constitute a major item of costs, varying with the chemical process as shown in Fig. 6-33. But to process steam must be added the requirement for building heat; a factor of 1.25 is used to account for the sum of process steam and building heat. To allow for radiation, line losses, building heat, and contingencies, the

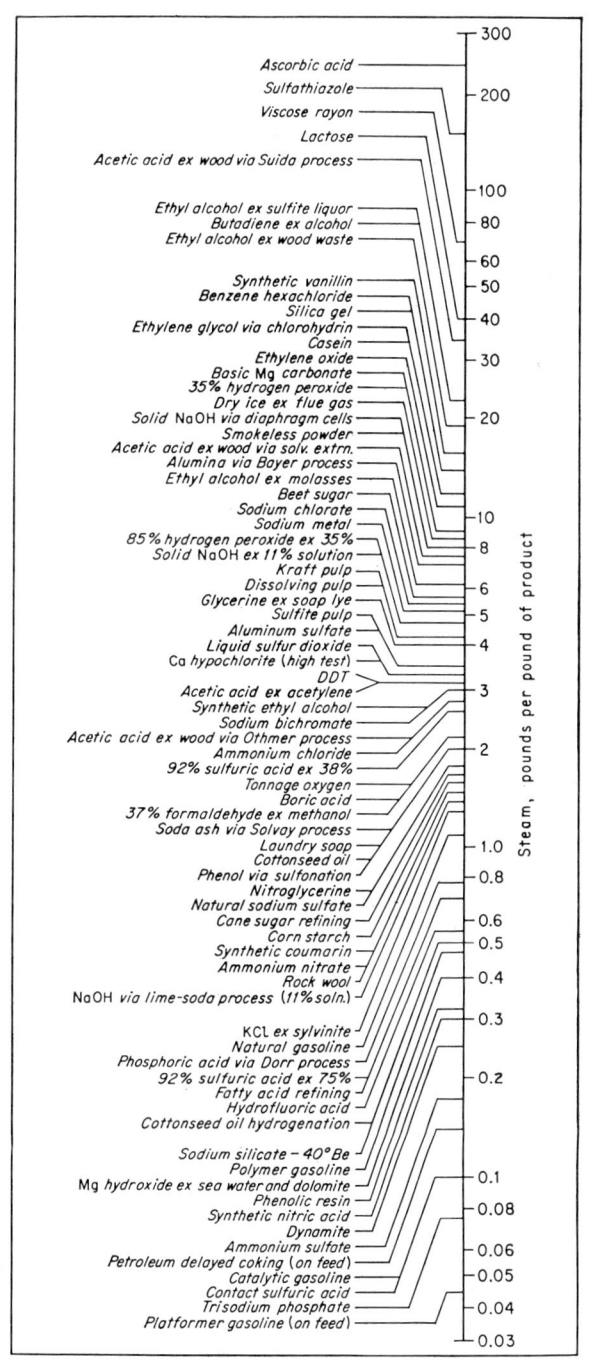

FIG. 6-33. Process steam requirements for chemical industries. [*Courtesy of C. H. Chilton, Chem. Eng.*, **58**(4): 111 (1951).]

230 CHEMICAL ENGINEERING PLANT DESIGN [CHAP. 6

factor of 2.00 is applied to the process heat requirements. A cost calculating chart is given in Table 6-12.

For capital investment purposes, a basis of $5 to $10 per 1,000 lb each hour is used for steam generation and distribution facilities.

TABLE 6-12. COST OF 1,000,000 BTU IN VARIOUS FUELS

Fuel	Heat content of fuel	Purchase price of fuel	Cost of 1,000,000 Btu
Fuel oil.........	152,000 Btu/gal	$0.023 per gal	$0.15
		0.030	0.20
		0.038	0.25
		0.046	0.30
		0.061	0.40
		0.091	0.60
Fuel gas........	1,000 Btu/ft^3	0.15 per 1,000 ft^3	0.15
		0.20	0.20
		0.25	0.25
		0.30	0.30
		0.40	0.40
		0.60	0.60
Fuel gas........	600 Btu/ft^3	0.25	0.15
		0.33	0.20
		0.42	0.25
		0.50	0.30
		0.67	0.40
		1.00	0.60
Coal...........	13,000 Btu/lb	3.90 per ton	0.15
		5.20	0.20
		6.50	0.25
		7.80	0.30
		10.40	0.40
		15.60	0.60
Steam..........	1,200 Btu/lb	0.18 per 1,000 lb	0.15
		0.24	0.20
		0.30	0.25
		0.36	0.30
		0.48	0.40
		0.72	0.60

Electric Power. Power requirements for process equipment and motors in specific chemical industries are summarized in Fig. 6-34. To allow for line losses and contingencies, these data should be multiplied by 1.11 to 1.25. The industrial power rates vary regionally and locally in the United States. Table 6-13 is a compilation of the average costs in individual states. In general, industrial power rates range from $0.008

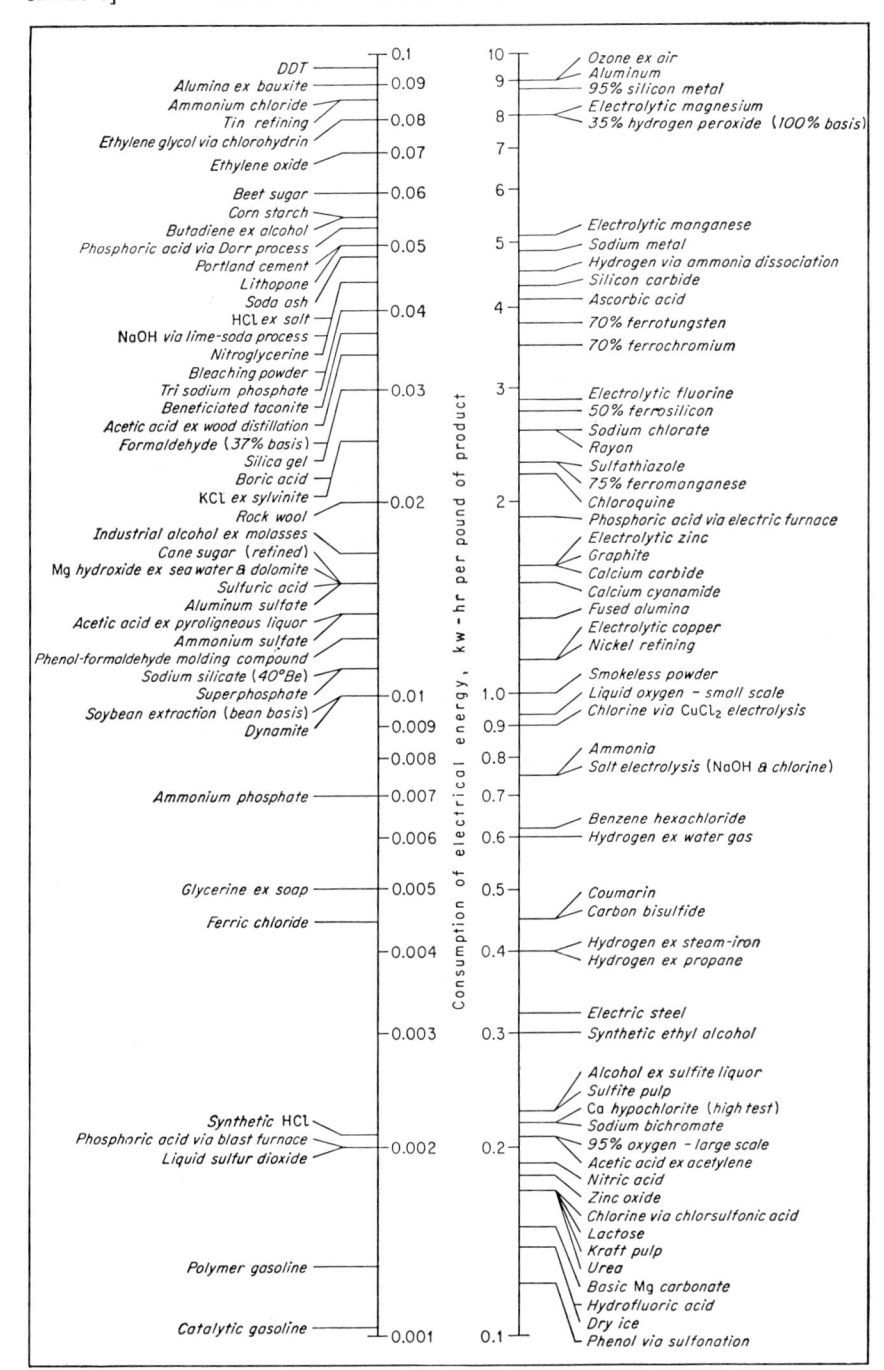

FIG. 6-34. Process power requirements for chemical industries. [*Courtesy of C. H. Chilton, Chem. Eng.*, **58**(3): 115 (1951).]

for locations having cheap power to $0.02 per kwhr in higher areas. For capitalization purposes $100 to $175 per kwhr used in 1 hr under average operating conditions is the share estimated for electric power generation and distribution facilities.

TABLE 6-13. ELECTRICAL POWER CHARGES BY REGION AND DEMAND FOR YEAR 1957

A. *Average power charges by region*
(Statistics for industries >50 kw demand)*

Location	Average charge, cents per kilowatt hour
United States	0.92
New England: Maine, New Hampshire, Vermont, Massachusetts, Rhode Island, Connecticut	1.72
Middle Atlantic: New York, New Jersey, Pennsylvania	1.21
East North Central: Ohio, Indiana, Illinois, Michigan, Wisconsin	1.02
West North Central: Minnesota, Iowa, Missouri, North Dakota, South Dakota, Nebraska, Kansas	1.31
South Atlantic: Delaware, Maryland, District of Columbia, Virginia, West Virginia, North Carolina, South Carolina, Georgia, Florida	1.13
East South Central: Kentucky, Tennessee, Alabama, Mississippi	0.51
West South Central: Arkansas, Louisiana, Oklahoma, Texas	0.93
Mountain: Montana, Idaho, Wyoming, Colorado, New Mexico, Arizona, Utah, Nevada	0.75
Pacific: Washington, Oregon, California	0.69

B. *Average power charge by demand and monthly consumption: United States average†*

Demand, kw	150	300	1,000	
Consumption, kwhr	30,000	60,000	200,000	400,000
Total charge, ¢/kwhr	2.07	1.87	1.64	1.07

* Abstracted from *Statistical Bulletin* 25, 1958, Edison Electric Institute, New York, N.Y.

† *Federal Power Commission Bulletins* R16 and R54, 1957, Washington, D.C.

Refrigeration. Refrigeration costs vary widely and depend on a number of factors, chiefly power, cooling water, depreciation, and repairs. Because of the interdependence of these factors and their variation from one installation to another it is necessary to estimate refrigeration costs for each installation if a reliable cost is required. The following approximate refrigeration costs are given for single-stage ammonia compression systems of roughly 10 to 500 tons in size, based upon power at $0.01 per kwhr, and 85°F cooling water at $0.02 per 1,000 gal. At evaporator temperatures of 34, 0, and −17°F the costs per ton-day approximate $0.47, $0.82, and $1.07, respectively. Steam jet refrigeration is widely used to obtain 50°F cooling water. If steam is taken at $0.25 per 1,000 lb and 85°F cooling water at $0.02 per 1,000 gal, cost of steam jet refrigeration will be approximately $0.52 a ton-day. Steam jet refrigeration is especially advantageous with cheap low-pressure exhaust steam and cooling water available. It is convenient to include the cost of refrigeration units in the manufacturing equipment and to charge the power, cooling water, repairs, and depreciation pertaining to refrigeration units under these items in the conversion cost, if refrigeration costs for the particular plant location are not available.

Installed costs for refrigeration on a ton-capacity basis can be obtained from Aries and Newton. The listed costs of equipment per ton of refrigeration effect are based upon the complete plant, with low-side equipment, ready to run, including reciprocating or centrifugal compressors, motor, starter, receiver, water and brine coolers, conditioner, water cooling coils, etc. If an evaporative condenser instead of a water-cooled condenser is used, add $35 per ton; for cooling tower, add $60 per ton.

Cold-storage holding costs can be estimated as follows:

For cold-storage holding		Dollars per cubic foot		
Capacity, ft³	Temperature, °F	Building	Insulation	Refrigeration equipment
>50,000	35	0.45	0.37	0.22
>25,000	0	0.45	0.50	0.38
>25,000	−20	0.45	0.67	0.53

Inert Gas. Nitrogen, carbon dioxide, argon, and helium are examples of inert process gas. Costs of commercially pure nitrogen delivered in trailers to compressed cylinder storage racks are:

Usage, cubic feet per month	Cost of nitrogen, dollars per 1,000 ft³
0–80,000	8.50
80,000–150,000	8.00
150,000–	7.50

Carbon dioxide can be purchased in cylinders for $3.80 per 100 lb compared to $1.30 per 100 lb if the inert gas is provided by combustion of natural gas.

Miscellaneous. Materials to be considered in cost estimation under the miscellaneous category are factory supplies, including such items as gaskets, lubricants, paint, test chemicals, janitor supplies, rags, etc. Unless demands for any of these items are excessively high, a cost allocation of 0.5 to 1.0 per cent of the plant machinery and equipment cost should be adequate.

Labor Costs

The cost of operation of chemical process equipment varies widely. Some equipment requires more skill, operators of better technical training, and more supervision than others (see Table 6-14). Engineering experience must be relied on to furnish most of this information. Available within each company are data on equipment operating labor costs, with typical values of 0.2 man per unit for continuous filters, evaporators, and columns, 0.5 man per unit for centrifuges, dryers, and continuous reactors, and 1.0 man per unit for batch reactors.

TABLE 6-14. TYPICAL JOB EVALUATIONS FOR CHEMICAL OPERATIONS*

Job title	Total points	Skill				Mental		Physical			Work		
		Mechanical ability	Dexterity	Application	Initiative	Material responsibility	Equipment responsibility	Application	Monotony	Working conditions	Hazards		
Reactor operation.........	19	2	0	5	3	5	2	0	0	1	1
Filter-dryer operator......	6	1	0	2	1	0	1	0	0	1	0
Still operator.............	19	3	0	5	4	5	1	0	0	1	0
Packer operator..........	6	0	1	0	1	1	0	2	1	0	0
Warehouseman...........	5	0	0	1	0	0	0	3	0	0	1
Furnace operator.........	13	4	1	2	1	2	2	0	0	1	0

* Job Evaluation, *Chem. Inds.*, **60**(5): 785 (1947).

There are several methods available for estimating operating labor. A practical experience method, also called rule of thumb, is to use one operating man for every ten process instruments in a well-instrumented and controlled process. A more accurate estimating method was devised by Wessel[1] and graphed in Fig. 6-35. The method is based on adding up the various principal processing steps on the flow sheets. A process step is defined as any unit operation, unit process, or combination thereof, which takes place in one or more pieces of integrated equipment on a repetitive cycle or continuously, e.g., reaction, distillation, evaporation, drying, filtration, and packaging. Take this total number of steps and multiply it by the daily man-hours per step at the specified daily plant

[1] H. E. Wessel, *Chem. Eng.*, **59**(7): 209 (1952).

product capacity as read from Fig. 6-35. This gives direct man-hours per ton. Chilton[1] summarizes direct man-hours per ton of operating capacity for 54 commercial processes with values ranging from 0.2 for tonnage oxygen and 0.25 for catalytic gasoline to 80 for cellulose acetate and coumarin production plants. A more precise estimate of the actual number and type of supervisors, repairmen, plant operators, clerks, etc.,

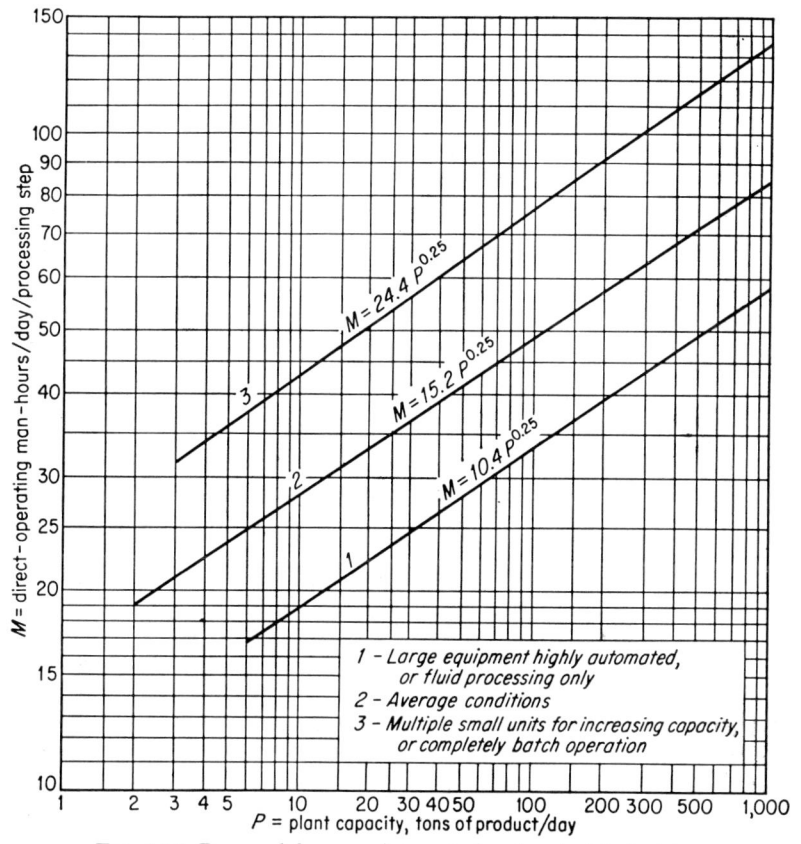

FIG. 6-35. Process labor requirements for chemical industries.

must be determined from the final operating instructions and preliminary time studies.

In determining labor costs, local conditions prevail, but some average must be obtained. A request to the local chamber of commerce will generally yield data on prevailing rates. Information from similar plants, labor agreements, job evaluations, and pilot plant data can be made the bases for labor estimates. Data on labor rates in each industry

[1] C. H. Chilton, *Chem. Eng.*, **58**(2): 151 (1951).

and craft can be obtained from labor boards in each region of the United States and from the Bureau of Labor Statistics, Washington, D.C. Table 6-15 gives data on hourly earnings in some chemical plants for the year 1955.

Approximate rates for preconstruction cost estimation purposes in 1958 are as follows:

	Rates per hour
Leaders and foremen	$2.65–$4.00
Skilled labor	$2.00–$2.70
Unskilled labor	$1.40–$1.90

An estimate is made of the number of hours of *supervision* required in normal manufacturing operations. The extra supervision needed during initial runs with the equipment is generally charged to start-up expenses. A supervisor's salary may range from $6,000 to $8,000 per year.

Payroll charges should cover such items as social security, vacations, pensions, and compensation insurance. They may be estimated as a percentage of labor plus supervision; the usual range is from 12 to 20 per cent of the sum of these items.

An estimate is made of the amount of *analytical work* that may be required and the time needed to complete this work. The average salary of an analyst for the time required plus 100 per cent for overhead will be a satisfactory charge in a preliminary estimate.

Indirect Labor Charges. The indirect expenses of employee service facilities are watchmen, locker rooms, dispensary, employment office, payroll preparation, cafeteria, yard, maintenance, general superintendence, purchasing and traffic, parking spaces, telephone service, etc. These costs are based upon direct labor and should be charged on the basis of labor costs. For preliminary estimation, the controllable indirect conversion expense can be approximated by either 40 to 60 per cent of the direct labor or 15 to 30 per cent of the direct conversion expense. Which of the two procedures will be used in a particular case depends upon the judgment of the estimator. A charge of $1 per week should be made for each man engaged in operating labor for clothing and laundry.

Maintenance and Repairs

Repairs. For preliminary estimates, repairs can be taken as a percentage of the installed plant machinery and equipment cost. Maintenance charges range from 2 to 10 per cent of the plant costs on an annual basis which corresponds to 1 to 4 per cent of the manufacturing costs per unit. A consideration of the type of equipment and operation involved will show whether repair charges should be high, low, or average.

TABLE 6-15. EARNINGS FOR WORKERS IN CHEMICAL INDUSTRIES, 1955
(Average straight-time hourly earnings* of workers in selected production occupations, United States and regions)

Number of workers and average hourly earnings in

Occupation†	United States		New England		Middle Atlantic		Border States		Southeast	
All production workers	153,647	$2.07	5,935	$1.85	44,887	$2.01	24,908	$2.03	5,047	$1.76
Selected production occupations:										
Carpenters, maintenance	1,443	2.36	47	2.09	400	2.35	252	2.22	54	2.16
Carboy fillers	121	1.79	—	—	29	1.88	—	—	—	—
Chemical operators, class A	17,001	2.29	702	1.99	4,718	2.27	2,562	2.34	358	1.98
Chemical operators, class B	16,447	2.14	353	1.82	4,749	2.00	2,668	2.21	680	1.87
Chemical operators' helpers	8,697	1.90	417	1.73	2,916	1.82	987	1.91	376	1.63
Compressors	735	2.31	—	—	46	2.15	140	2.33	—	—
Cylinder fillers	152	2.01	—	—	34	2.11	—	—	—	—
Drum fillers	458	1.92	—	—	184	1.89	8	2.02	27	1.35
Electricians, maintenance	2,764	2.41	55	2.17	633	2.36	420	2.31	84	2.21
Filling-machine tenders	856	1.89	—	—	195	1.89	140	1.72	20	1.41
Guards	1,733	2.02	69	1.83	458	1.90	273	2.04	51	1.86
Helpers, trades, maintenance	5,786	1.99	66	1.88	806	1.87	1,395	1.99	196	1.66
Janitors	3,183	1.74	162	1.69	864	1.68	570	1.70	88	1.47
Janitors (women)	249	1.66	—	—	103	1.65	13	1.49	—	—
Laboratory assistants	5,098	2.10	256	1.92	1,942	2.13	924	2.16	146	1.71
Laboratory assistants (women)	965	1.87	—	—	404	1.95	186	1.80	—	—
Laborers, material handling	8,267	1.67	235	1.69	2,006	1.68	1,544	1.55	630	1.41
Lead burners	221	2.58	—	—	66	2.63	38	2.56	—	—
Machinists, maintenance	2,745	2.41	56	2.22	564	2.32	467	2.29	80	2.10
Millers, class A	435	1.95	—	—	132	2.06	95	1.63	24	1.73
Millers, class B	642	1.82	—	—	244	1.75	100	1.57	30	1.60
Mixers, class A	658	1.94	—	—	265	1.87	34	1.71	64	1.34
Mixers, class B	992	1.80	—	—	548	1.81	92	1.60	74	2.15
Pipefitters, maintenance	4,231	2.41	108	2.18	1,119	2.37	731	2.35	29	1.94
Pumpmen	641	2.12	—	—	108	2.01	71	1.93	45	1.75
Stock clerks	1,405	2.07	35	1.84	337	1.96	247	2.03	95	1.60
Truckdrivers‡	1,915	2.03	84	1.88	444	2.14	351	2.00	—	—
Light (under 1½ tons)	400	1.99	—	—	68	1.92	56	1.88	—	—
Medium (1½ to and including 4 tons)	975	2.00	—	—	153	2.01	248	2.07	54	1.47
Heavy (over 4 tons, trailer type)	345	2.23	28	2.06	166	2.34	22	2.00	—	—
Heavy (over 4 tons, other than trailer type)	195	1.93	—	—	57	2.13	—	—	—	—
Truckers, power (forklift)	1,194	1.86	66	1.83	458	1.86	168	1.70	12	1.81
Truckers, power (other than forklift)	192	1.96	—	—	43	1.99	35	1.91	55	1.28
Watchmen	676	1.71	—	—	185	1.65	111	1.65	41	1.18

237

TABLE 6-15. EARNINGS FOR WORKERS IN CHEMICAL INDUSTRIES, 1955 (*Continued*)

Occupation	Number of workers and average hourly earnings in									
	Great Lakes		Middle West		Southwest		Mountain		Pacific	
All production workers	31,720	$2.10	4,623	$2.02	27,603	$2.25	2,593	$2.27	6,331	$2.15
Selected production occupations:										
Carpenters, maintenance	267	2.34	36	2.31	284	2.55	29	2.71	74	2.38
Carboy fillers	—	—	—	—	—	—	—	—	7	2.01
Chemical operators, class A	3,452	2.19	914	2.15	3,286	2.51	—	—	964	2.29
Chemical operators, class B	2,981	2.05	418	1.94	3,580	2.43	333	2.43	685	2.13
Chemical operators' helpers	1,828	1.92	245	1.88	1,392	2.11	113	2.08	423	2.00
Compressors	91	2.30	45	2.01	384	2.38	—	—	20	2.27
Cylinder fillers	21	2.00	—	—	—	—	—	—	20	2.12
Drum fillers	108	1.98	15	1.75	46	2.07	—	—	47	2.10
Electricians, maintenance	595	2.36	61	2.31	687	2.59	135	2.52	94	2.43
Filling-machine tenders	243	2.00	69	1.88	68	1.73	—	—	86	1.96
Guards	387	2.07	—	—	391	2.15	38	2.08	29	1.84
Helpers, trades, maintenance	1,340	2.00	146	1.87	1,582	2.07	75	2.31	180	2.08
Janitors	831	1.86	53	1.71	493	1.72	43	1.85	79	1.89
Janitors (women)	109	1.75	—	—	10	1.53	—	—	—	—
Laboratory assistants	761	2.00	169	2.03	716	2.23	40	2.15	144	2.15
Laboratory assistants (women)	166	1.65	—	—	121	2.02	—	—	—	—
Laborers, material handling	1,285	1.85	144	1.70	2,078	1.65	73	1.87	272	1.89
Lead burners	68	2.49	—	—	19	2.55	—	—	7	2.67
Machinists, maintenance	385	2.37	67	2.29	909	2.58	52	2.64	165	2.43
Millers, class A	130	2.05	—	—	32	2.04	—	—	15	2.07
Millers, class B	81	1.90	21	1.69	—	—	40	2.42	—	—
Mixers, class A	106	2.06	—	—	—	—	—	—	—	—
Mixers, class B	166	2.02	12	1.67	—	—	—	—	27	2.15
Pipefitters, maintenance	817	2.36	—	—	1,088	2.57	34	2.69	144	2.44
Pumpmen	148	2.10	25	1.97	210	2.31	—	—	26	2.08
Stock clerks	275	2.07	41	2.01	339	2.27	52	2.14	33	1.98
Truckdrivers‡	405	2.05	27	1.80	409	2.31	36	2.07	64	2.17
Light (under 1½ tons)	70	1.91	—	—	—	—	—	—	—	—
Medium (1½ to and including 4 tons)	238	2.03	24	1.77	196	2.01	24	2.26	23	2.05
Heavy (over 4 tons, trailer type)	73	2.23	—	—	29	2.04	—	—	30	2.27
Heavy (over 4 tons, other than trailer type)	24	2.11	—	—	105	1.98	—	—	—	—
Truckers, power (forklift)	215	1.99	59	1.94	12	1.70	—	—	68	2.08
Truckers, power (other than forklift)	53	2.04	—	—	47	1.76	—	—	20	2.10
Watchmen	199	1.95	26	1.45	—	—	—	—	53	1.67

* Excludes premium pay for overtime and for work on week ends, holidays, and late shifts. (Add 5–8¢/hr on second shift; 10–11¢/hr on third shift.)
† Data limited to men workers except where otherwise indicated. ‡ Includes all drivers regardless of size and type of truck operated.
Note: Dashes indicate no data or insufficient data to warrant presentation. SOURCE: Bureau of Labor Statistics, *BLS Report* 103.

Repairs on buildings may be about 5 per cent of their cost. For instruments, annual charges of 25 per cent of the original costs should suffice.

A large share of the maintenance required to keep a unit functioning is due to mistakes and accidents. Vessels, piping and tubing obsolescence, and pump maintenance are items of maintenance cost that can be approximated; even here, extra allowance should be made if the service is severe.

Leonard concludes from Table 6-16, wherein annual maintenance costs of specific equipment are listed, that the maintenance angle should be considered as fully as the operating aspects. A review of the preliminary plant design by a maintenance expert can save many dollars in the course of the operating life of the plant.

Common Denominator for Repair Costs.[1] A formula for budgeting repair costs should include two constants: repair labor index and repair material index; and two variables: pay with overhead per man-hour and kilowatthours consumed.

Such a formula may be written

$$c = x(a + by) \qquad (6\text{-}5)$$

where c = cost, dollars per year
x = kwhr per year ÷ 1,000
y = cost in dollars per man-hour with overhead
a = repair material index, as dollars per 1,000 kwhr
b = repair labor index as man-hours per 1,000 kwhr

Roughly, actual true repairs average 25 to 65 per cent, while the remainder is for repairs of supports, auxiliaries, etc.

Depreciation

Depreciation is the unavoidable loss in value of plant, equipment, and materials with lapse in time, caused by:

1. Chemical action or corrosion
2. Physical action:
   a. Decay
   b. Decrepitude
   c. Abrasion
   d. Normal wear
   e. Deferred maintenance or repair
3. Inadequacy
4. Obsolescence

In chemical plants obsolescence is high; process equipment rapidly declines in economic value relative to newer and more efficient alternatives available. The replacement of equipment or processes can be justified

[1] D. E. Pierce, A Common Denominator for Repair Costs, *Chem. Eng. Progr.*, **44**: 252 (1948).

TABLE 6-16. ANNUAL MAINTENANCE COSTS*

(Dollars per year)

Centrifugals:	
Continuous horizontals	
Small (24 in.)	$500
Large (54 in.)	3,000
Basket, underdriven and suspended	
Small (26 in.)	250
Large (60 in.)	400–500
Conveyors:	
Belt, pan, and roller per 100 ft	100
Screw per 100 ft	200
Moving slat per 100 ft	150
Bucket drag per 100 ft	1,200
Crystallizer:	
Batch, tank, per 2,000-gal unit	200
Continuous, rubber-lined 20- × 1½- × 1½-ft unit	100
Clarifiers, batch:	
Continuous thickeners	
Noncorrosive service, 20,000 gal	500–800
Corrosive service, 20,000 gal	2,000–3,000
Collectors, dust and mist	
Electrical precipitators, 10,000–15,000 cfm	5,000–8,000
Packed towers, per 1,000 cfm	100
Baffle spray, per 1,000 cfm	400–500
Bag filters, per 1,000 cfm	200–250
Cyclone, per 1,000 cfm	200–300
Distillation columns:	
Bubble plate, per ft^3 column space	5–50
Sieve plate, per ft^3 column space	3–10
Packed column, per ft^3 column space	5–15
Dryers:	
Continuous belt	
Short, 8 ft wide × 30 ft long	300
Long, 8 ft wide × 70 ft long	500
Continuous rotary, 6 ft dia. × 30 ft long	100–300
Spray	
Small	150
Long	350
Drum, per ft^2	3–4
Tray	
Atmospheric, per ft^2	1–2
Vacuum, per ft^2	2–3
Evaporators:	
Multiple effect, per 1,000 ft^2	500–600
Batch, atmospheric and vacuum per 1,000 ft^2	200–1,500
Pan, atmospheric and vacuum per 1,000 ft^2	800–1,000
Filters:	
Rotary vacuum, rubber and alloy units, per ft^2	9.00
Wooden units, per ft^2	11.50

* J. D. Leonard, *Chem. Eng.*, **58**(9): 149 (1951).

TABLE 6-16. ANNUAL MAINTENANCE COSTS (*Continued*)

(Dollars per year)

Pressure	
Large, per ft²	1.60
Small, per ft²	3.00
Intermittent vacuum	
Rubber and alloy, per ft²	2.00
Wooden, per ft²	4.25
Plate and frame	
Rubber and alloy, per ft²	0.85
Wooden, per ft²	1.10
Furnaces and kilns:	
Continuous rotary	
9-ft dia. severe hot end, per 100 ft	3,500–4,000
Plus severe cold end	8,000–10,000
Batch rotary, 10 ft dia. × 20 ft long	4,000–8,000
Mannheim for HCl, 10 tons per day	3,000
Nitric acid pan, 8 × 12 ft	800
Herreshof burner, 18 ft dia. × 20 ft high	3,000
Heat exchangers:	
Calandrias, per ft² surface	0.50–5.00
Condensers, per ft² surface	0.30–2.00
Jacketed pipe exchangers, per ft²	0.75–3.00
Falling film condenser, per ft²	0.20–0.75
Mixers and blenders:	
Continuous	
Light, per 50 ft³ capacity	100–200
Medium, per 50 ft³ capacity	200–300
Heavy, per 50 ft³ capacity	400–500
Batch	
Light, per 50 ft³ capacity	150–250
Medium, per 50 ft³ capacity	300–400
Heavy, per 50 ft³ capacity	500–600
Pumps:	
Centrifugal	
30 gpm 50 head ft	75–90
30 gpm 100 head ft	150–175
100 gpm 50 head ft	100–125
100 gpm 100 head ft	200–225
200 gpm 100 head ft	225–250
Diaphragm	
Up to 20 gpm	400–425
20–40 gpm	400–600
40–80	500–800
Piston	
Up to 20 gpm	125–150
20–40 gpm	150–175
40–80 gpm	200–250
Size reduction:	
Jaw and gyratory crushers	
Small	70
Large	400

242   CHEMICAL ENGINEERING PLANT DESIGN   [CHAP. 6

TABLE 6-16. ANNUAL MAINTENANCE COSTS (*Continued*)

(Dollars per year)

Pan-roller mills	
Small	2,000
Large	7,000
Jaw and gyratory crushers	
On TiO_2 per ton handled	0.60–0.71
On alum cake per ton handled	0.02
Ball-and-tube mills, per ton handled	0.08–0.12
Micropulverizers, per ton handled	0.04–0.07
Other impact mills, per ton handled	0.10–0.18
Screeners, vibrating:	
Dry, electrical, per ft^2 of screen	10–14
Mechanical, per ft^2 of screen	14–20
Wet, electrical, per ft^2 of screen	150–3,000
Mechanical, per ft^2 of screen	300–5,000
Mechanical sifter, per ton handled	0.10–0.20
Tanks, weak acid up to 50°C:	
Wood	
Small, per 1,000 gal	15
Large, per 1,000 gal	10
Rubber-lined	
Small, per 1,000 gal	30
Large, per 1,000 gal	40
Stainless steel, per 1,000 gal	5–10
Up to boiling	
Brick-lined, per 1,000 gal	25–40
Lead-lined, per 1,000 gal	200–250
Haveg, per 1,000 gal	5–10
Neutral or alkaline solutions, per 1,000 gal	10–15
Glass-lined, with agitator, per 1,000 gal	1,000
For storage, per 1,000 gal	800
Special, stainless steel, concrete, aluminum, per 1,000 gal	3–6
Plastic-lined	50–60
Piping, process, 1–6 in.:	
Mild steel, per 100 ft	10–50
Cast iron, per 100 ft	6–25
Brass, per 100 ft	8–35
Stainless steel, per 100 ft	7–28
Rubber-lined, per 100 ft	20–70
Glass, per 100 ft	22–80
Porcelain, per 100 ft	25–100
Carbon, per 100 ft	30–80
Silver, per 100 ft	50–90
Rubber hose, per 100 ft	6–36
Piping, service, 1–12 in.:	
Uninsulated, mild steel, per 100 ft	4–60
Cast iron, per 100 ft	5–42
Brass, per 100 ft	6–27
Insulated, mild steel, per 100 ft	7–11

economically when the savings in operating costs over the same period of time exceeds the cost of installing new equipment. Often new equipment will produce better products at lower costs. The cost of operating proposed facilities or procedures can be predicted accurately from suitable calculations and from experimental investigations. Pressure for better products at lower costs forces continual modernization of chemical plants. Reserves should be set aside out of current earnings to take care of technological antiquation of machines. This is depreciation.

From the probable useful life and from due consideration of obsolescence and economic life, the salvage values of the various types of equipment are established. Depreciation rates on equipment and machinery as fixed by the U.S. Bureau of Internal Revenue for tax purposes are presented in Table 6-17. A company may establish its own depreciation rates, but these must be approved by the U.S. Bureau of Internal Revenue. For preliminary estimates a rate equal to 10 per cent per year may be used for equipment and 3 per cent per year for masonry buildings and 4 to 5 per cent for frame buildings.

Since depreciation is a major fixed charge in establishing total product costs, the design engineer should become familiar with current Federal government regulations and internal accounting procedures in this area. Most chemical companies have recognized the advantages of increased depreciation allowances during the early years of a new process installation and pioneered internal company methods to accomplish this goal. It is possible for a plant to meet competition and still use a large write-off during the early years when prices are likely to be high and maintenance costs low. Replacement reserves thus grow rapidly at first, giving an early high recovery of the investment, yet profits and taxes are kept at an even level throughout the life of the process since prices generally decline in succeeding years. Under this system the cost of products during the first 3 to 5 years will be higher and in the next 5 years the cost will be lower than that computed by a straight-line method (e.g., one-tenth of the investment recovered each year for 10 years).

Methods of Determining Depreciation. There are nine principal methods of determining depreciation allowances which can be arbitrarily divided into two groups based on whether or not interest on investment is allowed. These methods and useful formulas are tabulated next. Consistent nomenclature is used throughout.

Group A (no interest on investment allowed):

1. Straight-line depreciation: Equal amounts are charged off over the useful life of the equipment.

$$d = \frac{V_o - V_s}{nV_o} \times 100 \qquad (6\text{-}6)$$

TABLE 6-17. PROBABLE ANNUAL DEPRECIATION RATES FOR CHEMICAL PROCESS EQUIPMENT

	Depreciation rate, per cent		Depreciation rate, per cent
Acids:		Chemical ware	50
Acetic:		Oil	5
Blow cases, cast iron and copper	33⅓	Water	25
Columns, fractionating	12½	Pots, condensing (earthenware)	14
Condensers:		Pumps and blowcases:	
Copper	10	Chemical ware lined	33⅓
Duriron	7	Rubber-lined blowcase	20
Lead	16⅔	Storage tanks (wooden, rubber lined)	7
Motors	7	Tanks, sulfuric acid storage (steel)	5
Pipes:		Tourills (silica)	10
Aluminum	33⅓	Towers, absorbing	10
Glass	20	Nitric:	
Glacial:		Blowers (stoneware)	20
Copper	10	Blow cases (earthenware)	50
Rubber	12½	Condensers (duriron)	8⅓
Aqueous	10	Condensers, S-bend (stoneware)	40
Pots	6	Elevators and conveyors (screw)	10
Pumps, vacuum	14	Flues, gas (duriron)	12½
Receivers, acid (stoneware)	7	Pans, niter cake (steel)	7
Scrubbers (stoneware)	7	Pipes and fittings (earthenware, duriron, lead)	50
Receivers, acid, for product (stoneware)	5	Pumps, sulfuric (iron), centrifugal	20
Stills:		Receivers (stoneware)	20
Cast iron	8	Retorts, 24-hr service	40
Refining, copper	7	Tanks (steel)	10
Refining, heating coil	33⅓	Towers, condensing	11
Tanks, storage:		Sulfuric (chamber):	
Steel	8	Air lifts, acid	10
Wood	4	Blowers, gas (lead)	6
Muriatic:		Blowcases	10
Air lifts (hard rubber)	10	Chambers	6
Cars, tanks	10	Coolers, acid (lead coil), for salt water	10
Coolers	10	Fans (cast iron)	10
Elevators, bucket	10	Pipes (lead)	10
Exhausters (rubber lined)	12½	Pots, niter	5
Flues (earthenware)	10	Pumps, acid	20
Furnaces, Mannheim	12½	Tanks (steel), acid storage, average weak and strong acid	5
Furnaces, pot and muffle	10		
Furnaces, retort	12½		
Grinders and coolers, salt cake	8⅓		
Motors	7		
Pipes:			
Acid (hard rubber)	14		

TABLE 6-17. PROBABLE ANNUAL DEPRECIATION RATES FOR
CHEMICAL PROCESS EQUIPMENT (*Continued*)

	Depreciation rate, per cent		Depreciation rate, per cent
Tanks, tower, acid distributing	12½	Storage (steel)	6
Towers, Gay-Lussac	5	Tank cars (steel)	8
Towers, Glover	5	Towers:	
Sulfuric (contact):		Absorbing	11
Air lifts	7	Cooler, cold scrub	8⅓
Blowcases (cast iron and steel)	20	Dry	10
Blowers:		Oleum	10
Connersville	14	Scrub	10
Sturtevant	6	Transferrers	11
Burners:			
Brimstone	10	Pulp and Paper and Paper Board	
Glens Falls	10	Absorbing system, milk of lime	10
Herreshoff	6⅔	Barkers, drum	10
Wedge, salt-water cooled	6⅔	Hand	6⅔
Wylde	7	Beaters	5
Coke boxes	6	Bins, storage, chip	3⅓
Combustion chambers, brimstone	10	Bleachers	6⅔
Compressors, air	6⅔	Burners, sulfur, acid plant	8⅓
Contact mass, including plates and supports	6	Calenders	4½
Converters	7	Chippers	5
Conveyors and elevators	10	Conveyors, inside	6⅔
Coolers:		Outside	8⅓
Drying acid	10	Cookers	5
Gas	7	Coolers	10
Gas, tower	10	Cutters	5
Product	16⅔	Cylinder machines, for paper and paper board	5
Dust chambers (brick)	7	Deckers	5
Filters, preliminary	9	Diffusers	4
Flues (iron)	7½	Digester linings	14
Gauges, meters, pyrometers	7	Digesters:	
Heaters, preliminary	7	Indirect	4½
Melters, brimstone	10	Rotary	5
Motors	6	Vertical, stationary	4
Pipes, acid	10	Drainers	3⅓
Platinum, in catalyst	2	Evaporators:	
Pumps, acid (iron)	14	Disk	6
Pumps, acid (lead)	12½	Multiple effect	4
Separators	7	Fourdrinier machines	5
Sublimers, brimstone	10	Furnaces, rotary	6
Tanks:		Grinders	5
Roasted ore storage (steel)	5	Jordans	5½
Storage (lead)	5	Knotters	6⅔
		Kollergangs	5
		Linings for blow pits	12½

TABLE 6-17. PROBABLE ANNUAL DEPRECIATION RATES FOR
CHEMICAL PROCESS EQUIPMENT (*Continued*)

	Depreciation rate, per cent		Depreciation rate, per cent
Melters, sulfur	12½	Devulcanizers, reclaimed rubber	6⅔
Pans:		Dipping machines	10
Causticizing	6⅔	Disintegrators	6⅔
Wash	4	Drums	8⅓
Pits, blow:		Dryers	7
Concrete	3⅓	Dusting machines, including	
Steel tank	5	chalking	6⅔
Platers	7½	Furnaces	6⅔
Pumps:		Grinders, pigment	6⅔
Acid	20	Mills, mixing or warming:	
Centrifugal	6⅔	Heavy duty	6
Plunger, duplex or triplex	5½	Light duty	6⅔
Pressure	6⅔	Mixers:	
Vacuum	6⅔	Large	6
Reels	6⅔	Small	6⅔
Rifflers:		Ovens	6⅔
Concrete	3⅓	Presses, cold	5
Wood	5½	Pulverizers	6⅔
Save-alls	5	Reeling machines	6⅔
Screens:		Refiners, roll type	7
Silver	8	Rolling machines	6⅔
Rotary	6⅔	Sealing machines	7
Shredders	8⅓	Separators	7
Slashers	7	Sheeters	6⅔
Smelters, sulfate process	14	Sifters	7
Stackers, pulpwood	7	Strainers	8⅓
Tanks:		Varnishing machines	6⅔
Causticizing	5	Vulcanizers	6⅔
Leaching	4½	Washers	6⅔
Mixing (wood)	8½	Winding machines	6⅔
Mixing (wood) for clay	5	Wrapping machines	8
Storage, acid	8⅓		
Storage or washing (concrete)	3⅓	Cement, Ceramics, Glass, Gypsum, and Lime and Limestone	
Storage or washing (wood)	6⅔		
Thickeners	5	Agitators	6⅔
Towers, absorbing system	6⅔	Augers	7
Washers, bleach or paper stock	5	Baggers	6⅔
Wet machines	5½	Beaters, tub	6⅔
		Blowers	6⅔
Rubber Goods		Blungers	8⅓
Autoclaves	10	Bottle machines	10
Boards, stock (wood)	25	Brickmaking machines	8
Calenders	5½	Burners	6⅔
Conveyors	6⅔	Calciners, continuous	6⅔
Covering machines	7½	Cars:	
Crackers, rubber	6⅔	Batch	8

Table 6-17. Probable Annual Depreciation Rates for Chemical Process Equipment (*Continued*)

	Depreciation rate, per cent		Depreciation rate, per cent
Dryer or kiln	6⅔	Pans, dry	6⅔
Mine:		Plungers	10
Steel	10	Polishers	8
Wood	20	Presses	5
Transfer	10	Pulverizers	6⅔
Casting and rolling tables	8	Pumps	6⅔
Charging machines	10	Pumps, clay	10
Controllers, temperature, automatic	10	Reels	5
		Riddles, gyratory	10
Conveyors	6⅔	Scales:	
Coolers	7½	Platform	5
Crushers	6⅔	Portable	6⅔
Cutoff machines	7	Screens	10
Cutting machines	7	Separators:	
Cutting machines, rock lath	5	Air	10
Disintegrators	12½	Magnetic	7
Drag lines:		Shovels, electric or steam	6⅔
Heavy	5	Sieves	12½
Light	10	Sifters, revolving	10
Medium	6⅔	Tables, drawing, grinding, or polishing	8
Dryers	9		
Dryers, rotary	6⅔	Tanks	5
Duster machines, bag	6⅔		
Dust collectors	5	**Oil and Gas Refining**	
Elevators:		Agitators	6¼
Bucket	6⅔	Carbon black plants	8
Screw	5½	Condensers	6⅔
Feeders	8	Exchangers, heat	6⅔
Filter presses	6⅔	Filtering plants	5
Furnaces	6⅔	Gasoline plants, natural gas	8
Furnaces, pot	8	Pipes, interunit lines, small diameter	6⅔
Grinders	6⅔		
Hydrators	6⅔	Pumps	6⅔
Jigs	10	Stills:	
Kettles	6⅔	Cracking	12½
Kilns	6⅔	Fire	6⅔
Lehrs	6⅔	Steam	6⅔
Loading machines	10	Tube or pipe	8⅓
Mills	6⅔	Vacuum	6⅔
Mixers	7	Tanks:	
Molds	20	Compounding	5
Molds, hydraulic	8⅓	Storage	5
Mud machines	8	Treating	6¼
Ovens, flattening	8	Towers, scrubbing	6⅔
Packers	10	Traps, gas and water	8⅓
Pallets and trays	12½	Wax plants	5

source: U.S. Bureau of Internal Revenue.

where d = annual per cent depreciation based on original investment
V_o = original investment value of installed equipment
V_s = final salvage value of equipment
n = allowable service life

2. *Multiple straight-line depreciation:* The straight-line method is applied over several periods, being short at first and then increasing as the life of the equipment is extended beyond the original estimate.

$$d_1 = \frac{V_o - V_s}{n_1 V_o} \times 100 \qquad d_2 = \frac{V_o - V_s}{n_2 V_o} \times 100, \text{ etc.} \qquad (6\text{-}7)$$

where d_1 = annual per cent depreciation during first period of years, n_1
d_2 = annual per cent depreciation during second period of years, n_2
$n_2 > n_1$

3. *Declining balance or fixed percentage method:* The annual depreciation is a fixed percentage of the current property value or remaining book value. Since the value is highest at the beginning, the depreciation is also highest.

$$d_d = 1 - \left(\frac{V_s}{V_o}\right)^{1/n} \qquad (6\text{-}8)$$

$$V_x = V_o(1 - d_d)^x \qquad (6\text{-}9)$$

where d_d = annual per cent depreciation based on remaining property value
V_x = property value after x years

4. *Double declining balance method:* This method is used to accommodate Federal laws of 1954 which permitted depreciation rates not more than twice the straight-line depreciation.

$$d_d = 2d \qquad (6\text{-}10)$$

Apply this in Eq. (6-9).

5. *Sum-of-the-years'-digits method:* This is an arbitrary method which gives the same type of depreciation trends as the declining balance method.

$$1 + 2 + \cdots n = \Sigma \qquad (6\text{-}11)$$

where $\dfrac{n}{\Sigma} = d_1$ = annual per cent depreciation on original investment for first year

$\dfrac{2}{\Sigma} = d_{n-1}$ = annual per cent depreciation on original investment for $(n-1)$st year

$\dfrac{1}{\Sigma} = d_n$ = annual depreciation on original investment for nth year

CHAP. 6] ECONOMIC EVALUATION OF THE PROJECT 249

6. Declining balance–straight-line method: This employs the declining balance during early years and the straight-line method in later years. It has some utility where service life of equipment is debatable, and yet allows faster early-life write-offs.
7. Unit of production method: This apportions depreciation over barrels of oil or tons of oil produced; it is limited to extractive industries.

Group B (interest on investment allowed):
1. Sinking fund method
2. Present worth method

Methods in this group are poor from the point of view of quick return of capital. They are only used by a few public utility concerns for depreciating any property that has long service life with little risk of obsolescence.

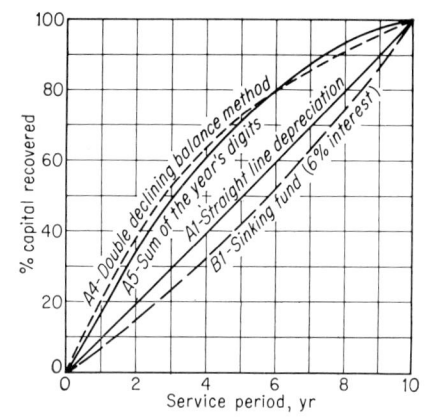

FIG. 6-36. A comparison of depreciation methods for capital recovery.

Figure 6-36 shows a composite plot of several capital recovery methods, illustrating the advantage of the rapid write-off techniques.

Other Fixed Charges

In addition to depreciation, other indirect manufacturing costs are local taxes, insurance on equipment and buildings, and general plant overhead. *Local taxes* are often 2 per cent of the investment annually, and *insurance* on the equipment and building is estimated at 1 per cent. In another method these two can be rated at about 6 per cent of the total manufacturing costs. *Social security taxes* and general plant overhead against individual products vary with the location. Estimates of 50 to 75 per cent of the operating and maintenance labor plus supervision have been used. The trend toward more fringe benefits may bring this up to 100 per cent in some cases.

Management and Marketing Expenses

General administration, selling, and research expenses are frequently based on average annual sales. In that order, based upon sales, estimates are made at 2 per cent, 2 to 10 per cent, and 2 to 5 per cent, respectively, depending upon the company. Research expenses cover the salaries and overhead for all technical personnel engaged in research and development work. Administrative expenses apply to the salaries and expenses of the officers of the company and to general expenses connected with company administration such as legal and auditing fees.

Summary of Total Product Costs

All the important items which make up the cost of a product have now been discussed. These can now be summarized by groups as shown in Table 6-18.

TABLE 6-18. SUMMARY OF TOTAL PRODUCT COSTS

1. Product ____X____ in _____ (drums, bags, tank cars, etc.)
2. Production rate 0,000,000 units (lb, tons, gal, etc.) per yr

	$/yr	$/unit
3. Manufacturing costs		
a. Raw materials	000,000	0.0000
b. Direct conversion	000,000	0.0000
c. Indirect conversion	00,000	0.0000
d. Packaging and shipping	00,000	0.0000
	0,000,000	0.0000
4. Management and marketing (6–15% of sales)	000,000	0.0000
5. Total product costs	0,000,000	0.0000

ECONOMIC ANALYSIS

Economic analysis in a broad sense is the determination of the relationship of income and expenses to the material welfare of the company. In previous sections the development of cost or expense data has been demonstrated. In its simplest form, the problem is now to establish what the income from sales will be, subtract the total product cost, and obtain a gross income. By subtracting income taxes, new earnings are obtained which must be linked to total capital investment to determine the attractiveness of the venture. These conclusions must be conveyed to management in one or more ways so that a sound decision can be rendered. One must recognize the fact that the project will have to compete with others for the investment money available and, in the final analysis, profitability will carry the most weight in the decision.

There are a wide variety of problems which will come up for economic analysis. Several typical ones would be:

Problem 1. Sales price already established by competition and plant capacity set by the marketing and sales group. Determine the profitability.

Problem 2. Sales price already established by competition. Determine profitability as a function of plant capacity.

Problem 3. A new venture—sales price to be established in terms of profitability.

Problem 1 is by far the simplest case and will be discussed more extensively than the other problems in terms of economic analysis for the benefit of the student.

CHAP. 6]   ECONOMIC EVALUATION OF THE PROJECT   251

Each company and each economist has one or more ways of determining profitability by economic analysis. It is not the purpose of this book to elaborate on these. Excellent books on chemical engineering economy are listed in the Additional Selected References. However, three of the more popular methods will be discussed: (1) return on investment, (2) pay-out time, (3) project present worth. To proceed with the economic analysis, net or new earnings must first be determined from selling price less costs.

New or Net Earnings

An examination of Table 6-19 will show quantitatively what is involved. Product costs have been established by methods outlined in Tables 6-8, 6-9, and 6-18. The other items in determining new earnings require further explanation.

Income from Salable Products. This covers the sale of the by-products as well as the principal product(s), which is primarily monitored by the market and sales group. The interrelationships among selling price, market demand and supply, production capacity, and investment return require careful economic analysis. In many cases, particularly new ventures, this can only be determined by a profitability analysis, e.g.,

TABLE 6-19. COST AND PROFIT SUMMARY FOR A PROPOSED PLANT

Type of plant: Lindane (99% γ isomer of benzene hexachloride)
Source of data: Table 6-24
Production: 240,000 lb/yr
Fixed capital investment: $1,220,000
Total capital investment: $1,430,000

Item	$/year
1. Total product cost	424,270
2. Product value	745,000
3. Gross profit or earnings (item 2 − item 1)	320,730
4. Income taxes (52% level)	166,700
5. Net profit or new earnings (item 3 − item 4)	154,030
Profitability analysis	
Annual per cent return on fixed capital	
Before taxes	26.1
After taxes	12.6

$$\text{Pay-out time} = n_b = \frac{\text{investment}}{\text{gross profit}}$$
$$= \frac{\$1,220,000}{320,730}$$
$$= 3.8 \text{ yr}$$

acceptable return on the investment money. The selling price must be set sufficiently high to justify a reasonable investment return and yet not be priced out of line. One relatively simple procedure is to determine

252   CHEMICAL ENGINEERING PLANT DESIGN   [CHAP. 6

investment returns for various selling prices, plot the data, and pick the selling price to match the company's requirement for investment return (Fig. 6-37). A modification of this is to establish the plant capacity and investment based on an anticipated sales price and required return on investment (Fig. 6-38).

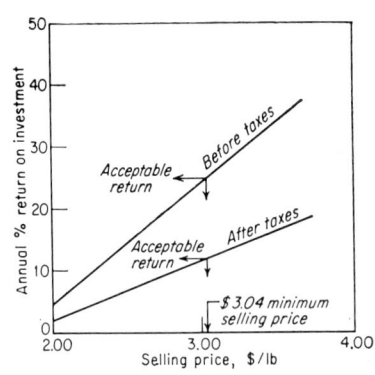

FIG. 6-37. Annual per cent return on investment versus selling price.

Type plant—Lindane (99% γ-BHC)
Annual capacity—240,000 lb
Reference: Table 6-24

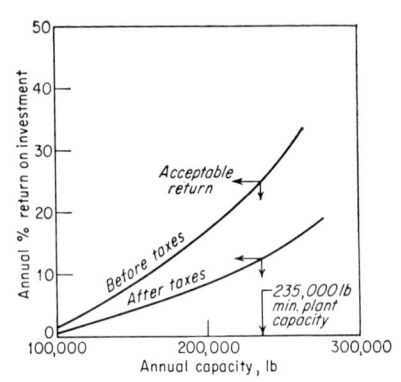

FIG. 6-38. Annual per cent return on investment versus plant capacity.

Type plant—Lindane (99% γ-BHC)
Selling price of product—$3.10/lb
Reference: Table 6-24

Income Tax Expenses. This represents a considerable expense to a profitable company. Table 6-20 shows that the average income tax for chemical companies is 52 per cent of gross earnings. The tax on earnings of the extractive, mining, and petroleum companies is somewhat less than this because of depletion allowances on their natural resource reserves.

TABLE 6-20. INCOME TAX RATES*

Type of company	Per cent tax on gross earnings
Chemical process	51.6
Pulp and paper, rubber, and synthetic fiber	50.5
Drugs	47.5
Extraction and mining	40.0
Petroleum	28.5

* Abstracted from *Chem. Eng. News,* **34**: 5647–5657 (1956).

Interest on Investment. Interest on the investment, usually at 6 per cent, is sometimes considered an element of operating cost. However, it cannot be so charged for income tax purposes, nor is it generally considered as operating expense by accountants. Although the question is debatable, it appears that the best opinion permits the use of interest on investment as a cost only when comparing the returns from two different investments, as, for example, two different processes for making the same or different products.

Application of Interest Charges to Project. Bullinger[1] holds that interest cost should be calculated on initial investment. His reasoning is that decline in plant value is always accompanied by the recovery of an equivalent amount of capital through depreciation, thus maintaining the original investment intact.

Grant,[2] on the other hand, maintains that interest charges should be computed on the average investment over the life of the project roughly half the initial investment on the basis that investment shrinks steadily as the plant depreciates.

Profit. Gross profit, usually on a yearly basis, is the difference between gross sales and product costs, which may include interest on investment. New earnings are obtained by subtracting income taxes from gross profits. They represent the sum of money that has been earned by the over-all investment of capital and labor to make the product available to customers.

Profitability Analysis

This analysis is the final judgment as to whether the project should be further financed. In its simplest form it examines net earnings in relation to capital investment. Several methods of analysis will be discussed.

Per Cent Return on Investment. This is the most widely used method since it very simply gives the annual rate at which earnings will return the investment. Formulas applicable to this method are:

$$r_b = \frac{P \times 100}{I} \qquad (6\text{-}12)$$

$$r_a = \frac{E \times 100}{I} \qquad (6\text{-}13)$$

where r_b = annual per cent return on investment before taxes
r_a = annual per cent return on investment after taxes
P = annual gross profit before taxes
E = annual new earnings = Px, where $x = [1 -$ fraction for income tax $(\cong 0.52)]$
I = investment, fixed or total, with r specified accordingly

---

[1] C. E. Bullinger, "Engineering Economic Analysis," 3d ed., McGraw-Hill Book Company, Inc., New York, 1958.
[2] E. L. Grant, "Principles of Engineering Economy," 3d ed., The Ronald Press Company, New York, 1950.

An accepted rate of anticipated return will vary with the degree of risk from obsolescence or competition. Typical criteria for acceptable returns in several industries are listed in Table 6-21.

TABLE 6-21. MINIMUM ACCEPTABLE RETURN ON INVESTMENT FOR SEVERAL INDUSTRIES

	Per cent return					
Industry	r_b, before taxes			r_a, after taxes		
	Low	Average	High-risk	Low	Average	High-risk
Chemical process	15	30	45	7	15	21
Drugs	25	43	56	13	23	30
Petroleum	18	29	40	12	20	28
Metals	10	17	25	5	9	13

Low-risk figures should apply only to those processes which have been well established commercially with firm sales markets. The high-risk values are applicable to pioneering ventures where scale-up and market conditions are uncertain.

Pay-out Time Method.[1] In making economic studies involving the purchase of new plants or equipment, it is frequently found desirable to estimate the so-called "pay-out period," i.e., the number of years n that will elapse before the investment has been completely recovered through savings or added earnings.

Pay-out time before taxes, n_b, is most commonly expressed, using a fixed investment I_f, as follows:

$$n_b = \frac{I_f}{P} \tag{6-14}$$

A comparison with Eq. (6-12) shows that n_b is $100/r_b$. Acceptable values of n_b can thus be calculated using the information in Table 6-21.

The effect of both taxes and interest can be included in a pay-out calculation by means of a stepwise calculation, which will determine, for each year, the net amount of capital recovered. This computation must be continued by years, until the entire investment is shown as recovered. Such a method is apt to be both laborious and time-consuming. The following formula, which achieves the same purpose, was developed to eliminate the use of the lengthy stepwise method:

$$n_i = \frac{\log Z - \log (Z - iI_f)}{\log X - \log Y} \tag{6-15}$$

[1] W. H. Buell, Calculating Payout Time for Equipment Investment, *Chem. Eng.*, **54**(10): 97 (1947).

where $X = 1 + i/2$
$Y = 1 - i/2$
$Z = P(1 - t) + tD = E + tD$
i = effective fractional rate of interest
t = fractional tax rate applicable to earnings
D = depreciation allowed for tax purposes

If the equipment to be installed replaces existing equipment, P is the annual profit made possible by the new equipment. If the plant under consideration is entirely new, P represents the annual profit of the plant before taxes and interest. The term i represents the effective rate of interest, i.e., the actual out-of-pocket interest cost after allowing for the effect on taxes. Since interest is chargeable against income and this reduces the taxes paid, the true interest cost will be less than the apparent interest. The term t represents the tax rate applicable to earnings and is the sum of the Federal and state tax rates.

Depreciation is not normally included with expenses in calculating a pay-out. However, depreciation is an allowable deduction in computing taxes. To cover the effect on taxes alone, a factor for depreciation must be included in the formula. Depreciation allowed for tax purposes may be different from depreciation charged on the books of the company or depreciation based on estimated years of life. This item, therefore, should properly be estimated by the tax department of the company.

Both P and t are assumed to remain constant for the entire pay-out period. If this period is expected to be long and if considerable changes in the values of these two items are in prospect, there is no alternative but to carry out a stepwise year-by-year calculation.

Formula (6-15) is usable only when i is greater than zero. If it is desired to neglect interest, the following simple formula should be used:

$$n_a = \frac{I_f}{Z} \qquad (6\text{-}16)$$

Project Present Worth Method. This method recognizes the time value of money and is more widely used by economists and accountants in presenting profitability reports to management, particularly when two processes require close comparison. Engineers should have an understanding of this method.

The time value of money depends on the earning rate at which money is reinvested. Reinvestment at high earning rates requires recovery of the initial investment as soon as possible. The method assumes that there are limitless opportunities to invest at a company average investment earning rate R. The receipts from the project under consideration are invested at this earning rate R, and the project evaluation involves a comparison of *over-all* company profit by investing money in the given project with the profit which would result were the money invested in projects yielding R per cent interest. Further details of this method can be obtained by studying suitable references.[1,2]

Profitability, Plant Capacity, and Sales

In order to provide management with a graphic representation of profitability analysis, working charts are extremely useful. Figures 6-37 and 6-38 are types which can be used. These were developed primarily to show the minimum size plant which should be built to assure

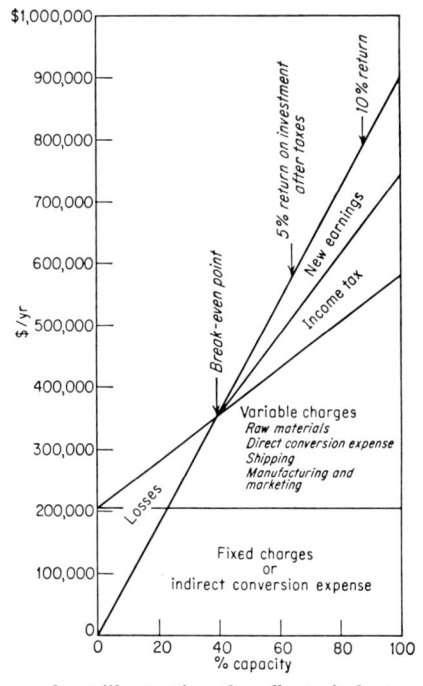

FIG. 6-39. Break-even chart illustrating the effect of plant capacity on earnings.

Type plant—Lindane (99% γ-BHC)
Annual capacity—240,000 lb
Reference: Table 6-24.

a reasonable return on the investment. Another concept is the effect upon costs and profits if a certain size plant is operated at less than full capacity. This effect can be illustrated by use of a type of graph known in economic parlance as a *break-even chart*. It is a plot of costs, sales, and

[1] J. C. Martin, "Economic Analysis," from J. J. Hur (ed.), "Chemical Process Economics in Practice," Reinhold Publishing Corporation, New York, 1956.
[2] A. Marston, R. Winfrey, and J. C. Hempstead, "Engineering Valuation and Depreciation," McGraw-Hill Book Company, Inc., New York, 1956.

profits as a function of the production level of operation of a plant (see Fig. 6-39). For a minimum size Lindane plant giving an allowable minimum return on the investment as established by company management, this graph shows that operation below 39% of full capacity will be done at a loss. The economic studies with this type of chart can be considerably expanded. The reader is referred to Aries and Newton.

REPORT ON ENGINEERING AND ECONOMIC EVALUATIONS

At this point a report should be written, presenting pertinent facts and conclusions reached, and distributed to all those who are likely to have any connection with the project. Since the evaluation is undoubtedly based on certain assumptions which may reflect the personal judgment and bias of the chemical engineer preparing the report, all assumptions and supporting experimental data should be clearly stated. Standard nomenclature should be used throughout the report. (See Appendix B.) The report must be well presented as the decision on the advisability of continuing the project may depend on how well the written presentation is made. A typical outline of a design report would be:

1. Letter of transmittal
2. Title page
3. Table of contents
4. Summary—points out results, conclusions, and recommendations for management to follow
5. Introduction
6. Survey of previous work
7. Proposed process description
    a. Laboratory data
    b. Assumptions
    c. Flow diagrams
8. Process design
    a. Flow diagrams
    b. Tabulation of equipment and specifications
    c. Process economics
        (1) Capital requirements
        (2) Profitability analysis
            (a) Product costs
            (b) Selling prices } as these affect investment return
            (c) Plant capacity
9. Recommendations
    a. Further research and development work, giving time and cost
    b. Further process design, giving time and cost
    c. Further plant design, giving time and cost
    d. Possible management decisions with alternatives
10. Acknowledgments
11. Summary of nomenclature used (optional)
12. Summary of references used

TABLE 6-22. PRECONSTRUCTION COST ESTIMATIONS FOR CRUDE BENZENE HEXACHLORIDE PESTICIDE PLANT—PROCESS EQUIPMENT COSTS AND CAPITAL INVESTMENT (1958)

	Item of equipment	No. required	Unit installed cost	Total installed cost
A-1.	Benzol transfer pump	2	$ 520	$ 1,040
A-2.	Benzol storage tank	2	4,000	8,000
A-3.	Benzol feed pump	2	280	560
A-4.	Water demineralizer	1	750	750
A-5.	Chlorine vaporizer	1	900	900
B-1.	Chlorinator	2	19,500	39,000
B-2.	Reflux condenser	2	6,000	12,000
B-3.	Vent gas separator	2	1,100	2,200
C.	Decanter	1	1,400	1,400
D-1.	Dilute acid receiver	1	1,900	1,900
D-2.	Acid still feed pump	2	500	1,000
D-3.	Acid stripping still	1	1,800	1,800
D-4.	Acid still condenser	1	1,000	1,000
D-5.	Conc. HCl receiver	1	1,300	1,300
D-6.	Conc. HCl pump	1	400	400
D-7.	Conc. HCl storage	1	9,500	9,500
D-8.	Acid recycle pump	2	500	1,000
E.	Crude product pump	2	300	600
F-1.	Caustic make-up tank	1	2,200	2,200
F-1a.	Caustic make-up tank agitator	1	500	500
F-2.	Caustic make-up pump	1	270	270
F-3.	Caustic storage tank	1	3,000	3,000
F-4.	Caustic feeder	2	500	1,000
F-5.	Neutralizer	1	1,400	1,400
F-6.	Spent caustic separator	1	700	700
G-1	Flash still feed pump	1	280	280
G-2.	Flash still	1	11,800	11,800
G-3.	Flash still condenser	1	2,900	2,900
G-4.	Solvent receiver	1	6,500	6,500
H-1.	Crystallizer feed pump	1	300	300
H-2.	Crystallizer	1	35,000	35,000
I-1.	Continuous centrifuge	1	8,600	8,600
I-2.	Recycle liquor receiver	1	2,600	2,600
I-3.	Wet crystal hopper	1	3,800	3,800
J-1.	Rotary vacuum dryer	1	20,000	20,000
J-2.	Vacuum condenser and receiver	1	2,500	2,500
J-3.	Vacuum pump	1	1,200	1,200
J-4.	Recovered solvent pump	1	280	280
K-1.	Dry BHC hopper	2	3,800	7,600
K-2.	BHC packaging	1	1,000	1,000
L-1.	Solvent still feed pump	2	300	600
L-2.	Solvent still	1	9,000	9,000
L-3.	Solvent still condenser	1	2,500	2,500
L-3a.	Condensate cooler	1	310	310
L-4.	Wet benzol receiver	1	500	500
L-5.	Wet benzol pump	1	300	300
M-1.	Monochlor pump	1	300	300
M-1a.	Monochlor cooler	1	280	280
M-2.	Monochlor storage	1	5,000	5,000
Total installed cost				$216,570
Building and land costs—60% of total				130,000

Capital cost calculations (see Table 6-7 on p. 222)
Fixed capital = 216,570 × 2.56 ≅ $556,000
Total capital = 216,570 × 3.00 ≅ $650,000

13. Appendix
    a. Calculations—all or sample calculations
    b. Detailed tables of data
    c. Pertinent research and development tests and data

SAMPLE PROBLEM

Economics of Benzene Hexachloride Production

As stated in Chap. 3, the ultimate goal of the process design was to present to management the comparative facts on crude BHC and Lindane production so that a decision could be made to commercialize one or the other process or abandon both. This section will present the important facts necessary to render a decision.

Crude BHC Process

Equipment specifications were prepared and illustrated in Chap. 4. Table 6-22 could thus be worked out, using individual equipment cost estimates from this book and others. The installed cost on a 1958 basis for the process equipment alone was $216,570. Table 6-24 was next prepared from figures available from the process flow sheet and other reasonable economic facts as discussed in here. Note that short-cut methods were employed where possible since only comparative economics were desired for this type of profitability analysis. Results show that a crude BHC plant at the production level recommended is unprofitable and definitely should not be considered further.

Lindane (99 Per Cent γ Isomer of BHC) Process

As explained in Chap. 3, the process design for this type of plant was not shown. It is to be left as Project 6-1. Equipment costs were developed for the process at two different capacities and the economic analysis is presented in Tables 6-24 and 6-27 and Figs. 6-37 to 6-39. The conclusions are that it is unprofitable to operate a Lindane plant at

TABLE 6-23. PRECONSTRUCTION COST ESTIMATIONS FOR CRUDE BENZENE HEXACHLORIDE PLANT—OPERATING LABOR AND SUPERVISION REQUIREMENTS

Employee	No.	Hourly rate	Annual salary per employee	Total cost per year
Shift supervisors	3	$3.00	$6,240	$18,720
Operators	6	2.25	4,743	28,458
Shipping clerk	1	1.80	3,742	3,742
	10			$50,920

TABLE 6-24. PRECONSTRUCTION COST ESTIMATION AND PROFITABILITY ANALYSIS FOR CRUDE BENZENE HEXACHLORIDE (12% γ ISOMER) AND LINDANE (99% γ ISOMER) PLANTS
Product cost estimations: See Tables 6-8, 6-9, 6-22, and 6-23

		Crude BHC in fiber containers 1,200,000 lb/yr of 6,000 hr, 250 days Plant location: Not determined Fixed capital: $556,000 (Based on Method 4) M+E: 426,000 Building and land: 130,000			Lindane in fiber containers 120,000 lb/yr of 6,000 hr, 250 days Not determined $820,000 (Based on Method 4) 646,000 174,000			Lindane in fiber containers 240,000 lb/yr of 6,000 hr, 250 days Not determined $1,220,000* (Based on Method 4) 960,000 260,000		
	Unit	Quantity/yr	Unit cost, $	Cost, $/yr	Quantity/yr	Unit cost, $	Cost, $/yr	Quantity/yr	Unit cost, $	Cost, $/yr
5. Raw materials:										
a. Benzene	Gal	69,500	0.36	25,000	69,500	0.36	25,000	139,000	0.36	50,000
b. Chlorine	Lb	1,057,000	0.031	32,700	1,057,000	0.031	32,700	2,114,000	0.031	65,400
c. Caustic soda	Lb	7,760	0.06	465	7,760	0.06	465	15,520	0.06	930
d. Methanol	Gal				14,800	0.27	4,000	29,600	0.27	8,000
e. Subtotal				58,165			62,165			124,330
f. Credit for by-products:										
γ isomers	Lb				1,050,000	0.05	52,500	2,100,000	0.05	105,000
Monochlorbenzene	Lb	261,250	0.08	21,000	261,250	0.08	21,000	522,500	0.08	42,000
Muriatic acid	Lb	229,750	0.02	4,595	229,750	0.02	4,595	459,500	0.02	9,190
Credit subtotal				25,595			78,095			156,190
g. Net raw material costs				+32,570			−15,930			−31,860
6. Direct conversion expense (DCE):										
a. Utilities:										
Steam (1,000 lb = M)	M	4,500	0.60	2,700	12,280	0.60	7,360	22,000	0.60	13,200
Electricity, kwhr	Kwhr	213,000	0.01	2,130	350,000	0.01	3,500	600,000	0.01	6,000
Water cooling (1,000 gal = M)	M	31,400	0.05	1,570	56,200	0.05	2,810	112,400	0.05	5,620
Water process (1,000 gal = M)	M	262	0.12	315	300	0.12	360	600	0.12	720
b. Labor		See Table 6-23		50,920			68,384			70,000
c. Supervision				12,000			14,000			14,000
d. Total direct labor charge		62,920 × 1.79		112,200	82,384 × 1.79		147,000	84,000 × 1.79		150,490
e. Laboratory expenses				15,000			20,000			20,000
f. Total DCE (6a + 6d + 6e)				133,915			181,030			196,030

260

TABLE 6-24. PRECONSTRUCTION COST ESTIMATION AND PROFITABILITY ANALYSIS FOR CRUDE BENZENE HEXACHLORIDE (12% γ ISOMER) AND LINDANE (99% γ ISOMER) PLANTS *(Continued)*

	Unit	Quantity/yr	Unit cost, $	Cost, $/yr	Unit	Quantity/yr	Unit cost, $	Cost, $/yr	Unit	Quantity/yr	Unit cost, $	Cost, $/yr
7. Indirect conversion expense (IDCE):												
a. Fixed charges and repairs		426,000 × 0.20		85,120		646,000 × 0.20		129,200		960,000 × 0.20		192,000
b. Building charges		130,000 × 0.05		6,500		174,000 × 0.05		8,700		260,000 × 0.05		13,000
c. Total IDCE				91,620				137,900				205,000
8. Bulk conversion cost (6*f* + 7*c*)				225,535				318,930				401,030
9. Bulk manufacturing cost (5*g* + 8)				258,105				303,000				369,170
10. Packaging and shipping costs:												
a. Containers				3,000				1,250				2,500
b. Packaging labor and overhead				6,000				5,000				8,000
c. Total				9,000				6,250				10,500
11. Total manufacturing costs (9 + 10*c*)				267,105				309,250				379,670
12. Management and marketing:												
a. Value of product		1,200,000 at $0.125/lb		150,000		120,000 at $3.10/lb		372,500		240,000 at $3.10/lb		745,000
b. Management and marketing costs		15,000 × 0.10		15,000		372,500 × 0.08		29,600		745,000 × 0.06		44,600
13. Total product cost (11 + 12*b*)	lb.		0.22	282,105	lb.		2.81	338,850	lb.		1.76	424,270
14. Gross profit (12*a* − 13)				−132,105 (loss)				33,650				320,730
15. Income taxes (52% of item 14)								17,500				166,700
16. Net profit or new earnings								16,150				154,030
17. Profitability analysis:												
a. Annual % return on fixed capital												
Before taxes				Loss		4.10%				26.1%		
After taxes						1.96%				12.6%		
b. Pay-out time before taxes						24.4 yr				3.8 yr		

* $\left(\dfrac{240,000}{120,000}\right)^{0.6} \times 820,000$. See also Table 6-25 for a more accurate estimate.

less than 235,000 lb/year production at current market prices. An even larger production with a continued research and development program should be recommended to avoid low investment returns in the face of competition and price drops. Final decision will have to be reconciled with a more detailed market analysis.

TABLE 6-25. FIXED CAPITAL ESTIMATES FOR A 240,000-POUND PER YEAR LINDANE PLANT USING MORE ACCURATE METHOD 3

Note: Equipment costs developed for this process are not shown. Students may be assigned this as a process design project. (See Project 6-1.)

Basis: Plant built adjacent to existing facilities.

Cost estimate by method 3

1. Delivered equipment costs....................................	$336,000
2. Installed equipment cost: 336,000 × 1.43.....................	478,000
3. Process piping: 478,000 × 0.20..............................	95,000
4. Instrumentation: 478,000 × 0.05.............................	23,900
5. Building and site development: 478,000 × 0.50................	239,000
6. Auxiliaries: 478,000 × 0.05.................................	23,900
7. Outside lines: 478,000 × 0.10...............................	47,800
8. Total physical plant cost (2 through 7).....................	$907,600
9. Engineering and construction: 907,600 × 0.30................	272,000
10. Contingencies: 907,600 × 0.10..............................	90,760
11. Size factor: 907,600 × 0.05................................	45,380
12. Total fixed capital cost (8 + 9 + 10 + 11).................	$1,315,740

PROCESS DESIGN PROJECTS

6-1. Lindane Plant

Design a 250,000 lb/yr Lindane (99 per cent γ isomer of benzene hexachloride) plant based on process information given in the current literature [e.g., see *Chem. Eng. Progr.*, **52**:281 (1956); *Chem. Week*, **78**:54 (1956); *Ind. Eng. Chem.*, **48**(10):41A (1956)]. Develop the process flow sheets, equipment specifications, and plant layout. Examine the economics of producing Lindane including (1) fixed and capital cost estimates by one or more of the methods outlined in Chap. 6, (2) profitability analysis showing return on investment and pay-out time, (3) break-even point analysis of the project, (4) economic estimates for a plant producing 500,000 lb/yr and 750,000 lb/yr.

6-2. Ferrous Sulfate Recovery Plant

Design a plant for the recovery of ferrous sulfate from the waste pickle resulting from the pickling of steel in a galvanizing plant.

A galvanizing plant uses 100 tons or 50°Bé sulfuric acid every 24 hr in the pickling of steel; before it is used this acid is diluted in 18°Bé. When exhausted in so far as pickling is concerned, the waste pickle liquor leaves the vats as a 25 per cent solution of ferrous sulfate containing 2 per cent sulfuric acid, and at a temperature of 175°F. This liquor is drained from the pickling vats through a header into a waste-liquor storage reservoir built underground.

From the pickle-liquor storage reservoir the liquor is to be pumped into neutralizer tanks containing scrap iron for neutralization of the acid; 48 hr at 170°F is considered sufficient to complete neutralization. The neutralized liquor is then to be pumped through filters to remove insoluble sludge before concentration in special evaporators. Crystallization to monohydrate is to be effected in the evaporators and the crystals are to be permitted to fall into a sump where temperature is maintained above 171°F. The wet crystals are to be removed by means of a pump or mechanical rake, centrifuged, and then cooled. The mother liquor from the centrifuge is to be returned to the sump. The dry crystals are to be conveyed to storage bins, preparatory to bagging, barreling, and transfer to freight cars.

The pumps, conveyors, agitators, and other motivating equipment are to be operated from a lineshaft, except that vacuum pump on the evaporator is to be steam-driven, the exhaust steam from this pump to be used in the first effect evaporator. If approved plant layout does not permit economical use of shafting for acid liquor pumping, a separate motor drive may be used for such purpose.

6-3. Deodorized Soybean Oil Plant

Design a plant for the deodorization of raw soybean oil obtained from an expressing plant handling 50 tons of raw beans, according to general practice; the deodorized oil is to be used for oleomargarine.

The deodorizing plant is to consist of a closed tank or deodorizer in which the oil is to be processed; equipment for heating the oil within the deodorizer while steam is blown through the oil; means for maintaining a high vacuum within the deodorizer; and equipment to cool and filter the oil after deodorizing.

The oil to be deodorized is to be heated by circulating it through a heater at 125°C; steam is to be injected into the oil; the deodorizing cycle is to be 8 hr. Cooling is to be accomplished by dropping the deodorized oil into a vacuum cooling tank equipped with cold-water coils.

6-4. Sodium Chlorate Plant

Design a chemical plant for the production of 4,000 lb of sodium chlorate per 24 hr by the Liebig method.

Salt from Louisiana will be shipped in, put into solution in wooden tanks, purified by sodium carbonate treatment to separate out the calcium and magnesium, neutralized with hydrochloric acid, and then electrolyzed using Nelson cells. The chlorine gas is to be piped to a mixing chamber, where it is to be mixed with the brine-caustic solution to form sodium hypochlorite; this solution is heated to 90°C to transform the hypochlorite to chlorate. The sodium chlorate–sodium chloride solution is then evaporated in triple-effect evaporators and the sodium chloride is removed. When evaporated to a concentration of 74.1 per cent sodium chlorate and a residual salt content of 0.41 per cent, the solution has a gravity of 1.65; then the solution is transferred to crystallizers and cooled. The crystallized sodium chlorate is centrifuged from the mother liquor and dried in a steam-heated rotary dryer before placing in storage. The separated sodium chloride is washed free from sodium chlorate by washing with sodium hydroxide solution and returned to the system for electrolyzing.

The average current efficiency of the Nelson cell is to be considered as 86 per cent, and the chemical efficiency rated at 80 per cent. All but 2 per cent of the sodium chloride is to be considered as recoverable. The Nelson cell is to operate under optimum conditions on a 26.4 per cent sodium chloride solution at 55°C, producing a solution of specific gravity 1.23 and containing 9.67 per cent sodium hydroxide and 14.52 per cent residual sodium chloride.

6-5. Butane Oxidation Plant

Design a plant for the production of solvents by the method of partial oxidation of butane.

A natural gasoline plant has available, as a by-product from the natural gasoline refinery, 60,000 ft^3 of butane per day, based on standard gas conditions. This by-product is to be utilized by oxidizing in the vapor phase the butane under controlled conditions to produce partial-oxidation products, totaling 159 lb/1,000 ft^3 of butane, the product consisting of aldehydes, alcohols, ketones, and acids, according to the following analysis:

	Per cent
Acids, calculated as acetic	4.0
Esters as ethyl acetate	0.6
Aldehydes:	
Acetaldehyde	18.5
Formaldehyde	15.0
Higher aldehydes	4.2
Alcohols, calculated as methanol	22.2
Ketones, calculated as acetone	9.5
Water	26.0

The conditions of the partial-oxidation process are the injection of a mixture of air and butane vapor, in a ratio of 10:1, into a stream of inert gas resulting from the elimination of the condensable vapors at 350 lb pressure. In order to carry the air-butane mixture through the system, a recirculation of inert gas is maintained at a ratio of 140:1, inert gas to butane. The temperature of the furnace must be maintained, so that the reaction coils heat the gas up to 720°F at the exit from the furnace. The reaction furnace is to be heated with butane; steam is to be generated by using butane fuel. Compressors are to be operated with internal-combustion engines using butane as the fuel. Water supplied for cooling is to be available at a range of 75 to 90°F.

The product obtained by oxidation of butane is to be subjected to fractionation, separating into three main fractions: (1) pure acetaldehyde, (2) a crude methanol fraction, and (3) a residue to be wasted. The second fraction is redistilled after treatment with caustic to polymerize the aldehydes, and the vapors are further treated by washing countercurrently in a 33 per cent caustic solution.

CHAPTER 7

Locating the Chemical Plant

In the previous chapter we have seen how an economic appraisal of a chemical process has been made. This was termed a preconstruction economic analysis at the process development level. The cost and profit figures were presented to management with fairly reliable costs on equipment and building requirements. An estimate on site development costs was included with no specific reference to the exact geographical location of the chemical plant.

Assume now that the sequential development of the plant design has reached the stage where management has surveyed the process development economics and given the approval to proceed with final plans for constructing and operating the plant. The final plans require detailed specification work and cost estimations on equipment and auxiliaries, building design, and site development. These topics will be discussed in Chaps. 8 and 9.

One of the most important parts of this final planning is the *site location*. If the plant is not located in the most economically favorable position, the competitive advantages of the process, so carefully engineered during research and development phases, can be wiped out. Without careful thought on all the factors which must be considered for optimum plant location, the plant may even be inoperable. A prize example of poor planning was the case of a plant site chosen without due consideration for all the water requirements. The project had proceeded to the point where materials of construction had been shipped to the site before the lack of water was apparent. Eventually a new site was chosen which best fitted all the requirements and the mistake was rectified but not without pain, strain, and additional expenditure of funds.

SUMMARY OF FACTORS IN PLANT LOCATION

Factors which generally apply to the economic and operability aspects of plant site location are classified into two major groups. The *primary factors* listed apply to choice of a region, whereas the *specific factors* are looked at in choosing an exact site location within the region. All factors are important in making a site location selection.

Primary Factors

1. Raw-materials supply:
   a. Availability from existing or future suppliers
   b. Use of substitute materials
   c. Distance
2. Markets:
   a. Demand versus distance
   b. Growth or decline
   c. Inventory storage requirements
   d. Competition—present and future
3. Power and fuel supply:
   a. Availability of electricity and various types of fuel
   b. Future reserves
   c. Costs
4. Water supply:
   a. Quality—temperature, mineral content, bacteriological content
   b. Quantity
   c. Dependability—may involve reservoir construction
   d. Costs
5. Climate:
   a. Investment required for construction
   b. Humidity and temperature conditions
   c. Hurricane, tornado, and earthquake history

Specific Factors

6. Transportation:
   a. Availability of various services and projected rates
      (1) Rail—dependable for light and heavy shipping over all distances
      (2) Highway—regularly used for short distance and generally small quantities
      (3) Water—cheaper, but may be slow and irregular
      (4) Pipeline—for gases and liquids, particularly for petroleum products
      (5) Air—for business transportation of personnel
7. Waste disposal:
   a. Regulation laws
   b. Stream carry-off possibilities
   c. Air-pollution possibilities
8. Labor:
   a. Availability of skills
   b. Labor relations—history and stability in area
   c. Stability of labor rates

9. Regulatory laws:
    a. Building codes
    b. Zoning ordinances
    c. Highway restrictions
    d. Waste-disposal codes
10. Taxes:
    a. State and local taxes
        (1) Income
        (2) Unemployment insurance
        (3) Franchise
        (4) Use
        (5) Property
    b. Low assessment or limited term exemptions to attract industry
11. Site characteristics:
    a. Contour of site
    b. Soil structure
    c. Access to rail, highway, and water
    d. Room for expansion
    e. Costs of site
    f. Site and facilities available by expansion on present company-owned property
12. Community factors:
    a. Rural or urban
    b. Housing costs
    c. Cultural aspects—churches, libraries, theaters
    d. School system
    e. Recreation facilities
    f. Medical facilities—hospitals, doctors
13. Vulnerability to wartime attack:
    a. Distance from important facilities
    b. General industry concentration
14. Flood and fire control:
    a. Fire hazards in surrounding area
    b. Flood history and control

Figure 2-2 shows another method of tabulating factors for an industrial plant location. No matter what method is outlined, a quantitative approach is usually required to establish the economic optimum site location.

One site location which should never be overlooked is that of present company-owned sites where new facilities can be added to present facilities at possibly lower costs. This type of site is subject to the same site location analysis as new sites; it is almost always included in the final economic comparison submitted to management (see Table 7-3).

A recommended procedure, which will be illustrated by an example in this chapter, involves a relative evaluation of the primary factors of raw materials and markets. This will serve to eliminate certain regions from the plant location analysis. For example, if neither the raw materials nor markets are located in the Rocky Mountain region, it would be foolish to set up a large tonnage operation in this region. The next step is to choose sites in one or more geographical regions of the United States, or possibly other parts of the world. Examples of regions are Southeastern, Gulf, Ohio Valley, Great Lakes, Northeastern, and West Coast of the United States; Latin American countries; etc. Then make a quantitative point evaluation or weighted scoring of all the primary factors and specific factors. An example of a quantitative relative point system of evaluation is shown in Table 7-2. This is presented at the end of this chapter in working out a sample problem on locating a site for the benzene hexachloride plant discussed previously in Chaps. 3 to 6.

ECONOMICS OF PLANT LOCATION

The final choice of the plant site usually involves a presentation of the economic factors for several equally attractive sites (see Table 7-3). The exact type of economic study of plant locations will vary with each company making a study. As explained in Chap. 6, it should include the following:

Investment
  Plant
    New money
    Existing facilities
Working capital
Annual sales
  Cost
    Manufacturing
    Distributing
    Selling, research

Annual earnings
  Operative
  Net, after taxes
Net annual return
  On total investment
  On new plant investment only

The limitations of preliminary plant location cost studies should be recognized and pointed out to management. No matter how carefully a survey is prepared, future trends such as population and market shifts, development of competitive processes, and the advent of new industries, services, and transportation facilities cannot be reliably predicted.

PLANT LOCATION FACTORS

The above presentation shows that a great deal of experience and knowledge is required to make a realistic plant location study. The next sections of this chapter present a more detailed discussion of the primary and specific plant location factors so that the student can gain some knowledge in this area. An actual plant location problem is then worked out at the end of the chapter. It is impossible to include a very complete compilation of specific information for a plant location study in this type of textbook. A listing of useful sources of information is presented in Table 7-1.

CHAP. 7]   LOCATING THE CHEMICAL PLANT   269

TABLE 7-1. SOURCES OF INFORMATION FOR PLANT LOCATION STUDIES

A. U.S. Government Agencies:*
  1. U.S. Department of Labor—cost of living, wage and hour statistics, price data
  2. U.S. Department of Commerce—manufacturing, distribution, population and business data
  3. U.S. Department of Interior—natural resources, seismological, and topographical maps
  4. Federal Trade Commission—trade laws
  5. Federal Power Commission—power policies
  6. Interstate Commerce Commission—shipping policies
  7. Federal Security Agency—employment statistics
  8. U.S. Public Roads Administration—transportation statistics
  9. U.S. Coast and Geodetic Survey—water transportation statistics
  10. U.S. Geological Survey—geological site information
B. State and Regional Agencies:
  Write to state office at capital of individual state requesting the information desired. This request will be forwarded to the correct agency. Information is generally available on possible industrial sites, resources, labor supply, unemployment, wage scales, state and local government taxes, and legal codes.
C. Private Information:
  1. Railroad companies—freight rates, shipping schedules, sites available on railroad routes
  2. Power companies—rates and economics for power demands
  3. Local chambers of commerce—for specific information on local area problems
  4. National Industrial Conference Board, 460 Park Ave., New York
  5. National Association of Manufacturers, 2 E. 48th St., New York
  6. Manufacturing Chemists' Association, 1625 Eye St., N.W., Washington, D.C.
  7. Moody's Investor Service, New York, industrial securities and manuals of security rating
  8. Standard and Poor's Corporation, New York, industrial surveys
  9. World Almanac and Book of Facts, *New York World Telegram and Sun*, New York (yearly)

* H. S. Hirshberg and C. H. Melinat, "Subject Guide to U.S. Government Publications," American Library Association, Chicago, Ill., 1947.

Raw Materials Supply

Probably the location of the raw materials of an industry contributes more toward the choice of a plant site than any other factor. This is especially noticeable in those industries in which the raw material is inexpensive and bulky and is made more compact and obtains a high bulk value during the process of manufacture. The supply of the basic raw materials should be controlled directly by the user; tonnage items like coal, salt, limestone, etc., should be available on company-owned property on the plant site or reached by not too distant pipeline, rail, or water transportation. Surveys should be made to chart definitely the quantity and quality of the basic raw materials; a minimum supply of 30 to 50 years is usually considered mandatory. The steel mills are located near the iron mines or at some intermediate point between the iron and the coal mines. The flour mills in the Middle West are near the wheat fields, and the cotton mills are in the heart of the cotton-growing section. The meat-packing industry in the United States is close to the great western fields upon which the herds are raised, and at the head of a transportation system feeding an extensive market.

The salt, gypsum, sodium sulfate, soda ash, carbonates, borax, natural nitrate, and many other industries that take their raw materials from saline residues are all, by necessity, located directly at the source of supply. The location in West Virginia of a company for the manufacture of synthetic ammonia was for the purpose of being near the coal fields that supply the necessary raw materials.

Physical distance is not the only controlling factor in the source of raw materials, for purchase price and buying expense, base point pricing, reserve stock, and reliability of supply are also determinants.

Markets and Transportation

The existence of transportation facilities has given rise to many of the greatest trade centers of the world. The character of a business will, however, determine the type of transportation used. The relation of railway to market is so close that no pains should be spared in making a careful investigation of freight rates before definitely deciding upon a plant location. A location should be chosen, if possible, which has several competing railroads and waterways in order that the competition will help to maintain low rates and give better service. The widespread use of motor trucking facilities, following the comparatively recent development of good interconnecting roadways, has supplemented and in some cases even supplanted the railroads. The formation of organizations that will pick up and deliver odd lots of freight has been a great help to isolated factories in the smaller towns. Transportation by airplane also calls for proximity to commercial airfields.

Industries that are national in scope find an advantage in locations where the transportation of a low-priced, bulky raw material is feasible, rather than where a high-priced material of small bulk must be transported. The location of the large sugar refineries along the seacoast is dictated by the fact that the raw, unrefined sugar is received by boats from the sugar-producing countries.

The freight rate probably plays a greater part in the success of chemical plants than in any other industry. The raw material is obtained from certain sections of the country, in some cases very much isolated, and price is regulated largely by the cost of transportation to consumers.

The rapid rise in freight rates during the past 10 years has been a strong contributing factor in causing many of the larger older plants, located in cities, to seek new locations. Oftentimes, a location is selected outside the city limits in order to have a railroad siding available and thus eliminate trucking costs to freight yards from the excessive cost of transportation. There will be more long-distance water transportation used in the future to reduce the cost of freight, with the spread between production cost and sales cost constantly narrowing.

Markets for Finished Product. The question of markets probably assumes greater importance for the intermediate and smaller industries, since such groups generally wish to deal directly with the market and dispense with the services of a middleman in disposing of the product. The concentration of industries in the larger cities is evidence of this fact.

The location of warehouses is largely a question of market. Large tonnages of steel are shipped by barge or lake boat to warehouses at the end of the water route, for final distribution of the material by rail. Grain is loaded direct from elevators at the head of the Great Lakes and shipped to Buffalo, where it is unloaded from boats to elevators for redistribution. Water shipments are economical where rail handling at the loading and delivery points can be kept down to a minimum and where the water haul is long enough to accumulate a saving as compared with all-rail freight.

The large oil refineries are located along the seacoast or near large cities where a market exists for the finished products. Crude oil is easily pumped by pipelines or shipped from the oil wells in the interior; cheap rates are thus secured for a crude, low-priced commodity, consumed in large quantities, while the finished products are made in the center of the market in order to lower the distribution cost.

Power and Fuel Supply

Fuel. The best plan is to locate near large coal fields if coal is the primary fuel or to tie up by long contracts sufficient gas and/or oil to ensure the continuity of operations over a long term. The Gulf Coast area has been built up in the past 15 years largely because of plentiful and cheap gas. When gas is the basic raw material, as in ammonia synthesis, synthetic gasoline, it is the controlling economic factor as it supplies the raw materials, heat, and power.

The cost of coal at the mines is constantly rising because of increased labor and other factors and this coupled with higher freight rates makes for an expensive fuel. The quality is slowly getting worse as the years go by and boilers should be built to burn the poorer grades of fuel over the long term.

Sources of Power. Power for chemical industry (Fig. 6-34) is primarily from coal, water, and oil; these fuels supply the most flexible and economical source, inasmuch as they provide for the generation of steam both for processing and for electricity production. Power can be economically developed as a by-product in most chemical plants, if the needs are great enough, since the process requirements generally call for low-pressure steam. The turbines or engines used to generate electricity can be operated noncondensing and supply exhaust steam for processing purposes.

Steam is generated from whatever fuel is cheapest, usually at pressure of 450 psig or more, expanded through turbines or other prime movers to generate the necessary plant power, and the exhaust steam is used in the process as heat.

Steam Power. The quantity of steam used in a process depends upon the thermal requirements, plus that to meet the mechanical power needs, if such power is generated at the plant. Steam-power-plant prime movers are well standardized and include (1) simple engines using saturated steam, (2) simple engines using superheat steam, (3) compound engines, (4) triple- and quadruple-expansion engines, (5) impulse turbines, and (6) reaction turbines. Other prime movers include water wheels, hydraulic turbines, and internal-combustion engines (gas, gasoline, and oil). Water wheels and hydraulic turbines require operation by specialists in these lines.

The chemical engineer may come into contact with such prime movers, but more often his plant equipment will be electrically driven, with electricity generated in the plant's own powerhouse or purchased from a public utility.

Process Steam Requirements. The data in Fig. 6-33 cover estimated process steam requirements for a number of various chemicals; these data for steam exclude that used primarily for mechanical drives and power generation. Data were selected and averaged.

Electric Power. The cost of electricity is the most important consideration in electrolytic and electrometallurgical processes, in many cases determining plant location over all other factors. An example of this is the location of the largest aluminum plant in the world, the 1,000 ton/day plant of the Aluminum Company of Canada at Arvida, Quebec; here 1,200,000 kw is available. In spite of the fact that the basic raw material, the bauxite ore, must be shipped from British Guiana, some 4,000 miles away, the Canadian plant produces probably the cheapest aluminum which can be shipped competitively to Europe and the United States. The favorable factors of plentiful, low-cost power and deep-water transportation overweigh all other considerations, including that of extremely severe winter weather making for difficult operating conditions. Similarly, the Buffalo–Niagara Falls region owes its growth since the turn of the century primarily to the ample supply and the low cost of hydraulic power generated at Niagara Falls. However, proximity to deep-water transportation and raw materials, plus a location on the Great Lakes in the heart of the East, is a combination not usually found and hard to beat.

In pulp and paper, portland cement, and glass and clay products, the source of fuel and power is of importance. Some of the largest generating and distributing companies in the country, however, have recently been developed in the South and offer power rates that compare favorably with the northern locations. Kanawha River Valley, West Virginia, and the region under the Tennessee Valley Authority are growing rapidly as industrial centers owing to proximity to the coal mines in the one case and cheap hydroelectric power in the other (see Table 6-13).

Over 70 per cent of the total electric power used in the chemical process industries is consumed by motor-driven machinery. It is particularly desirable for the process engineer to have a clear picture of the application of electric motors and controls because their characteristics have a very definite effect on the proper functioning of the associated process equipment.

The chemical engineering industries are the largest users of electric-power equipment among the industries today because the modern demand is for extreme flexibility that sometimes errs on the side of too many individual drives. Practically all modern chemical equipment, such as pumps, high- and low-pressure mixing vessels, dryers, high-speed pulverizers, attrition mills, compressors, and conveyors must be driven. All such equipment can be propelled either by individual electric motors or by systems of belting, shafting, and gearing.

Water Supply

Sources. Water for industrial purposes can be obtained from one of two general sources: the plant's own source or a municipal supply. If the demands for water are large, it is more economical for the industry to supply its own water. Such a supply may be obtained from drilled wells, rivers, lakes, dammed streams, or other impounded supplies. Before a company enters upon any project, it must ensure itself of a sufficient supply of water for all industrial, sanitary, and fire demands, both present and future. Data on temperature of water and on maximum, minimum, and average rainfall can be obtained from governmental agencies if surface water is to be impounded, or the data on stream flow of rivers can be acquired likewise. If wells are to be relied on, geologists and practical well drillers should be consulted.

The United States has abundant supplies of ground water and surface water based upon annual rainfall and provided adequate methods of preventing rapid runoff are installed. The 200 billion gpd water consumption in the United States is far less than that available from annual precipitation. All areas of the United States are not equally fortunate in the matter of rainfall or in the matter of annual runoff, as shown in Fig. 7-1. Two man-made problems, ground-water depletion and stream pollution, further aggravate the situation in congested industrial areas. Ground water accounts for 20 per cent of all water used—agricultural, industrial, and municipal. The distribution per capita has risen rapidly during the past quarter century because of the enormous quantities of irrigation water used; this amounts to a daily consumption of 10 billion gal, with prospects of rising rapidly with the practice of movable pipe irrigation now becoming widespread on Eastern, Southern, and Middle Western small farms. Recirculation, reuse, industry treatment of its own dilute industrial wastes, and the treatment of municipal sewage for

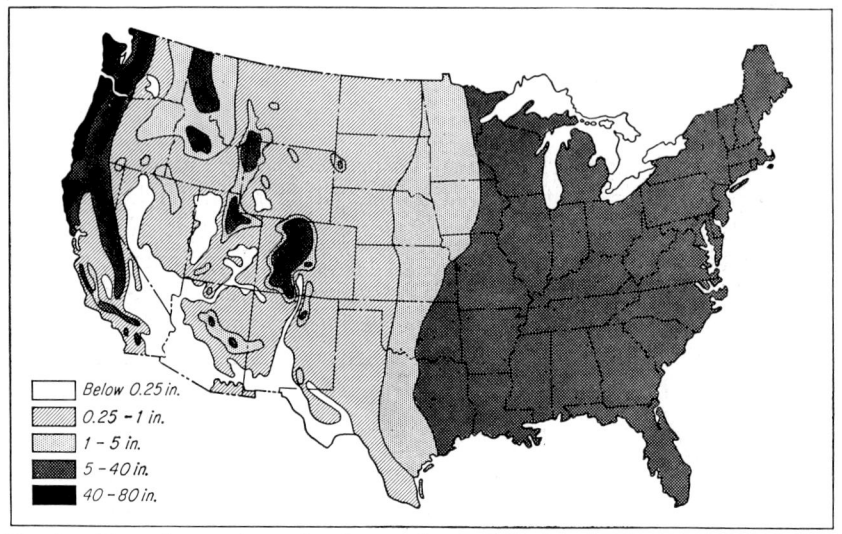

Fig. 7-1. Normal annual runoff. Stream flow and ground-water studies furnish data necessary for utilization and regulation for both industry and agriculture. [*Staff Report, Chem. Eng.*, **55**(1): 137 (1948).]

industrial water supply are methods adopted in specific areas to augment dwindling supplies or increasing demands.

Chemical Industries' Water Requirements. Water of sufficient quantity at the right temperature and quality, from either underground or surface sources, is the desire of a chemical engineer. Making chemicals consumes tremendous volumes of water. In most chemical industries, the quantity of water required is the greatest single item that goes into the plant.

In relation of water supply to chemical plant location, Powell and Von Lossberg[1] list the major items requiring consideration.

[1] S. T. Powell and L. C. Von Lossberg, Relation of Water Supply to Chemical Plant Location, *Chem. Eng. Progr.*, **45**: 289 (1949).

1. Availability of surface and underground water, and seasonal fluctuation in quality, quantity, and temperature
2. Data to show, statistically, the influence of prevailing meteorological conditions on the availability of the water supply under consideration
3. Chemical composition and physical characteristics of the water supply, including sufficient data to show the average, maximum, and minimum conditions
4. Quality of supply as revealed by microscopic and bacteriological analyses
5. Existing or predicted influence of industrial or domestic waste contamination
6. Estimated requirements for various needs of the proposed plant, such as water for processing, cooling, steam generation, sanitary uses, and fire protection
7. Evaluation of future requirements for the various services so as to provide for expansion
8. Availability and cost of using a municipal water supply compared with the development of a private supply
9. Design of heat-exchange equipment, recycling of cooling water, and provision of a cooling tower to provide for the conservation of the total volume of water required
10. Influence of the present rate of water consumption and the predicted results of industrial expansion in the area upon the continued availability of the supply and effect upon quality and temperature
11. Effect of industrial contamination on surface and underground waters, resulting from industrial activity in the area

The process requirements of water for specific chemicals vary considerably. Figure 6-32 lists 72 chemical industries and shows estimates of water requirements, based upon 1 lb of each chemical. It is understood in this estimate that any of the water going directly into the products must be treated to meet the specific requirements of the product being manufactured.

Quality. Water is a raw material; its natural quality is as diverse as the sites and regions from which it is obtained; for specific purposes it must be conditioned and treated; the variety of chemical industries consuming water is so great, and the specific physical, chemical, and bacterial requirements so diverse, that it is impossible to come up with standard specifications for water quality. Water-quality requirements in the same chemical plant change as processes are refined and new ones developed. Of greatest economic importance in determining the value of water to the chemical industries are temperature and chemical characteristics, such as organic and inorganic contaminants; the relative importance of each of these depends upon the service for which the water is wanted. Boiler feed water must meet the most exacting of water-quality requirements. The very-high-pressure boilers demand water from which almost all organic and inorganic salts (even traces of silica) have been eliminated. Ground water contains less suspended matter and is more apt to be free from contamination by sanitary and industrial wastes than surface water; but it often contains higher concentrations of soluble salts. Characteristics of surface waters vary widely. Lakes, especially the Great Lakes, are not influenced seriously by drought conditions. But rivers fluctuate considerably during wet and dry seasons; rivers have shown wide variations in hardness and salinity during high and low stages. Moreover, salt water penetrates upstream in tidal streams during droughts, increasing the salinity; on the James River sea water penetration upstream at these times has reached up to near Hopewell, Virginia. Plants on tidewater will always get water more or less salty; fresh water is always to be preferred over brackish water, because of the easier corrosion problem. Sea water, always available in volume, exhibits limited fluctuations in chemical quality compared with fresh-water surface sources. Along with the salt water there is ever present the problem of marine growth, such as barnacles, slimes, etc.

Temperature. There is probably no single water-supply factor of greater economic importance than the temperature of the water available. It governs design of equipment and the heat balances of many chemical processes. In most places, the temperature of the surface water tends to follow the atmospheric temperature throughout the year. Water from deep wells, therefore, is better than surface water for heat exhangers because of its more uniform temperature.

Cooling-water requirements of the chemical industries are enormous; for cooling, the volume available and the temperature far outweigh considerations of quality. An abundant supply of low-temperature cooling water can be evaluated as equivalent refrigeration tonnage. Sea water is fairly satisfactory for cooling; always available in volume, it exhibits limited fluctuation in temperature.

Ground waters, in specific areas, are an excellent source of uniform temperature if sufficiently available. The approximate temperature zones in isothermal bands throughout the United States are shown in Fig. 7-2.

Plant Measures for Conservation of Water. Use of water can be cut by means of cooling towers, recirculation, and other water-saving measures. In-plant economy, or water economy within the plant, calls for elimination of wasteful practices, for installation of multiple water systems, for recirculation and reuse of water, and for reclamation of usable water from plant effluents. *Progressive treatment of raw water* is possible. Water of different grades and temperatures can be used for different purposes. One big industrial plant in Ohio circulates cooling water progressively through heat exchangers requiring 54°F water and finally through units in which 85°F water is satisfactory. Process water may thus be utilized at increasing temperature levels. Separation of cooling water from waste water makes stream pollution control easier. *Recirculation of cooling water* over cooling towers brings enormous reductions in the volume of water required. With 160,000-gpm cooling towers, only 6 to 8 per cent make-up is required at the Texas City plant of Carbide & Carbon. Celanese, at its Bishop, Texas, plant, actually recirculates cooling water as many as fifty times before discharging it

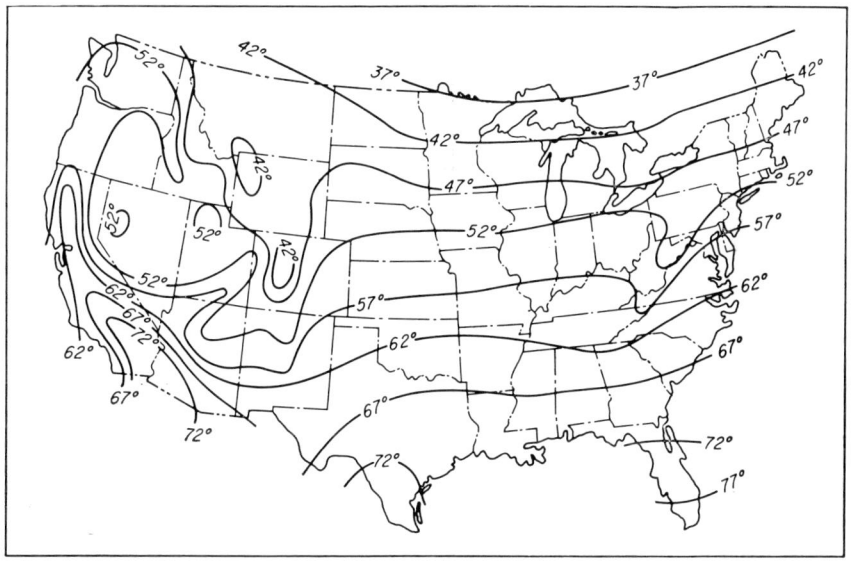

FIG. 7-2. Temperature of water from nonthermal wells at depths from 30 to 60 ft. It usually exceeds mean air temperatures by 3 to 6°F. [*Staff Report, Chem. Eng.,* **55**(1): 137 (1948).]

because of low quality. By *recirculation of process water*, Celanese cuts its requirements from 230 mgd to 4 mgd. A large West Coast steel plant recirculates cooling water and reuses effluent water from its sanitary and industrial waste treatment works. In starch manufacture the steep water is now recycled completely. Where water is used for extraction, it is advantageous to recycle it. Paper mills now recycle their white water; a board mill with efficient purification and recirculation of water can get by with 1,000 gal of fresh water per ton of product, while a pulp mill making bleached pulp and reusing little of the mill water requires as much as 50,000 gal/ton. A Texas refinery reclaims its water by treating its waste waters to avert stream pollution and thereby greatly conserves its limited fresh-water supply.

Water for industrial use is *reclaimed from sewage plant effluent* by the Sparrows Point plant of the Bethlehem Steel Company; with an option to take 100 mgd it now buys 40 mgd of sewage effluent from the city of Baltimore. It treats this sewage at a cost, exclusive of fixed charges, of $0.017/1,000 gal. As a result Baltimore has cut back its pumpage of ground water from 15,000 to 3,500 gpm. From a dangerous low of 150 ft in 1940, the ground-water level is back to a comfortable 80 ft.

Artificial Recharging of Ground Water. Artificial recharge of the ground-water supply has been made from the Ohio River at Parkersburg, West Virginia; Louisville, Kentucky; and Charlestown, Indiana. Water put back into the ground must be sanitarily uncontaminated and free from silt and minerals that might clog the aquifer and impair the efficiency of recharge wells. Putting too much warm cooling water back into the ground raises the temperature of the ground water. Before an artificial recharge program is undertaken, it must be ascertained whether the water returned to the ground at the plant will be confined in the immediate area or flow away to be withdrawn by other users. A thorough knowledge of geologic subsurface features is essential. Two wartime ordnance works, one on the Wabash River at Clinton, Indiana, and the other on the Ohio River at Charlestown, Indiana, used horizontal Ranney collectors to induce river water to flow into the aquifer next to the river, replenishing the underground supply. The aquifer also acts as a natural slow sand filter. Six Ranney wells at the Clinton plant yielded 72 mgd during peak operations; seven at Charlestown averaged 40 mgd. National Carbide at Louisville, Kentucky, has a single infiltration collector. In West Virginia, Connecticut, and elsewhere, other plants also employ river infiltration.

Lake Sources. To get water for its Whiting plant, Carbide & Carbon ran an intake pipe 700 ft out into Lake Michigan and put in an automatic pumping station. At Painesville, Ohio, Diamond Alkali takes 100 mgd and Industrial Rayon 30 mgd from Lake Erie. At Ashtabula, Ohio, Union Carbide's Electromet Division gets 7 mgd from Lake Erie, and National Distillers is pumping 5 mgd for various uses in its sodium plant.

Sea Water. To conserve fresh water, coastal chemical plants use sea water for cooling. That means higher fixed costs for corrosion-resistant equipment, but operating economies often compensate for this. Recirculated fresh water can be bottled up in a closed system and cooled with sea water. Conversion of sea water, either by ion exchange or distillation, is too costly for industrial use, approximating $0.80 to $1.30 per 1,000 gal.

Government Dams and Reservoirs. Long-distance transfer of water, use of storage lakes, and the building of dams and reservoirs are being undertaken by the municipalities, states, and Federal governments. Los

Angeles, for example, spent 220 million dollars to bring Colorado River water more than 300 miles from Parker Dam. Ohio, with some 400 reservoirs already, plans more. Texas has more than 40 major surface storage facilities with a total capacity of over 13 million acre-ft. The U.S. Bureau of Reclamation now proposes its 3 billion dollar basin development plan for California's Central Valley; applying the multiple-purpose philosophy to the entire basin, the bureau envisions 38 reservoirs, plus dams, power plants, transmission lines, pumping stations, and hundreds of miles of transfer canals. This plan would conserve the water resources of the San Joaquin and Sacramento Rivers and of the Central Valley Basin as a whole.

Municipal Water Supplies. A city water supply is an easy, if not an economical, solution to a water-supply problem. The quality of the water is easy to ascertain, as well as the size and condition of the supply mains, the normal and reserve supply, zonal distributive flow, the log of temperatures, and the pressure conditions.

Water Costs. (See Table 6-11.) The value of an abundance of good water is reflected in the selling prices of plant sites that have such supplies. Elaborate engineering techniques now required to procure, conserve, and treat water are significantly increasing costs. The rising cost of industrial-water procurement is also related to the substitution of surface water for inadequate ground-water supplies. Moreover, when surface water is substituted for ground water, treatment of process water becomes more expensive. Increased costs of water processing, which include that of treating the wastes discharged to the watercourse by municipalities and industries upstream, have made maximum use of the processed water essential; also, the high costs of constructing and operating a waste treatment plant have led to concentration of industrial wastes in the smallest amount of water, except where treatment processes require dilution.

Legal Restriction. State water laws, first passed in the West and Southwest, are now being enacted in the Middle West and East. Laws governing prior rights to water, withdrawal of ground water, and pollution of streams are already on the books. Pollution of ground-water reservoirs is serious too, because ground water moves so slowly that pollution, once introduced into a reservoir, may persist indefinitely. Many streams, especially in the East and Middle West, are so badly polluted that they are unusable. Several states have pollution control programs; the Federal pollution abatement law is in force. Arizona recently enacted a ground-water control law, thus becoming the eighth state to have such a statute. New Mexico's law calls for drilling permits and plugging of wells that leak. A New York law requires that for every well sunk to get water for air conditioning in the four Long Island counties, which include

Brooklyn, another must be used to return the water to the ground. New Jersey now allocates underground water to industries by granting or withholding drilling permits, with Pennsylvania having similar legislation. Virginia's new Water Control Act, permitting no new pollution and requiring reduction of present pollution levels, is creating great difficulty for industries already there and is halting the influx of new industries. In Indiana, a single user taking more than 200 gpm from the ground for cooling or air conditioning without a special permit must either circulate the water through cooling towers or other devices and use it over or return it to the ground through recharge wells.

Federal Water Pollution Control Act. Public Law 845 covers all interstate rivers and their tributaries; it authorizes the Federal government to study, advise, and help finance pollution control programs. And with the consent of the state concerned, it can eventually force industry to take whatever steps the courts decide. While the bill authorizes the Federal government to help defray the cost of municipal pollution control projects, it makes no such provision in the case of industry. However, 1 million dollars a year is authorized for grants to states for conducting research, surveys, and studies of the prevention and control of water pollution from industrial wastes. Under the Division of Water Pollution Control are to be 14 river basin offices covering the major drainage basins in the United States. Their job will be to conduct over-all studies, determine extent and sources of pollution, advise on corrective measures, and coordinate state and industrial programs.

Climate

Chemical plants as a general rule are rather difficult to insulate or to provide with artificial heat or conditioned air, except in the individual process units where air conditioning is essential. Excessive cold, deep snows, torrid heat, and excessive humidity reduce the productivity on the part of the workmen. Milder climates make for cheaper installations. In the South and West, many plants are built outdoors, with little or no buildings housing the main apparatus, making for relatively less expensive capital costs. Heating is much less of a problem than in the North and outdoor construction makes equipment easier in some cases to operate and repair. If the site selection is in a hot, damp climate adjacent to salt water, as on the Gulf, one must be prepared for much higher maintenance costs. Corrosion, particularly in the presence of acid and alkali fumes usually present around any large plant, increases maintenance costs. Essentially a process must be designed for *production* of a commodity or commodities under the most economic conditions; if outside plant climate does not fit into the picture of production, then the climate must be designed into the process and process equipment.

Labor

Before locating an industry in any particular locality, a careful study of the supply of available labor must be made. Factors to be considered in labor studies are supply, kind, diversity, intelligence, wage scales, regulations, efficiency, and costs.

Large corporations are decentralizing their production and building unit plants in localities throughout the country where labor supply is abundant.

The success of many an organization is dependent upon the means by which its laborers get to and from their work. A cheap site may have been chosen but no attention given to the housing facilities. Laborers sometimes live a great distance from the work. The workmen arrive tired at the start of a day's work and again must travel a great distance at its close. Thus what may have appeared to be a cheap location develops into a very expensive one because of inefficiency and high turnover.

Industrial housing, safety-first movements, welfare institutions, better sanitation, lunchrooms, etc., have all contributed to the solution of labor problems.

Labor surveys reveal the discrepancy in wage rates throughout the country and the industries is getting smaller because of the activities of the labor organizations and the "shortening" of distances between areas by modern methods of rapid transportation. The U.S. Employment Service obtains local and regional monthly data on the number of employables, currently employed and unemployed, listing of industrial shops and their active employment, and shifts and trends in employment. These data are supplemented in surveys by chambers of commerce and other civic organizations interested in sustaining or increasing industrial activities in definite areas. These data are also collected in general surveys by the U.S. Department of Labor, available to both labor and management.

Figure 6-35 gives an approximation of process labor requirements in terms of man-hours per process step per ton of chemical product. For firmer estimates it will be necessary to synthesize labor requirements on the basis of job analysis.

Community and Site Characteristics

In the selection of a definite plant site in a designated area there are many minor factors to which some attention should be given. The actual cost of the necessary *land* for the chemical industry is usually of minor importance, as is the item of *taxes*. Of great importance is the nature of the *subsoil;* the need of piling or other expensive foundations can materially affect construction costs (see Chap. 8). It takes real vision when locating a new plant to provide adequately for future *expansion*. No one can envision what the plant growth may be over a period of twenty-five to fifty years but one should err on the safe side and provide what appears to be more than sufficient space at the start. Land is relatively cheap compared with over-all plant costs and if provision is made for too much, it can be sold usually at a profit later on. When considering industrial plant locations in the light of possible *future atomic bomb attacks*, the National Security Resources Board states that most locations near cities of 50,000 population or less can be considered safe, that is, unless the industrial target should be a large important one such as a steel mill or a chemical plant making strategic war materials or the like.

The following site information is essential in the selection of a plot: (1) location with reference to adjacent areas, and restrictions imposed thereon; (2) codes (building, safety, and other—state and local); (3) labor regulations and conditions; (4) accessibility to freight and passenger transportation; (5) type of soil (rock, gravel, sand, silt, clay); (6) elevation above sea level, grade, and relation between grade and foundation tops and depths; (7) ground-water level; (8) drainage conditions, need for waterproofing of underground conduits; (9) atmospheric temperatures, wind velocities, rainfall; (10) special conditions such as earthquakes, cyclones, exceptionally severe or prolonged cold, heat, or humidity.

The ideal plant site is one which is at the crossroads of at least three or more railroads, has good hard-surfaced highways leading in all directions, and preferably has barge canal and deep-water transporation; also proximity to one or more commercial airports is a definite asset.

Chemical plants are not usually looked upon as desirable neighbors. They may be regarded as a source of danger from possible explosion, or, because of fumes, as a detriment to the health of the community and to its vegetation. The result of propaganda against the industry has been that many communities have passed *ordinances* against chemical manufacturing or certain classes of it. It is advisable to ascertain whether the attitude of the community is particularly unfriendly to chemical manufacturing; if this is the case, it is good policy to go elsewhere even though no restricting ordinances are in effect at the time. Such information can usually be obtained from the local chamber of commerce or board of trade, or by making inquiries at other plants in the locality.

Most large cities and many of the smaller towns are zoned, i.e., divided into residential, business, and unrestricted districts. A distinction is often made between so-called light manufacturing and heavy manufacturing. Chemical plants generally must be located in heavy or unrestricted zones. *Zoning regulations* must, therefore, be carefully investigated. Where such regulations are in force, a map is usually published outlining the various zones. As their boundaries are subject to small changes at frequent intervals, care must be taken to ascertain that the zoning map has been corrected to date. An unrestricted zone may even be slowly changing in character on account of the encroachment of residences, parks, or other civic developments.

If the industry for which a location is to be chosen is in any way hazardous or likely to be a nuisance owing to the possible escape of fumes or objectionable odors, a site should be selected at a considerable distance from houses and public institutions. This is a precaution that is frequently overlooked, which if not heeded may cause endless trouble and expense.

If there are other plants in the immediate neighborhood of a site under consideration, it is advisable to find out what products these plants are manufacturing and by what methods. For example, an adjoining plant manufacturing a highly flammable product would affect the insurance rate if its buildings were sufficiently close. The neighboring plant may also give out dust or fumes which may affect the product to be manufactured.

The introduction of *public improvements*, such as park and boulevard extensions and the conversion of swamps and unsightly public dumps, etc., into industrial areas, should be attended by a proper attitude of mind on the part of public officials, so that unnecessary and destructive limitations will not be placed upon such areas, nor the tax burden become prohibitive. Such improvements, on the other hand, add considerably to the upbuilding of the morale and welfare of the employees. Good health, good schools, good communities, and availability of adequate housing, with ease of purchase or rental, are all basic to good morale.

The availability of local *public utilities* is almost always an asset to any plant. Even when gas or electricity is made in the plant, an outside source is of considerable value for emergency purposes.

Fire apparatus provided in the average American city is for the most part adequate. So far as the location of the chemical plant is concerned, the important desideratum is that the fire apparatus be but a few minutes distant. If deep snow, congested traffic, bridge washouts, or stalled freight trains can prevent firemen from reaching a factory quickly, then a site subject to such conditions should be avoided. Often it is necessary to establish and maintain a company fire station and man it with volunteers or with full-time firemen.

In cities without a definite industrial district, the plant may become engulfed in a residential or a business area. A relatively permanent obstruction such as a railroad yard, a swamp, or a river will divert future community growth.

Industrializing of small cities is growing. Local chambers of commerce no longer cite cheaper labor. A recent survey made by the Chamber of Commerce of the United States in several Southern counties indicates that each 100 new factory workers in a rural community will result in the following additions:

   296 more people
   112 more households
    51 more school children
   174 more workers
     4 more retail establishments
   107 more passenger cars registered
   $590,000 more personal income per year
   $360,000 more retail sales per year
   $270,000 more bank deposits

These all make for greater local wealth in small communities.

Economic Relation to Other Industries. As a rule, a chemical plant will have a better chance of success if located near others of like nature. A group of plants can obtain favorable rail rates, better service from utilities, and plentiful supply of labor. Better banking and technical services are thus available, since familiarity with the needs of the industry results in segregation of knowledge and the establishment of sound credit relationships. Utilization of wastes from related plants or similarity of disposal offers unusual opportunities for economies.

Waste Disposal

Disposal of waste liquors and waste products is frequently a problem for the chemical plant and, therefore, must be given serious consideration in choosing a site. If there is a sewer in the street adjoining the property, the quantity of liquor to be disposed of should be estimated and the size of street sewer checked to determine whether it can take care of the liquor. If the waste liquor is acid or alkaline, contains solids, or has other objectionable features, it is advisable to learn from the local authorities whether the disposal of such liquor in the sewerage is permissible.

Plant Wastes and Waste Disposal. The proper disposition of liquid and solid plant wastes is a science in itself, particularly when a plant makes a wide variety of chemicals. The states are becoming more and more strict in this regard and an adequate waste-disposal system is an absolute necessity. Often plant wastes can be impounded and released to the river or ocean under controlled conditions in time of high water. Gaseous wastes and odors are also becoming of increasing concern to many plant operators, particularly those located in or near cities or towns. Many localities have passed laws to control at all times such things as fly ash, smoke from boilers, and the like, together with all kinds of noxious fumes and odors.

A policy toward waste disposal by any plant should include a statement to the effect that the company will abate all pollution which causes nuisance or results in conditions failing to conform to established standards or criteria of water quality. To control stream pollution effectively, an industry should avoid new sources of pollution through the construction and operation of new plants or additional units at existing locations, and accomplish necessary improvement in waste practices at existing locations. When a new product is being developed through research, methods for controlling the quantity of waste should be developed at the same time that the new process is being worked out. Designs covering new plants and expansion must include the necessary facilities for the proper disposal of waste. This applies particularly when considering the location for a new plant, since the degree of treatment will be influenced greatly by local conditions. Design of the waste-collection system will be influenced by the kind and amount of treatment required. Careful consideration of waste disposal is necessary throughout the design phase. The design engineer, therefore, should work in close cooperation with the company specialist in industrial waste disposal. Consideration should be given to the possibility of collecting the sanitary sewage and a part or all of the industrial waste for discharge into a municipal system. Where this can be done without causing damage to the sewerage system and without upsetting the municipal treatment plant, it may prove to be the most economical course. Before a final decision can be reached on this point, very accurate information on the quality and quantity of industrial waste will be required. Where satisfactory arrangements can be made with a municipality, pretreatment will often be necessary to prevent damage to the municipal system. Cost of doing this will have to be compared with the cost of improving waste quality sufficiently to discharge it directly into the receiving stream.

Another method of waste disposal is by seepage through the ground. If such a method is to be used, soil tests should be made to determine whether the soil is porous enough to permit the disposal of considerable quantities of liquor without accumulation. It is also advisable to check the topography of the area to determine where the liquor will seep in order to avoid trouble from neighboring plants or local authorities. Towns lower down the valley may draw their water supply from the drainage shed upon which the plant is situated.

Radioactive Wastes. With the exception that radioactive wastes cannot be destroyed chemically, the problem of radioactive waste disposal is similar in nature to the problem of chemical or sanitary waste disposal. Thus, discharged radioactive wastes can contaminate air or drinking water and, if absorbed by an organism, can cause biological damage. Health physicists and biologists have established the maximum concentrations of the various radioisotopes to be tolerated in air or water, and the problem of the engineer is to design the waste-disposal system so that these tolerance values are not exceeded. (See Chap. 10 for further information on this subject.)

SAMPLE PROBLEM

Site Location for a Benzene Hexachloride Plant[1]

Plant Design. A plant has been designed for an annual production of 240,000 lb of Lindane (99 per cent γ isomer of benzene hexachloride). It is estimated that ground area requirements are 1.75 acres minimum at an existing site, but at least 5 acres should be anticipated for landscaping, expansion, parking, and recreation facilities at a new site.

Raw Materials. The principal raw materials are crude benzene (99.5 per cent C_6H_6, 0.5 per cent or less $C_6H_5CH_3$) and chlorine. Only a minor amount of sodium hydroxide is required.

Markets. A market survey has shown that the principal use of Lindane is for an agricultural insecticide to control boll weevil and spider mites on cotton, and for the control of house flies, mosquitoes, and chiggers.[2]

Location of Possible Regions (Use of Raw Material–Market Proximity Method). One of the first principles to be applied here is to search for areas near *raw materials* and *markets*. A data source was located on *cotton* production by states and the information was recorded on a map of the United States (Fig. 7-3). This would constitute the most probable *market area* as a first approximation. A literature search for *benzene* production by states or regions failed to uncover any specific information. It was then necessary to look at the raw materials and processes used for benzene production, namely, coke-oven by-product tar distillation or petroleum refining operation side streams. The assumption is made that the major portion of benzene is produced at plants producing or processing these raw materials. The geographical production data on coke-oven by-products and crude petroleum processing were easily located and plotted on the same map of the United States on which market data were placed (Fig. 7-3). Fortuitously, the location of coal-consuming coke-oven plants and petroleum plants is also indicative of available fuel and power. Statistics by regions for chlorine, the other principal raw material, were directly available and also plotted in Fig. 7-3.

By the raw materials–market proximity principle, a study of Fig. 7-3 would indicate two probable regions for site location areas: (1) the South Central region comprising Texas, Louisiana, Mississippi, and Arkansas as possibilities and (2) the Western region with lower California as the best location. A quantitative procedure for several acceptable areas within these regions, using the weighted-score method, can be done next. (*Note:* For a student problem on plant location, the raw materials–market proximity method illustrated above, followed by a site location on a river or other reliable source of water near an industrial region having adequate power and fuel supplies, should be sufficient for the time allotted in most courses on plant design.) The detailed method described next is given to illustrate one method of how a final plant location report is prepared for management decision. It is obvious that a great

---

[1] See Chap. 6, pp. 259–262.

[2] E. R. DeOng, "Chemistry and Uses of Insecticide," Reinhold Publishing Corporation, New York, 1948.

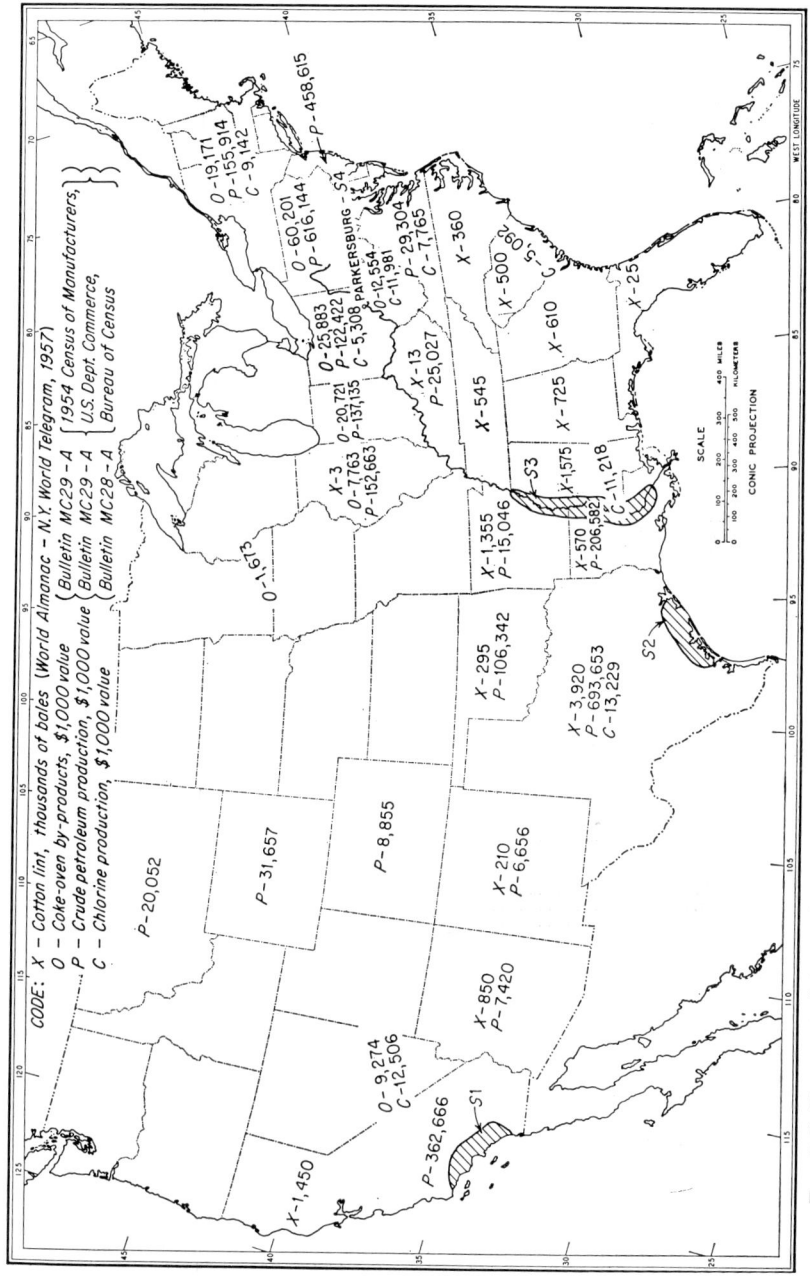

FIG. 7-3. Raw materials–market proximity map. (Blank map by permission of the University of Chicago Press.)

TABLE 7-2. EVALUATION OF SITE AREAS FOR A LINDANE PLANT BY WEIGHTED-SCORE METHOD

| Factor | Weighting item group | Plant location ||||
		Site 1 S. California	Site 2 Gulf Coast	Site 3 Mississippi River	Site 4 Parkersburg, W.Va.
Raw materials supply:					
Availability	50	30	40	35	40
Use of substitute materials	20	0	0	0	0
Distance	30	20	25	20	25
	100	50	65	55	65
Markets:					
Demand vs. distance	40	20	30	40	10
Growth or decline	30	15	20	30	25
Inventory storage requirements	10	10	10	10	5
Competition—present and future	20	20	10	10	20
	100	65	70	90	60
Power and fuel supply:					
Availability of electricity and various types of fuel	60	40	50	50	55
Future reserves	20	20	20	20	20
Costs	40	20	25	30	30
	120	80	95	100	105
Water supply:					
Quality	10	10	5	5	10
Quantity	40	5	5	20	30
Dependability	40	5	5	20	30
Costs	30	5	10	20	25
	120	25	25	65	95
Climate:					
Investment required for construction	20	20	20	20	10
Humidity and temperature conditions	20	15	10	10	10
Hurricanes, tornados, and earthquakes	20	5	10	5	15
	60	40	40	35	35
Transportation, availability and rates:					
Rail	30	20	25	25	25
Highway	30	25	25	20	20
Water	20	10	15	15	15
Pipeline	5	5	5	5	0
Air	5	5	5	5	5
	90	65	75	70	65
Waste disposal:					
Regulation laws	30	10	15	20	15
Stream carry-off possibilities	20	5	5	20	20
Air-pollution possibilities	20	5	10	15	15
	70	20	30	55	50
Labor:					
Availability of skills	40	20	25	20	15
Labor relations	30	25	25	25	20
Stability of rates	30	10	15	25	20
	100	55	65	70	55
Regulatory laws:					
Building codes	10	5	10	10	5
Zoning ordinances	10	5	10	10	5
Highway restrictions	20	10	15	20	15
Waste-disposal codes	10	5	5	10	5
	50	25	40	50	30
Taxes:					
State and local taxes	10	5	5	10	5
Industry exemptions	30	10	15	25	20
	40	15	20	35	25
Site characteristics—existing or new	60	40	40	40	60
Community factors	30	30	25	20	25
Vulnerability to wartime attack:					
Distance from important facilities	15	5	5	10	10
General industrial concentration	15	5	10	10	5
	30	10	15	20	15
Flood and fire control:					
Fire hazards in surrounding area	15	10	5	10	10
Flood history and control	15	15	15	5	5
	30	25	20	15	15
Total point score	1,000	545	625	720	700

deal of study by the engineering team approach is required to collect the detailed information and process it for economic evaluation.

Location of Possible Sites (Use of Weighted-score Method). The next step is to pick possible site areas within the geographical regions which look plausible and make a weighted-score analysis of the various primary and specific plant location factors outlined in the first part of this chapter. To illustrate the method, four site areas were studied (Fig. 7-3).

Area	Designation	Reason for choice
S1	Southern California	In established petroleum, industrial, and market area
S2	Gulf Coast	In established petroleum, industrial, and market area
S3	Mississippi River	In established petroleum, industrial, and market area
S4	Parkersburg, W. Va.	Company has existing facilities and available land at this site

A table of weighting factors for these areas was presented next (Table 7-2). *The valuations indicated for each factor for each area are not the result of careful study and are intended principally for illustration of the weighted-score method.*

The weighted-score analysis shows that a plant site along the lower Mississippi or the existing one at Parkersburg, West Virginia, should be further explored and an economic comparison made. The question of the remoteness of the Parkersburg site from the market with added transportation and distributing costs will have to be balanced against lower investment for site development and presently installed

TABLE 7-3. SUMMARY OF AN ECONOMIC ANALYSIS OF TWO SITE LOCATIONS FOR A LINDANE PLANT

Economic factors	Existing site at Parkersburg, W.Va.	New site at Natchez, Miss.
Investment:		
Plant		
New money	$1,020,000	$1,400,000
Existing facilities	200,000	
Working capital	183,000	210,000
Total investment required	1,403,000	1,610,000
Annual sales	745,000	745,000
Costs:		
Manufacturing	380,000	365,000
Distributing	24,600	21,000
Selling, research	20,000	25,000
Total costs	424,600	411,000
Annual earnings:		
Operative	320,400	334,000
Net (after 52% Federal tax)	154,000	160,500
Net return:		
On total investment	11.0%	9.95%
On new plant investment only	15.1%	11.5%

facilities. Intangible items such as diversification and geographical expansion, future unfavorable legal and tax regulations, etc., can only be decided at the management level.

Assume that further detailed work has located a new plant site at Natchez, Mississippi. Then a detailed economic comparison with the existing plant site at Parkersburg is made. The summary of results without itemized detail figures under investments and costs is listed in Table 7-3. These results are presented in report form to company management for a final decision. From an investment standpoint, the plant should be located at the existing site at Parkersburg, West Virginia.

CHAPTER 8

Site Preparation and Structures

One of the key phases in commercializing a process is the preparation of the site and erection of plant equipment and buildings. A typical problem in the final design stage may be to supply accurate specifications on land and building requirements for construction cost estimating in more detail than given in Chap. 6. A great deal of this work is handled by a *structural design group* consisting largely of architects and civil engineers. Consulting firms with personnel of similar background may also be brought into the picture. However, chemical engineers must have a basic understanding of this type of work and appreciate the effect of chemical plant layout as discussed in Chap. 5 on site preparation and structure designs. This chapter will present a general discussion of this area of plant design so that the chemical engineer will obtain a preliminary knowledge as a basis for further study where necessary.

SITE PREPARATION

Once the plant location area and plot plans have been chosen, as explained in Chaps. 5 and 7, the specific site must be selected and developed for installation of structures. Preliminary exploration of the subsurface conditions should be done prior to purchase of a piece of land, since foundation construction may be abnormally high if the bearing load of the underground strata is poor, even though the surface conditions appear satisfactory. Topographical maps of the site to establish plant grades and excavation, together with transportation facility locations, frost-line, and water-table information, are other important site data required. Some detailed aspects of these factors will be presented next.

Subsurface Evaluations

Since the support of every item in the plant ultimately depends on the soil within the region of the plant site, a thorough investigation of subsurface conditions is required for foundation design, as discussed later. In the final analysis the structural stability of a plant rests largely on the thoroughness with which the subsurface investigations were carried out and applied to foundation design. This type of work can best be done by firms specializing in subsurface exploration and analysis.

Soil Testing. The prime purpose of soil testing is to determine the load-bearing characteristics of subsurface soil. Representative test borings at different locations in the plant site and at varying strata are made to obtain information on many of the following soil and subsurface properties covered by ASTM standard tests:

1. Capillarity—measures rate of rise and height of water rise against gravitational forces
2. Compressibility—measures volume reduction under load
3. Density—measures natural consolidation of soil
4. Elasticity—measures shape recovery after release of applied load
5. Particle size and shape—gravel, sand, silt, clay, and colloids in order of decreasing particle size
6. Permeability—measures rate of gravitational water flow
7. Plasticity—measures ability of soil to change shape markedly under applied load
8. Shearing strength—ability to resist lateral flow
9. Water content and elevation of ground water—water close to the surface requires reinforcement of floors and underground equipment from water buoyancy forces and possible enhanced corrosion

Many of the properties can be grouped together in exploratory boring by means of a spoon-sampling penetration test. A standard 2-in. sampling spoon is lowered into the boring hole at any depth and soil compactness and load-bearing properties measured by the number of blows of uniform impact required to drive the sampler a given distance, usually 12 in. It is customary to carry *preliminary borings* to the refusal point at which the sampling spoon will not penetrate further or to a depth exceeding practical limits for foundation construction. Borings are often carried 5 to 10 ft beyond the refusal point to determine the type of rock present.

Preliminary boring analysis suffices for estimating foundation costs and may reveal (1) the undesirability of a plant location at the site, or (2) the nonuniformity of strata throughout the areas, requiring further boring work to get a reliable profile.

After the site location is definitely chosen, additional *deep borings* are made at points where heavy loadings will be placed or where dynamic loadings from reciprocating machinery will occur. *Shallow-type* borings are made under each major footing, using hand-driven equipment as a final check.

Bearing Load of Soil. The results of soil testing can be incorporated into quantitative unit loading values with suitable factors of safety, as evidenced by Table 8-1. Igneous rocks are best for foundations and will sustain the greatest weights; swamp loam and quicksands have the lowest values. In such cases it is necessary to add something to sustain the weight placed on the ground. Piles can be driven into the ground until the driving weight per unit distance is indicative of the load bearing characteristic of the piles from frictional and end-bearing forces.

TABLE 8-1. SAFE BEARING VALUES OF DIFFERENT FOUNDATION SOILS

Material	Tons per square foot
Granite rock formation	30
Limestone, compact beds	25
Sandstone, compact beds	20
Shale formation or soft friable rock	8–10
Gravel and sand, compact	6–10
Gravel, dry and coarse, packed and confined	6
Gravel and sand, mixed with dry clay	4–6
Clay, very dry and in thick beds	4
Clay, moderately dry and in thick beds	3
Clay, soft	1–1½
Sand, compact, well cemented and confined	4
Sand, clean and dry, in natural beds and confined	2
Earth, solid, dry, and in natural beds	4
Quicksand, alluvial soils, etc	1

Note: The foundations for a building housing heavy, vibrating machinery, such as steam hammers, shears, and grinding equipment, should receive some allowance for possible compression and rearrangements of soil owing to the vibrations transmitted through it.

Frost Line. The lowest depth to which frost penetrates the ground is important for foundation design. If a foundation is laid above the frost line, ground upheaval due to volume changes in freezing and thawing below the foundation will create an instability in structure foundations which must be avoided. Figure 8-1 gives the mean frost penetration lines in inches throughout the United States, and foundations should terminate below these distances. Practical embedment to prevent frost heaving and subsidence is generally 3 to 5 ft, depending on the climate. Foundations underneath buildings are not endangered by frost action except where artificially refrigerated conditions exist.

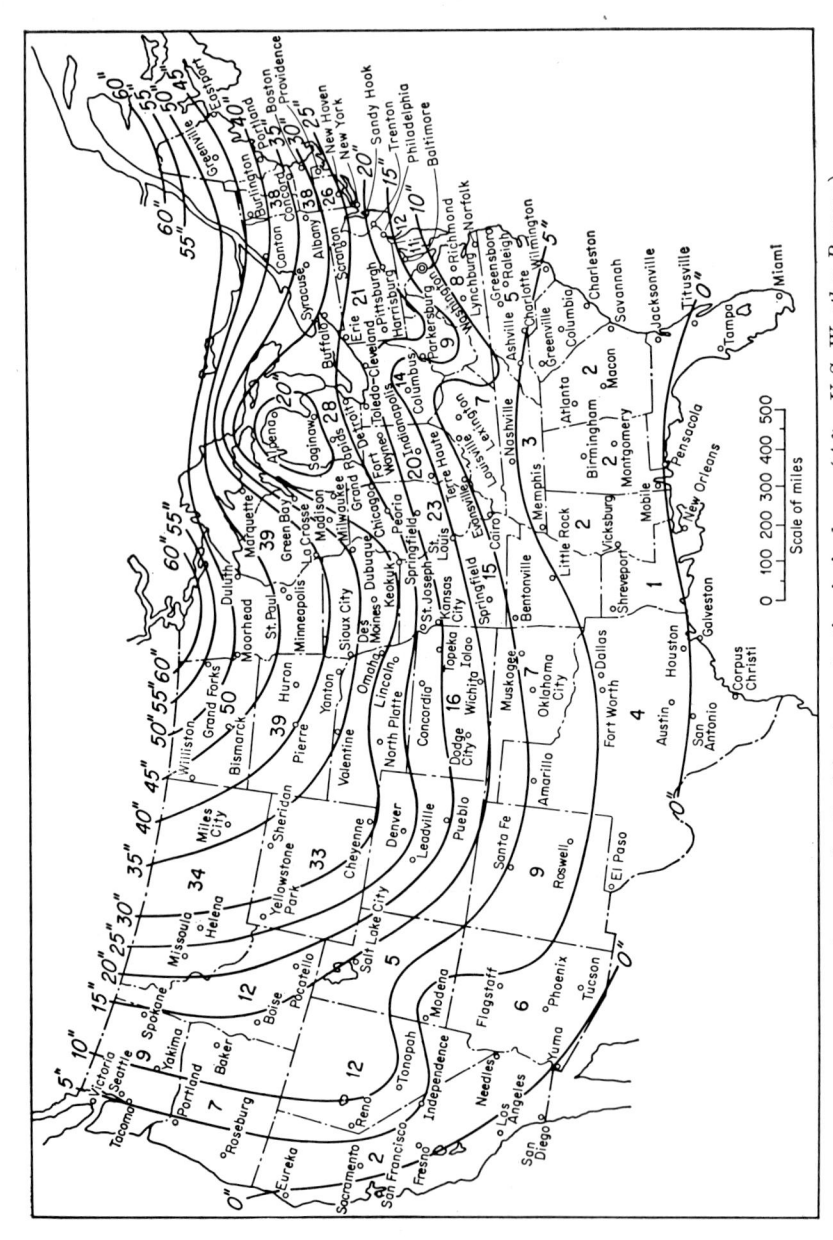

FIG. 8-1. Average depth of frost penetration in inches. (*After U.S. Weather Bureau.*)

Topography Problems

The local topography of the site should be mapped so that grading and drainage costs can be computed. The adequacy of slopes for gravity flow of materials and the layout of transportation facilities can also be determined from the information.

Reference Markers. Plant layout on the site requires a starting or reference point after the exact location of the property has been determined by permanent property markers already established or by independent survey of a local surveying firm. Markers are usually large-diameter capped steel pipe set in concrete, located outside the construction area, so that directional and elevation data can be taken relative to these permanent posts. One marker is called the "zero corner" point and all measurements start from there. "Plant north" and its relation to true north are next set up so that master plot plans can be oriented to the plant site and surveying stakes located in the working area.

Absolute *elevation* planes at the plant site are referred to sea level via bench marks erected by the U.S. Geodetic Survey or state surveys. Within the plant area, a reference datum level, generally at the lowest point on the site, is more useful in specifying plant elevations.

Grading and Excavations. A coordinated study of the topography maps, plant elevation requirements, and foundation designs determine the amount of grading, excavation, and filling required on the site. Machine operations encountered include bulldozers for leveling, compacting, and backfilling, trench diggers for underground piping and conduit layouts, and drag-line cranes for heavy construction excavations.

Construction on fresh fills should be carefully planned since fills do not generally settle to a relatively permanent datum plane until after the third year. Machine tamping, water consolidation of the fill soil, and use of heavy gravel or weak concrete materially aid the permanent settling characteristics of the fill. In other cases buildings are designed to settle uniformly on the fill, but irregular settling can occur with the attendant problems of equipment misalignment, unsightly building cracks, etc.

FOUNDATIONS

The purpose of foundations is to distribute the loading from structures and equipment so that perpetual settlement of the load-bearing soil will not cause excessive maintenance or impair the usefulness of the plant. The selection of a suitable type of foundation or soil loading support structure depends on the loads to be transferred to the foundation, on the material on which the foundation rests, and on the method of placing the foundation as dictated by the subsoil conditions.

In designing foundations the following items should be taken into consideration:

1. Grade level and elevation of foundation
2. Effects of frost on the foundation
3. Bearing load of the soil
4. Gravitational load to be supported by the foundation
5. Intermittent loading from overturning moment produced by winds
6. Dynamic loading from reciprocating or rotating machinery, or infrequently from earthquakes
7. Shape and distribution of load
8. Type and shape of foundation
9. Type of equipment support and anchorage
10. Materials of construction
    a. Plain concrete
    b. Reinforced concrete
    c. Steel
    d. Wood
    e. Combinations of the above materials

The first three points have already been discussed under subsurface evaluations. Some of the important points in connection with the other items in the list will be given next.

Type and Shape of Foundations

The type of foundation chosen will depend on the load to be supported and the bearing capacity of the soil, i.e., the unit loading or pressure. The intensity of pressure on the soil should ideally be uniform over the entire foundation. This is possible to achieve on hard rock, gravel, or good sand, and foundations of the *footing* or *mat* type are specified. In the case of compressible soils, such as mud, silt, or clay, differential settling occurs and *piling* must be used.

Footings. This is the oldest and least expensive type of foundation, which is designed in several ways, depending on the physical and economic conditions. Figure 8-2 pictures five variations of footings.

The foundation for a light building may be a downward continuation of the concrete wall, resting on hard pan or rock. It is termed plain *nonspread footing* and its use is only possible where a suitable bearing stratum is near the surface or near the line of basement excavation (Fig. 8-2e).

With soils of lower bearing capacity it is necessary to increase the bearing area under the support wall or column by means of *spread footings* in order to reduce pressure and avoid undue settlement. Plain spread footings (Fig. 8-2a and b) of either square or octagonal shape are most generally used because of the low-cost forming and reinforcing required. The projection b should not be greater than the depth d of the footing unless transverse steel reinforcement is used. Where a considerable projection is required and reinforcing is not used, some saving in concrete costs can be made by using stepped or trapezoidal spread footings (Fig. 8-2c and d).

Mats. In designing foundations for some structures in low-load-bearing soils the combined area of spread footings approaches that of the supported structure. It is then more practical to use one large reinforced concrete slab or mat under the entire structure. In Fig. 8-3a typical mat construction for support of walls, columns, and flat-bottomed storage tanks is illustrated. A ventilated concrete mat and vertical support foundation for high-temperature supported structures is shown in Fig. 8-3b.

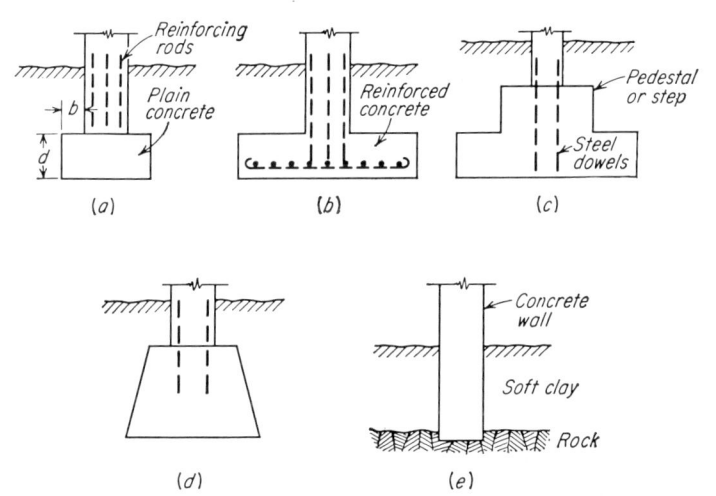

FIG. 8-2. Typical foundation footings. (a) Plain spread footing; (b) plain reinforced spread footing; (c) stepped spread footing; (d) trapezoidal spread footing; (e) plain nonspread footing.

A "floating" foundation can sometimes be designed on compressible soil. This is a boxlike, hollow foundation which displaces as much weight of soil as the applied dead load. For example, in the basement excavation of Fig. 8-3a the weight of soil displaced had a pressure on the supporting stratum of 1,200 psf, or 1.2 kips/ft². The resulting foundation loading was also 1.2 kips/ft²; hence there was no increase in applied load to the soil and the foundation floats with no reason to settle. It is obvious that changes in soil weight due to water flooding or changes in mat loading will create instability in floating mat construction.

Piles. The need for sites adjacent to water-transportation facilities creates a problem of supporting structures on soft clay, silt, and mud of low-bearing capacity (see Table 8-1). Piles must be used as foundation supports in this case.

298   CHEMICAL ENGINEERING PLANT DESIGN   [CHAP. 8

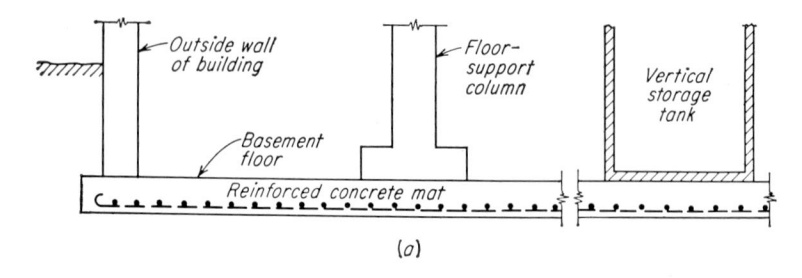

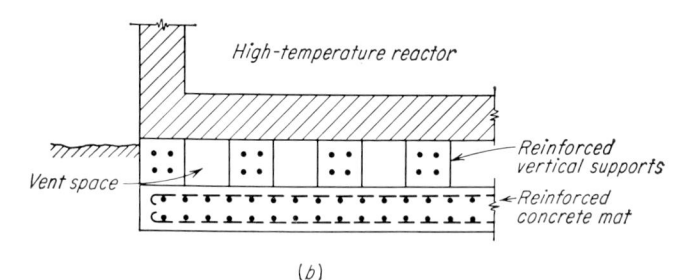

FIG. 8-3. Types of mat foundations. (a) Floating mat; (b) mat for supporting high-temperature equipment.

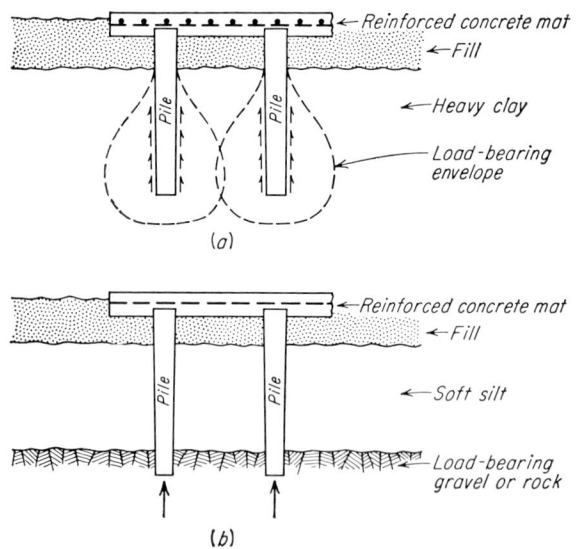

FIG. 8-4. Typical load-bearing action of piles. (a) Friction piling; (b) end-bearing piling. Combinations of friction and end-bearing loading on pilings frequently occur.

There are basically two types of load reaction of piling construction which can be advantageously used. A *friction pile* transfers its support load to the surrounding soil by means of skin friction along its surface (Fig. 8-4a). Design values for skin friction will vary with soil conditions and depth, ranging from 50 to 150 psf in soft silt to 300 to 600 psf in stiff clay or fine wet sand. *End or point bearing* piles are sunk deep enough to rest on hard stratum (Fig. 8-4b). They behave as equally distributed loaded columns in calculating total safe loads.

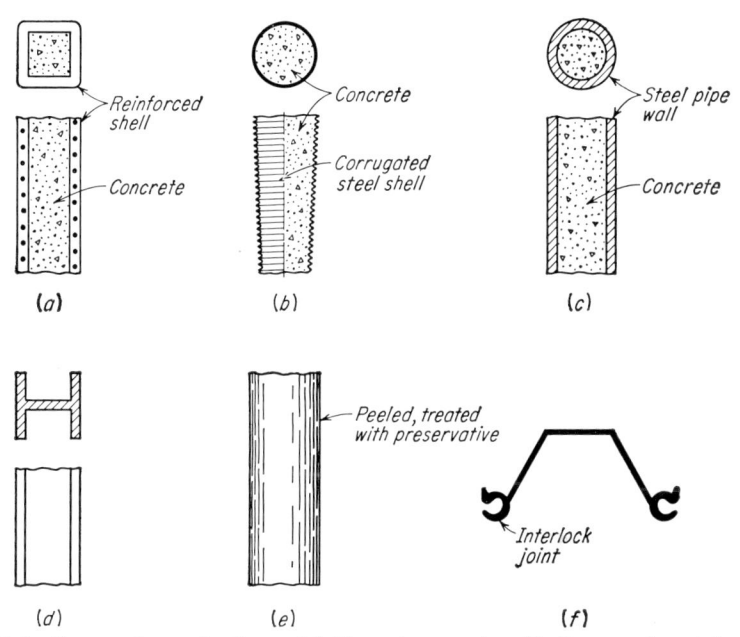

FIG. 8-5. Commonly used piles. (a) Precast concrete; (b) Raymond cast-in-place pile (may have spread footing by ramming concrete); (c) steel pipe pile (may have spread footing by ramming concrete); (d) steel W-beam pile; (e) wood; (f) steel sheet piling.

Some of the types of piling available are shown in Fig. 8-5. Comments which will aid in selection of piling are desirable:

1. *Precast concrete* (Fig. 8-5a). Difficult to handle and transport; shatters when driven into hard strata unless water jets are used; strong as columns and end-bearing supports since they are steel-reinforced and/or prestressed; useful where in-place casting is difficult under water; corrosion resistant.

2. *Cast-in-place concrete* (Fig. 8-5b and c). Easy to transport and drive; tapering or corrugation useful for friction loading; adjustable to any length; piling may be damaged by adjacent driving if concrete setup

time is not allowed; can be rammed while casting to produce an end-bearing spread footing; heavy-wall pipe piling useful for deep-driven end-bearing piling.

3. *Steel W beam* (Fig. 8-5d). Adaptable for long piling in dense soil; corrosion a problem in underground flowing-water conditions.

4. *Wood* (Fig. 8-5e). Light and easily handled; can be floated to many jobs; lowest first cost; increasingly difficult to obtain in lengths over 40 ft; difficult to drive in dense soil; suitable only for friction piling since they are weak in compression; subject to decay unless placed under the water table.

5. *Sheet pilings* (Fig. 8-5f). Useful mainly as a retaining wall, not as a structural vertical load-bearing member.

6. *Composite piling.* For economy a combination of wood and concrete piling is used for deep bed work. The wood section is driven below the elevation of the lowest ground-water elevation to avoid decay, and then the upper section is poured concrete in a metal shell.

Machinery and Equipment Foundations

When erecting equipment foundations one must consider the use to which the equipment will be subjected and the possible effects of vibration and shock (dynamic loading) on the foundations, equipment, and supporting soil.

Foundations for Dynamic Loading. Machinery with moving parts, such as compressors, centrifuges, and turbines, when placed on poorly designed foundations, may produce vibrations sufficient to render the building and foundations subject to early deterioration. Also, outside interests may desire to be free of annoying vibrations. Allowable unit soil loading for a dynamic load is one-third to one-half that of a static load. The natural frequency of the soil and foundation must be at least $1\frac{1}{2}$ times the frequency of exciting forces created by the moving parts of the equipment to prevent resonance. Sufficient mass is provided in foundations to absorb the energy and limit the amplitude of vibration to 0.0025 in. or less. This gives rise to the massive mat construction pictured in Fig. 8-6. Where more than one machine is to be installed, it is generally desirable to place all of the machines on a common mat. Shallow foundations well spread out are preferred to a deep foundation of the same mass. Sometimes it is better to remove subgrade soil and replace it with compacted material of more favorable load-bearing and natural-frequency characteristics.

An alternative solution for lighter equipment, such as blowers and motors, is to dampen the vibration by use of supported elastomers, heavy springing, compressed cork mats, or wooden timbers in tying the equipment to the floor.

Anchoring Equipment. A secure fastening of equipment to the foundation is provided by anchor bolts, such as diagrammed in Fig. 8-6. The removable anchor type is preferable for replacement reasons, since raised bolting can be sheared off during the installation of heavy equipment or during the tightening of the bolting.

Since cement contractors cannot finish foundations to exact levels, it is customary to provide a rough finished machine foundation 1 to 2 in. below the required level. The equipment is temporarily wedged and then positioned by adjusting nuts. The frame is then "grouted" in place by use of a mortar fill which adheres to the rough concrete surface (Fig. 8-6c).

Steel-tank Supports. The design of the foundation for a steel tank depends upon the type and design of the tank, which in turn depends

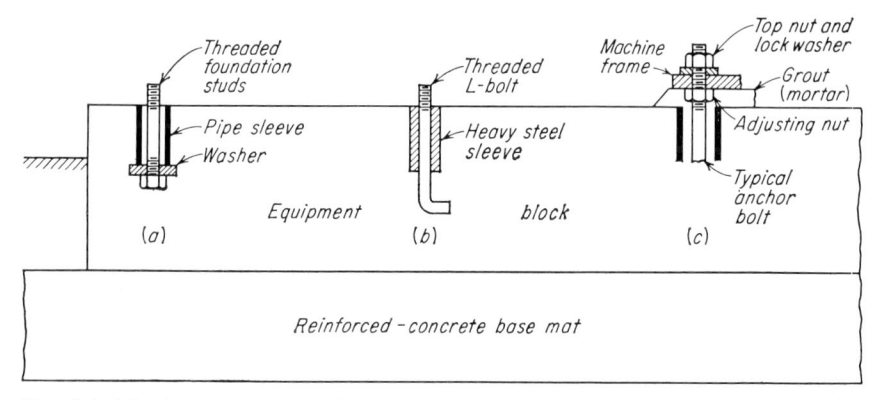

Fig. 8-6. Massive mat construction for heavy vibration-producing equipment showing three types of anchors: (a) removable anchor; (b) fixed anchor; (c) grouting.

entirely upon service. There exist tanks with flat and dished, concave and convex heads; tanks that are horizontal or vertical; tanks for pressure, atmospheric, and vacuum service. In the design of the necessary foundations and supports for a steel tank to store tar, fuel oil, or benzol, the tank to be placed outdoors, one must first consider the over-all dimensions and the shape of the ends, whether flat, dished, or concave. The conventional radius r generally employed for the ends is taken equal to the diameter of cylinder or tank. If $r = \frac{1}{2}$ diameter of cylinder, the head reaches maximum strength for a minimum quantity of steel.

The next consideration is the lugs on which the tank rests upon the supports. These lugs or brackets (Fig. 8-7) are either welded or riveted to the tank, may be either cast or welded, and should be designed to minimize eccentricity.

The foundations should also be as wide as the tank, built below the frost line, and of a size dependent on the load. The load is a dead load,

FIG. 8-7. Riveted support lugs.

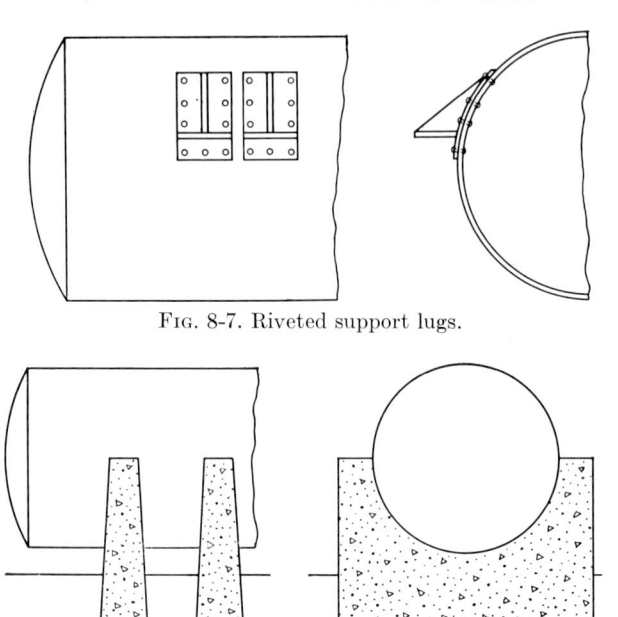

FIG. 8-8. Solid piers.

and the foundations must be built to carry the maximum expected weight—that of the filled tank. The piers likewise should be built to carry a uniform load. Precautions in construction must be taken to provide for creep and expansion, which will vary considerably with temperature.

Solid Piers. Styles of foundation vary considerably. One type is the solid concrete or bricked-up pier, shown in Fig. 8-8. This type must be properly insulated from the tank to prevent the development of corrosion areas on the tank at points of contact with the pier. Wood, rubber, or asphalt can be used for this purpose.

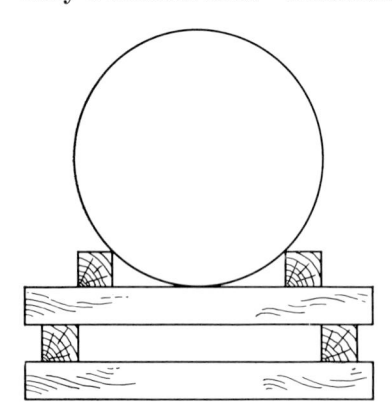

FIG. 8-9. Cribbing support.

Cribbing Piers. Timbers can be cribbed, or if the storage tank is small, timbers can be bolted together, the tank resting in the cradle thus provided (see Fig. 8-9). Such supports provide for movement and shock absorption.

Supporting I Beams Resting on Piers. This type of support, shown in Fig. 8-10, permits variation of temperature in the equipment without rupture of the piers by creeping. Usually the tank is anchored at one

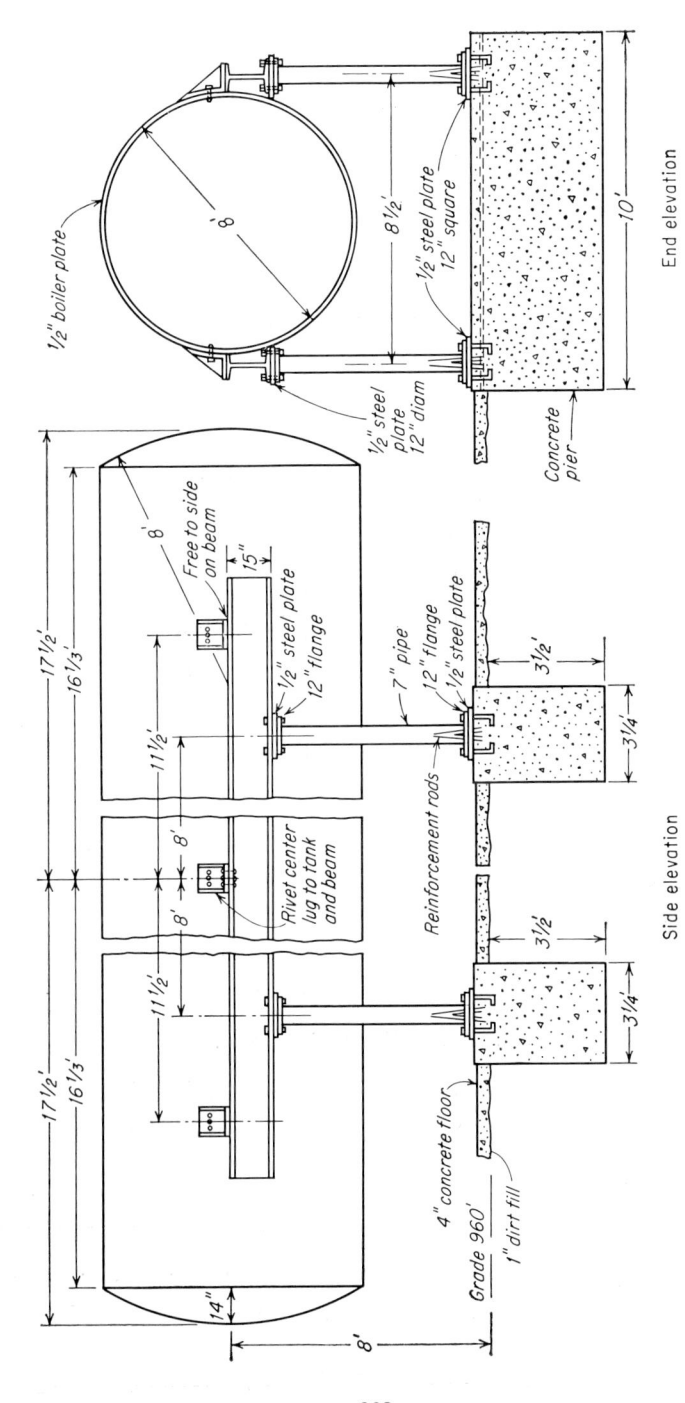

FIG. 8-10. Horizontal tank foundation and support.

point, while the other supports or lugs on the tank are free to move on the I beam. The I beam is firmly anchored to the supports and the supports to the concrete foundations. The supports can be either standard piping or H columns.

Vertical Tank Supports. In order to provide for variations in length or movement of the tank owing to change in temperature, or if storage floor area is limited, vertical tanks may be used with supporting lugs located in the mid-section area, permitting expansion to take place both

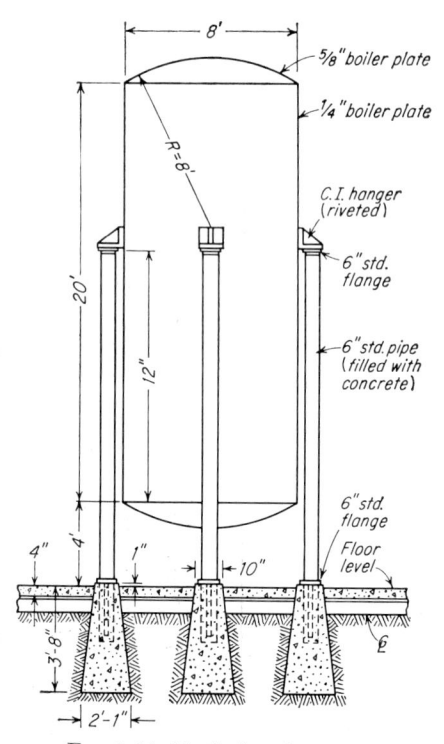

FIG. 8-11. Vertical tank support.

up and down. A support of this type appears in Fig. 8-11. Such installation is often used in extraction batteries and stills. A common manner in supporting steel tanks, especially on concrete floors, is to weld three legs onto the lower periphery and slightly under the tanks. The fourth leg on tanks provides more stability, but the load should be calculated for three legs.

Solid-block Tank Foundations. Vertical tanks of some importance are oftentimes placed upon single huge blocks of concrete and the tank properly supported by dunnage strips to relieve any possible strain on the chime. The entire weight of the tank contents, therefore, rests upon the bottom. Where several vertical tanks of similar size and shape are to be erected side by side, the monolithic block foundation is built sufficiently large to support the entire assembly. Oftentimes the concrete foundation is extended above the ground several feet more than drainage requirements call for, for the additional concrete costs less than a short section of structural steel. One practice for flat-bottom tanks, 20 ft in diameter by 45 ft deep, is to use a circular pier of 1:2:3 concrete around the outer edge, filling the central well with sand, then grouting between the tank bottom and the pier where needed when the tank is filled.

Wooden Tank Foundations. Poor foundations are a common cause for leakage of wooden tanks. In the design of good ones, there are three cardinal principles to be observed:

1. The weight must be supported on the bottom only. The staves of wooden tanks must not carry any of the load, and where the tank is to rest on a level surface, it is best to use dunnage strips or subjoists that will support the bottom and raise the ends of the staves (chime) from the foundation.

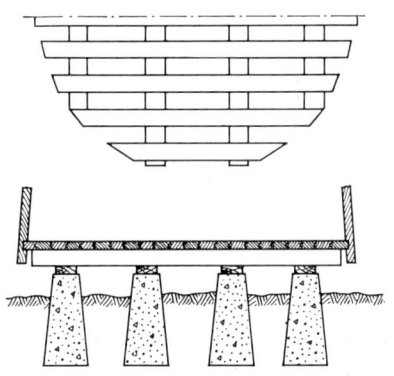

FIG. 8-12. Wooden tank foundation.

2. The supporting pieces under the bottom must not be spaced over 18 in. apart (preferably less), and the bottom boards of wooden tanks must run across the dunnage strips or joists supporting them.

3. Concrete foundations, both monolithic and separate piers, must extend below the frost line when on the ground.

Figure 8-12 shows a standard foundation for tanks on the ground. It consists of concrete walls with wood joists placed across them.

Concrete—Plain and Reinforced

Concrete is important in both foundations and buildings as a material of construction. It has fair compressive strength (1,000 to 4,000 psi) but little tensile or shear strength unless reinforced with steel. Design and construction principles and codes have been developed largely through the efforts of the following organizations:

1. American Concrete Institute (ACI), 22400 W. Seven Mile Rd., Detroit, Mich.

2. American Society for Testing Materials (ASTM), 1916 Race St., Philadelphia, Pa.

3. Concrete Reinforcing Steel Institute (CRSI), 38 S. Dearborn St., Chicago, Ill.

4. Portland Cement Association (PCA), 33 W. Grand St., Chicago, Ill.

Numerous publications on concrete can be obtained from these organizations. A few of the important concepts on plain and reinforced concrete will be discussed next with further information given in a subsequent section on reinforced concrete building structures.

Definition. Concrete is a mixture of cement (portland, natural, or special cement), fine aggregate (sand), coarse aggregate (gravel), and water. Concrete mixtures are defined by a volume ratio, i.e., 1:2:4 concrete contains 1 part of cement, 2 parts of sand, and 4 parts of gravel. Additive materials, such as lime and calcium chloride, are used to prevent freezing, improve workability, and accelerate the setting time.

Strength. The strength of plain concrete depends on several factors: proportioning of ingredients, particularly water-cement ratio, time, temperature, and method of curing. A rich mixture (1:1½:3 or 1:1:2) develops a compressive strength of approximately 4,000 psi for use in thin and heavily reinforced concrete structural beams and columns. Floor slabs and structural members requiring 2,000 to 3,000 psi concrete are made from a 1:2:3½ mixture, while a lean mixture of 1:3:6 is used for massive concrete abutments or fills.

Water-Cement Ratio. This is defined in the trade as U.S. gallons of water per 94-lb or 1 cu ft bag of cement (x). This ratio varies from 5 to 8. For average jobs, the compressive strength (S_c in psi) can be given as

$$S_c = A(B - Cx) \qquad (8\text{-}1)$$

where A, B, and C depend on the type of dry concrete mix. Thus, high-water-content cements are weaker but have increased fluidity for pouring.

Curing. Concrete hardens because of a very slow hydration reaction which takes place between cement and water to produce a bond among aggregate particles. Its strength, under favorable curing conditions, will increase rapidly during the first 7 days, taper off to 28 days, and then very slowly increase during the ensuing years. It is important that concrete should be protected against premature drying or freezing for 7 to 21 days, the period being longer at lower temperatures.

Methods of Testing. ASTM tests for various concrete masonry units are available. These consist of compressive strength and sometimes per cent moisture and rupture modulus tests at a 28-day period. Compressive tests are often made by core-sampling actual construction work during early stages. To avoid waiting, 7-day tests are made and related to 28-day tests as follows:

$$S_{c28} = S_{c7} + 30 \sqrt{S_{c7}} \qquad (8\text{-}2)$$

where S_{c28} = estimated compressive strength at 28 days
S_{c7} = tested compressive strength at 7 days

Reinforced Concrete. This is more commonly used in structural work than plain concrete to improve the strength. Square and round bars of steel (20,000 psi allowable design tensile strength) with protrusions or indentations for increasing adhesion are spaced in high-strength concrete according to coded specifications for structural beams and columns. Expanded metal or wire is used for reinforcing floor slabs, walls, and other thin sections.

Typical Placement Specifications for Reinforcement. The minimum clear distance between parallel bars should be 2 times the side dimensions for square bars and $1\frac{1}{2}$ times the diameter for round bars. Reinforcement of footings and columns should be sealed with at least 3 in. of plain concrete on the ground contact surface. Surfaces exposed to weathering should have at least a 2-in. protective layer of plain concrete. Structures subject to fire hazards should have a fire-resistant coating of concrete 1 in. thick for slabs and 2 to 4 in. thick for structural members.

Design Principles. The design strength of concrete members must be worked out from structure calculations which are beyond the scope of this book. Study references include:

1. "Building Construction Handbook," sec. 4-5, McGraw-Hill Book Company, Inc., New York, 1958.
2. L. C. Urquhart, C. E. O'Rourke, and G. Winter, "Design of Concrete Structures," 6th ed., McGraw-Hill Book Company, Inc. New York, 1958.
3. Publications of ACI, CRSI, and PCA.

STRUCTURES—ENCLOSED AND UNENCLOSED

Industrial operations are commonly conducted inside of buildings in order to afford shelter to workmen, the manufacturing equipment, the control instruments, and the processing materials. Many large chemical manufacturing operations, however, have little need for buildings for the following reasons:

1. The number of workmen per unit of factory space is generally small.
2. The equipment is of such a nature that it can be protected more economically without a building and can be operated via automatic control instrumentation from a small central control building.
3. Materials needing protection are not exposed except at one or several points where specifically designed structures can be erected, thus giving the most economical total plant.

Buildings are required for enclosing chemical manufacturing operations not readily adaptable to outside weather conditions, particularly solids-handling processes. Auxiliary facilities, such as administration, laboratories, maintenance shops, storage warehouses, and personnel accommodations such as cafeterias, change houses, and medical first-aid centers, will require housing irrespective of climatic conditions.

308   CHEMICAL ENGINEERING PLANT DESIGN   [CHAP. 8

It is seen, therefore, that the chemical engineer in plant design work will be faced with all types of unenclosed and enclosed structures, and should have a basic familiarity with the design problems in this area.

Outdoor Plants

The outdoor plant, as represented by Fig. 8-13, consists mainly of unhoused or unenclosed equipment structures, and has an economic advantage in initial investment cost as well as operating and maintenance costs. This fact is pointed out in Tables 6-5 and 8-7. In temperate

Fig. 8-13. Unhoused plant except for small control building on right—Amoco Refinery, Yorktown, Va. (*Courtesy of The M. W. Kellogg Co. and American Oil Co.*)

climates where the weather is inclement only a short time and lack of shelter is only a transitory inconvenience, outdoor plant construction is particularly popular. The trend is not solely confined to mild climate regions with the advent of improved outdoor engineering construction.

Engineering Design Features. Equipment such as pumps and motors have been designed for outdoor exposure; pipelines are well insulated and weatherproofed, jacketed or steam-traced to take care of widely fluctuating temperatures. Weather-resistant metals, materials, and finishes (stainless steel, plastic coatings, impregnated wood, etc.) are significant factors.

Plant layout studies are made by template and model methods, as explained in Chap. 5. Ample space is provided for access by crane or fork truck to heavy equipment requiring repair or replacement and for snow- and ice-removal equipment. Shelters of some kind must be provided for equipment requiring protection from rainfall and snow and/or frequent maintenance and operating attention. Examples of equipment of this nature are centrifuges, compressors, and pumps. A relatively cheap shelter of structural steel, designed for overhead crane and monorail support, with corrugated sheet-metal roof and partially open sides suffices. An example of an indoor-outdoor plant with equipment shelters is shown in Fig. 5-6.

Modern instrumentation design is one of the principal keys to the success of the outdoor plant. As explained in Chap. 9, the error-detecting and final control elements can be located on outdoor equipment and signals transmitted by electronic wiring or pneumatic tubing to a process control center which generally consists of a small air-conditioned building. The transmitting equipment must be protected from ice, snow, and wind loading, but this is a relatively simple problem. In very cold climates the pneumatic air must be kept sufficiently dry so that its frost point is below the lowest ambient temperature.

Heating and ventilation problems are largely eliminated. From a safety standpoint, outdoor plants are superior since combustible fumes are quickly dispersed and rarely accumulate to explosive limits. Twelve- to twenty-four-inch-high fire walls of concrete or impacted dirt are placed around all storage equipment containing flammable materials, unless separate emergency drain and sewage systems are provided. The reader is advised to check Perry's "Chemical Engineers' Handbook," 3d ed., sec. 30, pp. 1848–1879, for a complete discussion of safety and fire-protection measures.

Wind loading on outdoor structures, particularly tall stacks, towers, and pipeline suspensions, require special attention. The upsetting wind moment on vertical structures must be allowed for in foundation design. Oscillations due to von Kármán vortices at low, steady wind velocities can be removed by designing the equipment with a natural frequency out of phase with the frequency of the von Kármán vortices that is likely to occur owing to the prevailing steady winds of that locality.

Enclosed Structures

The need for housing certain functions of plant operation has been pointed out. The design engineer will usually have the assistance of architects to provide an over-all plan for buildings, open structures, and landscaping. Plot plans and models of process equipment areas are developed and then architectural designs are submitted for the final

FIG. 8-14. Example of an indoor-type plant—Central Soya Inc. and McMillen Feed Mills Plant, Chattanooga, Tenn. (*Courtesy of Factory Management and Maintenance.*)

selection based on functionality, pleasing appearance, and economy. Figure 8-14 is an example of a typical indoor-type plant with a predominance of enclosed structures. Several types of building construction which can be selected will be discussed next.

Selection of Building Types. The chemical plant has relatively simple building requirements which can be classified as follows:

1. Process buildings
2. Process auxiliary buildings
    a. Powerhouse
    b. Shops
    c. Warehouses
3. Plant laboratories
    a. Control
    b. Research
    c. Development or pilot plant
4. Administration and personnel
    a. Offices
    b. Cafeteria
    c. Recreation
    d. Medical and first-aid center
    e. Change facilities containing locker and shower rooms
    f. Guardhouses

There are many building types which can be specified by the architect-design group to meet the above requirements, but these fall arbitrarily into three categories based on functionality and in the order of increasing cost: (1) prefabricated; (2) custom-designed factory type, principally for housing equipment; (3) custom-designed buildings for laboratory, administration, and personnel requirements. It should be pointed out that there can be some overlapping of functionality in these three classifications, e.g., a prefabricated building can be used for housing control laboratory requirements. Some of the features of these three types of buildings will be pointed out.

Prefabricated Buildings. In a large chemical plant there are usually a number of buildings that have no special design requirements and it is possible to effect considerable savings by use of prefabricated buildings. These buildings are available from suppliers in a wide range of sizes and types of construction, and are field-assembled in accordance with the manufacturer's standard instructions. An example of this type of construction in a rigid-frame design is shown in Fig. 8-15, and representative costs are given in Table 8-6. Buildings of this type have walls and roofs of light gauge metals, such as corrugated steel or aluminum, and/or corrugated asbestos cement. This construction will not provide the length of service that a custom-designed heavy construction building would. To improve the appearance, a brick veneer is sometimes placed on the side most frequently visible to visitors.

Prefabricated buildings are most suitable for storage, maintenance, shops, boiler rooms, equipment shelters, and temporary construction, particularly in warmer climates.

Custom-designed Factory Type. Buildings of this type are designed functionally along conservative lines, using inexpensive materials to handle nearly all process requirements. Typically, these buildings have concrete foundations and floors; the side walls are of masonry or light insulated panel construction, and the framing is of structural steel with roofs of varying type and material design. The buildings which house equipment must be designed for crane, monorail, elevator, or other services peculiar to a particular process.

Typical equipment and industries requiring this type of enclosure are (1) solids-handling, weighing, and packaging machinery; (2) solids-separating equipment, such as filters and centrifuges; (3) pharmaceutical and food processing; (4) instrumentation and control; (5) any facilities requiring a high degree of operating attention by plant personnel.

It is possible to incorporate within these buildings a number of the personnel requirements, such as locker, toilet, and shower rooms, first-aid, and cafeterias, by judicious use of partitioning, particularly in the areas next to outside walls. However, modern architectural design will usually specify more expensive facilities as discussed next.

Custom-designed Buildings for Laboratory, Administration, and Personnel Requirements. These buildings of more substantial and costly construc-

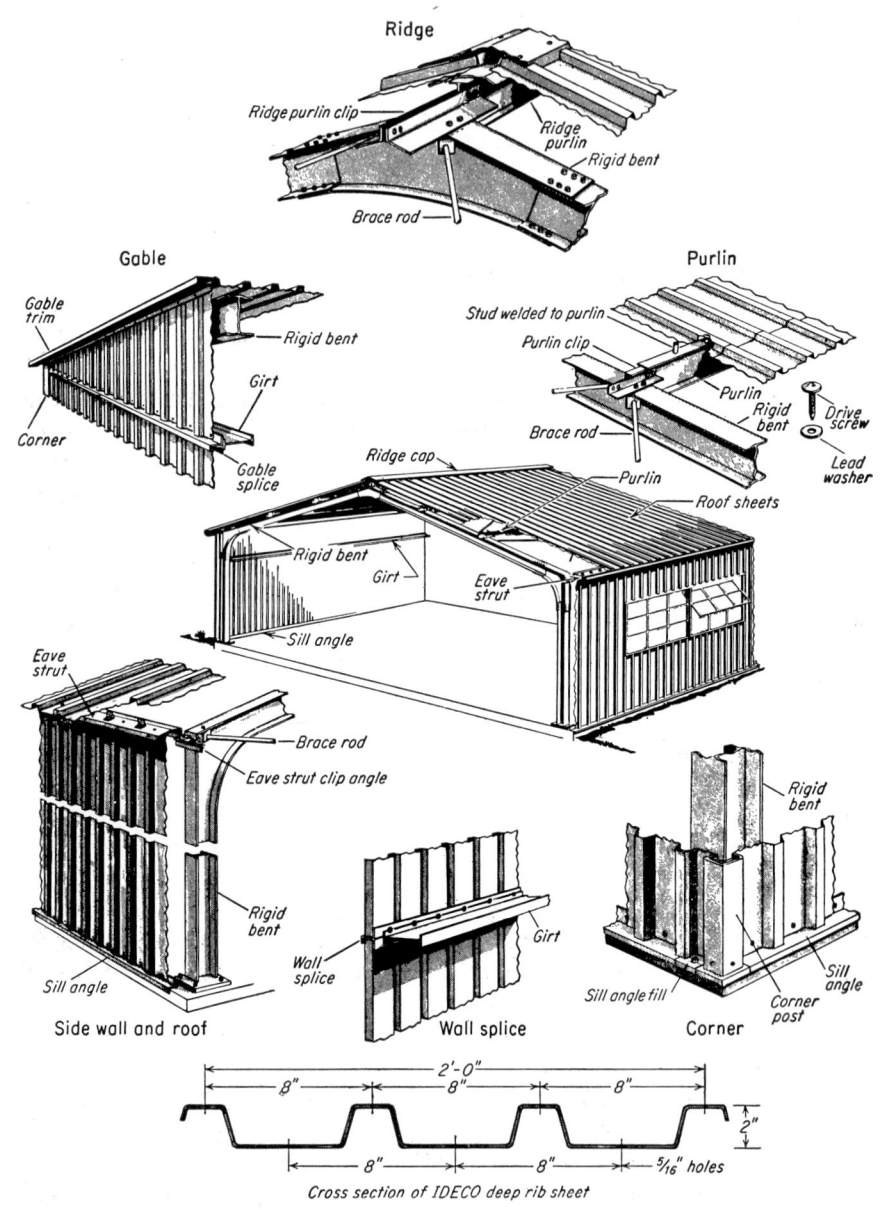

Fig. 8-15. Example of prefabricated construction showing rigid framing. (*Courtesy of Dresser-Ideco Co., Columbus, Ohio.*)

tion are provided for the optimum efficiency of the plant staff from a physical and psychological standpoint. Employees will have better morale if they are provided with surroundings of pleasing appearance. These buildings are often located at the plant entrance and the impression that many people get of the plant is created by the exterior appearance. Visitors entering the buildings are also impressed by well-designed interiors. Thus, the additional cost for improvements to appearance and work efficiency is money well invested.

Buildings of this type vary markedly in design. Architects take advantage of modern materials to create pleasing yet useful designs. Liberal use is made of glass sheet and block, aluminum, and stainless-steel paneling in addition to the more conventional finishes of stone and brick.

The many details which must be included in any set of building designs will be discussed under building design principles.

BUILDING DESIGN PRINCIPLES

Engineers engaged in designing chemical plants must take into consideration location, climatic conditions, and the nature of the chemical industry which the plant is to house in deciding on the type or types of building construction to adopt. The process and the materials handled will indicate the general design requirements. Careful attention must be given to the type of floor, structural frame, walls, roof, fume handling, drainage, ventilation, heating, and foundations of both building and equipment. Although experienced engineers can determine the proper building practices, information regarding relative merits and costs of different forms of structure is valuable for preliminary survey purposes for the chemical engineer.

Types of Building Designs

Height Classification. Usual industrial buildings are of two types: (1) multistory (Fig. 8-16) and (2) single-story (Fig. 8-18). In certain of the chemical engineering industries the multistoried building (or, at least, a tall single-story building) is required where gravity flow from point to point in the process is absolutely necessary. In such a case, equipment must be supported at different levels, either on different floors of a multistoried building or on working platforms independent of the building wall and reached by open stairways.

Where plant construction must be made in congested areas, space for single-story construction may not be available except at prohibitive cost. The general tendency of industry in recent years, to move from congested central districts to outlying communities, has been attended by a great increase in the number of single-story buildings, built where land is cheap and taxes low. However, modern industry is coming to find out that the single-story type is preferable, almost without regard to taxes and land values. For many industries, the single-story building is the most efficient and economical, whether land cost is $200 or $2,000 an acre.

That *multistory buildings* are less expensive per square foot than single-story buildings is a common misconception, based on the fact that the structural floor of a multistory building serves also as a ceiling for the story directly beneath it (Fig. 8-21). In contrast to this, the single-story building naturally must have a floor and roof for every square foot of floor space. However, there are factors that more than offset this. For example, practically all multistory buildings have basements that must

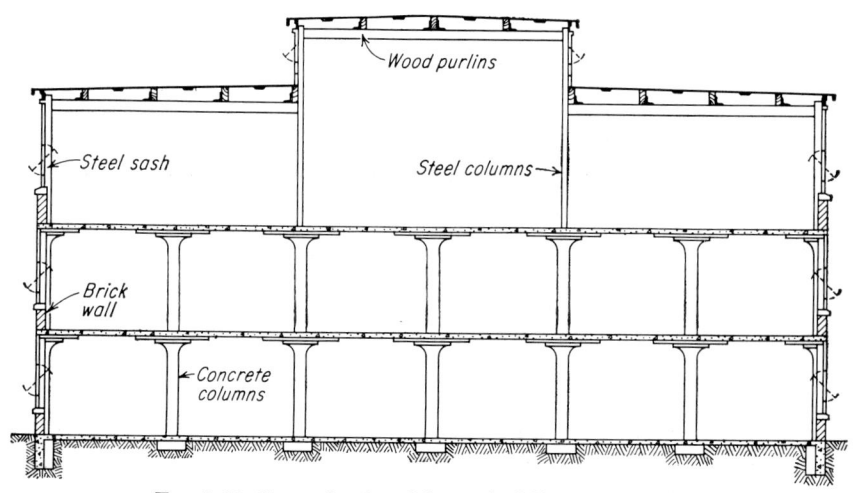

FIG. 8-16. Example of multistory building construction.

be excavated. Basement walls are heavier than other walls in all buildings, and considerable expense for waterproofing of the basement walls and floor is entailed. The cost of stairs, elevators, approaches, outside and interior walls, and wall columns runs high in the multistory building. Cost of elevators and their commonly required enclosing walls is a large item, and most multistory buildings of fair size require either two freight elevators and one passenger elevator or one of each type. Building codes usually call for at least one stairway at each end of a multistory building, and these must be enclosed in fire walls at added expense. The frequent heavy live-load requirements put on multistory building floors often increase the cost of multistory construction.

Single-story buildings, with some few exceptions, are less expensive than the multistory type in original cost, maintenance cost, and operating cost. The increasing intensity of industrial competition, making economical and efficient straight-line production virtually a matter of self-preservation for the manufacturer, requires the use of the single-story plant in the great majority of cases. In the single-story building, the load the floor will carry usually is limited only by the bearing value of the soil on which the floor rests. Most soils have a bearing value of many thousands of pounds per square foot.

Flexibility is another important advantage of the single-story building. If a manufacturer finds it necessary to effect radical changes in product or method, he may find a multistory building unsuited to the new production scheme, whereas a single-story building lends itself to almost any kind of arrangement, because of the absence of large columns and the uniformly high floor-loading capacity.

Roof Classification. Industrial buildings can often be classified into a few simple designs according to roof structure (Fig. 8-22):

1. Flat-roof buildings may be either single- or multistory and are often used where it is advisable to house a number of tanks or other vessels or equipment on or above the roof. Although there are a good many flat-roof buildings in use in the chemical industry as well as in many other manufacturing uses, it is usually advisable to eliminate traffic of any sort over roof surfaces in order to avoid possible mechanical damage or punctures that will result in leaks. Many plant operators feel, too, that it is advisable to get water and snow off the roofs as soon as possible and, therefore, frequently insist upon pitched roofs with fairly steep angles.

2. Pitched-roof buildings may be of single- or multistory design. In either case this type of roof is often combined with a monitor that may be designed to give additional interior daylighting in the building or to be equipped with louvers for ventilation. In many cases ventilated types of sash provide both ventilation and interior illumination.

3. Saw-tooth buildings are usually best located with the glass in the saw teeth facing directly north. Although somewhat expensive, there are distinct advantages in lighting and visibility in a building having the *saw-tooth* type of construction. For certain kinds of plants, notably those in which macroscopic examination of processed materials is important, as in dyeing, it is recommended. Direct sunlight is excluded, and the time of use of artificial lighting during the day is considerably reduced. There is a uniform diffusion of light to all parts of the space, which makes all space working space. The disadvantages lie in the roof construction, and special care must be exercised in design. Leaks, excessive loss of heat, excessive condensation on the underside of the roof, and poorly controlled ventilation are commonly encountered in saw-tooth construction. The cost of construction is greater owing to the cost of windows, glazing, special flashing, condensation conductors for skylights, and higher cost of heating.

Fire Code Classification. Industrial buildings are usually insured against fire loss, so that the building code recommended by the National Board of Fire Underwriters, 85 John St., New York, should be consulted. It is based on the fire resistance or degree of fire stability as measured by the time in hours that the various materials of construction will stand a specified amount of heat under fire exposure as determined by a recognized test (Standard Method of Fire Tests for Building Construction and Material, ASTM E119-50). Based on this test, a time duration scale ranging from 1 to 4 hr serves to classify materials of construction used in buildings. A summary of the fire code classification for building structures in terms of increasing order of fire resistance of the various components of the structure is:

1. *Poor Fire-resistant Construction (Frame Construction).* The exterior walls or portions thereof of wood; also a building with wooden framework veneered with brick, stone, terra cotta, or concrete, or covered with stucco or sheet metal.

2. *Moderate Fire-resistant Construction.* Buildings having masonry walls with heavy timber or unprotected metal support structures.

3. *Good Fire-resistant Construction.* Buildings with masonry, protected steel, or reinforced-concrete construction.

Materials of Construction on Classification. A last and most important classification of industrial buildings is arbitrarily divided into two areas: (1) structural-steel and (2) reinforced-concrete. Other materials, such as wood and aluminum, have been adapted by structural designers, but building construction is based largely on structural steel and reinforced concrete or combinations thereof. Hence, further discussion of these two types is required.

Structural-steel Building Design

Steel is well suited for use in large plants having long spans, heavy loads, and large clearance heights; it can also be equally well adapted to use in small structures. This is a valuable asset because of the variety of shapes and sizes, as well as flexibility, needed to house chemical plants. It accounts for the success of prefabricated building construction for many industrial housing requirements.

Steel is noncombustible, but it is not classified as highly fire-resistant in structural design because of its reduction in ultimate strength with increasing temperature and its high coefficient of expansion; this causes sagging and deformation after short-time exposure to temperatures above 900 to 1100°F created by inside fires. When required by code, an increase in fire resistance can be accomplished by use of a 2- to 4-in.-thick coating of concrete.

Structural-steel design is based on the specifications of the American

Institute of Steel Construction (AISC).[1] The steel framework is planned with use of commercially available sizes and shapes (Fig. 8-17). Structural members are joined together in shop fabrication by riveting or welding. Final connections of large shop-assembled sections are made in the field, using high-tensile bolting tightened by calibrated torque wrenches to produce a tension equal to 90 per cent of the minimum test loading of the bolt. This prevents slipping by the high frictional coefficient developed in the joined members. Field welding is economical when a large number of bolts are required to develop the proper connection in the framework. Beam seats and temporary bolting are required for positioning the members to be welded.

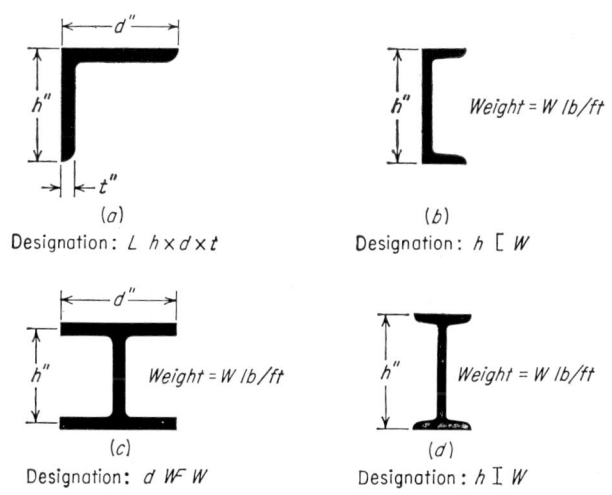

FIG. 8-17. Commercially available structural-steel shapes: (a) angle; (b) channel beam; (c) wide flange beam; (d) I beam.

The variety of shapes, sizes, and details of industrial structures framed with steel are numerous, depending on individual building needs.

Mill Building. A mill building denotes a single-story structure having one or more relatively wide aisles of considerable length and large clearance. They require long-roof rigid frames or trusses to span the aisles, and substantial cross and knee bracing to withstand wind loads on sloped roofs and walls. Figure 8-18 shows examples of mill-building framing, while Fig. 8-19 gives a composite view of a typical mill-building structure. Both these figures show truss framing in the top of the building to allow lighter top chord steel beams over long spans, but this design reduces working headroom. Rigid frame design, as shown in Fig. 8-15,

[1] Steel Construction Handbook, 5th ed., American Institute of Steel Construction, 101 Park Ave., New York.

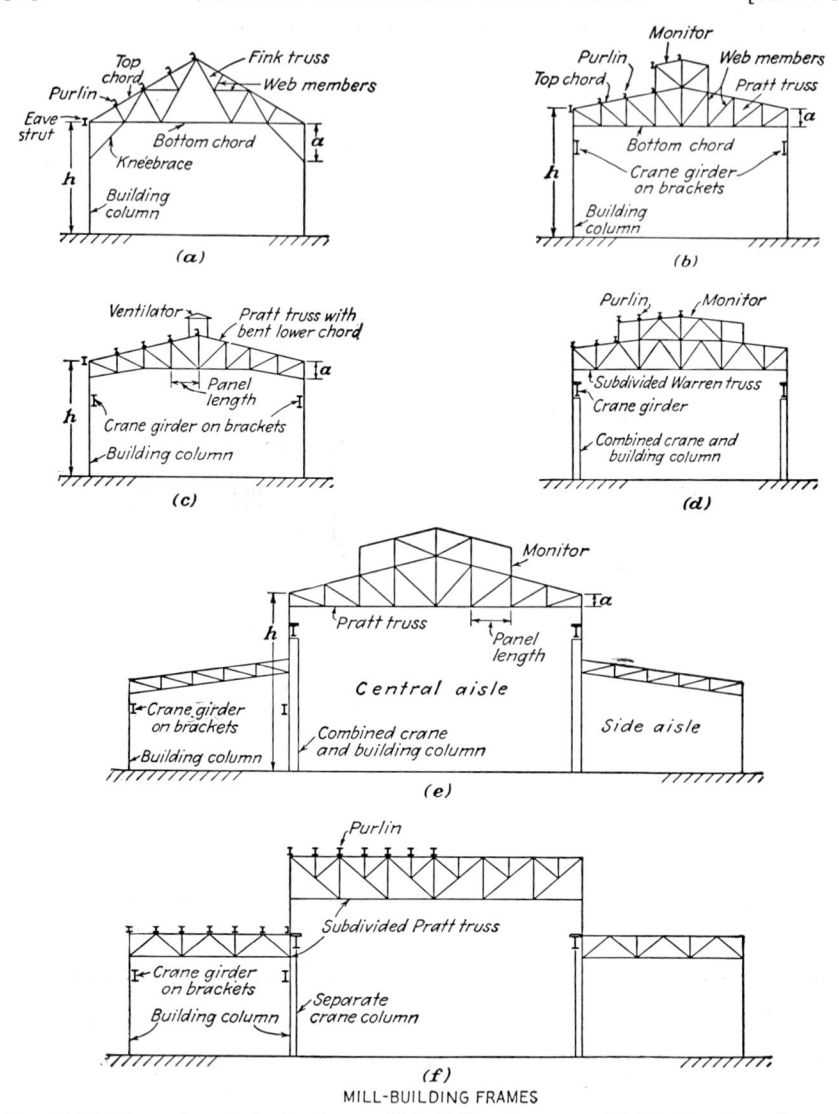

FIG. 8-18. Examples of single-story mill building frames. (*C. W. Dunham, by permission from "Planning of Industrial Structures," McGraw-Hill Book Company, Inc., New York, 1948.*)

solves this problem, but span lengths are reduced and foundations must be designed to take shear loading at the frame-foundation joint.

Multistory Steel Building. Industrial structures of more than one story in height are built with structural-steel framing, but reinforced-concrete construction is a strong competitor. Columns and framework for floor support must be provided in multistoried building construction

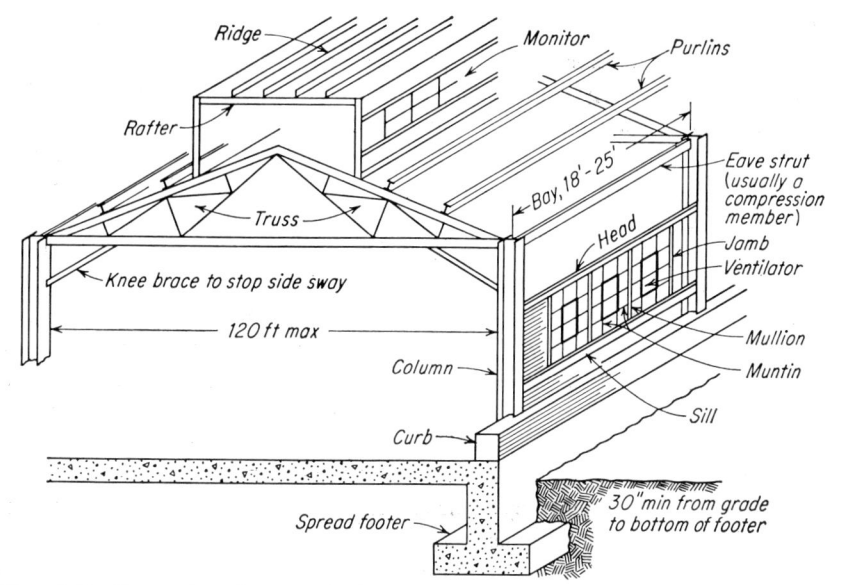

Fig. 8-19. Composite view of mill building construction showing frame, wall, and window nomenclature. (*Courtesy of R. W. Parkinson, Engineering Drawing Department, The Ohio State University.*)

in comparison with single-storied structural-steel design. Other facilities which must be provided include elevator shafts, stairways, chimneys, chutes, duct systems, and bays.

Reinforced-concrete Building Design

Reinforced concrete is an excellent building material if properly used. It should be specified where fire-resistant buildings of rigid construction are required. In order for reinforced-concrete construction to be competitive with structural steel, the building should have:
1. Limited rigid span lengths (less than 100 ft)
2. Repeated module or form size requirements
3. A minimum of large concentrated loads in multistoried structures
4. Height requirements limited to ten stories but more than one story

These requirements will rule out reinforced-concrete framing for most chemical process buildings.

Some of the important properties and considerations in using reinforced concrete were listed in a previous section on foundation design. Planning of concrete structures should incorporate, where possible, such inherent features as monolithic or continuous framing and high compressive strength in specifying beams, columns, and flooring systems.

Beams and Slabs. Reinforcing is particularly important for these structural members since the bottom section is under tension where plain concrete has no strength. Rectangular beams are used for girders, while T beams are cast monolithically with floor and roof slabs to take advantage of the length of the slab. *Precast* beams are competitive with poured form beams since they eliminate forming and pouring except at junctions. *Prestressed* beams can be purchased for improved strength-weight ratio characteristics. Here wires or rods pulled under high tension are cast into the beam. After solidification, the end forces on the steel are released and the tension is transferred in part as a compressive build-up in the surrounding cast concrete.

Columns. Two types of columns generally used are (1) square and rectangular columns with longitudinal reinforced bars stayed laterally with tie rods; (2) round and octagonal columns with circumferential longitudinal bars tied by a continuous spiral of wire $\frac{1}{4}$ to $\frac{5}{8}$ in. in diameter wound on the outside of the bars. Columns can be monolithically joined to concrete beams or flat slabs either by direct pouring into column and girder forms or by poured junction connection to *precast* concrete columns with their *possible economic advantage*.

Formwork. Forms must possess adequate strength to maintain the true shape of concrete and should be well braced and tight. They should be designed for reuse several times on the same job, as forms are an expensive part of reinforced-concrete construction. Use standard lumber and plywood sizes to avoid cutting; use heavy-fiber cylindrical tubes made specifically for concrete column work.

Concrete Flooring Systems. Floors should be designed for the *live* loading tolerance specified by the building code governing the construction site. Typical live loads in pounds per square foot for chemical plants, as distinguished from the dead weight of floor and surfacing, are: process areas, 150 to 200 psf; warehouses, 500 psf or greater; offices and laboratories, 100 psf on first floor, 75 psf on upper floors. To accommodate these loads, four different types of reinforced-concrete flooring may be used: (1) beam-girder; (2) flat-slab; (3) ribbed floors, with ceramic or steel-tile fillers; (4) steel- or precast-concrete-joist floors.

Beam-girder Floors. Parallel beams transmit the reinforced-concrete slab load to girders run at right angles supported on columns. Substitution of beams for girders gives a two-way slab construction with approximately square spans not exceeding 18 ft. When parallel beams only in one direction are used, this *one-way* construction is suited for rectangular panels with a width not exceeding 12 ft.

Flat-slab Floors. A flat-slab floor is a reinforced-concrete slab built monolithically and using a top-spread column support without the aid of beams and girders. In Fig. 8-16 the second and third floors are flat-slab construction. It is ideal for industrial buildings with live loads of 100 psf and square panel spans of 18 to 30 ft, but the flooring must be more heavily reinforced. In the *two-way system* the reinforcing bars are placed parallel to lines of columns in both directions with a maximum allowable spacing not greater than 1½ times the thickness of the slab, or about 12 in. The *four-way* system consists of parallel and diagonal reinforcing bar braces.

Ribbed Floors. For light loads over long spans, such as the loads occurring on office or laboratory floors, 4- to 12-in. ceramic or steel square tiles are located on supporting beam-girder frames and spaced as fillers to create an adjacent series of monolithically cast concrete T beams. When terra-cotta or gypsum tiles are used, the tile blocks are left in place, giving a flat ceiling for plastering. Steel cores can be either permanently placed or removable for reuse. Metal lath, welded to reinforcing rods projecting from the bottom of floor beams, supports the ceiling plaster for the lower floor. Ribbed construction is also adaptable to roofs.

Steel- or Precast-concrete-joist Floors. This is another light-load type of floor. The floor joists are spaced 12 to 30 in. on center and support a steel mesh on which a 2- or 3-in. concrete slab is poured. The joists span 4 to 32 ft and are supported by concrete or steel beams or girders.

Concrete Mill Buildings. Rigid framing of single-storied buildings with reinforced concrete is not competitive with structural-steel buildings except where favorable costs on precast beams can be obtained. Even then, buildings are limited to clear bays of 100 ft or less and moderate vertical clearances of less than 35 ft.

Flooring

Over 90 per cent of recently constructed plants have concrete floors. The types of finishes for this concrete base flooring are listed in Table 8-2, making possible a selection of one or a combination of materials for chemical plant flooring requirements. In many cases where new chemicals are being handled, laboratory tests will have to be run to select suitable flooring finishes.

Walls

Steel or wood framework of buildings can be filled in with a number of combinations. *Concrete block* is probably the best material to use for walls in a building of permanent construction, although *brick* is nearly ideal where the size of the project warrants the investment. Relative fire resistances of walls and partitions for brick, concrete, hollow tile, and plaster are given in Perry's "Chemical Engineers' Handbook," 3d ed., table 6, p. 1852.

Sheet-metal construction of any kind has the advantage of cheap and quick erection. It will not be found desirable in the colder climates as

TABLE 8-2. CHARACTERISTICS OF FLOOR FINISHES*

Characteristic	Cement	Mastic	Wood block	Wood, laminated	Brick	Rubber	Plastic tile	Linoleum	Mastic tile	Mastic block	Cork	Magnesite
Cost index.............	3[a]	7[b]	[c]	7[d]	8[e]	6	5	1	2	1	3	4
Ease of repair..........	3	1	1	1	4	1	1	2	1	1	1	1
Fast installation—days...	5[f]	2	1	1	1	1	1	2	2	1	1	2–5
Compressive strength....	2	4[g]	5	6	1	8	3[h]	8	4	3	7	2
Tensile strength.........	2	4	8	5	8	1	1	7	3	8	6	1
Toughness.............	2	3[j]	1	3	7[k]	5	1	6	3	2	4	3
Resistance to denting....	2	4	4	5	1	2	3[g]	5	4	3	5	2
Abrasion resistance (wear), in.	0.025[l]	0.007–0.032	[p]	0.002[m]	[q]	0.006	0.002	0.002	0.007	0.007–0.032	0.011[n]	0.006
Sparkproof[s]...........	N[x]	N[x]	Y	Y[r]	N	Y	Y	Y	Y	N[x]	Y	Y
Self-healing............	N	1	N	N	N	N	N	N	N	2	N	N
Resistance to moisture[t]...	3	2	N	N	1	4	2–4	N	4	3	N	N
Easy wheeling..........	1	3	3	3	2	[v]	2	[v]	[v]	3	[v]	1
Nonskid—walking[z]......	3	2	2	3	2–4	3	3	2[w]	2	2	1	2
Resilient, comfortable....	11	9	6	4	12	1	7	3	10	8	2	5
Variety of color.........	4	N	N	N	N	1	1	1	1	N	2	3
Quiet.................	11	9	5	4	12	2	6	3	10	7	1	8
Odorless (after 7 days)...	4	3	3	2	2	2	1	1	1	1	1	1
Acid- and alkali-proof....	9	2	7	8	1	6	3	12	5	4	10	11
Fire resistance..........	1	6	7	11	1	9	3	10	4	5	8	2
Nominal thickness, in....	4–6	1½	2–3	2½ 4½	1⅜ 4½	⅛ ¼	⅛ ¼	⅛ ¼	⅛ 3⁄16	½ 3	⅛ ½	⅜ ¾
Weight, psf†...........	5 60	15–20	2½ 9	3 12	2¼ 25–35	3⁄16 0.8	3⁄16 1.5	3⁄16 0.9	3⁄16 2.0	2 16–20	½ 1.58	½ 3–6

Note: Lowest number indicates highest quality; in cost index, the lowest cost. Y = Yes. N = No or None.

[a] Separate topping, metallic hardener.
[b] With acidproof membrane under mastic.
[c] Creosote-impregnated end-grain lug block (Kreolite).
[d] Tar-Rok.
[e] Acidproof 2¼-in. brick.
[f] 24 hr for Lumnite cement.
[g] Can be hardened if necessary, but increases tendency to crack or chip.
[h] For hard type.
[j] Soft base, tough top.
[k] Hard to classify for toughness. Will stand considerable mechanical abuse.
[l] Based on 1:2:4 mix. Regular 1:1:2 mix with fluosilicate wash should be approx. 0.012.
[m] Based on maple finish.
[n] High density; 5⁄16 in. dark = 0.026 in.; 5⁄16 in. light = 0.011 in; ½ in. light = 0.019 in.
[p] No available tests—probably about 0.035 in.
[q] No available tests, probably about 0.005 to 0.010 in. for shale.
[r] Concealed or copper nails.
[s] Cement, mastic, rubber, linoleum, mastic tile, and magnesite may be purchased with conductive quality to prohibit sparking from static discharges.
[t] Moisture to extent that material cannot dry between periods.
[v] Not recommended.
[w] Without wax or lacquer.
[x] Can be obtained on special order.
[z] Dry surfaces, designed for traffic.

* Courtesy of R. D. Rodgers; reproduced by permission from W. Stanier (ed.), "Plant Engineering Handbook," McGraw-Hill Book Company, Inc., New York, 1950.
† For thickness indicated.

this type of building is hard to heat unless insulation is used on one side of the sheet metal. However, for foundries and warehouses, sheet metal is very largely used since heating is unimportant and low initial cost is a major factor. Where corrosive conditions will not permit the use of unprotected metal such as steel or aluminum, it is possible to use an asphalt-asbestos-cement covering which will resist corrosion.

Hollow tile offers a compromise in cost between metal and heavy masonry. The tile walls are easily, quickly, and cheaply erected and are not unduly expensive to maintain. However, it should be understood that they are primarily of temporary construction and subject to rather rapid deterioration because of vibration or shock. This type of construction is very popular for plants erected in the Northern states. It has a further value in explosives plants in that it will disintegrate easily upon shock. The combination of hollow-tile walls with concrete pilasters is commonly used and results in a high-grade building of low initial cost.

Wooden walls are cheap, noncondensing, quickly erected, and easily altered. A study of the fire hazard, corrosive action of fumes, and insurance rates will generally be enough to show whether a wooden building is desirable for the industry under consideration. In the Southern states wood is subject to splitting due to high humidity and temperature changes and is subject to rapid disintegration due to termite attack.

Glass in the form of sheet-glass or metal-glass block prefabricated panel units is used to provide large areas for transparent or diffuse lighting and an aesthetic appearance. Over 50 per cent of the total wall area in many industrial buildings has glass in one form or another. Ventilation can be provided in the sash design (Fig. 8-19). In cold climates, glass can be used to an advantage in radiant heating, but the panels should be insulated by use of a double thickness with an intervening air gap. Transparent glass panels should not be used in structures where closely controlled air conditioning is essential.

Roof

The first requirement for the roof in the average chemical plant is that it should have a high degree of resistance to corroding fumes; and the second, that it should be noncondensing, or as nearly so as possible. The problem is again the choice of a suitable material to meet the manufacturing conditions. A roof for a building in which an explosive process is carried out should be light and capable of disintegration upon shock.

The problem may be to resist heat, give maximum light, exclude the weather, prevent condensation, furnish ventilation and fire protection, and to combine any or all of the above with as strong and well-appearing a roof as possible. Appearances, however, rarely enter into a factory design. The *steel roofs* of the train sheds in our larger cities are a good example of roofs that combine strength and lightness with unusually good appearance.

Wood shingles or the various *tar and gravel-specification roofs* are easy to erect, cheap, and of reasonably long life. In some cases, excessive fumes may cause a rapid disintegration of the woodwork. In such cases a more resistant roof should be installed. Otherwise, a wood-and-tar roof is probably better for factory purposes than concrete or steel, as its thermal conductivity is lower, thus improving working conditions, reducing condensation, and lowering the heating costs.

Slate is noncorrodible, quickly laid, and unusually durable. It is, however, somewhat expensive for factories and is also subject to heavy condensation. The same objections can be applied to a *tile roof*. Scaffolding is necessary for installation, and repairs are hard to make.

Table 8-3. Minimum Design Dead Load on Roofs*

Roof coverings	Psf
Asbestos-cement corrugated or shingles	4
Asphalt shingles	2
Copper or tin	1
Corrugated iron	2
Clay tile plus mortar	22–26
Cement tile plus mortar	26
Composition:	
4-ply felt and gravel	5.5
5-ply felt and gravel	6
Slate, $3/16$ in.	7
Skylights, $5/16$-in. glass	5
Wood shingles	3

* As recommended in American Standard A 58.1-1955.

Tin, copper, lead, aluminum, and *zinc roofs* are long-lived. They are, however, expensive and subject to corrosion in specific cases and are, therefore, not generally applicable to chemical plants.

Concrete, as precast slabs, makes an unusually good roof. It is not so subject to corrosion as most other types and is fireproof. The objections of higher installation cost, excessive condensation, and greater weight per unit of thickness may be enough to swing the decision in favor of some of the other types.

Corrugated iron is especially suited for buildings of a temporary character but has nothing to recommend it for permanency. An improvement results when the corrugated iron is covered with a heavy tar paint and in some cases asphalt and asbestos. Such a roof is cheap, lasting, easily erected, quite resistant, and largely self-supporting. Only a light steel framework is required.

An unusually durable and fireproof roof is obtained by the use of *asbestos and concrete corrugated board,* which can be installed with the ease of sheet iron. Such a roof does not need paint, will not rot, and is practically permanent. This material can also be obtained in the form of flat shingles which can be placed over an old shingle roof. Thus an actual advantage is obtained by securing a roof which is thicker than usual, better insulated, and, therefore, noncondensing.

The perfect adhesion of *built-up roofing felts and insulation* to the steel-deck type of roof, on which such coverings are often installed, enables the roofer to make a positive guarantee of roof service. Roofing felts do not fracture when laid on insulated steel roof decks, as is the case with roof decks that expand and contract during temperature changes. When built-up roofing is applied to concrete roof slabs, a priming coat before the laying of built-up roofing is required to ensure a good bond, whereas wood roof decks require the nailing of built-up roofing to ensure bond.

Table 8-4. Combined Snow and Wind Live Loads, Psf

Section of United States	Slope of roof, deg					
	60	45	30	20	Flat
Southern and Pacific Coast	25	23	21	18	28
Central and Rocky Mountain	25	24	22	27	32
Northwest and New England	26	25	22	32	38

Loads on Roofs. There are two types of load on roofs which must be taken into consideration in a chemical building: (1) dead load, consisting of structural load, such as roof surface, trusses, and purlins, and (2) live load, consisting of snow and wind load. The design of roofs, columns, and foundations is the task of the architect, and to him the chemical engineer must go to obtain the correct design for the building. However, oftentimes the chemical engineer is interested in rough estimates on roof loads when some situation arises where an architect cannot be consulted and a shelter of some type must be provided. Especially is it essential to the chemical engineer to be able to answer a question on additional loads on structures should replacement or new construction be considered in an old structure. Table 8-3 contains dead-load data for roofs, excluding steel-truss work.

For live loads, consideration must be given to the added force from wind and snow. Simplified data for estimating these live loads appear in Table 8-4. Additional data on weather can be obtained from the U.S. Weather Bureau. A useful summary of temperature range and wind characteristics for 88 cities in the United States is given in Perry's "Chemical Engineers' Handbook," 3d ed., p. 46.

Safety and Fire Protection

Possible loss of life, property, and productivity makes the hazard aspects of building design one of the major study areas for the plant design group. Safety problems have already been discussed at length in Chap. 2. The reader is referred to Table 2-3 as a source of regulatory groups and information on design principles and codes for all types of hazard protection. The National Fire Code published by the National Fire Protection Association, 60 Batterymarch St., Boston, should be studied. The code is available in volumes:

Vol. I, Flammable Liquid and Gases
Vol. II, Combustible Solids, Dusts, Chemical and Explosives
Vol. III, Building Construction and Equipment
Vol. IV, Extinguishing Equipment
Vol. V, Electrical
Vol. VI, Transportation

Another excellent reference for study is Perry's "Chemical Engineers' Handbook," 3d ed., sec. 30, pp. 1847–1879.

Safety recommendations in building design can be summarized in terms of a safety check list:

1. Separate buildings to avoid spread of fumes and fires; use 50 ft or more as the separation distance.
2. Use outdoor plant construction wherever possible.
3. Use single-story rather than multistory buildings, if possible, with fire wall cutoffs or subdivisions for highly combustible occupancies.
4. Use as high a degree of fire-resistant construction as is economically feasible, considering all possible losses in a disaster analysis.
5. Provide at least two exits from all confined areas, and two exits from each floor; use emergency chutes and escape doors in hazardous areas.
6. Explosion-hazard areas should be designed for rapid pressure relief; one or more walls or ceilings should be easily pushed out by an explosion pressure wave; score window panels for easy rupture.
7. Eliminate all open doorways and elevator shafts.
8. Ventilation should be designed for positive elimination of harmful and combustible fumes and for elimination of the spread of fire.
9. Provide the correct lighting to eliminate accidents.
10. Design electrical systems according to National Electric Code classification.
11. Provide safety and alarm devices, such as fire-alarm systems, combustible-vapor detectors, flame arresters, pressure-relief venting of equipment, flame-failure controls for oil- and gas-fired equipment.
12. Design adequate disposal systems for unwanted and hazardous residues, flammable or dangerously reactive liquid or sludge waste.
13. Provide floor drainage to properly vented sewers or emergency holding storage areas for accidental spillages.
14. Employ proper fire-protection equipment:
    a. Water spray or fog automatic sprinkler systems
    b. Reliable source of water at accessible locations

    c. Well-designed fire hose and hydrant systems in terms of water-discharge rates
    d. Adequate emergency fire-extinguisher system (see Perry's "Chemical Engineers' Handbook," 3d ed., p. 1855)
15. Place proper guards and warning devices on moving equipment, stairways, ladders, platforms.
16. *Last but not least*—study and design according to *all codes* which relate in any way to plant design, construction, and operation.

Illumination

Correctly designed lighting throughout an entire chemical plant and its auxiliary facilities will pay off in increased productivity and comfort of the personnel and in reducing accidents. Good illumination requires lighting of sufficient intensity, measured in foot-candles, without glare or shadows. An excellent section on lighting requirements and characteristics of various lighting fixtures is found in Perry's "Chemical Engineers' Handbook," 3d ed., pp. 1755–1759.

Lighting in Process Plants. Chemical plants call for proper lighting at strategic points near equipment where physical and chemical hazards exist. Recommended values of 15 to 30 foot-candles should be used. In the control areas where lighting is needed to enable the operators to make observations and adjust controls more accurately, the intensity should be increased to 50 to 75 foot-candles of diffuse lighting. The spotting of the outlets and location of desired points of illumination cannot be entrusted entirely to an illuminating engineer. The chemical engineer must assist him in pointing out illumination in terms of operating needs. Each plant should be custom-designed in terms of lighting requirements.

Details of location for conduits, complete code specifications, and other such details are the responsibility of the illuminating engineer. Types of switches, outlets, lights, and reflectors are chosen by him only after consultation with the chemical engineer relative to the health, mechanical, chemical, and fire hazards incidental to the processing.

The design of a complete factory-illumination layout for a chemical plant is not a difficult task with standard factory electrical equipment. Pamphlets issued by manufacturers of electrical equipment, containing specifications, details, codes, and quotations, are available and serve as excellent guides to the designing chemical engineer. Conventional symbols (see Fig. 8-20) for indicating various pieces of electrical equipment should be used in all sketches submitted to the illuminating engineer by the chemical engineer to convey to the former the requirements for the chemical plant.

Painting. The color of surfaces within an enclosure is considered a part of the fundamental design of plant illumination. Painted surfaces are essential in reflecting and absorbing light, particularly on the ceiling

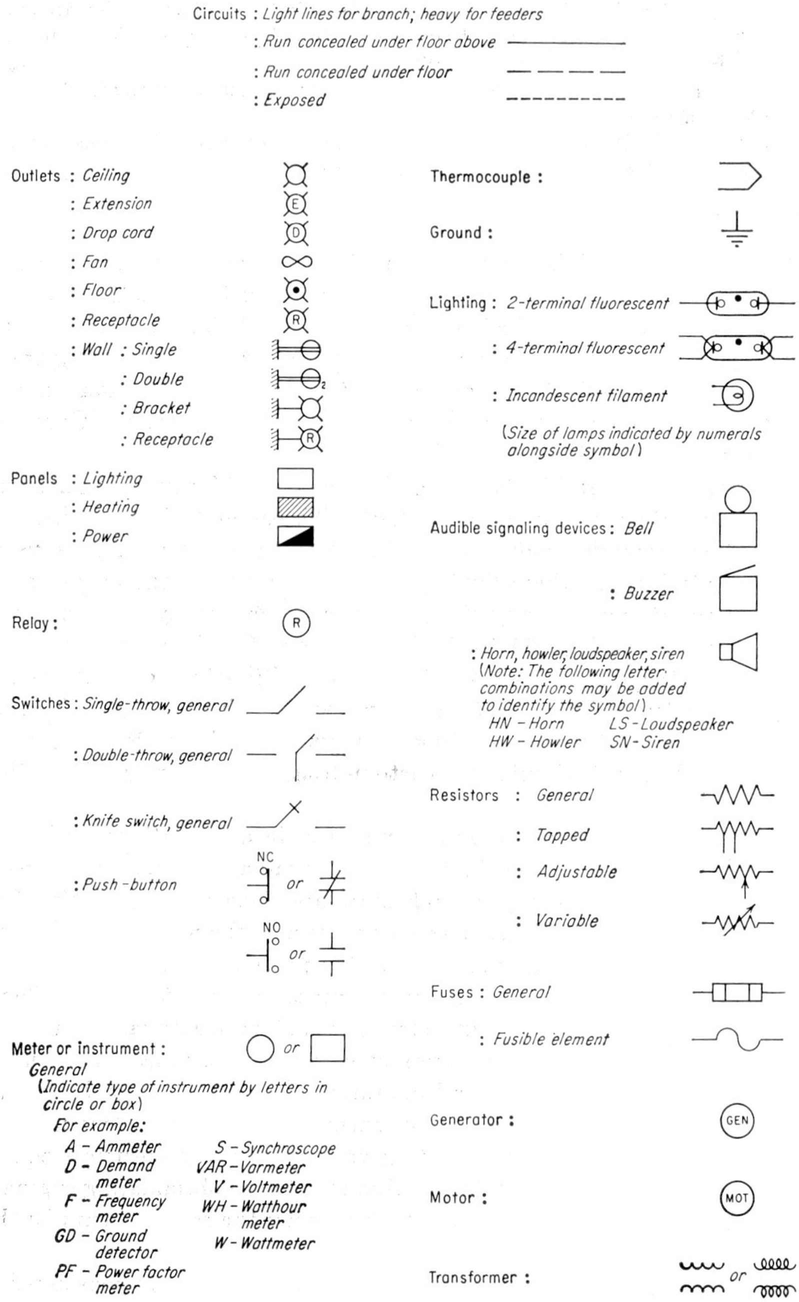

328   CHEMICAL ENGINEERING PLANT DESIGN   [CHAP. 8

Circuits : *Light lines for branch; heavy for feeders*
: *Run concealed under floor above*
: *Run concealed under floor*
: *Exposed*

Outlets : *Ceiling*
: *Extension*
: *Drop cord*
: *Fan*
: *Floor*
: *Receptacle*
: *Wall* : *Single*
: *Double*
: *Bracket*
: *Receptacle*

Panels : *Lighting*
: *Heating*
: *Power*

Relay :

Switches : *Single-throw, general*
: *Double-throw, general*
: *Knife switch, general*
: *Push-button*

Meter or instrument :
General
(*Indicate type of instrument by letters in circle or box*)
For example:
A – Ammeter   S – Synchroscope
D – Demand   VAR – Varmeter
  meter   V – Voltmeter
F – Frequency   WH – Watthour
  meter   meter
GD – Ground   W – Wattmeter
  detector
PF – Power factor
  meter

Thermocouple :

Ground :

Lighting : *2-terminal fluorescent*
: *4-terminal fluorescent*
: *Incandescent filament*
(Size of lamps indicated by numerals alongside symbol)

Audible signaling devices : *Bell*

: *Buzzer*

: *Horn, howler, loudspeaker, siren*
(Note: The following letter combinations may be added to identify the symbol)
HN – Horn   LS – Loudspeaker
HW – Howler   SN – Siren

Resistors : *General*
: *Tapped*
: *Adjustable*
: *Variable*

Fuses : *General*
: *Fusible element*

Generator :

Motor :

Transformer :

FIG. 8-20. Electrical symbols.

and upper part of the wall. Good light-reflecting paints are the white, light yellow colors, or aluminum paints (65 per cent or greater reflectance). The light green or light gray colors which are popular for equipment paints have intermediate reflecting properties (35 to 65 per cent reflectance). Dark green, blue, brown, and black have very poor reflectance and should be avoided as a predominating color. All potentially hazardous areas should be painted with a standard Safety Color Code (see Perry's "Chemical Engineers' Handbook," 3d ed., p. 1853).

Air Conditioning—Heating and Ventilation

Industrial process air conditioning is broadly defined as the engineering control of the temperature, relative humidity, ventilation, cleaning, and movement of air within a process area. A study is conducted by the plant design group and a decision is made to control all or only a portion of the factors governing air conditioning. For example, a pharmaceutical plant may require a complete air-conditioning system in the packaging area to provide the customer with a clean, free-flowing hygroscopic powder. Yet a filter press room where aqueous inorganic slurries are being handled will require only space heaters for providing a comfortable temperature level in the winter time and windows for cross-flow natural ventilation.

Ventilation. Chemical processes inherently contain toxic and flammable materials which require a careful study of ventilation requirements for each installation. In its simplest form, a knowledge of the rate of release of hazardous vapors and dusts and the allowable safe concentration in a known volume of the enclosed area will determine the ventilation requirements for clean air input.

Safety Standards. The *toxicity tolerance* in terms of maximum allowable concentration (generally in parts per million on a volume basis) has been published for a number of chemicals.[1,2] The tolerance for many new chemicals may not be established, and safe standards will have to be worked out in cooperation with public health authorities.

Fire and explosion hazards are controlled by designing according to the *lower limit of flammability*, which is the least amount of a substance capable of sustaining flame propagation when mixed with air. Values for individual materials are listed by the NFPA.[3]

---

[1] A. D. Brandt, "Industrial Health Engineering," John Wiley & Sons, Inc., New York, 1947.

[2] N. I. Sax (ed.), "Dangerous Properties of Industrial Materials," Reinhold Publishing Corporation, New York, 1957.

[3] National Fire Protection Association, "Fire Hazards of Flammable Liquids, Gases and Volatile Substances" and "Prevention of Dust Explosions in Industrial Plants," 60 Batterymarch St., Boston.

Dust-explosion control is more difficult to evaluate, as it depends on the particle size, hygroscopicity, surface, and combustibility of the dust. The lower explosive limit of air-borne dust generally ranges from 0.01 to 0.5 oz/ft^3 of air. More detailed information is available from the NFPA.

When confronted with the problem of establishing the explosive limit of mixtures of several kinds of vapors, experimental testing according to ASTM standards is required. A reasonable estimate can be made according to the following equation:

$$S_m = \frac{1}{x_a/S_a + x_b/S_b \cdots} \qquad (8\text{-}3)$$

where S_m = lower flammability limit of the mixture as per cent in air
x_a, x_b, etc. = volume fraction or mol fraction of combustible vapors a, b, etc., on an air-free basis
S_a, S_b, etc. = corresponding lower flammability limit of vapors a, b, etc., as per cent in air

Design of Ventilation Systems. Air in controlled area ventilation is supplied by a positive- or negative-pressure system. The *positive-pressure system* is generally used for complete air conditioning in non-hazardous areas. A controlled volume of clean heated or cooled air is blown into the room through non-draft-inducing distributors at velocities from 400 to 1,500 fpm. Air is removed by suction ducts placed to avoid short circuiting. The exhaust air is generally cleaned and recirculated with make-up air to save on heating and cooling requirements. Air losses are always outward with no infiltration of ambient untreated air of a possible hazardous nature. This system is often used in explosion-proofing small rooms or equipment.

The negative-pressure system is used to prevent leakage of hazardous fumes from an enclosure to cleaner surrounding air. A suction system is provided which draws in clean ambient air, often over heating or cooling fins, and discharges it through the blower to treating equipment and/or dispersion stack in this once-through system. Air should be treated by cyclone separators, washing towers, air filters, or electrostatic precipitators where air pollution is a possibility or where recovered fumes would have some economic value (see Perry's "Chemical Engineers' Handbook," 3d ed., pp. 1013–1050). Intake and exhaust ducts should be placed to avoid short circuiting of the fresh air required for dilution and mixing purposes. Safety codes for installation in hazardous areas are available from the National Fire Protection Association and the National Board of Fire Underwriters.

The *air-change method* for determining fresh-air-input requirements in general room ventilation is commonly used but is often misleading. The air-inlet rate is established either from recommended values of air changes per hour in the enclosure (Table 8-5) or by material balance calculations, using allowable safe concentration limits when dealing with hazardous vapors.

TABLE 8-5. RECOMMENDED AIR CHANGES

Type of building	Minimum air changes per hr, C
General offices	4–6
Mill buildings	4–6
Toilets (restricted use)	5
Toilets (public use)	10
Conference rooms	8–10
Machine shops	8–10
Power houses	10–15
Chemical process enclosures	10–30
Chemical laboratories	10–30

By use of Table 8-5 the air-inlet rate is established as

$$R = fVC \qquad (8\text{-}4)$$

where R = air rate, cfh
V = enclosure volume to be ventilated, ft^3
C = minimum air changes per hr (see Table 8-5)
f = factor of safety (always greater than 1)

When dealing with a hazardous material, the air-inlet rate can be calculated as follows:

$$r_a = fr_h \left(\frac{100}{S} - 1\right) \qquad (8\text{-}5)$$

where r_a = air rate, mols/hr
r_h = hazardous material input within the entire enclosure, mols/hr
S = *lowest* limit of hazard, considering both flammability and toxicity; mol per cent hazardous material per mol of air plus hazardous material if a vapor
f = factor of safety (always greater than 1)

It is often possible to reduce air-input requirements by removing the hazardous material at the point of discharge by *local ventilation*. This lowers the r_h value in Eq. (8-5), which assumes possible disposal of hazardous material within the entire enclosed volume of the enclosure being ventilated. Hoods and exhaust ducts are placed over such equipment as open filter presses, pulverizers, open tanks, and over laboratory benches and equipment to catch the maximum amount of vapor or dust without interfering with normal operation and maintenance. Local air velocities in the region of pickup will depend on density of the hazardous material or its particle size if a dust or fume. Air velocities greater than 200 fpm are usually employed for industrial operations, while chemical laboratory fume hoods range from 70 to 125 fpm when fully opened.

Temperature and Humidity Control. Ventilation requirements must be met to create a safe working environment. Comfort and process requirements in terms of temperature and humidity of the air must also be fulfilled. The quantity of comfort controlled air required on a once-through system is often prohibitive if ambient air conditions deviate from room requirements to any great extent. In the winter the air would be heated; in the summer the heat duty on an air-conditioning or -cooling system, where specified, would be excessive unless a recirculation-treating system were used. The summer internal heat load for a complete air-conditioning system is the load-determining factor, as winter requirements are invariably smaller.

Heat Load. The heat load within an enclosed structure consists of the following major factors:

1. *Conductive heat transmission* through the building structure: the maximum plausible temperature difference between the enclosure and the surroundings is used in conjunction with heat-transfer coefficients for various materials and areas of building structure—walls, windows, partitions, floors, ceilings, and roof. Heat-transfer coefficients for building and insulation materials can be found in Perry's "Chemical Engineers' Handbook," 3d ed., pp. 457–458.

2. *Solar transmission* through the building: sun effects are difficult to calculate; one practical method is to use conductive heat transmission and increase the temperature difference 25°F on roofs and 15°F on side walls in northern climates with 35°F on roofs and 20°F on walls in southern climates.

3. *Heat load of occupants.* Heat and moisture output from personnel depends on activity of the individuals and ambient room conditions; use 500 Btu/hr per person for a design basis.

4. *Heat load from lighting.* Use 3.41 Btu/hr for each watt of lighting in the conditioned space.

5. *Heat load from motors.* Use 2,545 times actual brake horsepower divided by the fractional motor efficiency (0.80 to 0.85) for all motors within the enclosure. This gives motor heat load in Btu/hr.

6. *Heat load from other equipment.* This can be a major factor; use conventional heat-transfer calculations.

In the summertime the heat load with humidity control is removed by cooling systems based on mechanical refrigeration or naturally cold water (see Perry's "Chemical Engineers' Handbook," 3d ed., pp. 758–797). Make-up heat and humidity control in the winter is incorporated in the air-conditioning system, if available. Separate heating systems use steam from the power plant, or small gas- or oil-fired heat. Enclosed finned heat exchangers are popular in both forced draft and natural circulation systems with small hot-blast or unit heaters provided in many applications where local heating in large buildings or complete heating of small enclosures is required.

Personnel and Service Facilities

In planning toilets, locker rooms, and other facilities for the service and convenience of employees, factors that have a direct effect on operating and maintenance costs may be even more important than those governing initial costs. Proper layout and distribution of such facilities, for instance, may save enough of employees' time to offset any differential in initial costs, while the selection of materials and equipment that require a minimum of janitorial services and maintenance will show corresponding economies.

First-aid and hospital facilities, required to keep pace with industrial safety programs, can be met economically with compact unit layouts using standardized hospital equipment. Substantial savings can be made by anticipating the need for private treatment rooms and multiple facilities at the outset.

Facilities for in-plant food service can be held to a minimum cost by careful planning and the adoption of staggered schedules. When it is advisable to provide large central cafeterias, their cost may be reduced by designing them for efficient alternate uses in employee training, recreational, and community functions.

In office buildings and research laboratories, clear spans of 50 to 60 ft permit a degree of flexibility that makes it possible to maintain efficient layouts in the face of changing needs. As in production areas, permanent walls are held to a minimum and removable partitions are installed wherever privacy is essential. The character of work and personnel represented by the occupancy of such buildings frequently makes it advisable to include air conditioning for maximum control of working conditions.

The wide range of service facilities normally required by chemical research and development laboratories can frequently be met at maximum economy by providing central pipe galleries and service trenches. Additional service facilities which may be required and for which industrial building contractors must provide are listed in the following:

Electrical:
  Transmission line to site
  Cost of any changes to substation from which transmission lines are run
  Power generation
Steam:
  Boiler house, including buildings, equipment, piping, and other accessories
  Steam line to site
  Yard distribution piping

Water:
　Water intake and supply facilities
　Water storage
　Water treatment
　Cooling towers
　Line to site
　Yard distribution piping
Fuel gas:
　Line to site
　Yard distribution piping
Fuel oil:
　Fuel-oil storage and handling
　Line to site
　Yard distribution piping
Compressed air:
　Air compressors
　Line to site
　Yard distribution piping
Sewers:
　Lines from site
　Yard sewer piping
Fire protection:
　Water line to site
　Elevated water-storage tank, fire pumps, etc.
　Yard distribution piping
　Fire-fighting equipment
　Sprinkler installations
　Other fire protection facilities
Miscellaneous:
　Office furniture and equipment
　Laboratory furniture and equipment
　Cafeteria furniture and equipment
　First-aid equipment, etc.

COST ESTIMATING FOR PLANT SITES AND STRUCTURES

A simplified approach for estimating site and building costs in terms of installed equipment costs was given in Chap. 6 (see Table 6-5). A more detailed analysis of these costs is given in the sections which follow.

Site Costs

Site Search and Investigation. A search for a plant site involves the selection of the plot of ground best suited to the production and distribution problems of any given manufacturer. It is a painstaking problem whose many ramifications cannot be overlooked; in many instances, it is the most significant step in the plant construction program. This is especially true in the chemical industries, because of their great dependence on such factors as the sources of raw material and the power and water supply (see Chap. 7).

The price range of costs of searching for and investigating sites depends entirely on how exhaustive the plant owner wants to make it. Sites without too specialized qualifications can be found at relatively low cost, sometimes for less than $5,000. If highly specialized requirements are a factor, the cost may be many times more.

Land Values. Surveying a plant site is a negligible cost factor for the average chemical plant. Generally, sites are confined to 20 acres or less, and even the most detailed surveys can be completed within a week. Land values vary according to location. In rural areas, land can be purchased for as little as $200 an acre; in industrialized areas, costs run as high as $10,000 an acre.

Preparation and Landscaping. Preparation and grading of plant sites is influenced by the terrain and the locality of the site. In hilly areas, site preparation is costly. The same is true if existing structures must be demolished. The cost of this part of a building project usually is slightly less than one-half of 1 per cent of the total cost of the project in ordinary cases.

Landscaping costs may vary too. If elaborate landscaping is desired, experts are called in and even very small plots will require expenditures of several thousand dollars. If only a grass covering is desired to keep dust from blowing into equipment and buildings, costs will be considerably less.

Roads, Parking, and Yard Service. Parking areas, with a black top surface, generally range from $3 to $5 per square yard, depending on the terrain. A similar yardstick may be applied to roads. Railroad sidings usually range from $10 to $15 per lineal foot in today's market; these figures include grading, ballast, ties, track, and an average number of switches. Another consideration that may be counted on is fencing. The chain-link type of fence is almost standard. Its cost may be computed at the rate of $3 to $4 per foot, depending on the number of gates and on site conditions.

Building Costs

The decisions affecting building costs involve the number of buildings and stories. Here the fundamental approach should be that of over-all economy, with the efficiency of the process, manufacturing operations, or other functions taking precedence over first cost.

Equal efficiency may be possible with a number of layouts, some of which will permit minimum building costs, while others would needlessly increase these costs. Thus, although sound economic reasons may often dictate a functional layout that requires an expensive building, the important point is to avoid these higher costs if they contribute nothing to the operational efficiency.

Recognition of certain basic principles of building layout can help the design engineer make judicious decisions. The fundamental principle is to secure minimum total building costs (amortized first cost, plus maintenance and operating cost of building services). The total required

TABLE 8-6. COST ESTIMATES FOR BUILDINGS (1958 PRICES)*

I. Type of structure	Erected cost of bare shell including foundation (add required services and equipment costs from Sec. II), $/ft² of floor area
Single-story:	
Mill buildings for chemical processing	
Prefabricated; steel frame, side walls and roof; concrete floor	6
Custom-designed, steel frame, masonry side walls with glass sash, concrete floor, fire-resistant roof	14
Warehousing	
Prefabricated mill building; steel frame, side walls and roof; concrete floor, 2,500 psf loading	4
Reinforced concrete, highly fire-resistant	9
Laboratories, 10-ft ceiling height	
Custom-designed; steel frame; masonry, glass and/or metal panel side walls; concrete floor; fire-resistant roof	17
Offices, 10-ft ceiling height	
Custom-designed; steel frame; masonry, glass and/or metal panel side walls; concrete floor, fire-resistant roof	15
Multistory:	
Chemical processing, 14-ft ceiling height	
Structural-steel framing, masonry walls, concrete floors, fire-resistant roof	18
Reinforced concrete, masonry walls, concrete floor and roof, 2–10 stories high	19
Warehouses, 2,000 psf loading on floors, 16-ft ceiling height	
Structural-steel frame, walls and roof	6
Reinforced concrete, highly fire-resistant, 2–10 stories high	11
Laboratories, 10-ft ceiling height	
Custom-designed; steel framing; masonry, glass and/or metal panel side walls; concrete floors and roof	21
Custom-designed, reinforced concrete, highly fire-resistant	22
Offices, 10-ft ceiling height	
Custom-designed; steel framing; masonry, glass and/or metal panel side walls; concrete floors and roof	18

* For further information see the following:
Figures 8-21 and 8-22.
R. S. Aries and R. D. Newton, "Chemical Engineering Cost Estimation," McGraw-Hill Book Company, Inc., New York, 1955.
Engineering News-Record, Oct. 24, 1957, p. 79.

TABLE 8-6. COST ESTIMATES FOR BUILDINGS (1958 PRICES) (*Continued*)

II. Additional costs of services and equipment	$/ft² of floor area
Air conditioning (including heating)	4.00
Electric lighting:	
Process buildings and laboratories	1.25
Offices	0.90
Warehouses	0.70
Heating (excluding main heating plant)	1.30
Plumbing (excluding process piping)	3.00
Fire-prevention equipment: alarms, extinguishers, sprinkling system	0.80
Office equipment	4.00
Laboratory equipment:	
Control laboratory	12.00
Research laboratory	15.00–25.00

plant floor area should be provided in the minimum number of buildings and with the minimum number of floors. The ultimate preference is the single-story structure if land is available.

Costs of Single-story vs. Multistory Buildings. For the purpose of comparison, it is customary to express building costs on a square foot or cubic foot basis. These unit costs are influenced by many factors: the type of construction; local labor, material, and contracting costs; the size of the building; the ratio of wall perimeter to floor area, etc. A unit cost for one building should not be used in estimating the cost of another unless the conditions are similar. The only reliable method of obtaining building costs for a final cost estimate is to have a set of prints made up by the architects and get a firm bid price. For estimating building costs from plant layout plans, as discussed in Chap. 5, a fair degree of accuracy can be achieved by use of information such as given in Table 8-6. Comparative costs for different types of construction and layout are shown in Figs. 8-21 and 8-22. From these data it is obvious that multistory buildings are 20 to 40 per cent more expensive than a single-story building with the same basic construction materials.

Foundation and Excavation Costs. One can safely figure piling (wood, concrete, or steel) at $2 to $7 per linear foot in place in ground which will permit straight and easy driving. Excavation costs of soft dirt or clay for the average small foundation for machinery or buildings are of the order of $3 to $5 a cubic yard. Wooden forms for concrete can be placed at $0.50 to $1 a square foot. For heavier foundations these figures will average as follows: excavation, $2 a cubic yard; forms, $0.75 a square foot; and concrete, $20 a cubic yard. A concrete conduit tunnel will cost approximately $20 per running foot, whereas a larger walking tunnel averages $45 per foot.

In a preliminary estimate, it is not customary to give separate estimates for the foundations required for each unit of equipment. An estimate for all the equipment foundations located in the yard and all the foundations situated in each building is sufficient. Equipment foundation costs may be estimated either as a percentage of the cost of all equipment supported on the foundations or from the number of cubic yards of concrete required and by applying a unit price to this quantity. Similar considerations apply to equipment supports. The square feet of platform required for equipment access and the approximate weight of the

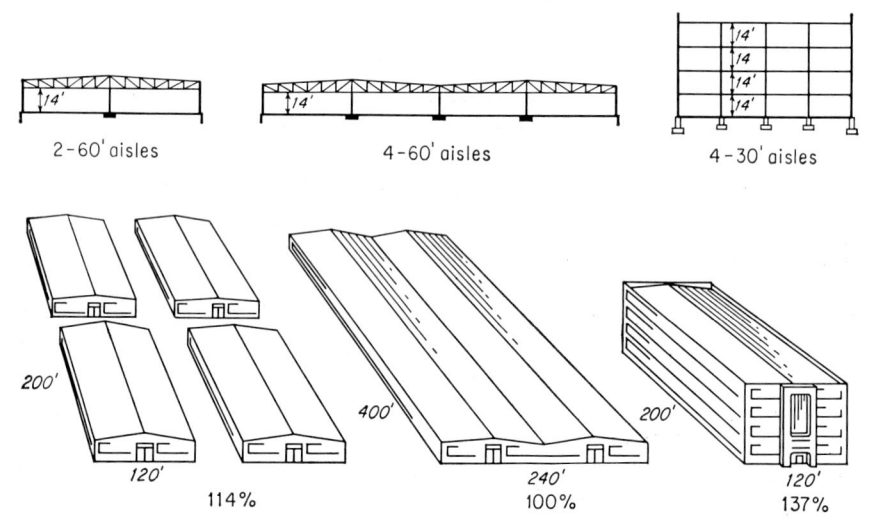

Fig. 8-21. Comparative construction costs for the same gross floor areas as provided by three different building layouts. [*Courtesy of G. A. Bryant, Chem. Eng.*, **54**(5): 114 (1947).]

equipment support on mat structures are estimated from preliminary arrangement drawings, and a unit price applied to the mat area. Supports that are tied into the mat framing may be estimated separately as a percentage of the cost of equipment requiring individual supports.

Indoor versus Outdoor Construction Costs

A breakdown of costs of two plants, each designed for indoor and outdoor construction, is shown in Table 8-7 to illustrate the distribution of costs in the two particular cases.

The structural cost differential between indoor and outdoor plants normally approximates the difference in over-all plant cost. Where the amount of equipment surface requiring insulation increases greatly when the plant is installed outdoors, detailed knowledge of the process is required to obtain realistic cost figures.

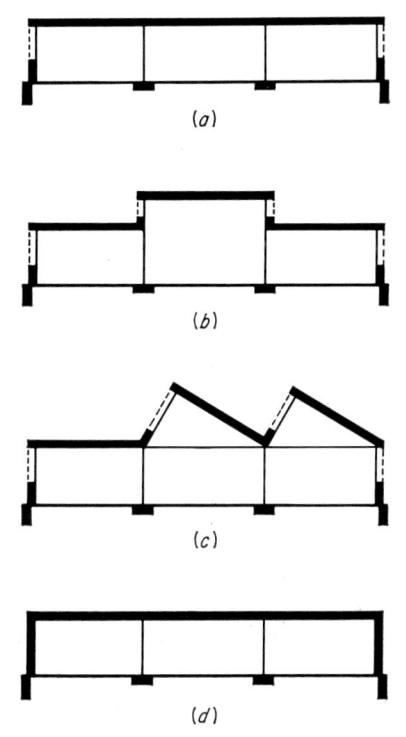

Fig. 8-22. Comparative construction costs for similar buildings as affected by variations in fenestration and roof structure. (*a*) Continuous wall sash and flat roof, 100 per cent; (*b*) continuous wall sash and monitor, 109 per cent; (*c*) continuous wall sash and saw teeth, 111 per cent; (*d*) no sash except vision panels, 105 per cent. (*Courtesy of G. A. Bryant, Chem. Eng.*, **54**(5): 115 (1947).]

TABLE 8-7. COMPARISON OF INVESTMENT COST DISTRIBUTION FOR INDOOR AND OUTDOOR PLANTS

| | Dollars per $100 of indoor plant cost ||||
| | Plant A || Plant B ||
	Indoor	Outdoor	Indoor	Outdoor
Process equipment..........	51.6	51.6	42.5	42.5
Piping....................	16.5	16.6	15.8	15.8
Structures................	19.5	11.4	32.6	20.3
Insulation................	6.0	6.4	2.4	5.0
Electrical.................	4.8	4.6	4.5	4.2
Miscellaneous.............	1.6	1.4	2.2	1.7
	100.0	92.0	100.0	89.5

CHAPTER 9

Process Auxiliaries

As shown in Chap. 6, preconstruction economics does not require a thorough study of such important design areas as piping, instrumentation, control, and power systems. These areas of chemical plant design must be considered in detail when finalizing the plans for putting a chemical process into commercial production. This chapter will highlight the design principles of these process auxiliaries which by and large are the highly specialized areas of chemical engineering and require a backlog of education and experience to do this type of design work.

PIPING

Piping is used as a general term referring to any type of closed conduit for transportation of fluids. In chemical process industries, piping can be considered as the arteries and veins of the plants supplying the life-giving fluids essential to keep the plants running. Even in those plants handling mostly solids, finely divided solid material is quite as readily handled by piping as are ordinary liquids and gases. In Chap. 6 it was pointed out that piping costs can run as high as 50 to 70 per cent of the cost of equipment to which the piping is connected.

Some of the usual piping design problems facing the chemical engineer are:

1. Choosing the correct material according to experience and established codes

2. Selecting the most economical pipe size to handle a given flow

3. Layout of pipelines for ease of accessibility, proper drainage, and minimum stress and fatigue

4. Selection of the best valves for specific service conditions

CHAP. 9]   PROCESS AUXILIARIES   341

5. Selection of suitable means for joining piping and sealing joints
6. Selection of anchors, hangars, and other supports to provide proper installation and life service requirements
7. Specifying an economic insulation where required
8. Determining proper outside pipe colors for identification
9. Preparing detailed and accurate cost estimates of piping systems

In the past several decades, piping design was a relatively simple matter. The engineer set the pipe size from flow considerations, picked the pressure classification from simple formulas or rating tables, and provided the draftsmen with flow sheets from which to make piping layouts. Thermal expansion was provided by installing loops of pre-established movement if necessary. Details of branch connections, attachment of structural supports, and anchors were left to the construction crews. This way of handling piping still is generally workable where mild service conditions exist and low-carbon steel piping can be readily adopted.

Today the demands of a highly competitive and expanding economy have led to complex processing conditions which add a host of new problems to piping design. Higher pressures require either heavier pipe walls or stronger materials. The increased strength might be produced by use of new materials, alloying, heat treatment, or cold work in the case of extended pipe. Broadening the temperature regime introduces problems of creep at high temperatures and notch brittleness at low temperatures. More severe corrosion conditions, economic demands for fuller utilization of material strength, and installation to avoid fatigue are other design complexities.

Where flammable, toxic, lethal, or otherwise dangerous fluids are to be conveyed, safety in piping design is a prime consideration. Striking an economic balance between piping costs and loss of life or property resulting from minimum design standards cannot be met by a simple set of design rules. Specifications have been worked out by experienced designers in the form of *codes*, many of which have legal and, therefore, mandatory statutes.

Explanation of Codes

Any body of published piping design requirements which is national in scope of application is defined as a code, though it may bear the title of regulations, rules, or specifications. Though chemical engineers may not be deeply involved in design by code specifications, it is important from a supervisory standpoint to have a basic familiarity with codes related to piping design. Tables 9-1 and 9-2 summarize the principal codes available. Since these codes are constantly being revised in the light of new materials and research findings, engineers specializing in

TABLE 9-1. SOURCES OF INFORMATION ON STANDARDS AND CODES*

ASA   American Standards Association. Standards for pipe, pipe flanges, and fittings. Published by the American Society of Mechanical Engineers, 29 W. 39th St., New York.

ASME   American Society of Mechanical Engineers, New York. Standards for pressure vessels and all ASA standards.

ASTM   American Society for Testing Materials, 1916 Race St., Philadelphia. Standards relating to metals.

AISA   American Iron and Steel Institute, 350 Fifth Ave., New York. Steel products manual.

MSS   Manufacturing Standardization Society of Valves and Fittings Industry, 420 Lexington Ave., New York. Standard practices.

API   American Petroleum Institute, 1205 Continental Building, Dallas. Oil-piping design standards.

FSSC   Federal Specifications. Superintendent of Documents, Washington, D.C. Metallic piping standards.

USN   U.S. Navy, Bureau of Supplies and Accounts, and/or Bureau of Ships, Washington, D.C. Specifications, standards, plans, and publication legally required for USN work.

* An excellent summary of codes and pipe material specifications is given in "Piping Engineering," Tube Turns, Division of National Cylinder Gas Co., Louisville, Ky.

piping design should obtain the latest code information by writing to the various organizations listed in Table 9-1.

Selection of Piping

ASTM specifications are available for well over 150 different types of materials which can be used in the manufacture of pipe and tubing. Perry[1] lists properties and size of pipe and tubing for 47 different materials and the reader is advised to study this section of the handbook thoroughly. However, an explanation of piping nomenclature and fabrication methods provides the necessary background requirement for studying the working in the field of piping design.

Methods of Fabrication. The methods used in making tubular products are welding, casting, extrusion, forging, cupping, and piercing.

Welded steel pipe is either butt-welded for pipe under 2 in. or lap-welded for larger pipe. In the butt-welded process, long narrow strips of steel called "skelp" are heated in a furnace to 2600°F and drawn through a bell die and welding machine which bends them into a circular shape and simultaneously welds the butted edges together. In lap welding the edges are overlapped by cutting the edges of the strip at a scarfing angle before bending the skelp. Welding is done by an electric resistance method, where rolls are employed to press the scarfed edges together

[1] J. H. Perry (ed.), "Chemical Engineers' Handbook," pp. 413–454, 1650–1651, McGraw-Hill Book Company, Inc., New York, 1950.

TABLE 9-2. IMPORTANT CODES FOR PIPING DESIGN

ASA B 31.1–1955	Code for Pressure Piping
	1. Power Piping Systems
	2. Gas and Air Piping Systems
	3. Oil Piping Systems
	4. District Heating Piping Systems
	5. Refrigeration Piping Systems
	6. Fabrication Details
	a. Pipe Hangers, Supports, Anchors, Sway Bracing and Vibration Dampeners
	b. Pipe Joints Other than Welded
	c. Expansion and Flexibility
	d. Welding of Pipe Joints
	e. Welding Branch Connections and Fabricated or Case Specials
	7. Materials—Their Specifications and Identification
USN 21Y	Code for Installation of Power Plants
API 600	Valve Standards
ASA B 2.1–1945	Pipe Threads
ASA B 16b–1944 R1953	Cast-iron Pipe Flanges and Flanged Fittings, Class 250
ASA B 16.5–1953	Steel Pipe Flanges and Flanged Fittings
ASA B 16bl–1931 R1952	Cast-iron Pipe Flanges and Flanged Fittings (for 800-lb hydraulic pressure)
ASA B 16.9–1951	Steel Butt-welding Fittings
ASA B 16.11–1946 R1952	Steel Socket-welding Fittings
ASA B 16.20–1950	Ring Joint Gaskets and Grooves for Steel Pipe Flanges
ASA B 16.21–1951	Non-metallic Gaskets for Pipe Flanges
ASA B 26–1925 R1947	Fire-hose Couplings (Screw Thread)
ASA B 36.19–1952	Stainless Steel Pipe
ASA B 57.1–1953	Compressed Gas Cylinder Valve Outlet and Inlet Connections
ASA B 60.1–1950	Refrigerant Expansion Valves
ASA (no symbol)	Tolerances for Cylindrical Fits (a primer—not a standard)

under high pressure. Lap welding gives a stronger joint and, hence, is specified for larger pipe.

Seamless pipe is stronger than welded pipe because its wall is uniform throughout. It is made by piercing a circular steel billet at a sufficiently high temperature to maintain plastic flow. This is done on a piercing mill which contains two compression rolls to compress the billet radially with the hole size being controlled by an inner piercing mandrel. The product from this operation is put through dies to adjust diameter and wall thickness to final tolerance specifications. Short lengths of heavy-wall seamless pipe are made by forging a central opening into a hot circular billet with forge hammers.

Thin-wall tubing of various materials can be made by cold-drawing crude seamless pipe from piercing mills through a mandrel-die machine.

Ductile materials such as copper, lead, and plastic can be made directly by extrusion of solid rod material.

Casting is generally applied to brittle materials. The molten metal is poured into vertical or horizontal stationary sand molds. By rotating the mold, a centrifugal casting is produced with a stronger homogeneous wall than that produced by stationary casting.

Pipe Strength and Wall Thickness. The American Standards Association has set up specifications for wall thickness by *schedule number* based on the following formulas:

$$\text{Schedule number} = 1{,}000 \frac{P_s}{S_s} \tag{9-1}$$

$$\text{Schedule number} = 2{,}000 \frac{t}{D_m} \tag{9-2}$$

where P_s = internal working pressure, psi
S_s = allowable stress, psi
t = pipe wall thickness, in.
D_m = mean diameter of pipe, in.

Allowable stresses for various ASTM code steel pipes at different temperatures are listed in Perry's "Chemical Engineers' Handbook," 3d ed., pp. 1650–1651, and ASA B 31.1 Code for Pressure Piping. For example, S_s is 6,500 for butt-welded and 9,000 for lap-welded carbon-steel pipe at temperatures up to 250°F.

Ten schedule numbers are used: 10, 20, 30, 40, 60, 80, 100, 120, 140, and 160. Only schedule number 40, 80, 120, and 160 are stock items for steel pipe under 8-in. size.

Nominal Pipe Size (NPS). Piping is specified in terms of a *nominal pipe size* or diameter which is related to an outside diameter independent of the schedule number or wall thickness. This is required to ensure interchangeability of fittings. The nominal *pipe* size of steel pipe ranges from 1/8 to 30 in. For 14-in. or larger steel pipes, the nominal pipe size is the outside diameter. The nominal value is slightly less than the outside diameter for 3- to 12-in. pipe but for very small pipe the actual outside diameter is much larger (see Perry's "Chemical Engineers' Handbook," 3d ed., pp. 415–417). Pipe of other materials is also made with the same outside diameter as steel pipes for the same NPS to permit interconnecting of pipelines and substitution of alloy fittings and valves when necessary.

The size of tubing is indicated by the outside diameter. The wall thickness is specified by the BWG (Birmingham wire gauge), ranging from 24 (very light) to 7 (very heavy). Sizes and wall thickness of tubing are listed in Perry's "Chemical Engineers' Handbook," 3d ed., pp. 417, 424–438.

Ferrous Pipe. Ferrous materials account for a large part of all the piping in process industries. Carbon steel remains the heavy-tonnage piping material, but the chrome-nickel and high-chrome alloys have been moving rapidly ahead. Commercial steel pipe, or black iron, is made of a low-carbon steel specified under ASTM specification A53. It is made in "random lengths" averaging 22 ft and "double random lengths" averaging 44 ft. It commands a premium for specified lengths. Specification A120 covers a similar steel when galvanized. For higher-temperature service, lap-welded and seamless steel pipe are produced from a low-carbon steel containing under 1 per cent manganese (ASTM specification A106). Line pipe specified by the American Petroleum Institute is of quite similar composition.

A large number of *alloy steels* can be secured in the form of piping, some primarily for temperature and creep resistance, some for corrosion or oxidation resistance. Small percentages of molybdenum or molybdenum and chromium are added for temperature and creep resistance. Increasing the percentage of chromium in such steels tends to increase the corrosion resistance. Such steels also contain small percentages of silicon. Many of them are covered by ASTM specification A335. In a somewhat similar range of alloys specified under ASTM A333, the substitution of moderate percentages of nickel for the chromium and molybdenum permits the use of the steel for low-temperature service.

Austenitic stainless alloys containing 8 per cent or more of nickel and 16 per cent or more of chromium are produced in many types. In the form of pipe or tubing, among the available austenitic stainless alloys are AISI types 304, 321, 347, 316, 317, 309, and 310. The lighter schedules 5 and 10 are more commonly used than the heavier schedules 40 and 80. In addition to the austenitic stainless, the high-chrome ferritic steel type 430 (16–18 Cr, no Ni) finds considerable piping use for high-temperature oxidation resistance, for example, in tube stills and catalytic cracking.

Cast-iron pipe is made by pit casting and centrifugal casting. The latter gives a denser, stronger product by eliminating blowholes and impurities. Cast-iron pipe has not yet been fully standardized. The American Water Works Association has had regional standards for water pipe and fittings for many years, as has the American Gas Association for gas piping. ASA now has approved standards for centrifugal-cast pipe. Pit-cast pipe made under the AWWA standards is produced in eight pressure classes for hydrostatic heads from 100 to 800 ft and in nominal diameters from 8 to 84 in. The AGA standards apply to centrifugally cast pipe for pressures from 100 to 250 psi and in sizes from 3 to 24 in. nominal diameter. Cast-iron pipe is also produced in iron pipe size, with the same outside diameter as standard steel pipe in sizes from 1½ to 12 in., and for pressures of 125 and 250 psi. Either cast-iron water and gas pipes are made with plain ends for use with special mechanical couplings which will be described later, or they are provided with cast-on bell-and-spigot ends for caulked joints or with cast-on flanges for bolted and gasketed joints.

The most commonly used special cast irons for process applications are the high-silicon cast irons. These alloys, duriron and durichlor, are too hard for machining and so must be cast and ground. They are made in inside diameters from 1 to 8 in. using cast-on flanged ends, and from 1½ to 12 in. with bell-and-spigot ends for use in drainage systems.

Nonferrous Pipe and Tubing. Nonferrous pipe and tubing in general come in size ranges from ⅛ to 12 in. in diameter and specified wall thicknesses, and in quite a variety of metals and alloys. Commercial copper and red-brass pipe conform in outside diameter to steel pipe, although the inside diameters differ; weights include both standard and extra-strong wall pipe; such pipe may be threaded.

The copper and alloys group includes materials known as commercial copper, brass, bronze, and cupronickel. Usual red *brasses* for pipe and tubing have 15 per cent zinc. The term *bronze* generally includes a group of alloys in which zinc is largely or entirely displaced by other metals such as tin, aluminum, or silicon, with copper usually making up 90 per cent or more of the composition. Cupronickel or *copper nickel* covers a range of high-nickel alloys with 10 to 30 per cent nickel. *Copper and brass type B* tubing have outside diameters equal to those of NPS pipe but wall thicknesses close to that of the intermediate grade of water tubing. Tubing is also made in outside-diameter sizes with wall thicknesses expressed in decimals of an inch or by the BWG or Stubbs gauge system.

Nickel, monel, and *Inconel* are produced in both NPS and outside-diameter tubing types, generally in sizes from about ½ to about 4 in. and in a variety of tempers, depending on the type of use. Some of the other high-nickel alloys such as the Hastelloys can be secured as pipe, generally produced by casting. Hastelloy cast pipes come in ⅛ to 4 in. NPS size, with wall thicknesses closely corresponding to extra-strong steel pipe. Some of the Hastelloys are made in thin-walled tubing by rolling and welding.

Aluminum seamless pipe and tubing are produced in a variety of compositions, including such alloys as 3S, 61S, and 63S. Pipe is made to the same dimensions and thicknesses as standard and extra-strong iron pipe; tubing sizes are rated by outside diameter and BWG wall. Both drawing and extrusion are used in making aluminum tubular products. *Aluminum bronze* alloys of the Ampco group are made in schedules 40 and 80 pipe from ½ to 4 in., schedule 20 in larger sizes. Condenser tubes come in ⅝- to 4-in. OD sizes.

Lead pipe is produced by extrusion in several compositions including *chemical lead, antimonial lead,* and *tellurium lead.* The first is most commonly used, except where the greater mechanical strength of the alloy materials is needed. Pipe sizes are rated by actual inside diameter and by wall thickness. From four to six wall thicknesses are produced in each size. For applications where the lead needs reinforcement, various weights of *lead linings* are installed in standard steel pipe, in NPS sizes from ½ to 12 in. Three lining methods are used: (1) low-cost method of expanding the extruded lead sleeve into the steel pipe, without bonding; (2) bonding the lead sleeve by heat and pressure to the pipe through use of a bonding alloy applied to both the lead sleeve and the pipe interior; and (3) casting the lead in place with the pipe and around a mandrel.

Tantalum may be formed into tubular products by lap or butt welding or by seamless drawing. Such tubes are thin-walled, in a variety of thicknesses, and cannot be threaded. Lap-welded tubing is made in inside diameters from ⅜ in. up, and butt-welded from ½ in. up. Seamless tubing is produced in a range of wall thicknesses in inside or outside diameter from $\frac{1}{16}$ to 2 in.

Nonmetallic Piping and Tubing. For many years the process field has used numerous nonmetallic materials for pipe, valves, and fittings, among them ceramics, glass and fused silica, glass-lined steel, carbon, asbestos-cement, rubber, and plastics. Chemical stoneware is used for corrosive service in pipe with bell-and-spigot, conical flange, cemented-on metal flanges, and cast-on threaded plastic sleeve joints. Chemical stoneware fittings, lubricated and nonlubricated plug cocks, stoneware-body diaphragm valves, and a variety of low-pressure dampers are available. Porcelainware pipe is available with cemented-on metal flanges and conical flanged joints, ells, tees, crosses, caps, reducers, sight-glass fittings, and packed expansion joints. In *Lapp Y valve* construction, a combined plug and stem is used which passes through a deep stuffing box equipped with a lantern gland, sometimes packed with Teflon wedge rings, sometimes armored with fiber-glass bonded cloth for critical locations. Armoring for stoneware pipe, fittings, and valves is becoming common in the stoneware and porcelain industry.

Industrial glass (*Pyrex*) pipe and fittings come with specially heat-treated ends said to be 2½ to 3 times stronger than ordinary glass. Lengths of pipe and fittings are connected by conical flanged joints with an interface gasket and molded inserts. Available glass fittings include spacers, slip joints, ells, tees, crosses, laterals, return bends, reducing fittings, angle spigots, and straight-through plug cocks. Adapters are available for connecting to metal pipe and tubing and to glass-lined pipe. Glass pipe armored with glass-fiber tape and epoxy resin is a recent development.

Standard Pfaudler glass-lined pipe is made in NPS sizes from 1½ to

12 in. for 150 psi working pressure. Joints are a modified Van Stone joint with forged low-carbon steel flanges with gaskets. Fittings include ells, tees, crosses, return bends, reducers and reducing flanges, hose connectors, adapters, and valves which use a porcelain disk and seat, although stainless and Hastelloy can be supplied for these parts and for the stem.

Impervious *graphite* or *carbon* pipe is manufactured in nominal diameters from 1 to 10 in., with threaded, cemented, and flanged joints; the joint assemblies are cast-iron split flanges with through bolts and a flexible packed coupling using neoprene seal rings. Fittings include ells, tees, crosses, plugs, caps, couplings, nipples, collars, reducers, all-carbon globe valves with Teflon packing, and diaphragm valves, with neoprene or Teflon diaphragms.

Standard *Transite* pressure pipe for water and process use is furnished with a sulfur-base cemented cast-iron flanged hub. Fittings are fabricated of steel and lined with transite, produced in equivalent flanged ells, tees, crosses, laterals, packed expansion joints, and reducers used with steel pipe.

Soft rubber is supplied in the form of flexible rubber pipe with rubber-flanged ends, often internally wire reinforced if used in suction service. Rubber linings ranging from semihard to soft can be installed in fittings and ferrous pipe. Rubber-lined pipe, fittings, and valves have flanged ends with the lining carried out over the flange. Hard-rubber and hard-rubber-lined screwed or flanged pipe, fittings, cocks, and gate and diaphragm valves are available. Available rubbers include natural rubber and various synthetic rubbers.

Plastics. Types of plastics used in piping are cellulose acetate butyrate (*Tenite II*), polyvinylidene chloride (*Saran*), reinforced phenolics (such as *Haveg*), polyethylene, rigid polyvinyl chloride (unplasticized), polyester–glass fiber products, and glass-reinforced resin. *Saran* tubing, rigid Saran pipe, fittings, and diaphragm valves, Saran linings swaged into steel pipe, cast-iron and cast-steel fittings, and in plug and diaphragm valves, are available in some types up to 6 in. *Polyethylene* pipe, valves, and fittings come in sizes of ½ to 2 in. Rigid polyvinyl chloride pipe and fittings come with threaded, flanged, and cemented joints. Tenite II pipe and fittings are joined by solvent cementing. Glass-reinforced *Permanite* furane resin pipe and fume ducts can be joined with glass fabric and resin cement wrappings, or the pipe can be supplied with enlarged ends to take a special cast-iron split flange.

Criteria for Selection of Materials. The optimum selection is generally an economic one based on experience of the design engineer backed by code specifications. Alloy steel and nonferrous and nonmetallic materials have become standard for specific services. Essentially the same type of study of materials selection made for process equipment is applicable to piping (see Chap. 4).

Fabrication and erection costs must be considered in selecting piping materials. Some inexpensive material might require expensive installation and joining costs as well as high maintenance costs. On the other hand, expensive alloy pipe material lasting well beyond the expected obsolescence period of the process would not be justifiable except where product quality is lowered by corrosion product contamination.

Pipe Sizing by Internal Diameters. The types of problems in fluid flow which usually confront the piping engineer are:

1. A new line or system is to be designed for handling a given rate of flow. Find the optimum diameter of pipe.

2. The capacity of an existing line is to be increased or other conditions affecting flow are to be altered. Find the pressure drop as a function of flow rate.

3. Both line size and flow are established; find pressure drops over the entire line or between certain points as a check on available pump or compressor capacities.

Students in a plant design course will have had a basic understanding of the fundamentals of fluid flow and will have undoubtedly solved problems under each of the above categories. A summary of conventional methods can be found in any undergraduate chemical engineering unit operations textbook. Perry's "Chemical Engineers' Handbook," 3d ed., pp. 369–396, gives an excellent treatise on fluid flow, and the following tables and nomographs from this handbook are of particular use in piping design:

	Pages
Nomograph for solving pipe flow problems	379
Compressible flow charts	380, 381
Fanning friction factors	382
Economic pipe diameter	385, 386
Friction loss in screwed fittings	390

Since construction of new plants will generally involve problems of type 1 and since this is probably least emphasized in undergraduate courses, the economic approach will be discussed further.

In choosing the inside diameter of pipes to be used, the selection should be generally based on costs of piping and pumping. Small-diameter pipe costs less but pumping costs are excessive. Conversely, larger-diameter pipelines will have too high a fixed capital charge even though pumping costs are reduced. This suggests an economic balance as a first approach. A complete discussion of this economic balance method is given in Perry's "Chemical Engineers' Handbook," 3d ed., pp. 384–385, with a nomograph for turbulent flow conditions on p. 386.

The design equations for ordinary industrial process conditions using steel pipe are: [1,2]

Type of flow	Inside diameter, D_i	Design equation for D_i	
Turbulent	≥ 1 in.	$D_i = 2.2 w^{0.45} \rho^{-0.31}$	(9-3)
Turbulent	< 1 in.	$D_i = 2.5 w^{0.49} \rho^{-0.35}$	(9-4)
Viscous	≥ 1 in.	$D_i = 1.9 \left(\dfrac{w}{\rho}\right)^{0.36} \mu_c^{0.18}$	(9-5)
Viscous	< 1 in.	$D_i = 2.1 \left(\dfrac{w}{\rho}\right)^{0.40} \mu_c^{0.20}$	(9-6)

where w = mass flow, thousands of pounds per hour
ρ = density, lb mass/ft^3
μ_c = viscosity, centipoises

Representative values for an economic fluid velocity expressed as feet per second in various size pipes are given in Table 9-3.

TABLE 9-3. ECONOMIC FLUID VELOCITY, FPS

Pipe diameter, NPS	Viscous flow, viscosity, centipoises					Turbulent flow, density, pcf				
	0.01	0.1	10	100	1,000	1	10	40	50	60	70
1	38	10	3.1	1.0	0.3	10	4.8	3.3	3.1	2.9	2.7
2	40	21	...	1.5	0.5	11	5	3.4	3.2	3.0	2.9
4	45	24	...	2.5	0.8	13	6	4.1	3.8	3.6	3.4
8	54	28	...	4.1	1.3	15	7	4.5	4.3	4.1	3.9

For unusual conditions such as expensive alloy piping, high installation and maintenance costs, low hourly operating time per year, a separate study should be made [see Perry's "Chemical Engineers' Handbook," 3d ed., p. 385, eq. (32)].

The considerations which lead to an economic pipe diameter are not always realistic, based on plant experience factors. Future changes in demand, corrosion and fouling, and absence of pumping costs in gravity feed lines are factors not taken into account. In many instances the selection should be made on practical experience factors as illustrated in Table 9-4.

The values are representative, not mandatory, as unusual conditions may dictate velocities well outside the indicated ranges. The direction will generally lie toward lower velocities and larger pipe sizes. This is

---

[1] R. P. Genereaux, *Ind. Eng. Chem.*, **29**: 385 (1937).
[2] B. R. Sarchett and A. P. Colburn, *Ind. Eng. Chem.*, **32**: 1249 (1940).

TABLE 9-4. PRACTICAL FLUID VELOCITIES IN PIPELINE

Type of fluid	Type of flow	Velocity, fps, D_i = inside diameter, in.
Thin liquids, water, alcohol, etc., $\mu_c < 10$ cp	Pump inlet Pump discharge Pipeline, pumping Pipeline, gravity	$1.3 + 0.16D_i$ $4 + 0.5D_i$ 5–7 0.5–1
Viscous liquids, heavy petroleum fractions, $\mu_c > 10$ cp	Pump inlet Pump discharge Pipeline, pumping	$0.2 + 0.05D_i$ $0.5 + 0.1D_i$ $1 + 0.5D_i$
Gases*..................	Low pressure < 100 psig High pressure > 100 psig	50–100 100–200

* For better values for steam, see Fig. 9-1.

particularly true for small pilot plants where the use of small line sizes based on economic methods would result in almost immediate clogging by dirt, sediment, and corrosion products generally present in this type of work.

Choosing the Final Pipe Size. When the inside diameter for flow has been determined, the next problem in piping design is to select a pipe material and a size which closely approximates the desired inside diameter and has a safe wall thickness to withstand internal working pressure. The schedule number is found by Eq. (9-1). If severe corrosion is anticipated, a larger schedule number should be used than calculated by the stress formula. Pick a nominal pipe size with the specified schedule number which will give an inside diameter for flow the same or slightly larger than obtained by fluid-flow calculations or experience factors as described above. This method will be illustrated by an example.

Example 9-1. Specify the correct pipe size and material for a high-pressure water line operating at 250°F and handling 20,000 lb of water per hour at 1,500 psia. Seamless piping of ASTM Serial Designation A106 is required by state code.

Calculation of schedule number:
$S_s = 11,000$ (Perry's "Chemical Engineers' Handbook," p. 1650)
$P_s = 1,500$

$$\text{Schedule number} = \frac{1,000 \times 1,500}{11,000} = 136 \cong 160 \quad \text{[from Eq. (9-1)]}$$

Calculation of D_i:
Density of water = 58.8 lb/ft³ @ 250°F
From Table 9-4, velocity in pipe = 6 fps

$$0.7854 \left(\frac{D_i}{12}\right)^2 \times 6 = \frac{20,000}{3,600 \times 58.8}$$

$D_i = 1.68$ in. ID

The closest stock pipe is a 2-in. nominal pipe size, 160 schedule, 1.689 in. ID (Perry's "Chemical Engineers' Handbook," 3d ed., p. 415).

If the economic pipe diameter method is used, then, by Eq. (9-3),

$$D_i = 2.2 \left(\frac{20{,}000}{1{,}000}\right)^{0.45} (58.8)^{-0.31} = 2.38 \text{ in. ID}$$

This would require a 3-in. nominal pipe size. Experience factors here would dictate the smaller 2-in. NPS for this installation.

Process Steam Piping

Steam-pipe design problems are common in refineries, chemical processing plants, and power plants. Since steam is a compressible gas handled at high temperatures and pressures, fluid-flow calculations are somewhat more difficult to handle. This section will discuss methods of sizing steam pipelines.

Steam is delivered from power plants at pressures above 150 psig, but many chemical plants operate with pressures of 15 psig or less. Black-iron (steel) pipe is ordinarily used for steam, except where specially constructed equipment may be necessary, such as thick cast iron to withstand corrosion. When oil is the heat-transfer medium for high-temperature processing, equipment not especially constructed for high-pressure work can be used.

Steam-reducing and Regulating Valves. Jacketed equipment of many kinds, for which the safe working pressure is 15 to 30 lb, is common in chemical plants. In order to bring steam down from the line pressure, a reducing valve is necessary. This is placed in the service line between the feed and the equipment. To assist in repairing a reducing valve, which often leaks or becomes unseated, a bypass should always be made around the reducing valve. Such a bypass is best provided with a globe valve to permit manual control of pressure if the reducing valve is out of service. In order that this can be accomplished, the reducing valve should have valves on both inlet and outlet sides, with the pressure gauges beyond the bypass connections. Valves of all kinds leak, and leaks when the plant is not operating would permit the building up of pressure in the service lines and equipment. Such conditions might lead to destructive consequences should the equipment be unable to withstand the pressure. Therefore, service lines close to equipment should be provided with pop safety valves adjusted to safe pressures in accordance with the equipment in use. A popping valve, hence, indicates leakage at the reducing or shutoff valves and the need for repair.

Selection of Steam-pipe Size. Compressible flow calculations for gases such as steam in pipes are complex. In commercial practice the designer usually does not have time to derive his own formulas, nor can

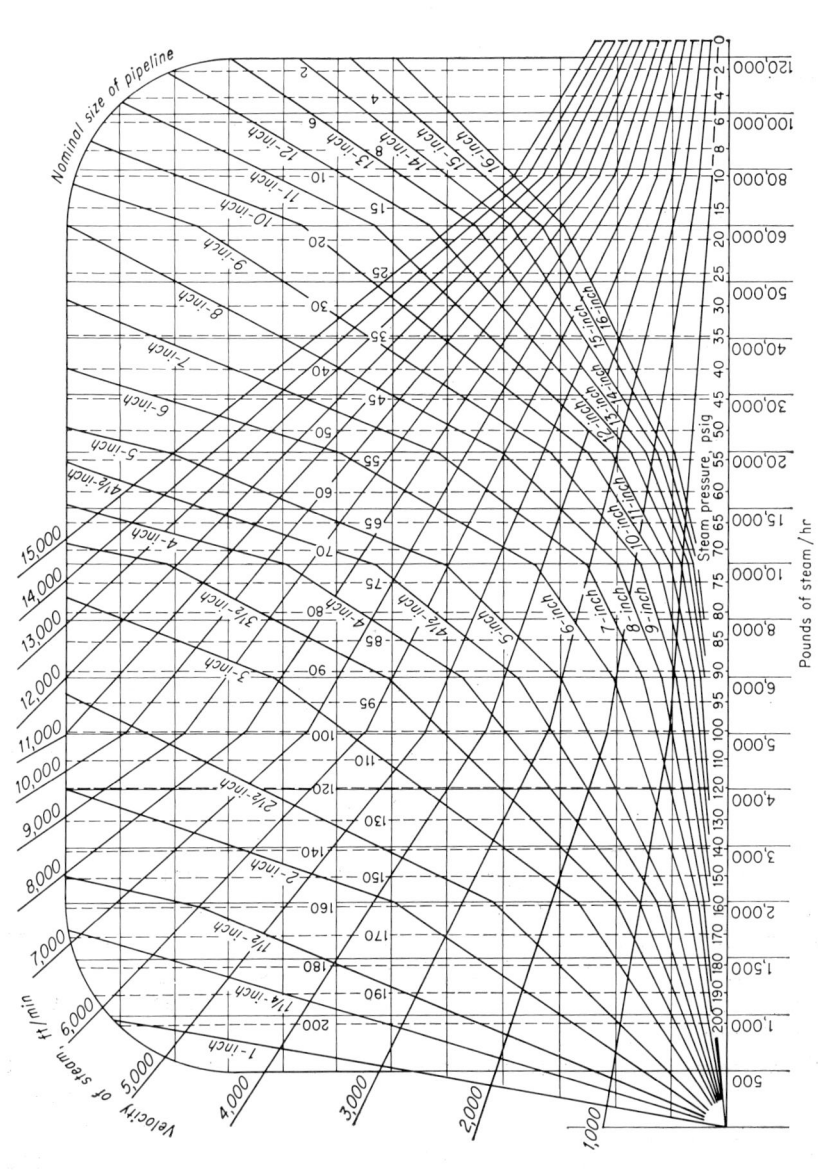

FIG. 9-1. Steam velocity chart. (Courtesy of J. M. Spitzglass, Republic Flow Meter Co., Chicago, Ill.)

he usually spare time to make tests or trace back through extensive mathematical transpositions for proofs. Instead he must select a standard formula, apply it to the work at hand, and check the results against those of some other well-known formula. Or, if tables and graphs are available, he will use such of them as will apply to the problem in question (see Fig. 9-1).

The velocity chart (Fig. 9-1) is a great timesaver in calculating velocities, discharge, and size of pipe required for given conditions of flow of *steam*.

Example 9-2. Allowing a velocity of 5,000 fpm, what size of pipe is necessary to deliver 8,000 lb of steam per hour at 120 lb gauge?

Solution. Trace the 5,000-ft velocity line to 120 lb gauge on the chart. From the intersection, follow horizontally to 8,000 lb of steam per hour. Read the nearest size of pipe, namely, 4 in.

The probable drop or loss of pressure is dependent upon the velocity of flow, length of line, number of turns in fittings or valves, and the covering of the pipes.

Pressure-relief Systems

The loss of operating control of a process will sometimes occur, particularly on start-up of a new process, failure of the cooling-water system, or accidental fire exposure. These conditions develop excessive pressure in the piping and process equipment which cannot economically be designed for containment of fluids under such abnormally high pressures. Therefore, a pressure-relief system must be designed for adequate fluid discharge from a system building up abnormal pressures. Piping codes specify design requirements; the ASA Code for Pressure Piping states: "The combined discharge capacity of the pressure relief or safety devices and their location shall be such that the maximum allowable working pressure of the piping system shall not be exceeded by more than 10 per cent when the pressure relief or safety valves are blowing. The relief or safety valve should be located adjoining, or as close as possible to, the reducing valve."

Pressure-relief Devices. Spring-loaded pressure-relief valves and rupture disks of thin sheet material are available from manufacturers (see "Chemical Engineering Catalog"). When using pressure-relief disks, a strictly noncorrosive material must be specified because of the requirement of maintaining strength in very small thickness.

Design of Pressure-relief Systems. Discharge rates are calculated by charts, tables, or equations supplied by manufacturers. In general, the method is to establish the discharge-pressure rates by code requirements and use fluid-flow calculations for the pressure-relief devices, which behave as nozzles or orifices, and the associated piping. Discharge coefficients for pressure-relief devices can be approximated as follows:

	Discharge coefficient, C	
	Relief valve	Rupture disk
Vapor.........	0.95	0.80
Liquid........	0.40	0.60

Atmospheric venting close to the equipment is sometimes possible on nonhazardous fluid operations or at heights where the contaminant would be satisfactorily air-dispersed, e.g., a relief valve on top of a tall fractionating tower. Safe disposal of large quantities of fluids requires such devices as burning pits, water-quenching towers, or 100- to 200-ft-high flare towers containing a pilot flame. The discharge piping from the relief device to the disposal equipment should have a line size at least as large as the relief valve outlet and a total pressure drop not exceeding 15 per cent of the discharge pressure. Further references for study of pressure-relief systems are available.[1]

Other Piping Hardware

Pipe lengths are generally limited to 22 ft so that long runs of pipe or short lengths with directional change require connection by use of threaded fittings (up to 3 in. NPS) or flanges. Recent piping practice employs bending and circumferential butt welding of the piping at the plant site. Other auxiliaries which require connection to pipelines are valves, flowmeters, steam traps, side-line drawoffs. All these fittings are rated by ASA code on internal system pressures of 25, 125, 250, and 800 psig for cast iron and 150, 300, 400, 600, 900, 1,500, and 2,500 psig for steel. Adequate description of these hardware items is given in Perry's "Chemical Engineers' Handbook," 3d ed.:

	Pages
Pipe joints, flanges, and fittings.........	441–447
Process valves.......................	447–451
Control valves.......................	1326–1328
Gaskets.............................	451–453
Flowmeters..........................	397–412

Piping Layout

The pipe designer has been given process and engineering flow sheets and equipment layout drawings with or without scale models. He has chosen pipe sizes and materials according to the methods discussed in

---

[1] N. E. Sylvander and D. L. Katz, The Design and Construction of Pressure Relieving Systems, *Univ. Mich. Eng. Research Inst.*, Bull. 31, 1948.

previous sections of this chapter. The next job is to design a layout of this piping to meet the following requirements:
1. Easy and economical installation
2. Operation accessibility
3. Ease of inspection, maintenance, and replacement
4. Protection of the piping system from physical or thermal shocks
5. Minimum stress and fatigue from thermal or vibrational environment by providing well-designed flexibility
6. Adequate support and anchorage without handicapping accessibility to equipment or piping
7. Safety codes followed to the letter

Piping Flow Sheets. All the process and utility piping is shown on engineering flow sheets. A typical type of engineering flow sheet is drawn in Fig. 3-9. Lines are coded according to a system which makes piping design identification and procedures easy to follow. The code system shown in Fig. 3-9 is:

O – 102 – 3″
Fluid designation   Line number   Nominal pipe size

Fluid designations are convenient letter symbols for the various fluids handled in the process. Lines are numbered from one piece of equipment to another. This system allows recording of various lines on forms for identifying such items as flow rate, pressure, temperature, class and rating of line and fittings, material and labor costs take-offs, etc. Valves can be identified by a suitable code number on both the flow sheet and specification sheets.

Layout by Scale-model Method.[1] In Chap. 5 the use of models for equipment layout was emphasized. This method is highly recommended for piping layout as well; experience with this method has shown that piping design costs can be reduced as much as 40 per cent and piping installation costs by 5 to 10 per cent. Piping layout drawings are made after some juggling, using wires on the three-dimensional scale model, has been done.

Some piping layout rules based on design experience are:

1. Keep the piping above ground if possible. These lines are easier to install and repair and are free from underground corrosion. Buried lines are hard to maintain and create a hazard since leaks are hard to detect.
2. Headroom requirements should be approximately 20 ft over railroads, 15 ft over roadways, 9 ft above grade, and 7 ft above floors or platforms.
3. Piping which must be located below the minimum 7 ft of headroom should be located at floor level or in shallow trenches with walkways provided.
4. Avoid overhead layout for piping less than 1 in. (NPS) for structural reasons.
5. For below-ground piping in trenches, provide fire stops and drains.

[1] W. N. Troy, *Petroleum Processing*, **9**(2): 224–226 (1954).

6. Allow space for extra piping which is invariably required after operations start; provide for future plant expansion if possible.
7. Allow space for other utilities such as service lines, instrument line ducts, and electrical conduit lines.
8. Valving design and layout considerations are:
    a. All valves, particularly large ones, should be readily accessible from floor level or platforms.
    b. Valves on lines discharging into open tanks should be located so that the operator is not exposed to splashing or fumes.
    c. Use a blocking valve of the gate or globe type behind check valves which invariably leak.
    d. Use double block valves on lines where service on a portion of the line must be supplied while equipment is being removed for repair on another part of the line.
    e. Locate emergency dump valves and discharge lines a safe distance from the operating area outside of process buildings.

The most economical layout usually results in a lane or shadow design in which all of the north-south piping is located at one elevation and all east-west piping at another elevation. This allows for future expansion or modification without interference. Key lines and large expensive piping are positioned first for the shortest, most direct route. Smaller pipes and auxiliary systems, such as steam tracing or jacketing to keep process lines hot, are then fitted in. After all piping and auxiliaries are installed on the model, representatives from various design groups study the model for corrective suggestions. Whereas only one or a few ways of rerouting piping can be seen on two dimensions even by experienced piping designers, a multitude of ways are apparent on the scale model and all of these ways can be tested in a matter of minutes. Accessibility can very easily be checked by this procedure.

A general agreement on piping layout can be reached speedily by this method of attack. Then piping drawings can be made up for use in cost estimating, field fabrication, and installation.

Piping Drawings. Chemical engineers should be familiar with the general systems used in making piping drawings. *Diagrammatic* drawings are most frequently used to represent piping layouts for architectural plans, plant layouts, or small-scale drawings or sketches. Fittings are shown by approved ASA symbols (Fig. 9-2) and piping by a single line, regardless of diameter. An example of a diagrammatic piping drawing is given in Fig. 9-3. For more exact layout work, diagrammatic drawings in more than one view are required (Fig. 9-4). The usual orthographic projection in plan and elevation is at A. By swinging all the piping in one plane as at B, a clearer view is often obtained. Isometric or oblique drawings as at C are most often employed for a clearer understanding of the layout.

The general rules for dimensioning drawings hold for piping plans, but there are additional points that may be mentioned. Always give

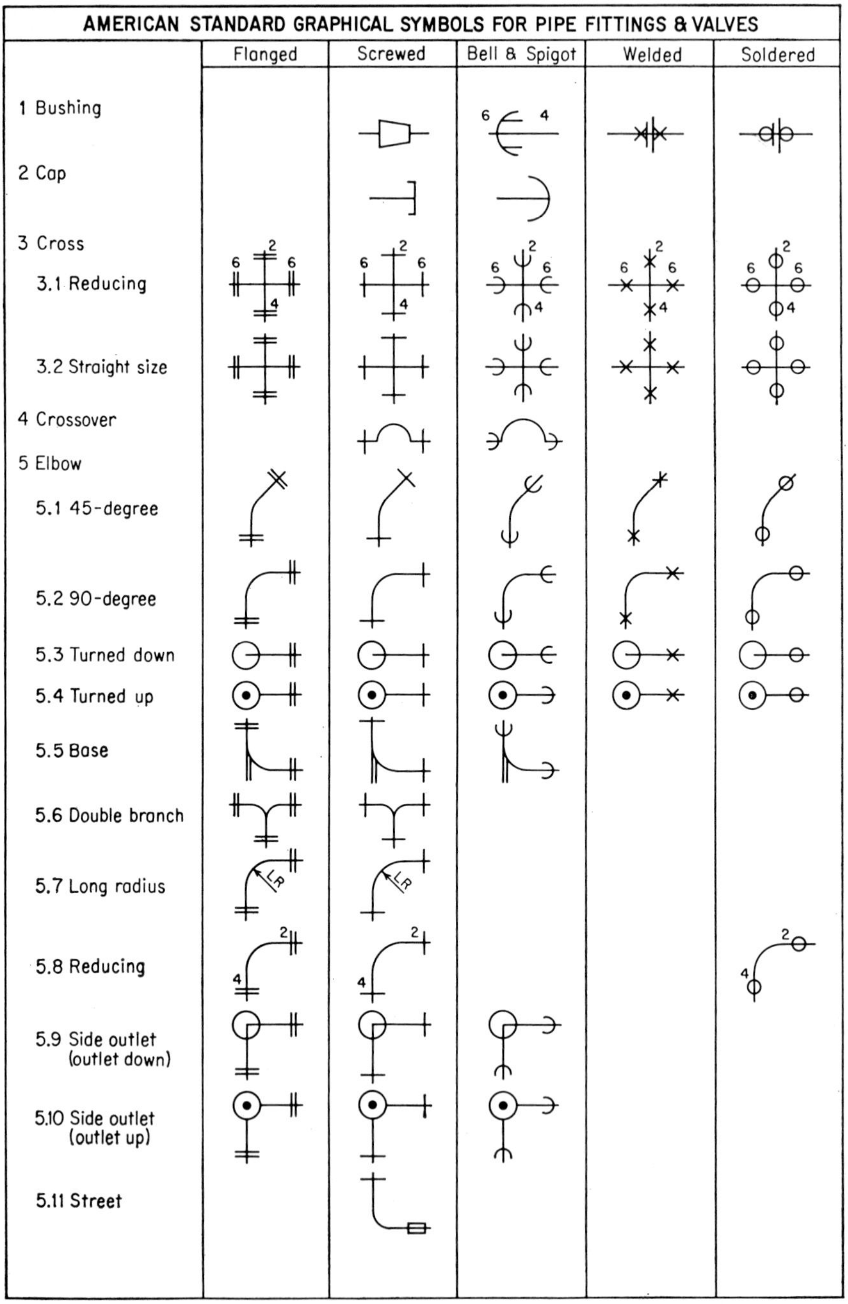

358   CHEMICAL ENGINEERING PLANT DESIGN   [CHAP. 9

AMERICAN STANDARD GRAPHICAL SYMBOLS FOR PIPE FITTINGS & VALVES					
	Flanged	Screwed	Bell & Spigot	Welded	Soldered
1 Bushing					
2 Cap					
3 Cross					
3.1 Reducing					
3.2 Straight size					
4 Crossover					
5 Elbow					
5.1 45-degree					
5.2 90-degree					
5.3 Turned down					
5.4 Turned up					
5.5 Base					
5.6 Double branch					
5.7 Long radius					
5.8 Reducing					
5.9 Side outlet (outlet down)					
5.10 Side outlet (outlet up)					
5.11 Street					

Fig. 9-2. Graphic symbols for piping systems. (*Extracted from American Standard publisher, The American Society of Mechanical Engineers, New York.*)

AMERICAN STANDARD GRAPHICAL SYMBOLS FOR PIPE FITTINGS & VALVES

	Flanged	Screwed	Bell & Spigot	Welded	Soldered
6 Joint					
6.1 Connecting pipe					
6.2 Expansion					
7 Lateral					
8 Orifice flange					
9 Reducing flange					
10 Plugs					
10.1 Bull plug					
10.2 Pipe plug					
11 Reducer					
11.1 Concentric					
11.2 Eccentric					
12 Sleeve					
13 Tee					
13.1 (Straight size)					
13.2 (Outlet up)					
13.3 (Outlet down)					
13.4 (Double sweep)					
13.5 Reducing					
13.6 (Single sweep)					
13.7 Side outlet (outlet down)					
13.8 Side outlet (outlet up)					

Symbols for Piping Systems, Z32.2.3-1949, reaffirmed 1953, with permission of the

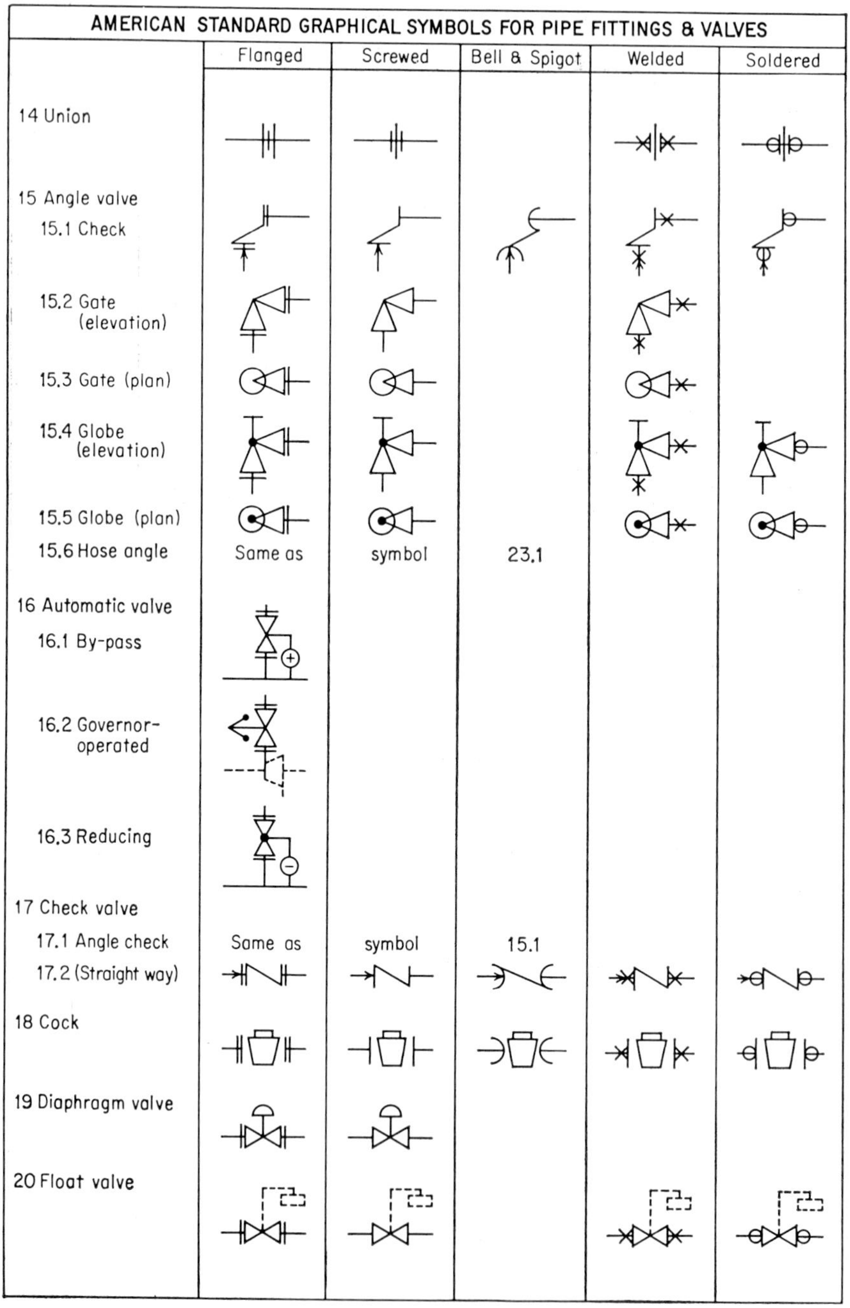

360 CHEMICAL ENGINEERING PLANT DESIGN [CHAP. 9

AMERICAN STANDARD GRAPHICAL SYMBOLS FOR PIPE FITTINGS & VALVES					
	Flanged	Screwed	Bell & Spigot	Welded	Soldered
14 Union					
15 Angle valve 15.1 Check					
15.2 Gate (elevation)					
15.3 Gate (plan)					
15.4 Globe (elevation)					
15.5 Globe (plan) 15.6 Hose angle	Same as	symbol	23.1		
16 Automatic valve 16.1 By-pass					
16.2 Governor-operated					
16.3 Reducing					
17 Check valve 17.1 Angle check 17.2 (Straight way)	Same as	symbol	15.1		
18 Cock					
19 Diaphragm valve					
20 Float valve					

Fig. 9-2 (*continued*).

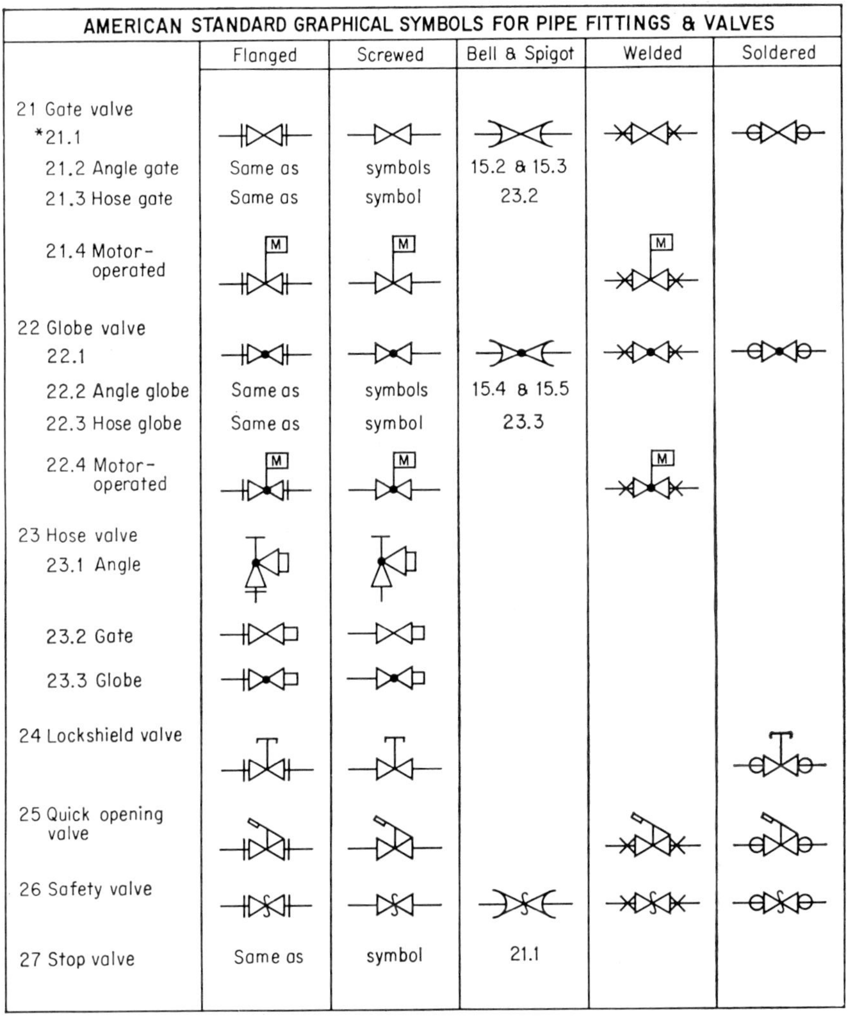

CHAP. 9] PROCESS AUXILIARIES 361

AMERICAN STANDARD GRAPHICAL SYMBOLS FOR PIPE FITTINGS & VALVES						
	Flanged	Screwed	Bell & Spigot	Welded	Soldered
21 Gate valve *21.1 21.2 Angle gate 21.3 Hose gate	 Same as Same as	 symbols symbol	 15.2 & 15.3 23.2		
21.4 Motor- operated					
22 Globe valve 22.1 22.2 Angle globe 22.3 Hose globe	 Same as Same as	 symbols symbol	 15.4 & 15.5 23.3		
22.4 Motor- operated					
23 Hose valve 23.1 Angle					
23.2 Gate					
23.3 Globe					
24 Lockshield valve					
25 Quick opening valve					
26 Safety valve					
27 Stop valve	Same as	symbol	21.1		

*Also used for general stop valve symbol when amplified by specification

FIG. 9-2 (continued).

AMERICAN STANDARD GRAPHICAL SYMBOLS FOR PIPING

	Air conditioning	
28	Brine return	— — — BR — — —
29	Brine supply	———— B ————
30	Circulating chilled or hot-water flow	———— CH ————
31	Circulating chilled or hot-water return	— — — CHR — — —
32	Condenser water flow	———— C ————
33	Condenser water return	— — — CR — — —
34	Drain	———— D ————
35	Humidification line	— · — H — · —
36	Make-up water	— · — · — · —
37	Refrigerant discharge	———— RD ————
38	Refrigerant liquid	———— RL ————
39	Refrigerant suction	— — — RS — — —
	Heating	
40	Air-relief line	— — — — —
41	Boiler blow off	——— ——— ———
42	Compressed air	———— A ————
43	Condensate or vacuum pump discharge	—o——o——o—
44	Feedwater pump discharge	—oo——oo——oo—
45	Fuel-oil flow	———— FOF ————
46	Fuel-oil return	— — — FOR — — —
47	Fuel-oil tank vent	— — — FOV — — —
48	High-pressure return	—#—#—#—
49	High-pressure steam	—#——#——#—
50	Hot-water heating return	— — — — — —
51	Hot-water heating supply	————————
52	Low-pressure return	— — — — — —
53	Low-pressure steam	————————
54	Make-up water	— · — · — · —
55	Medium pressure return	—+—+—+—
56	Medium pressure steam	—+——+——+—
	Plumbing	
57	Acid waste	———— ACID ————
58	Cold water	— · — · — · —
59	Compressed air	———— A ————
60	Drinking-water flow	— · — · — · —
61	Drinking-water return	— ·· — ·· — ·· —
62	Fire line	—— F —— F ——
63	Gas	—— G —— G ——
64	Hot water	— ·· — ·· — ·· —
65	Hot-water return	— ··· — ··· — ··· —
66	Soil, waste or leader (above grade)	————————
67	Soil, waste or leader (below grade)	— — — — — —
68	Vacuum cleaning	—— V —— V ——
69	Vent	— — — — — —
	Pneumatic tubes	
70	Tube runs	════════
	Sprinklers	
71	Branch and head	——o——o——
72	Drain	—— S — — — S ——
73	Main supplies	———— S ————

FIG. 9-2 (*continued*).

CHAP. 9] PROCESS AUXILIARIES 363

figures to the centers of pipe, valves, and fittings, and let the pipe fitters make the necessary allowances. Dimensions of fittings can be found in Perry's "Chemical Engineers' Handbook," 3d ed., pp. 432–450. If a pipe is to be left unthreaded, it is well to place a note on the drawing calling attention to the fact. If left-hand (LH) threads are wanted, this

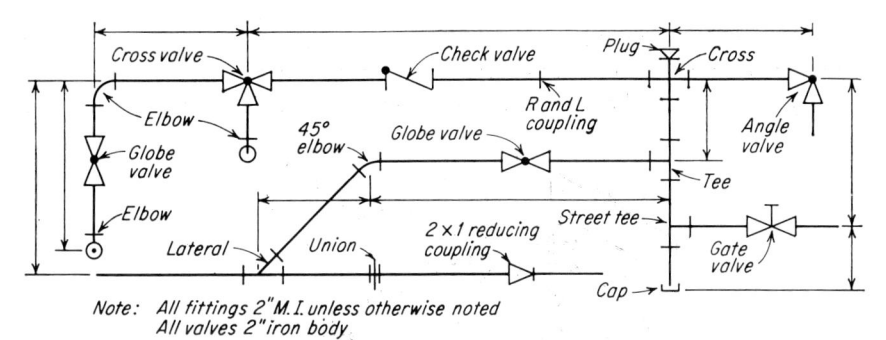

FIG. 9-3. Piping drawing, diagrammatic. (*T. E. French and C. J. Vierck; by permission from "A Manual of Engineering Drawing," McGraw-Hill Book Company, Inc., New York.*)

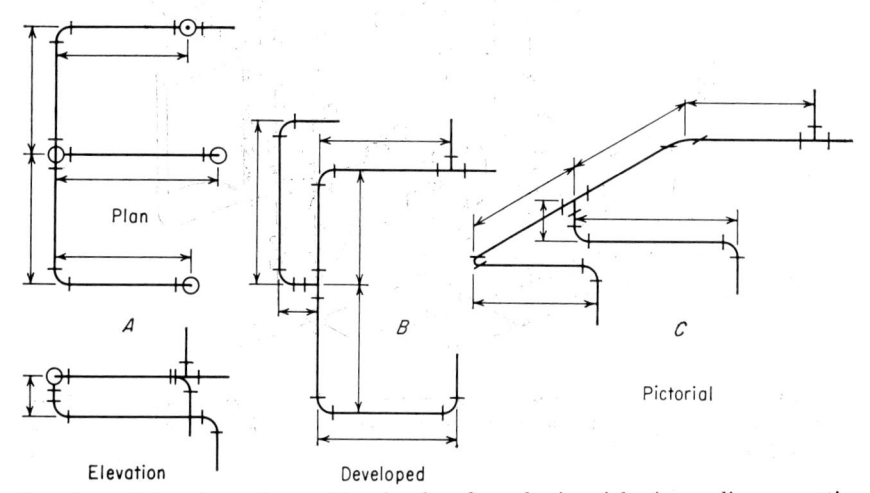

FIG. 9-4. Piping in orthographic, developed, and pictorial views, diagrammatic. (*T. E. French and C. J. Vierck; by permission from "A Manual of Engineering Drawing," 8th ed., McGraw-Hill Book Company, Inc., New York.*)

fact should be noted. Wrought pipe sizes can generally be given in a note using the nominal sizes.

Flanged valves, when drawn to large scale, may have the over-all dimensions given, including the distance from the center to the top of the handwheel or valve stem when open and when closed, the diameter of wheel, etc. Separate flanges should be completely dimensioned, as

364   CHEMICAL ENGINEERING PLANT DESIGN   [CHAP. 9

should all special parts. It is necessary to locate the piping, so that the parts of the building containing the piping must be shown and must be accurately dimensioned. The relative location of apparatus and pipe connections should be given by measurements from the center lines of machinery, distances between centers of machines, heights of connections, etc.

Final drawings should be made after the equipment, boilers, and other machinery have been decided upon, for they can then be drawn completely and accurately. At least two views should be drawn, a plan and

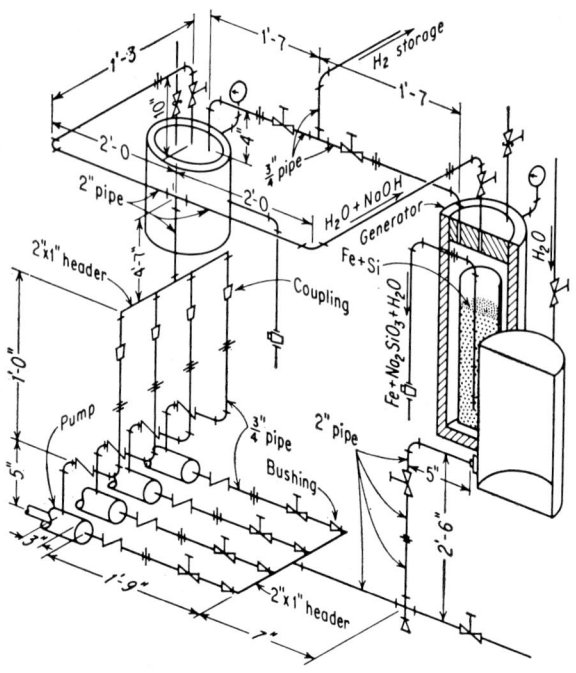

Fig. 9-5. Schematic dimensioned piping drawing. (*T. E. French and C. J. Vierck; by permission from "A Manual of Engineering Drawing," 8th ed., McGraw-Hill Book Company, Inc., New York.*)

an elevation. Often extra elevations and detail drawings are necessary. Every fitting and valve should be shown. A scale of $3/8$ in. to the foot is desirable for piping drawings when it can be used, since it is large enough to show the system in detail.

A typical diagrammatic sketch with dimensions for the piping in a section of a chemical processing plant is shown in Figs. 9-5 and 9-15.

When lines carry different materials, they are identified by ASA coded symbols (Fig. 9-2). These can be color-coded, using colored ink to correspond to ASA color-codes (see Table 9-12).

CHAP. 9] PROCESS AUXILIARIES 365

A second type of piping drawing is often prepared where large-diameter, close-fitting piping layouts are required. These are called piping *scale layout* drawings. They are also used where parts are cut, threaded, and shipped to the job from a remote source of supply. Figure 9-6 illustrates a typical scale drawing. Dimensions of fittings can be found in Perry's "Chemical Engineers' Handbook," 3d ed., pp. 432–450, or any engineering drawing text.

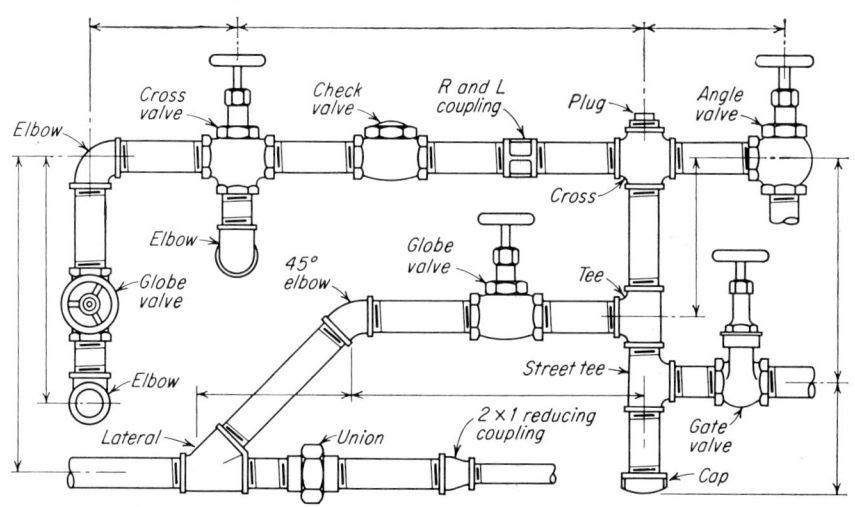

FIG. 9-6. Scale layout piping. (*T. E. French and C. J. Vierck; by permission from* "A Manual of Engineering Drawing," *8th ed., McGraw-Hill Book Company, Inc., New York.*)

Piping Stress Design

Stresses which develop in process piping and the attached equipment under operating conditions are the result of two effects: first, pressure and gravitational forces exerted internally, externally, or inherent in the fabrication and erecting process; second, thermal expansion and contraction of the piping and equipment materials. For example, a 1,000-ft length of carbon-steel pipeline, 16-in. NPS schedule 40, carrying oil at 750°F, would expand 76 in. from 100°F ambient temperature conditions before start-up. The compressive stress developed in the pipe, if it were considered a rigid beam fixed at both ends, would be 170,000 psi, or 260 psi/°F. This is well above the yield strength of carbon steel and a failure would occur unless some method of stress relief were provided in the design. Furthermore, the total weight of piping and liquid is 220,000 lb, which requires a support design. The manner in which forces of such magnitude are to be reduced and handled safely and economically justifies

366   CHEMICAL ENGINEERING PLANT DESIGN   [CHAP. 9

a thorough analysis in most cases. Piping design engineers are responsible for limiting the magnitude of compressive and tensile forces and bending and torsional moments on process equipment and associated piping.

A complete presentation of stress and flexibility analysis and design in piping systems is beyond the scope of this book. This section will serve as an introduction to the subject, and references for more detailed study will be included.

Internal or External Fluid-pressure Stresses. A safe design can be made by use of ASA B 31.1 Code as explained in the section on Pipe Strength and Wall Thickness [see Eqs. (9-1) and (9-2), p. 344].

Departure from normal steady-state operation can be anticipated. A few such conditions are:

1. Pulsations from reciprocating or unbalanced centrifugal pumps
2. Water hammer due to a sudden valve-off or plugging
3. Overpressure of boiling liquid resulting from fire or loss of process control

The design pressure is usually set slightly higher to anticipate such overloads, and incorporation of pressure-relief valves on associated equipment is required design practice.

Piping-material Stresses. These stresses remaining in the material by fabrication or erection can be relieved by annealing or change of procedure. Purchasing from a reliable manufacturer who uses modern manufacturing and inspection methods is the best way to eliminate the fabrication problem. Construction practices for welding, fastening, and inspection as outlined in the Code should be strictly adhered to.

Thermal Stresses. The problems created by temperature stresses are much more difficult to analyze than the other piping stresses because of the many assumptions to be made and the inherent complexity of rigorous force and stress calculations. Detailed references on the subject are available for further study.[1-6]

Thermal Expansion Effects. Changes in pipeline temperature cause expansion or contraction. When the pipe is not restrained, dimensional changes occur with no new stresses added (Fig. 9-7a). If the pipe is

---

[1] "Piping Engineering," sec. 4.0, published by Tube Turns, A Division of National Cylinder Gas Co., Louisville, Ky.

[2] ASA B 31.1 Code for Pressure Piping, American Society for Mechanical Engineering, New York.

[3] The M. W. Kellogg Co., "Design of Piping Systems," 2d ed., John Wiley & Sons, Inc., New York, 1956.

[4] *Catalogue* 53, Crane Co., Chicago.

[5] F. E. Walosewick, *Petrol. Refiner*, **30**(2): 69 (1951).

[6] H. A. Wert and S. Smith, "Design of Piping for Flexibility with Flex-Anal Charts," Blaw Knox Co., Pittsburgh, Pa.

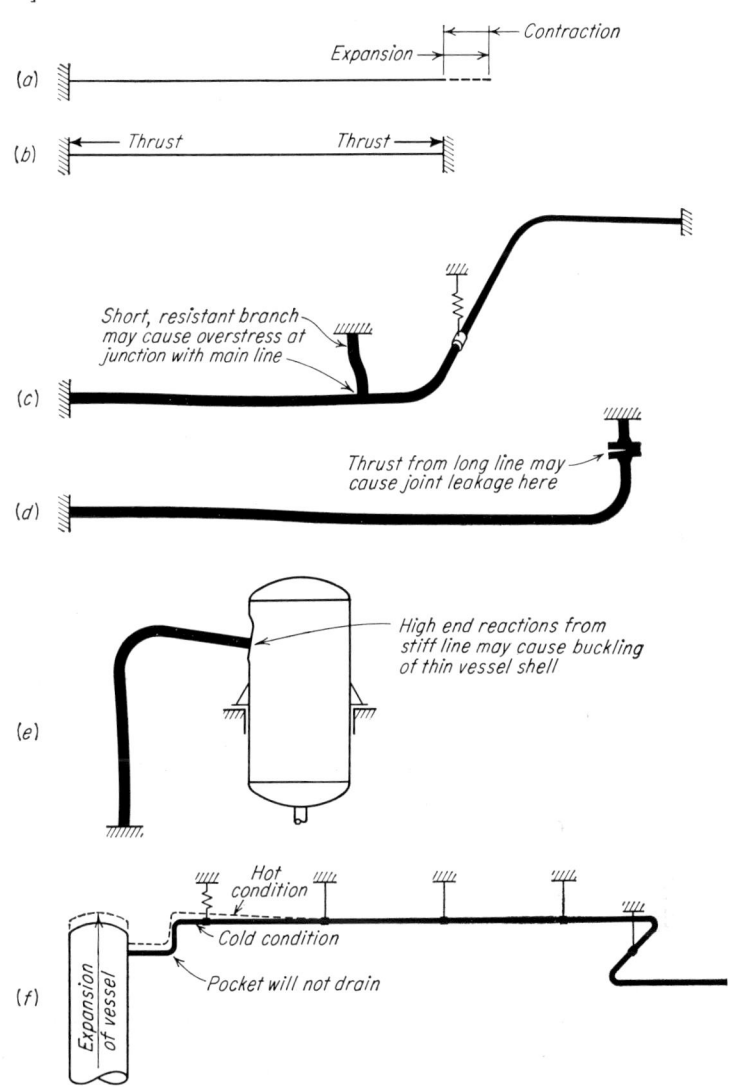

Fig. 9-7. Typical conditions caused by thermal stresses in piping without proper stress relief design. (*Reprinted by permission from Tube Turns' Piping Engineering Paper 4.01—"Introduction to Piping Flexibility Problems," copyright 1950.*)

restrained from movement, large stresses are set up with the pipe and its supporting structure (Fig. 9-7b). The free expansion in terms of inches per 100 ft relative to 0°F can be obtained from Perry's "Chemical Engineers' Handbook," 3d ed., table 1, p. 413. The coefficient of expansion, or strain, e, is derived from the table by dividing the expansion per degree by 1,200.

Example 9-3. Compute the free expansion of 1,000 ft of a schedule 80 steel pipe in heating from 100 to 750°F. From Perry's "Chemical Engineers' Handbook," 3d ed., table 1, p. 413,

$$\text{Free expansion} = \frac{(8.36 - 0.76)(1,000)}{100} = 76 \text{ in.}$$

Example 9-4. Compute the coefficient of expansion in Example 9-3.

$$e = \frac{8.36 - 0.76}{(750 - 100)(12 \times 100)} = 9.7 \times 10^{-6} \text{ in./(in.)(°F)}$$

Stresses and forces developed in restrained piping are computed as follows:

$$S_e = Ee \, \Delta T \tag{9-7}$$
$$F = S_e A_p \tag{9-8}$$

where S_e = compressive pipe stress, psi
 E = modulus of elasticity, psi. (see Perry's "Chemical Engineers' Handbook," 3d ed., table 4a, pp. 1539–1541)
 e = coefficient of expansion, in./(in.)(°F) (see Example 9-4)
 ΔT = temperature change, °F
 A_p = cross-sectional area of pipe wall, in.²
 F = expansion thrust or force, lb

Typical calculations will show that temperature differences of 170°F will create forces about equal to the yield strength of carbon steel, whereas temperature differences less than this value will cause the undesirable condition pictured in Fig. 9-7c to f.

Increasing pipe-wall thickness will not remedy overstress as it would in the usual fixed-load static problem; in fact this tends to increase the forces and moments at the ends [see Eq. (9-8)]. Undesirable thermal stress conditions can only be relieved by proper flexibility design.

Flexibility Design. Flexibility in piping systems can be provided by two basically different methods:

1. Expansion absorbing devices in straight-line layouts between two fixed terminals
2. Piping layout to provide departure from straight-line connections between two fixed terminals

Expansion joints which cause the piping to behave as two free-end systems are illustrated in Fig. 9-8a. These units are expensive and are generally used only where space is limited. An additional feature of the corrugated or bellows joint is that it does not transmit vibration. Slip joints are rarely used because nearly perfect alignment of the moving parts must be maintained to avoid cocking and jamming. Packing selection and maintenance under elevated temperatures or severe corrosive conditions are other limitations. The bellows-type joint has poor corrosion resistance in the thin-walled, highly stressed corrugations. Flow resistance and erosion can be partly alleviated by use of liners.

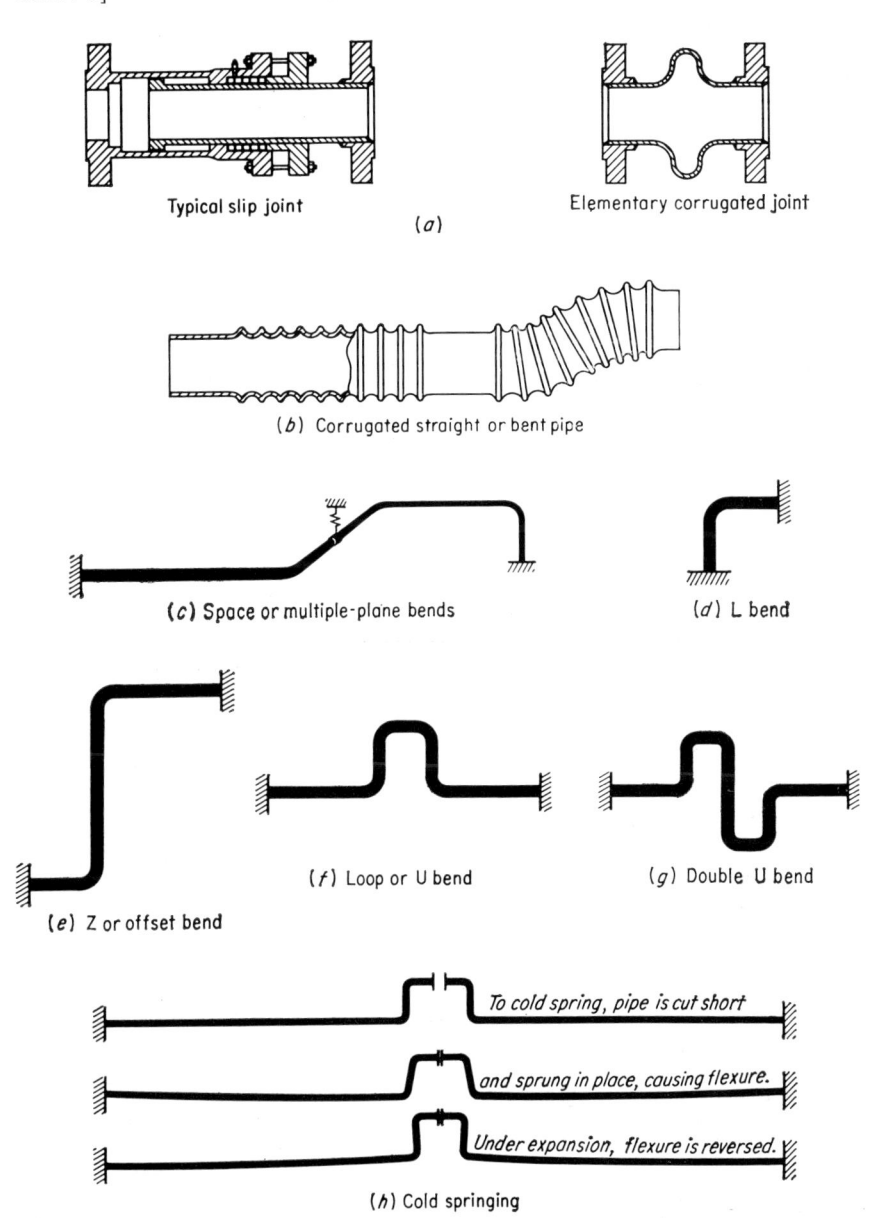

Fig. 9-8. Methods of providing flexibility and thermal stress relief in piping layout design. (*Reprinted by permission from Tube Turns' Piping Engineering Paper 4.01— "Introduction to Piping Flexibility Problems," copyright 1950.*)

Corrugated-pipe designs shown in Fig. 9-8b behave structurally as an intermediate between a bellows joint and straight pipe. They are also intermediate in price and require tie-rod support to keep the corrugations from straightening out under creep conditions. Such expansion designs are limited to short-expansion, straight-line requirements.

The most common and economical method of providing flexibility in one or more non-straight-line layouts is pictured in Fig. 9-8c to h. These designs are based on the rule that departure from a straight-line neutral axis connecting two fixed points creates an improvement, with the tie members, such as welding elbows, furthest removed from the neutral axis contributing most to the flexibility. By running lines in more than one plane (Fig. 9-8c) a 30 per cent increase in flexibility is achieved, since the tie members are in torsion rather than bending.

A comparison of the controlling stresses in the various uniplanar bends shown in Fig. 9-8d to g with the basic square L bend dimensions gives their relative effectiveness for stress relief as follows:

Square L bend	1.0
Z bend	0.84
U bend	0.12
Double U bend	0.04

Offsetting the improved effectiveness of the U bends is the resulting increase in pressure drop.

A method allowable under the Piping Code is known as *cold springing* (see Fig. 9-8h for one typical example). The pipe is initially placed under tension by deliberately cutting one or more joints short and closing by springing. The final compressive stress of heating the pipe to operating temperature is thus reduced. The Code allows a cold-springing credit equivalent to one-third the total expansion.

Mathematical methods of analyzing stresses in complex piping layouts are completely rigorous, time-consuming, and costly, despite adaptation to high-speed computers. The short-cut methods with charts used in most design firms are still too complex to be included in the scope of this book and the reader is referred to the detailed references listed on page 366.

Model testing using strain gauges is often used by large pipe design firms to solve complex piping designs and verify mathematical methods and assumptions when a high degree of accuracy is required.

Piping Installation

Since piping systems are usually not self-supporting, piping design must necessarily include methods of supporting, restraining, and bracing the piping. A convenient glossary of terms and principles of piping installation has been set forth by The M. W. Kellogg Co.[1]

[1] The M. W. Kellogg Co., "Design of Piping Systems," 2d ed., chap. 8, John Wiley & Sons, Inc., New York, 1956.

Principles of Layout to Facilitate Support and Maintenance. Some of the principles of piping layout to effect an economic support design are:

1. Group pipelines to minimize the number of pipe supports, restraints, and braces.

2. Locate lines near other required supports such as building wall and flowing supports.

3. Make the piping system self-supporting as far as practical within flexibility requirements.

4. Additional supports or restraints may be required to avoid excess amplitude of movement or vibration (e.g., vertical pipelines which require only a bottom support for sustaining the weight).

5. Route piping from compressors or vacuum pumps which are prone to vibrate on supports independent of other process piping and use rests or sliding supports which provide some damping capacity rather than hangers.

6. Pipeline should be sufficiently close to the support point so that connection has adequate rigidity.

7. Piping from upper connections on vertical vessels is advantageously supported from the vessel to minimize relative movement between supports and piping; hence such piping should be routed next to the vessel and supported close to the connection.

8. Route piping beneath platforms near major structural members favorable for adding loading of the piping.

9. Provide sufficient space for accommodating the support assembly and for its maintenance.

10. Route very large lines to make short and direct runs, even though it means skewing at a sacrifice in appearance, to save on piping and support costs as well as operating expenses.

11. Provide adequate headroom and drainage.

Some of these points will be discussed further.

Typical Supports for Piping. A high percentage of the piping in modern chemical process plants is carried on structural steel pipe supports. Figure 9-9 is an example of a typical outdoor support for multiple group piping using horizontal steel beams supported by structural steel stanchions. Similar indoor construction under flooring and platforms can be designed by use of a hanger as pictured in Fig. 9-10d. Examples of other manufactured hangers and supports are shown in Fig. 9-10. These are generally cheaper than field fabricated hangers for single pipe support.

Design of Overhead Piping. In designing such piping structures, the clearance and gradient or slope are primary considerations. The headroom required over railroad tracks usually is 20 ft and may need to be more to accommodate locomotive cranes. Plant roadways and power lines also require headroom consideration. Overhead piping should be

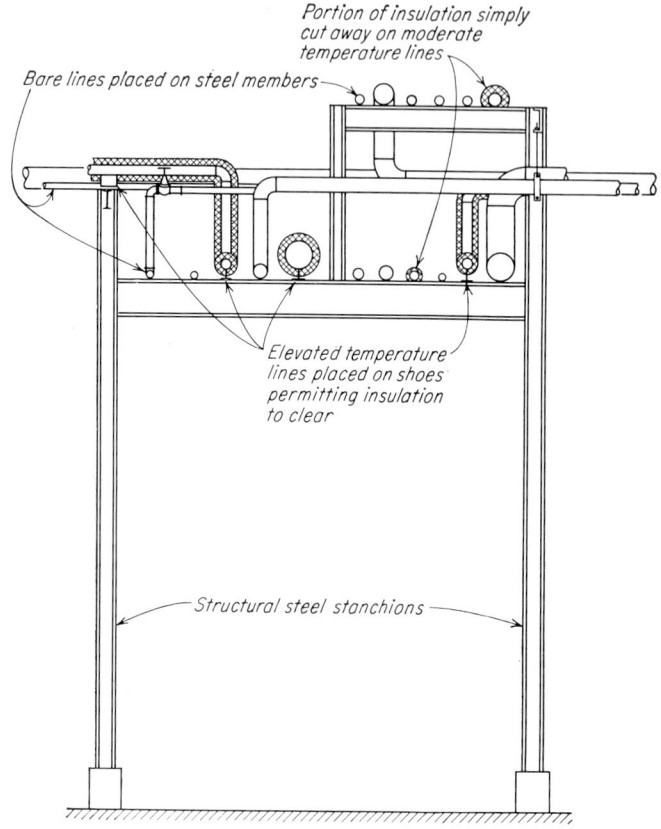

Fig. 9-9. Section through typical outdoor overhead pipe rack showing arrangement of north-south runs at two elevations and east-west runs at an intermediate elevation. (*Reprinted with permission from The M. W. Kellogg Co., "Design of Piping Systems," 2d ed., John Wiley & Sons, Inc., New York, 1956.*)

set on a grade so that it will drain to avoid air pockets or low places which may freeze or permit the deposition of sediment.

When the nominal height of the piping above ground has been decided, the proportioning of spans is the next factor, i.e., whether to use more posts and shorter spans or fewer posts and longer spans. Sweeney[1] discusses this design problem in terms of weight of pipe and its contents and wind loading, together with drainage-deflection conditions. Allowable span length can be computed from a formula for a continuous beam with maximum bending moments at the supports:

$$L = \sqrt{\frac{2SI}{WD_o}} \qquad (9\text{-}9)$$

[1] R. J. Sweeney, *Chem. Eng.*, **63**(3): 199 (1956).

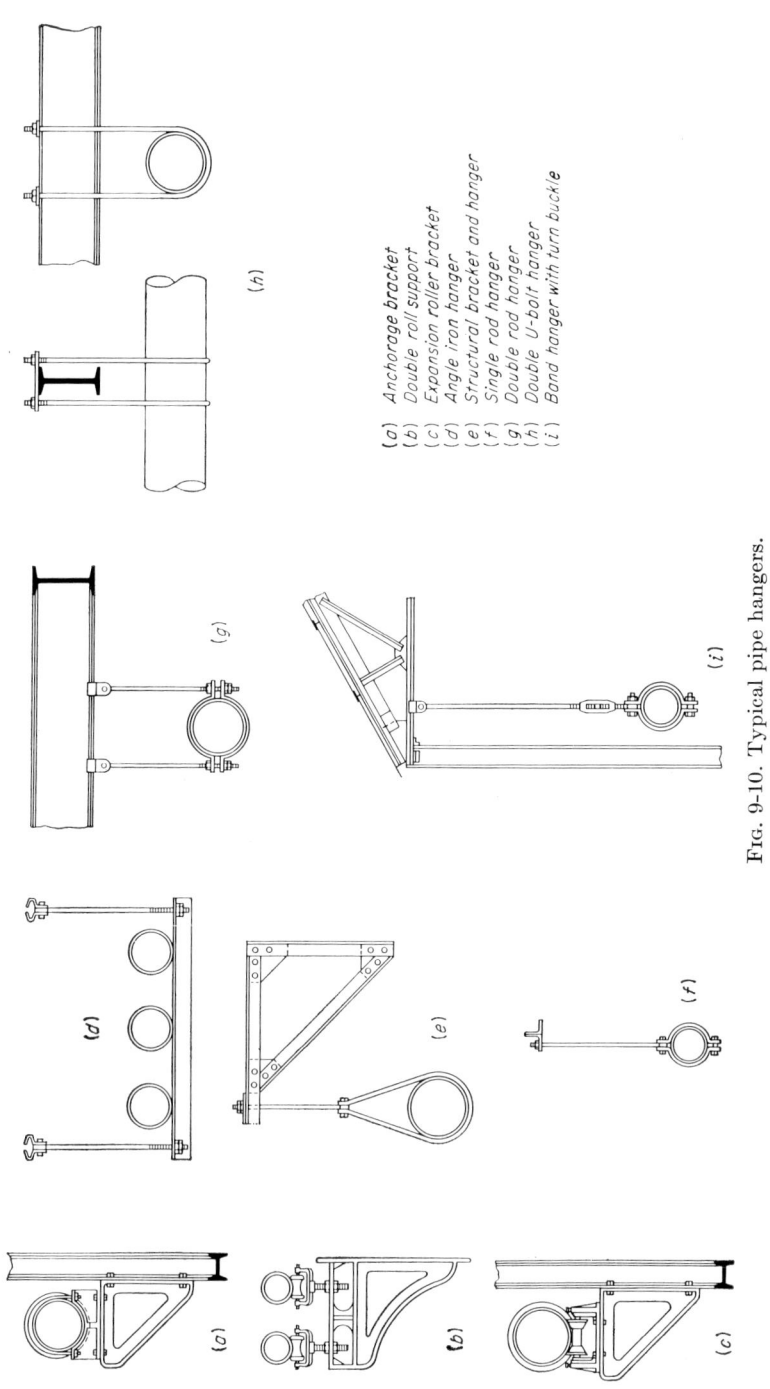

(a) Anchorage bracket
(b) Double roll support
(c) Expansion roller bracket
(d) Angle iron hanger
(e) Structural bracket and hanger
(f) Single rod hanger
(g) Double rod hanger
(h) Double U-bolt hanger
(i) Band hanger with turn buckle

Fig. 9-10. Typical pipe hangers.

where L = span length, ft
 W = total loading per unit length, lb/ft
 S = maximum fiber stress, psi (approx. 20,000 psi for carbon-steel pipe)
 I = moment of inertia, in.^{4}
 D_o = outside diameter of pipe, in.

The elevation over span length L required for proper drainage is

$$e = \frac{2.26 W_p L^4}{IE} \qquad (9\text{-}10)$$

where e = elevation over span length L, ft
 W_p = weight of empty pipe, lb/ft
 E = modulus of elasticity, psi (approx. 3×10^7 for carbon-steel pipe)

These formulas are satisfactory for pipe sizes larger than 2 in. Table 9-5 summarizes the important data for designing overhead piping, using carbon-steel schedule 40 pipe filled with water, and wind loading in excess of 100 mph. Pipes smaller than this show larger deflections due to plastic deformation at point of support (see Catalog 53, Crane Co.). It is best not to use overhead design for piping less than 1 in. If the support pole elevation for long lines becomes excessive on level terrain, it is better to drop the line periodically to the low limit and add a low-level drainage connection.

TABLE 9-5. DESIGN INFORMATION FOR OVERHEAD PIPING*

Nominal pipe size, in.	OD, in.	I, in.^{4}	Pipe W_p	Water	Wind, horizontal component	Total W	Span length L, ft	Elevation between supports, ft	Load at support poles, lb
2	2.375	0.666	3.65	1.45	3.01	8.11	37.2	0.82	302
3	3.500	3.017	7.58	3.22	3.46	14.26	49.2	1.14	700
4	4.500	7.233	10.79	5.50	3.91	20.20	56.5	1.18	1,140
6	6.625	28.14	18.97	12.54	4.69	36.20	68.6	1.14	2,480
8	8.625	72.49	28.55	21.70	5.15	55.30	78.0	1.12	4,320
10	10.75	160.7	40.48	34.13	5.31	79.92	86.6	1.08	6,910
12	12.75	279.3	49.56	48.70	5.74	104.0	92.0	0.98	9,560

* Abstracted from R. J. Sweeney, *Chem. Eng.*, **63**(3): 199 (1956).

Design of Underground Piping. Laying a pipeline in a tunnel (1) reduces the heat losses and insulation costs, (2) does not occupy valuable overhead space, and (3) protects the piping from mechanical injury and freezing. However, it is a good rule to consider some of the disadvantages before installing pipe underground. Leakage, which would be difficult to locate underground, is virtually certain to occur in chemical lines sooner or later. This would require frequent inspection and would make repairs difficult. When piping must pass under certain obstacles, e.g., under a concrete roadway or a shallow stream, it should be installed in such a manner as to permit it to be withdrawn readily for replacement. Underground piping is also subjected to the usual underground corrosion. The tunnel should be built below the frost line to avoid movement of the line. Pipe supports installed in an underground trench or tunnel are usually of simple design—the roller, wedge, or common pipe or bar.

Steam lines and returns, all hot-water lines, and any hot chemical lines that run outside the power plant or between buildings can be run underground. These lines may be insulated by a special system of underground insulation.

Erection of Piping and Supports. The design engineer is frequently called in to assist with and observe the adequacy of the piping installation. For this reason he should familiarize himself with field practice methods.

The permanent pipe-support structures are usually positioned first to minimize temporary rigging. A large majority of the process piping is shop or field welded and tested so that long sections can be dropped in place to avoid overhead welding or other difficult handling in the construction area. Minor plastic deformation of piping and structural supports is tolerated, particularly on cold-springing design for flexibility stress relief. Pipe is first aligned at the more critical locations; then the intermediate stations are adjusted by pulling into alignment with or without heat treatment at locations away from girth welds. Ring or muffle gas burners are used advantageously for the required uniform heat treating, whereas torch heating must be applied by at least two continuously moving torches to give even heating.

During start-up, the performance of all supports, restraints, and braces should be checked and adjustments made to avoid fatigue failure and unusual distortion with possible leakage.

Insulation

The object of insulation is to prevent an excessive flow of costly heat to the surroundings from piping and equipment in which heat is generated, stored, or conveyed at temperatures above the surrounding temperature. A second object is to provide for protection of personnel against skin damage from high- or low-temperature surfaces. Insulation is just as useful in preventing an excessive flow of heat from the outside to materials which must be kept at temperatures below that of the surroundings.

The student has had a good background in heat transfer so that this section will give the applied principles for correctly selecting, installing

TABLE 9-6. CHARACTERISTICS OF THERMAL INSULATION MATERIALS*

Material	Type†	General forms	Strength	Resistance to water	Resistance to vapor	Composition	Advantages	Disadvantages	Temp. limits, °F Min	Temp. limits, °F Max	Conductivity, Btu/(hr)(ft²)(°F/in.) 32	70	212	500°F	Density, pcf
1. Air cell	SR	Laminated corrugated felts formed into pipe insulation and sheets	Little compressive strength; moderate tensile strength	Poor	Poor	Asbestos felts cemented together	Light weight, low cost	Will not withstand severe abuse	100	300	....	0.490	0.550	....	9
2. Asbestos	SR	Laminated felts formed into pipe insulation, blocks and curved segments	Little compressive strength until compressed approximately 20%; moderate longitudinal strength	Moderate	Poor	Asbestos sheets separated by spongy cellular nodules or by indentation	Fair tensile strength	Sags and becomes soft when wet	100	700	....	0.390	0.450	0.560	23
3. Asbestos fiber (standard)	R	Processed into pipe insulation and block	Good flexural and excellent tensile strength	Excellent	Fair	Long amosite fibers blended with binders	Excellent for bridging and for use in steam tracing‡	....	33	750	....	0.320	0.390	0.538	14
4. Asbestos fiber (super)	R	Processed into pipe insulation and block	Good compressive and flexural strength; limited tensile strength	Fair	Moderate	Hydrous calcium silicate blended with long asbestos fibers	High shearing resistance; high compressive strength	Not suitable for bridging under heavy load; nonflexible	212	1200	....	....	0.42	0.56	18
5. Cellular glass	R	Rigid, fabricated pipe insulation, lagging, segments and fitting covers	High compressive strength; fair tensile strength	Excellent	Excellent	Inorganic glass containing microscopic, hermetically sealed cells	Easily shaped and fitted; does not depend on vapor barrier for its vapor resistance; incombustible‡	Should be protected against abrasion	−400	800	0.35	0.39	0.46	....	10
6. Cellular silica	R	Rigid cellular structure block	High compressive strength (130–210 psi)	Excellent	Excellent	Foamed silica	Resistant to thermal shock; resistant to acids except hydrofluoric and hot phosphoric‡	Must be protected against abrasion	−300	2200 (intermittent) 1600 (continuous)	....	0.44	0.62	1.0	10–12

376

7. Cement	C	Mineral wool-water mix; semithermo-setting	Resistance to impact normally good	Water will soften the dried cement	Moderate	Nodulated mineral wool and asbestos fibers with inorganic binders	Good troweling characteristics; medium smooth finish; good adhesion and cohesion	Slow setting; will not stand beating rain	100	1800	..... 0.690 0.820 1.058	26
8. Cement	C	Mineral wool-water mix; hydraulic setting	Impact resistance very good	Will not soften when wetted	Moderate	Nodulated mineral wool and asbestos fibers and inorganic binders	Fair troweling characteristics; smooth finish, good adhesion and cohesion; quick setting	Cannot be reworked	32	1200	0.46 0.525 0.610 0.840	49
9. Diatomaceous earth	R	Molded and formed pipe insulation, segments and blocks	Good compressive strength; limited flexural and tensile strength	Fair	Moderate	Diatomaceous earth blended with long asbestos fibers	Handles well, used as inner layer under 85% magnesia where operating temperature exceeds 575°F	Not suitable for bridging under heavy load; nonflexible	22	1900	..... 0.66 0.720	23
10. Glass fibers	BF	Felted into rolls	Soft, flexible; little compressive resistance after application	Good	Moderate	Fine glass fibers lightly bonded	Flexible; fibers respond to repeated compressing; good thermal cushioning	Pronounced depression under heavy loading; depends on vapor barrier	−300	600	0.23 0.25	1-3
11. Glass fibers	SR	Formed into pipe insulation and block	Soft, flexible fibers; will not break down under flexing	Excellent	Moderate	Glass fibers bonded	Flexible; light fibers will not break down under moderate impact; fair tensile strength	Application limited under certain conditions; depends on vapor barrier	−300	600	0.270 0.273 0.320	8
12. Glass fibers	BF	Felted glass fiber blanket with various types of facings	Soft, flexible; little compressive resistance after application	Good	Moderate	Felted glass fibers fabricated with metal mesh or other facing	Flexible; light fibers will not break down under moderate impact; fibers respond to repeated compressing	Application limited under certain conditions due to binder	−300	1000	0.21 0.280 0.320 0.600	6

377

TABLE 9-6. CHARACTERISTICS OF THERMAL INSULATION MATERIALS* (Continued)

Material	Type†	General forms	Strength	Resistance to water	Resistance to vapor	Composition	Advantages	Disadvantages	Temp. limits, °F Min	Temp. limits, °F Max	Conductivity Btu/(hr)(ft²)(°F/in.) 32	70	212	500°F	Density, pcf
13. Hairfelt	BF	Felted into blanket roll	Standard density, very spongy; heavy density, very firm	Little resistance	Little resistance	100% cattle hair	Excellent emergency low-temperature insulation; can be formed over any size or shape pipe or vessel	Requires careful vapor treatment; depends on vapor barrier for its service in low temperatures	−150	200	0.25	0.260			8.25 (standard)
										200	0.26	0.29			17.25 (heavy)
14. Hydrous calcium silicate	R	Molded and formed insulation, segments and blocks	Good compressive and flexural strength; limited tensile strength	Good	Moderate	Hydrous calcium silicate blended with long asbestos fibers	High shearing resistance; high compressive strength‡	Nonflexible	100	1200		0.37	0.41	0.52	12
15. Hydrous calcium silicate— extra-high temperature	R	Molded block and pipe insulation	Good compressive strength	Good	Moderate	Special process hydrous calcium silicate with asbestos fibers	High shear strength; low shrinkage at high temperature‡	Nonflexible; limited bridging strength	100	1800		0.38	0.43	0.57	11–12
16. Magnesia (85%)	R	Molded and formed pipe insulation, segments and blocks	Good compressive strength; limited flexural and tensile strength	Fair	Moderate	85% hydrated magnesium carbonate	High shearing resistance; high compressive strength	Not suitable for bridging under heavy load; nonflexible	212	575		0.410	0.460		11
17. Mineral cork	R, BF	Formed pipe insulation and blankets	Moderate compressive and tensile strength	Good	Moderate	Mineral wool fibers with asbestos binders	Easily fitted at atmospheric temperatures	Requires careful vapor treatment; depends on vapor barrier for its service in low temperatures	−300	150	0.305	0.325			15

378

18. Mineral wool	R	Molded into blocks and lagging	Limited compressive and tensile strength; fair flexural strength	Excellent	Moderate	Mineral fibers blended with binders	Flexible and withstands considerable expansion and contraction	Will not resist excessive mechanical abuse without indentation	100	1700	..... 0.390 0.420 0.420	18
19. Mineral wool	BF, R	Semirigid blanket, etc., formed into pipe insulation, block and lagging	Moderate compressive and tensile strength	Fair	Moderate	Mineral-wool fibers bonded	Easily fitted at atmospheric temperatures	Requires careful vapor treatment; depends on vapor barrier for its service in low temperatures	−400	250	0.300 0.320	14
20. Perlite, asbestos fibers and binders	R	Molded block and pipe insulation	Good compressive strength	Excellent	Good	Formed Perlite asbestos fibers and binders	Good compressive strength; water resistant‡	Nonflexible	32	1200	..... 0.38 0.43 0.58	10–12
21. Polystyrene (expanded)	R	Block and pipe insulation	Moderate compressive strength; fair tensile strength	Excellent	Excellent	Polystyrene with air cells	Easily fitted; flexible, clean and light-weight‡	Depends partly on barrier for effective service at low temperatures, combustible	−300	175	0.23 0.26	2–2.3
22. Polyurethane (expanded)	R	Molded in blocks and pipe insulation	Fair compressive strength	Good	Good	Foamed diisocyanate and polyester resin	Light weight; low conductivity	Limited upper temperature	−50	230	..... 0.17 0.35	3
23. Reflective preformed metal	R	Factory formed into flat and curved pipe and fitting insulation	Good compressive strength; high tensile strength; will bridge gaps	Excellent	Excellent	Factory formed into rigid forms of the number of reflective sheets required	Light weight; fast application; can be removed and replaced without damage; nonabsorbent	Fittings and special shapes must be factory preformed	−200	1100	Depends on number of reflective sheets	5–6 (average)
24. Vegetable cork	R	Molded pipe insulation, blocks, lagging and fitting covers	Compressible under light load, but does support heavy loads	Good	Fair	Granulated vegetable cork bonded with natural cork resin	High vibration resistance, fair tensile strength‡	Should be fireproofed in critical areas	−300	200	0.285 0.300	11

TABLE 9-6. CHARACTERISTICS OF THERMAL INSULATION MATERIALS* (Continued)

Material	Type†	General forms	Strength	Resistance to water	Resistance to vapor	Composition	Advantages	Disadvantages	Temp. limits, °F Min	Temp. limits, °F Max	Conductivity Btu/(hr)(ft²)(°F/in.) 32	70	212	500°F	Density, pcf
25. Wool felt...	BF, R	Laminated sheets formed into pipe insulation	Good compressive and tensile strength	Moderate	Moderate	Bonded sheets of rag felt	Low cost	Not to be used in critical areas	32	225	0.510				18
26. Loose and fill types	.....	Numerous materials, over 23 in number, are available under this general classification, consisting of powders, fibers, and granulations in addition to other compositions. The information in adjacent columns shows property ranges and usefulness of these types of materials; their *specific* properties should be evaluated from manufacturer's specifications					Low conductivity; lightweight construction; easy to apply to cavities	Will pack down; little mechanical strength	−300 to 200	200 to 2500	0.10 at 32°F to 2.35 at 1600°F				¾ to 30
27. Reflective sheets or foil	.....	Reflective foil, with and without lamination to other sheet insulation, comprises this general classification. Examples would be aluminum foil, with or without lamination to corrugated asbestos paper. The information in adjacent columns shows property ranges and usefulness of these types of materials; their specific properties should be evaluated from manufacturer's specifications					Incombustible; moisture resistant if sealed properly; lightweight construction	Low tear and mechanical strength; may require prefabrication to fit; needs noncorrosive atmosphere to maintain reflectivity	−325 to 32	200 to 1100	Depends on radiation properties for effective heat barrier				

* By permission of R. Thomas and W. C. Turner, *Chem. Eng.,* **60**(6): 222 (1953); supplemented by information from W. C. Turner, Carbide and Carbon Chemicals Company.
† R, rigid; BF, blankets or felts; C, cements; SR, semirigid.
‡ Suitable for shop prefabrication.

and maintaining insulation. (For review, see Perry's "Chemical Engineer's Handbook," 3d ed., sec. 6.)

Types of Insulation Available. There are numerous types and forms of insulation available. These can be classified according to temperature levels of operation:

1. Low temperature—below room temperature; used in refrigeration service (see Table 9-6)

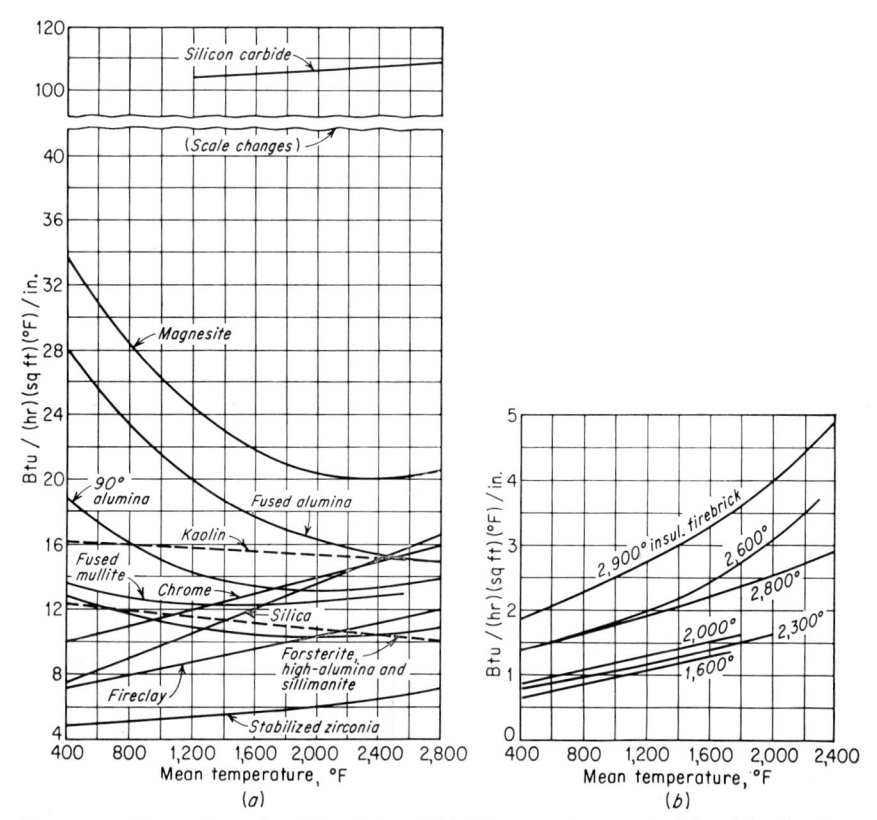

Fig. 9-11. Thermal conductivity data of high-temperature materials: (a) refractory brick and (b) insulating firebrick. [*Courtesy of C. L. Norton, Jr., Chem. Eng.,* **60**(6): 216 (1953).]

2. Medium temperature—from room temperature to 1900°F; most piping insulation for heat savings falls into this classification (see Table 9-6)

3. High temperature—above 1900°F; refractory ceramics available in molded forms or can be cast in place. Metallic refractory sheets are available for reflective insulation. See Tables 9-6 and 9-7 and Fig. 9-11.

Up to 600°F, 85 per cent *magnesia* is the most commonly used insulation.

TABLE 9-7A. CHARACTERISTICS OF HIGH-TEMPERATURE INSULATION. CLASSIFICATION AND USE PROPERTIES OF REFRACTORY BRICK

Name	Composition, per cent	Melting point, °F	Normal use limit, °F*	Approx. $/M 9-in. equiv.	True specific gravity	Bulk density, lb/ft³	Source
Alumina-silica:							
Fireclay base							
First quality	35–40 Al_2O_3; 54–60 SiO_2	3,090–3,175	2,400–2,700	95	2.60–2.70	125–140	Fireclay
Semi-silica	70–80 SiO_2	2,940–3,060	2,400–2,700	100	2.40–2.45	125–140	Fireclay
Superduty	41–45 Al_2O_3; 51–55 SiO_2	3,175–3,200	2,500–2,800	115	2.65–2.75	130–145	Flint fireclay
High burned superduty	41–45 Al_2O_3; 51–55 SiO_2	3,175–3,200	2,500–2,800	155	2.65–2.75	130–145	Flint fireclay
50% alumina	50 Al_2O_3	3,200–3,245	2,500–2,800	160	2.75–2.85	130–145	Fireclay, diaspore, bauxite
60% alumina	60 Al_2O_3	3,245–3,310	2,700–2,900	200	2.90–3.05	130–145	Fireclay, diaspore, bauxite
70% alumina	70 Al_2O_3	3,290–3,335	2,700–2,900	235	3.15–3.25	135–150	Diaspore, bauxite
80% alumina	80 Al_2O_3	3,335–3,390	2,800–3,000	260	3.35–3.45	140–155	Diaspore, bauxite
90% alumina	90 Al_2O_3	3,390–3,425	2,800–3,200	1,175	3.55–3.65	172	Calcined alumina
Kyanite base	$Al_2O_3 \cdot SiO_2$ 58–68 Al_2O_3	3,250–3,300	2,800–3,000	500–680	3.00–3.06	140–150	Kyanite
Kaolin base (high-fired)	$Al_2O_3 \cdot 2SiO_2$; 44–45 Al_2O_3	3,190	2,800–2,900	170–335	2.65–2.75	135–145	Kaolin
Fused mullite base	$3Al_2O_3 \cdot 2SiO_2$; 72–75 Al_2O_3	3,325–3,350	2,900–3,200	685–1,370	3.08–3.25	150–160	Bauxite and synthetic
Silica:							
Standard	SiO_2	3,142 (pure)	3,000	95–100	2.30–2.38	100–105	Quartz sand, ganister
Superduty	SiO_2	3,142 (pure)	3,000 plus	110	2.30–2.38	100–105	Quartz sand, ganister
Magnesite:							
Burned	MgO plus SiO_2, Fe_2O_3, Cr_2O_3	5,070 (pure)	3,000–4,000	520	3.40–3.60	160–165	Magnesite, sea water brines, brucite
Chemically bonded	MgO plus FeO, Cr_2O_3		2,900–3,100	465	3.60–3.80	170–175	Magnesite, sea water brines, brucite
Chrome:							
Burned	$FeO \cdot Cr_2O_3$ pure; plus some Al_2O_3, MgO, Fe_2O_3, SiO_2	3,540–3,990	2,800–3,200	430	3.60–4.10	185–190	Chromite ores
Chemically bonded	$FeO \cdot Cr_2O_3$ plus MgO	3,540–3,990	2,900–3,100	450–485	3.90–4.10	170–180	Chromite ores
Forsterite	$2MgO \cdot SiO_2$	3,461 (pure)	3,000	530	3.30–3.40	150–155	Olivine or synthetics
Silicon carbide	SiC	Dissociates at 4,082	2,800–3,200	1,520	3.19	155	Synthetic
Fused alumina	Al_2O_3	3,722 (pure)	3,400	2,000	3.70–3.90	175–195	Bauxite
Zircon	$ZrO \cdot SiO_2$	4,532 (pure)	3,400	1,210	4.70	205	Zircon sands
Zirconia (stabilized)	ZrO_2	4,870 (pure)	4,300	9,430	5.75	275	Zircon sands
Carbon (graphite)	C	6,330	4,000 (reduc. atm.)	1,600	2.25	137	Carbon

* Note: "Normal-use-limit" temperatures are approximate only because service conditions sometimes will change allowable temperatures by several hundred degrees. Such conditions may include heavily reducing atmospheres and unusual loading.

From C. L. Norton, Jr., Refractories for Every Use, *Chem. Eng.*, **60**(6): 217 (1953).

Diatomaceous-silica type insulation is suitable for temperatures up to 1900°F and is used as an inner temperature-reducing layer in combination with 85 per cent magnesia or other lower-temperature insulations. *Mineral wool* has an excellent temperature utility range (−150 to 1700°F) and is generally used where insulation must be removed frequently since it is held together by metal fabric.

Prefabricated molded pipe insulation is specified by an ASTM Dimensional Standard system which provides great flexibility in the choice of thickness. The actual thickness of any given pipe insulation is one-half

TABLE 9-7B. CHARACTERISTICS OF HIGH-TEMPERATURE INSULATION. CLASSIFICATION OF INSULATING FIREBRICK

Type*	Composition, per cent	Normal use limit, °F†	Approx. $/M 9-in. equiv.	Bulk density, lb/ft³
Group 16....	15–37 Al_2O_3; 30–60 SiO_2 plus TiO_2, Fe_2O_3, alk.	1,600	115	21–37
Group 20....	26–38 Al_2O_3; 45–61 SiO_2 plus TiO_2, Fe_2O_3, alk.	2,000	125	26–45
Group 23....	25–42 Al_2O_3; 45–67 SiO_2 plus TiO_2, Fe_2O_3, alk.	2,300	150	27–47
Group 26....	40–46 Al_2O_3; 47–55 SiO_2 plus TiO_2, Fe_2O_3, alk.	2,600	195	43–64
Group 28....	45–53 Al_2O_3; 42–52 SiO_2 plus TiO_2, Fe_2O_3, alk.	2,800	260	45–65
Others......	45 Al_2O_3	2,900	285	52
	65 Al_2O_3	3,000	480	69
	90 Al_2O_3	3,250	750	81

* Group number is std. ASTM classification, indicating normal use limit (×100).
† See note under Table 9-7A.
From C. L. Norton, Jr., Refractories for Every Use, *Chem. Eng.*, **60**(6): 217 (1953).

of the difference between the OD of two successive pipe-covering layers minus the clearance. This insulation is prefabricated for pipe sizes from ½ to 30 NPS so that insulation will have nominal thicknesses of 1, 1.5, 2, or 2.5 in. By selecting one, or a nest of two or more coverings, it is possible to obtain thicknesses up to 10 in. in increments of ½ in. Prefabricated insulation coverings for fitting, flanges, and valves are also available. Flat or curved block, blankets, molded brick and cast cements are other forms available for insulation of both piping and equipment. As pointed out in item 23, Table 9-6, a reflective, preformed insulation is now available in a composite construction consisting of many layers of reflective metal sheets.

Factors Governing Selection of Insulation. The *proper* selection of insulating material is an engineering problem which cannot be treated by rule of thumb. Certain principles can be listed for guidance:
1. Select the material to withstand working temperature range.
2. Choose an inexpensive low-thermal-conductivity type.

TABLE 9-7C. CHARACTERISTICS OF HIGH-TEMPERATURE INSULATION. CLASSIFICATION OF REFRACTORY CASTABLES
(Refractory aggregates plus hydraulic cement)

Name, standard types	Composition	Normal use limit, °F*	Approx. cost, $/T, CL	Bulk density, installed and fired, lb/ft³
Alumina-silica:				
Fireclay type........	Al_2O_3, SiO_2; plus CaO, Fe_2O_3, TiO_2, alk.	2,400–2,700	60–75	100–140
Superduty type......	Al_2O_3, SiO_2 plus CaO, Fe_2O_3, TiO_2, alk.	2,700–3,000	100–170	100–140
Chrome..............	$FeO \cdot Cr_2O_3$ plus Al_2O_3, CaO, SiO_2	2,700–3,100	90–150	160–190
Insulating types:				
Expanded mica......	....	1,600–1,800	....	40–50
Diatomaceous base...	SiO_2 plus Al_2O_3, impurities	1,800–2,000	....	55–70
Bloated clay base.....	Al_2O_3, SiO_2 plus CaO, Fe_2O_3, TiO_2, alk.	2,000	....	70–100
Porous refractory base.	Al_2O_3, SiO_2 plus CaO, Fe_2O_3, TiO_2, alk.	2,000–2,500	90–130	50–90

* See note under Table 9-7A.
From C. L. Norton, Jr., Refractories for Every Use, *Chem. Eng.*, **60**(6): 217 (1953).

3. Choose a type and form which has sufficient durability and structural strength to withstand severe conditions. This includes resistance to moisture and the chemical environment.

4. Specify the proper methods of application to ensure retention of original form and insulating value.

5. Pick the most economic insulation from a fixed charges–heat savings

standpoint (see Fig. 9-12, Tables 9-8 to 9-11, and Perry's "Chemical Engineers' Handbook," 3d ed., pp. 479–480).

Application of Pipe Insulation. Prefabricated insulation is applied to a pipe with joints tightly butted and pointed up. Each section of insulation is wired to the pipe with not less than three loops of 16-gauge annealed iron wire on pipe up to and including 6 in. and with not less

TABLE 9-7D. CHARACTERISTICS OF HIGH-TEMPERATURE INSULATION. CLASSIFICATION OF REFRACTORY PLASTICS AND RAMMING MIXES

Name	Composition	Melting point, °F	Normal use limit, °F*	Approx. cost, $/T, CL	Bulk density, installed and fired, lb/ft³
Alumina-silica:					
Fireclay base	Al_2O_3, SiO_2, plus Fe_2O_3, TiO_2, alk.	2,850–3,100	2,500–2,700	50–70	125–140
Superduty base	Al_2O_3, SiO_2, plus Fe_2O_3, TiO_2, alk.	....	2,700–3,000	60–80	130–140
Kyanite base	$Al_2O_3 \cdot SiO_2$, plus Fe_2O_3, TiO_2, alk.	....	3,000–3,100	150–175	140–150
Fused mullite base	$3Al_2O_3 \cdot 2SiO_2$ plus Fe_2O_3, TiO_2, alk.	....	3,000–3,200	175	150–160
Silica	SiO_2	3,142 (pure)	2,800–3,000	....	100–105
Magnesite:					
Grain	MgO	5,070 (pure)	2,900–3,100	70–100	150–160
Mixtures	....	....	....	....	145–165
Dolomite	$CaO \cdot MgO \cdot 2CO_2$	4,650–5,070 (pure)	2,900–3,100		
Chrome	$FeO \cdot Cr_2O_3$ plus Al_2O_3, MgO	3,765 (pure)	2,900–3,100	95–110	180–200
Silicon carbide	SiC	Dissociates at 4,082	2,800–3,150	....	150–170
Fused alumina	Al_2O_3	3,722 (pure)	2,900–3,400	....	175–195
Zircon	$ZrO \cdot SiO_2$	4,532 (pure)	3,200	....	205
Zirconia (stabilized)	ZrO	4,870 (pure)	3,800	....	275

* See note under Table 9-7A.
From C. L. Norton, Jr., Refractories for Every Use, *Chem. Eng.*, **60**(6): 217 (1953).

than four loops for larger pipe sizes. All asbestos-sponge felt insulation 2 in. and greater in thickness, 85 per cent magnesia of 3 in. and double standard in thickness, and combination insulation, should be applied in two layers with both circumferential and horizontal joints staggered and with each layer securely wired in place as previously described. The insulation on bends should be given a thin finishing coat of cement to present a smooth, even surface. Insulation of flanges and fittings over 4 in. in diameter should be the same as the insulation on the line, surfaced with

½ in. of cement applied in two layers, the first coat being left to dry with a rough surface before the application of the smooth finishing coat. Canvas is stretched tightly over the cement and pasted neatly. On lines under 4 in., the fittings and flanges are insulated entirely with cement to the same thickness as the adjacent insulation. Outdoors, the flanges and fittings are waterproofed with Insulkote or a similar preparation, applied in place of the second coat of cement, the canvas being omitted. All insulation on piping indoors should be finished with an extra jacket of

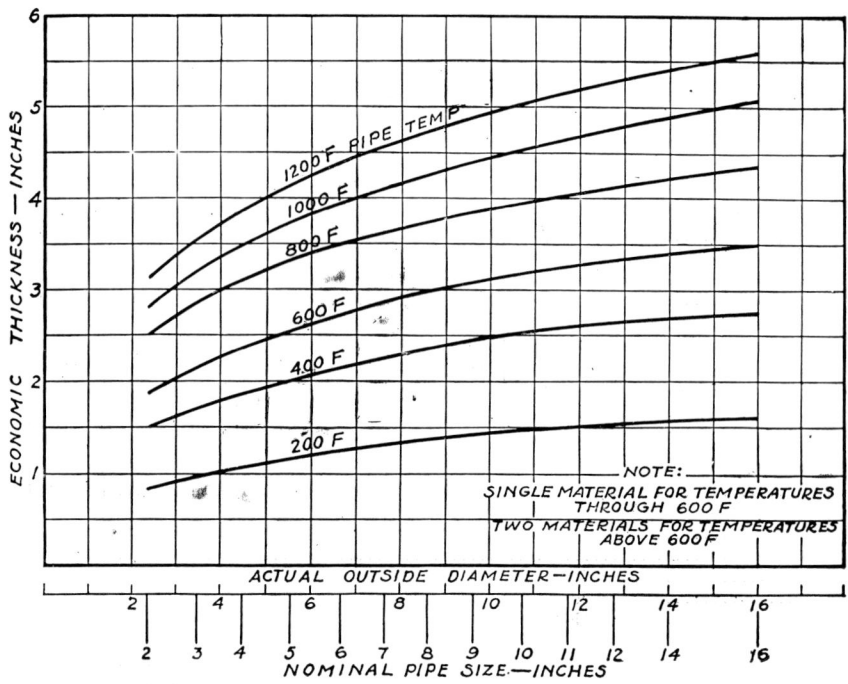

Fig. 9-12. Economical pipe insulation thickness. [*By permission from* "*Plant Engineering Handbook*," *W. Staniar* (ed.), *McGraw-Hill Book Company, Inc., New York,* 1950.]

8-oz canvas, sewed on over rosin-sized paper or asbestos paper and, if desired, sized with glue and painted with two coats of lead and oil paint. All insulation on piping outdoors, or exposed to the weather, should be protected with a waterproof jacket, the canvas being omitted. Pipe insulation located close to the ground or where there is possibility of mechanical injury should be protected with a metal jacket.

Weatherproof and Fire-resisting Pipe-insulation Jackets. For outdoor lines, particularly those of some length, the most satisfactory insulation is asbestos-sponge felt with an integral, waterproof jacket. Asbestos-sponge felt not only has high insulation value but also, owing to its construction, is not damaged by fall or blow and maintains its efficiency on hot lines over a long period of time. The integral jacket provides complete weather protection and obviates the labor of a separate application. Where insulation other than asbestos-sponge felt with integral

TABLE 9-8. COST OF PIPE INSULATION (1958 PRICES)
(Basis: Installed cost per 100 ft of straight pipe. For approximate cost of insulating valves and fittings as well as straight pipe, multiply values in table by 1.35. "M" denotes 85 per cent magnesia; "S" denotes high-temperature block insulation; number immediately following denotes thickness in inches.)

Nominal pipe diameter, in.	Temperature range, °F							
	200–270	270–350	350–400	400–500	500–600	600–700	700–800
2	M–1 $135	M–1 $135	M–1½ $164	M–2 $187	M–2 $187	S–1½ + M–1½ $344	S–1½ + M–2 $426
3	M–1 $149	M–1 $149	M–1½ $182	M–2 $217	M–2 $217	S–1½ + M–1½ $395	S–1½ + M–2 $466
4	M–1 $162	M–1 $162	M–1½ $197	M–2 $260	M–2 $260	S–1½ + M–1½ $465	S–1½ + M–2 $550
6	M–1 $190	M–1½ $247	M–2 $320	M–2 $320	M–2½ $425	S–1½ + M–2 $570	S–1½ + M–2½ $680
8	M–1¼ $230	M–1½ $280	M–2 $382	M–2½ $483	M–2½ $483	S–1½ + M–2 $668	S–1½ + M–2½ $792
10	M–1¼ $274	M–1½ $328	M–2 $455	M–2½ $573	M–2½ $573	S–1½ + M–2 $751	S–1½ + M–2½ $1040
12	M–1½ $362	M–2 $516	M–2½ $641	M–3 $728	M–3 $728	S–1½ + M–2 $848	S–1½ + M–2½ $1120

Adapted from R. M. Braca and J. Happel, *Chem. Eng.*, **60**(1): 183 (1953).

TABLE 9-9. COST OF INSTALLING MAGNESIA BLOCKS
(Basis: Dollars per square foot—1958 prices)

Thickness, in.*	Installation height		
	Below 15 ft	15 to 25 ft	Above 25 ft
1	$1.35	$1.50	$1.74
1½	1.45	1.60	2.00
2	1.55	1.70	2.10

* With additional ½ in. cement finish.

waterproof jacket is used, one of the best forms of weather protection for insulated pipelines out of doors consists of a waterproof asbestos jacket with all joints lapped at least 3 in. and all horizontal laps located on the side of the pipe, turned down to shed rain. Rings of heavy copperweld wire are applied on 4-in. centers to hold the jacket in place over the insulation. Sheet metal is another popular form of weatherproofing. Where exposed to fire hazard only, it is good practice to apply an asbestos fire-retarding jacket. In such cases the application of asphalt-saturated roofing jackets is inadvisable, since flame may be carried along exposed piping when a fire occurs adjacent to lines so protected. The fire-retarding jacket consists of one sheet of asphalt-saturated asbestos felt over which is cemented an unsaturated felt. It will not drip asphalt, carry flame, or support combustion.

Cold Insulation. Refrigeration cannot be efficient without insulation of the fittings, valves, and flanges, as well as the straight piping. Fitting

Table 9-10. Cost of Installing Cork Lagging
(Basis: Dollars per square foot—1958 prices)

Thickness of single layer, in.	Inside installation height		Outside installation height	
	Below 15 ft	Above 15 ft	Below 15 ft	Above 15 ft
1	$1.10	$1.38	$1.60	$2.00
1½	1.10	1.38	1.60	2.00
2	1.10	1.38	1.60	2.00
3	1.10	1.38	1.60	2.00
4	1.25	1.60	1.78	2.25
5	1.35	1.70	1.85	2.32
6	1.45	1.85	1.95	2.45
7	1.53	1.92	2.05	2.52
8	1.60	2.00	2.12	2.60
10	1.68	2.07	2.20	2.75

covers are made from cork granules compressed and baked in molds, of exact size to fit snugly and leave no space between the cold surface and the insulators where frost could accumulate. Built-up covers have a lower material cost, but a higher cost of labor on the job erases this possible saving; and there is always danger that the built-up job may not be well fitted and result in trapping frost in the hollow spaces, later resulting in pushing the insulation out of place. Covers for fittings are made in the same thicknesses as the corresponding straight pipe covering, carefully sized to join the sections of the pipe covering without a break. For the insulation of equipment in general refrigeration, for cold-storage work, and for roof insulation, there is standard corkboard in sizes 12 by 36 in., 18 by 36 in., 24 by 36 in., and 36 by 36 in. Table 9-11 is a guide for thickness and size of cork insulation.

Cold insulation is more expensive than heat insulation. For example, a Houdry cracking unit required 9.1 per cent of the total project cost for insulation, while a large-scale refrigeration job required 21.4 per cent of the total project cost.

Preinsulated Pipe. A number of concerns are able to supply preinsulated pipe in special or standard 21-ft lengths for both underground and overhead installation, in both high- and low-temperature insulations.

TABLE 9-11. GUIDE FOR THICKNESS AND SIZE OF CORK INSULATION*

Temperature, °F	Pipe or equipment diameter, in.						
	Pipe ¼–1½	Pipe 2–6	Pipe 8–20	Pipe and equipment, 20–54	Pipe and equipment, 54–120	Equipment and flat surfaces over 120 in.
	Insulation type and thickness, in.					
−300 to 250	6–8, C	9–11, C	12–14, C	14–16, L	16–18, B	22, B
−250 to 200	5–7, C	8–10, C	11–13, C	13–15, L	15–17, B	18, B
−200 to 150	5–6, C	7–8, C	9–11, C	11–12, L	12–15, B	15, B
−150 to 100	4–5, C	6–7, C	8–9, C	9–10, L	10–12, B	12, B
−100 to 50	4, C	5–6, C	7–8, C	8–9, L	8–10, B	10, B
− 50 to 20	4, C	4–5, C	5–6, C	6–7, L	7–9, B	9, B
− 20 to 0	STB, C	STB, C	5, C	5–6, L	6–8, B	8, B
0 to 20	BT, C	STB, C	STB, C	4, L	4–6, B	6, B
20 to 50	BT, C	BT, C	BT, C	3, L	3–4, B	4, B
50 to 100	IWT, C	IWT, C	IWT, C	2, L	2, B	2, B

KEY: C, sectional pipe covering; L, lagging; B, board; STB, special thick brine sectional pipe covering, 2.63 to 4 in. thick; BT, brine thickness sectional pipe covering, 1.70 to 3 in. thick; IWT, ice-water thickness sectional pipe covering, 1.20 to 1.93 in. thick.

* F. C. Otto, *Chem. Eng.*, **54**(5): 118 (1947).

One or more pipes may be installed and insulated in the same conduit. For underground use the pipe (or pipes) is packed in insulation within a 16-gauge corrugated iron conduit which is then covered with asphalt-saturated felt. For overhead use the insulated assembly is wrapped in copper or aluminum foil (min. 0.003 in. thickness) bonded with asphalt and asphalt-impregnated felt. In all such preinsulated assemblies the pipe ends protrude to allow welding or other methods of joining adjacent lengths of pipe. After joining the lengths the gap is insulated and sealed.

Color Coding[1]

The identification of piping and equipment by color has been standardized through the efforts of the American Standards Association (ASA). See Table 9-12 for typical color codes. All process piping should be first classified according to fluid content as safe, hazardous, or protective and then color-coded accordingly. The colors may be used over the entire surface of the pipe covering or color stripes and bands, placed next to valves, junctions, and other critical locations are used for economy reasons. Self-adhesive color tapes are available for this purpose. Often the name of the fluid and direction of flow arrows are stenciled on the colored bands or pipe covering. Recommended ASA stencil letter sizes in inches for various pipe sizes are: $3/4$ to $1\frac{1}{4}$ NPS, use $1/2$; $1\frac{1}{2}$ to 2 NPS, use $3/4$; $2\frac{1}{2}$ to 6 NPS, use $1\frac{1}{4}$; 8 to 10 NPS, use $2\frac{1}{2}$; 12 NPS or greater, use $3\frac{1}{2}$.

The *cost of painting* (material and labor) for bare steel pipe, including sandblasting, can be estimated as $0.35 per square foot. If painting over canvas-coated insulation, reduce this to $0.20 per square foot.

Drainage Piping Design

Gases released from drainage systems in a chemical plant, owing to defective design or installation of the waste-disposal piping, endanger the health and even the lives of the workmen. Such gases may arise from both the sanitary and the chemical waste systems. Still greater danger and discomfort may exist when the two types of waste, in intermixing, act upon each other to accelerate or decelerate reactions that might occur normally or abnormally. In other cases, such as in food-products plants, the product may be contaminated by gases released from improper waste-disposal systems.

Plumbing Codes in Chemical Plants. The system for drainage and sewage, together with water-supply and water-heating equipment and accessories, is installed by the plumber. This is usually done in accordance with specifications prepared by the architect; in chemical plants, it will be in accordance with the wishes and desires of the design engineer, since drainage systems are connected directly with equipment. The specifications are also subject to official municipal regulations, because the local authorities assume responsibility for the installation of an efficient and safe system of sanitary plumbing. Systems for handling liquid wastes in chemical plants are not clearly defined in plumbing codes, the solution to the problems being left to the processor (until such time as damage to others may bring the case to the attention of the courts).

Effluent. The reference point—the lowest point in a drainage system—will in general be one of two types: (1) an effluent pipe into a sand pit,

[1] D. E. Garrett and W. A. Jordan, *Chem. Eng.*, **63**(9): 195 (1956).

TABLE 9-12. COMPARISON OF PIPING IDENTIFICATION COLOR SCHEMES*

Material	ASA classification	General Motors	Du Pont	Monsanto	American Cyanamid	Union Oil	American Potash & Chemical (proposed)
Steam, high pressure	Dangerous—yellow or orange	Aluminum	Alert orange	Yellow, 2 black stripes	Orange	Aluminum, yellow band	Orange, 1 black stripe
Steam, low pressure	Safe below 212°F—green	Aluminum	Alert orange	Yellow, 1 black stripe	(Safe below 130°F)	Aluminum, yellow band	Orange, 2 black stripes
Gas (fuel)	Dangerous—yellow or orange	Yellow	Yellow	Yellow band	Yellow	Yellow	Light orange
Dangerous process fluids (acids, etc.)	Dangerous—yellow or orange	Orange	Yellow	Yellow	Yellow	Orange	Yellow
Fresh water	Safe—cold Dangerous—hot Green	Green	Green	Green	Green (yellow stripes above 130°F)	White (yellow band when hot)	Green (yellow stripe above 140°F)
Brackish water	Safe—green	Green	Green	Green, 1 blue band	Green	White, black, or brown band	Green, 1 black stripe
Compressed air	Safe to 300 psig—green or achromatic	Blue	Yellow	Green, 1 orange band	Green (yellow stripes above 75 psig)	Blue (white on low pressure)	Gray (yellow stripe above 50 psig)
Vacuum	Safe—green or achromatic	None	Gray	Green, 1 white band	None	Green	Gray, 1 orange stripe
Sprinkler (water) and fire equipment	Fire protection Red	Red	Red	Red	Red	Red	Red
Electric conduit	None	Black	Black	Orange	None	None	None
Radioactive	Purple (tentative)						

Note: A stripe is generally less than 1½ in. in width, while a band is wider, e.g., Union Oil Co.'s 1st color is three pipe diameters wide, while the band is one diameter wide, centered in the first color.

* D. E. Garrett and W. A. Jordan, *Chem. Eng.*, **63**(9): 196 (1956).

stream, pond, or river, or (2) a tap into another sewer. An effluent of the first type must be well protected from washing by the construction of a concrete support and spillway, with such a curvature as to direct the drain water out into the stream to reduce backwashing and undermining. An effluent of the second type, into a larger drain, is either the connection to the drain at a preexisting tap, or a cut made into the drain and the terminal line cemented into place.

Where the natural drainage is shallow and where rainfall may convert a tract of land into a shallow lake, drainage is difficult. Usual practice is to drop the sewage into a well, then pump from this reservoir to a higher level and at a distance to permit gravity disposal or long-distance pumping. Common practice for chemical sewage is to impound the waste and permit diffusion through the soil or diversion of the impounded waste through a treating plant or bed before it is run off into the natural drainage system.

Capacity. The first point to consider in determining the location and size of drains of a sewage system is the service to be rendered. The proper size of drain pipes for chemical plants is subject to considerable variance of opinion. Plumbing codes specify 4 in. as a common minimum for buried piping and 2 in. as the minimum where suspended. The sizes of drains vary with the number of floors. In a chemical plant, however, the question is not one of connecting sanitary branches, but one of disposal of wastes, either corrosive or containing suspended matter. A small pipe allows scouring and prevents the deposition of solids on the sides of the pipe. Calculations of flow must be made if the time required for emptying is an important item in removing wastes from equipment or from an area. Floor drains cannot be considered as all being in full capacity at the same time unless the specific plant process calls for it. A flow of liquid to fill the pipe but half full should be considered as the carrying capacity of a drainage pipe. In Table 9-13 will be found data on the carrying capacity of different sizes of sewer pipe for various drops per 100 ft. The pitch of a waste-disposal system is a matter of local health code specifications, but ordinarily this amounts to about $\frac{1}{4}$ in./ft.

Tile. Drain tile and pipe are made of a variety of materials including wood, fiber (orangeburg), concrete, steel, unglazed vitrified clay, plastic, terra cotta, salt-glazed vitrified clay, Transite, or cast iron. Systems installed underground are preferably constructed of bell-and-spigot or hub cast-iron pipe, calked at the joints with lead. In a large number of the cities in the United States this is a requirement. The closing of cast-iron pipes is usually accomplished with oakum and lead, and of ceramic pipes with oakum and cement. Cast-iron pipe has fewer joints than tile pipe since its sections are generally 5 ft long, against 2- and 3-ft lengths of tile. Tile, furthermore, is more susceptible than cast iron to breakage, owing to settling. Terra cotta tile and pipe come in standard sizes, shapes, and grades; these are sold locally in 3-ft lengths. Cast iron is manufactured in three grades: light (or standard), heavy, and extra heavy. For chemical plants the standard grade is unsatisfactory. The extra heavy is costly, but where corrosive liquids are dumped into the disposal system, the added life of the extra heavy more than offsets the initial high cost. Heavy pipe is ordinarily considered the normal grade to use where corrosion is not an item. Wrought iron and steel can be used but are considered as highly unsatisfactory in chemical plants. As added protection against corrosion, cast-iron pipe is heavily coated inside and outside with asphalt.

TABLE 9-13. CARRYING CAPACITY OF SEWER PIPE*
(Gallons per minute)

Size, in.	Inches fall per 100 ft							
	1	2	3	6	9	12	24	36
4	27	38	47	66	81	93	131	163
6	75	105	129	183	224	258	364	450
8	153	216	265	375	460	527	750	923
9	205	290	355	503	617	712	1,006	1,240
10	267	378	463	755	803	926	1,310	1,613
12	422	596	730	1,033	1,273	1,468	2,076	2,554
15	740	1,021	1,282	1,818	2,224	2,451	3,617	4,467
18	1,168	1,651	2,022	2,860	3,508	4,045	5,704	7,047
24	2,396	3,387	4,155	5,874	7,202	8,303	11,744	14,466
27	4,407	6,211	7,674	10,883	13,257	15,344	21,771	26,622
30	5,906	8,352	10,233	14,298	17,714	20,204	28,129	35,513
36	9,707	13,769	16,816	23,763	29,284	33,722	47,523	58,406

* F. E. Kidder and H. Parker, "Architects' and Builders' Handbook," 18th ed., p. 1751, John Wiley & Sons, Inc., New York, 1931.

Large-sized pipe is commonly made of concrete, cast on the job, or a concreted tunnel is constructed. The smaller tile can be either straight joint or of the bell-and-spigot type. Building and sanitary codes of municipalities control the type and size of drain tile permitted, especially where the sanitary sewage of the plant is handled in the municipal system. In general, main soil pipe cannot be less than 4 in.; soil pipe for more than five water closets not less than 6 in.; drain soil lines not less than 4 in.; and main waste pipes not less than 2 in.; when service is required for more than five sinks, urinals, or showers, the size should be not less than 4 in. Vent pipes should not be less than 2 in.

A knowledge of building and drainage codes is necessary to specify the construction of the system, from the digging of the trench in the proper place and at the correct drainage depth to the connection of the reference-point effluent into its ultimate objective.

Cleaning. Provision for the easy cleaning of sewers must always be made. This is best accomplished by the construction of wells or manholes, properly bricked up or concreted, from which inspection of the lines can be carried on. At every bend, change in elevation, or terminal point, means must be supplied for inspection and cleaning.

Fittings. Standard catalogues of fittings should be consulted to determine the standard types and sizes of all parts of the drainage system, because the construction of special equipment is costly. Plumbing and sewage contracting is provided with an endless choice of materials. Plumbers' handbooks and trade catalogs are the best sources of information on different types of equipment available, giving service requirements, capacities, construction details, essential improvements, and comparative costs. The large number of such sources of information available in manufacturers' literature makes it unnecessary that detailed descriptions of such equipment and service be given here.

Both the type and location of service urinals and stools for both men and women employees require prime consideration. It is important that the sanitary service be somewhat removed from the plant processing in order to simplify fittings underground. Process equipment service may interfere with sanitary service in drainage, since equipment drainage may overtax the system at periodic intervals. Therefore, each piece of equipment should drain directly into the main laterals.

Floor Drains. Floor drains should be liberally used in a chemical plant. A pitch to the floor of 1 in. in 8 ft is not sufficient for the rapid runoff so essential to chemical plants, and a pitch of from four to six times as much will repay the designer in efficient floor drainage. Floor drains are made with integral traps and cleanouts. If a simple trap is used, a deep seal trap and a cleanout flush with the floor should be provided. Drains for floors not on the ground should be provided with floor seals to prevent leakage seeping around the setting to the floor below.

Pipe Joints. Pipe joints of the hub type, or bell and spigot, are sealed, first, by packing with oakum, driven in tight, and following this with molten lead. Incidentally, the fact that jute costs five times as much as lead makes a calked joint rather costly. Each joint for a 2-in. pipe requires 4 lb of lead and 0.1 lb of oakum. For 4-in. pipe the quantities are 7.5 lb of lead and 0.2 lb of oakum, while a 6-in. pipe requires 10.25 lb of lead and 0.3 lb of oakum. All joints should be given proper attention for perfect sealing. Where small cast-iron piping is used, the joints may be screwed, but these should be specially threaded and the fittings provided with raised internal parts that eliminate the recesses and dams found in the water and steam types of standard pipe fittings. Fittings of this sort give free flow of liquids and reduce the tendency to clog.

Vents. Where local health codes require that waste-disposal piping systems be ventilated, a survey should be made of the specific code to ascertain the correct number and placement of the vents. Oftentimes, the code does not apply to chemical plant wastes; in such cases ample venting should be provided to eliminate the possibility of breaking trap seals on account of any rush of waste when a piece of equipment is voided.

Traps. Every fixture and each piece of equipment should be separately trapped by a valve or a water-sealing trap placed as close to the fixture outlet as possible, and no trap should be placed more than 2 ft from any fixture. The sizes of traps for chemical equipment depend upon the desired drain-off, but for sanitary purposes the minimum size is stipulated in the code. Closed traps must be of 4-in. size, while those for sinks, urinals, showers, and the like must not be under 2 in.

Traps are known as full S, $\frac{3}{4}$S, $\frac{1}{2}$S, or P, V, bag, running-Y, and grease traps, and combinations, depending upon the use and configuration. Floor drains usually incorporate easy cleanout traps in their construction to reduce joints and to add serviceability.

Suspended Piping. Suspended piping should be hung below the ceiling to permit ready examination of all connections. It should be provided with Y branches at the changes in direction of flow, to permit easy access to the piping for rodding and cleaning. Substantial support should be given all suspended piping in order that vibration and the weight of the pipe may not cause a slow sagging and opening of the pipe at the joints.

Drains under Processing Equipment. Detail drawings of drainage under equipment must show all construction and sewage lines up to equipment. Large openings and rapid runoff are desirable, so no traps are provided. Cleanout ferrules are always used to permit unscrewing the cap and securing ready access to the drain. Brass traps and ferrules are preferable because cast iron "freezes" on long standing.

A sudden rush of liquids from equipment into the drainage system would cause compression of entrapped gases in the sewer ahead of the stream and a vacuum behind the stream. Such pulsation drains all the traps and permits permeation of odors and sewer gases. The breather system, by proper installation of vents, is always installed according to code. In order to prevent the back rush of liquids, a ball check is installed ahead of a condenser. Both check valve and subsequent trap should be accessible for cleanout. The bell of the trap extends down from the perforated or gridded floor screen and covers a portion of the drain pipe extending above the bottom of the drain pit. More commonly the floor drains, drain screen, bell, and outside case of the pit are cast iron.

Sanitary Drainage. If complex systems of traps and check valves are to be avoided, the sanitary sewage and process wastes must run into a separate drain and sewer system. However, in municipalities this cannot be done and both types of sewage are mixed. Urinals, stools, washbasins, showers, and fountains are recommended for personnel in industrial plants on a basis similar to those proposed in Table 9-14. Sanitary codes control in a measure what type of service must be provided and, hence, must be consulted. Cleanliness and clean appearance, with accessibility, are as important in the plant as in office toilets.

TABLE 9-14. TOILET FACILITIES

Water faucets	1 per 3 persons
Showers	1 per 25 persons
Urinals	1 per 20 persons
Stools	1 per 15 persons
Fountains	1 per 50 persons

Testing Drainage Systems. The necessity for testing waste-disposal systems should be evident to chemical engineers. Not only are suspended pipes a source of danger and hazard, but high leakage in underground pipe through which chemicals may be flowing, insufficiently diluted by other streams, may seriously affect the structure or neighboring land. The water test is easy of application. The engineer should not permit the vertical stack, which connects the first and second floors, to be sealed until he can have the downcomer pipe plugged and all the openings closed except the one to the roof via the vent. Water should then be pumped in and the joints tested for leaks. Either a drop in the water level or visible evidences of leaks at the joints are sought. Additional tamping of the lead usually seals the leak. Plugging of the outlet and filling the underground system with water, with subsequent checking on the water level, enable one to ascertain to what extent underground leaks may exist. Such leaks are difficult to find. Consequently, the engineer should insist on an inspection of the waste piping before covering in the case of underground systems.

Design of a Drainage System. The points to be taken up in the design of a drainage system, in the order of their consideration, are as follows:

1. Quantity of service demanded
2. Quality of service demanded
3. Reference point
4. Pitch of drain
5. Effluent protection
6. Wells and manholes for cleanout purposes
7. Location of equipment
8. Location of traps with cleanouts
9. Location of check valves
10. Layout in building
11. Provision for vents
12. Toilet facilities
13. Equipment drains
14. Fixtures

Symbols for use in drawings are given in Fig. 9-13.

General Rules of Drainage Design. Even though the best quality of pipe material be used, the sewage system will give unsatisfactory service unless certain precautions are taken in the design and installation. All underground pipe should permit as direct a run of sewage as possible; all changes of direction should be with long sweeps or curves; and branch connections should be Y branches (Fig. 9-14). Ample provision should

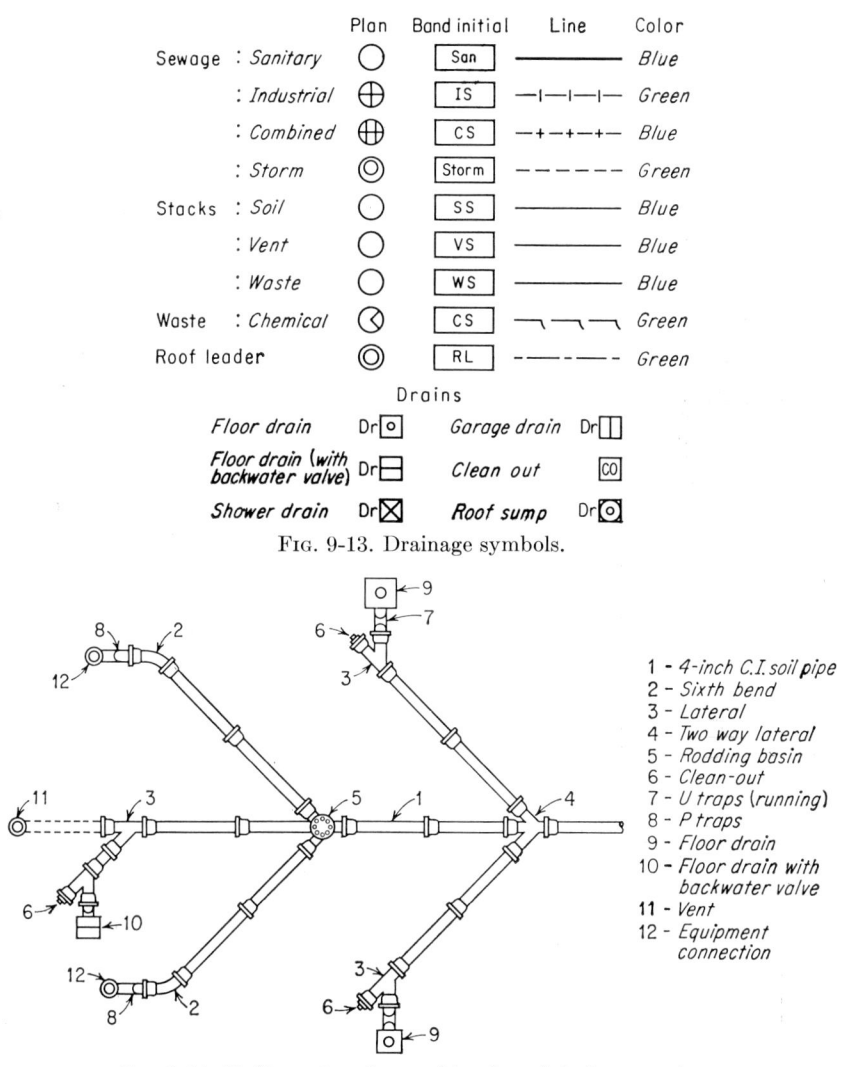

		Plan	Band initial	Line	Color
Sewage	: Sanitary	○	San	————	Blue
	: Industrial	⊕	IS	—I—I—I—	Green
	: Combined	⊕	CS	—+—+—+—	Blue
	: Storm	◎	Storm	— — — —	Green
Stacks	: Soil	○	SS	————	Blue
	: Vent	○	VS	————	Blue
	: Waste	○	WS	————	Blue
Waste	: Chemical	⊗	CS	⌐⌐⌐	Green
Roof leader		◎	RL	-——-—-	Green

Drains

Floor drain	Dr ○	Garage drain	Dr □
Floor drain (with backwater valve)	Dr ⊟	Clean out	CO
Shower drain	Dr ⊠	Roof sump	Dr ○

Fig. 9-13. Drainage symbols.

Fig. 9-14. Undimensioned assembly plan of drainage system.

1 - 4-inch C.I. soil pipe
2 - Sixth bend
3 - Lateral
4 - Two way lateral
5 - Rodding basin
6 - Clean-out
7 - U traps (running)
8 - P traps
9 - Floor drain
10 - Floor drain with backwater valve
11 - Vent
12 - Equipment connection

be made for cleaning out all lines, for chemical precipitates and sludges may settle in the lines and cause no end of trouble. Y branches should be used at changes in direction of the lines and each branch provided with a calked-in ferrule, sealed with a close-fitting, screw-in brass plug. If

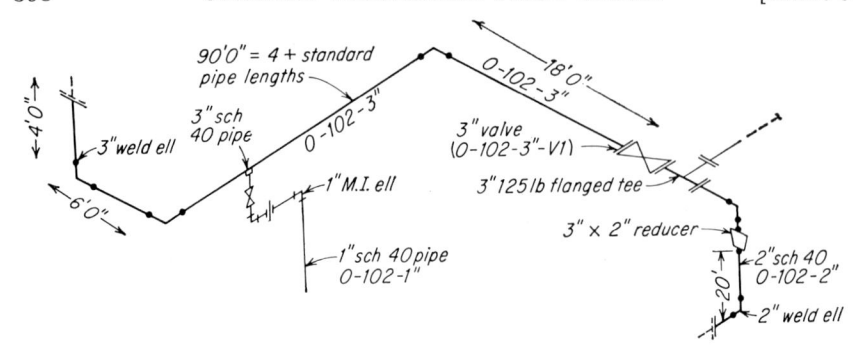

FIG. 9-15. Dimensioned diagrammatic piping drawing for cost estimating. [*Courtesy of W. G. Clark, Chem. Eng.*, **64**(7): 243 (1957).]

rodding basins are permitted, these should be placed at connecting points of branch sewers to provide access to danger points. Iron rodding basins should be mounted flush with the floor and provided with sealed covers, to be removed only when the plugging of lines calls for cleaning out. After cleaning, these should be resealed. A layout of sewer and drainage lines should be made so that proximity to piers or foundations is avoided, in cases where the weight of the pier or foundation would be likely to cause crushing or disruption of the lines.

Estimating Piping Costs

Piping cost estimating as a percentage of the purchased or installed equipment cost was outlined in Chap. 6. This method suffices for preliminary or order-of-magnitude type estimates. When an estimate to within 10 per cent or less is required, a more accurate procedure using piping flow sheets, models, and drawings is in order. Materials and labor take-off to get accurate cost estimates by the diameter-inch method of Clark[1] will be explained by means of a sample problem.

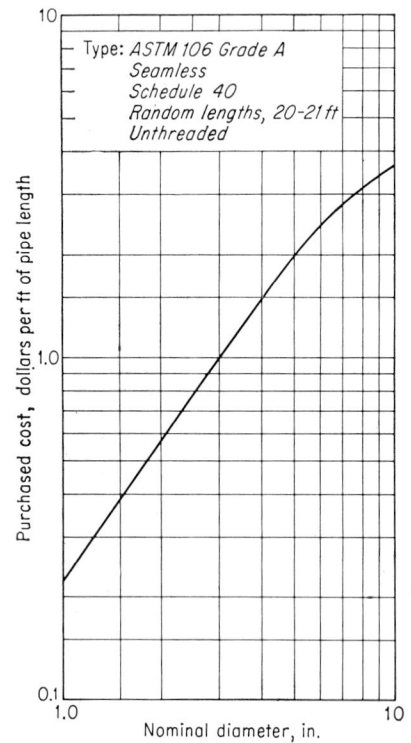

FIG. 9-16. Base price for carbon-steel (black) pipe (1958 price).

[1] W. G. Clark, *Chem. Eng.*, **64**(7): 243 (1957).

Table 9-15. "X" Factors for Estimating Piping-material Costs
(Based on steel pipe; reference: Fig. 9-16)

	"X" factor	
Nominal pipe size (NPS)	¾–2″	3–8″
Pipe:		
Carbon steel, schedule 40, seamless, random length	1.0	1.0
Carbon steel, schedule 40, butt- or lap-welded	0.82	0.93
Carbon steel, schedule 40, threaded on both ends and coupled	1.06	1.10
Carbon steel, schedule 80	1.42	1.42
Carbon steel, schedule 10	0.72	0.50
Type 304 stainless steel, schedule 5	6.3	4.1
Type 304 stainless steel, schedule 10	8.0	4.8
Type 316 stainless steel, schedule 5	7.1	5.2
Type 316 stainless steel, schedule 10	10.2	6.5
Galvanized steel, schedule 40	1.4	1.3
Saran-lined steel pipe	7.8	4.6
Aluminum	4.0	2.7
Brass	7.1	6.2
Pyrex glass	4.5	2.9
Lead	5.3	2.9
Cast iron	......	3.0
Asbestos-cement	1.50	0.71
Valves, all types:		
Carbon steel, 150-lb class	78	101
Carbon steel, 300-lb class	125	162
Carbon steel, 600-lb class	272	354
Type 304 stainless steel, 150-lb class	284	284
Type 316 or 347 stainless steel, 150-lb class	375	362
Brass, 150-lb class	83	137
Aluminum, 150-lb class	375	362
Cast iron, 125-lb class	120	86
Monel, 150-lb class	164	200
Fitting, 90° ell, reducer or coupling:		
Carbon steel, schedule 40	3.2	4.7
Carbon steel, schedule 80	5.2	6.7
Type 304 stainless steel, schedule 10	14.7	18.0
Type 316 or 347 stainless steel, schedule 10	21.6	23.0
Cast iron, 125-lb class	36.0	17.8
Cast steel, 150-lb class	90.0	47.3
Fitting, tee or unions:		
Carbon steel, schedule 40, weld	14.3	13.0
Carbon steel, schedule 80, weld	13.5	18.8
Type 304 stainless steel, schedule 10, weld	24.6	26.0
Type 316 or 347 stainless steel, schedule 10, weld	30.5	32.3
Cast iron, 125-lb class	75	24
Cast steel, 150-lb class	120	58
Fitting, stub end:		
Type 304 stainless steel, schedule 10	9.3	6.0
Type 316 or 347 stainless steel, schedule 10	12.0	8.1

TABLE 9-15. "X" FACTORS FOR ESTIMATING PIPING-MATERIAL COSTS
(*Continued*)

Nominal pipe size (NPS)	"X" factor	
	¾–2"	3–8"
Flanges:		
Forged steel, slip-on, 150-lb class	6.1	3.2
Forged steel, slip-on, 300-lb class	10.0	6.7
Forged steel, slip-on, 600-lb class	24.4	13.7
Forged steel, welding neck, 150-lb class	9.2	4.8
Forged steel, welding neck, 300-lb class	15	10
Forged steel, welding neck, 600-lb class	36	21
Type 304 stainless steel, slip-on, 150-lb class	25	13.5
Type 304 stainless steel, slip-on, 300-lb class	40	28
Type 304 stainless steel, welding neck, 150-lb class	41	21
Type 304 stainless steel, welding neck, 300-lb class	67	45
Type 316 or 347 stainless steel, slip-on, 150-lb class	28	15
Type 316 or 347 stainless steel, slip-on, 300-lb class	45	31
Type 316 or 347 stainless steel, welding neck, 150-lb class	46	24
Type 316 or 347 stainless steel, welding neck, 300-lb class	75	50

Cost of Materials and Labor for Piping and Fittings, Including Fabrication and Erection. A dimensioned piping drawing or scale model should be available. Figure 9-15 will be used for the example. Material price estimates, exclusive of the costs of pipe supports, hangers, and gaskets, are given in Fig. 9-16 and Table 9-15. To price an NPS fitting, obtain the cost per foot of black pipe of the same NPS (Fig. 9-16). Then multiply by the correct "X" factor for the fitting, as given in Table 9-15. (For a higher degree of accuracy, up-to-date manufacturers' catalogs and quotations on all job items are necessary.)

Table 9-16 was prepared using this information and the diameter-inch method. All threaded and welded connections are counted and multiplied by the corresponding nominal pipe diameter and this product multiplied by the correct labor factor listed in Table 9-17. Bolting of flanges is accounted for by a separate labor factor in Table 9-17. Piping labor costs vary with the job location; use $3 to $4 per hour.

The method properly gives little consideration to linear length of pipe since the fabrication and erection costs are practically identical whether a pair of elbows are 3 ft or 15 ft apart. However, when standard lengths of pipe are joined together, this is counted as a connection and totaled in the inch-diameter column for labor charges.

Cost of Pipe Supports, Hangers, and Gasket Sets. A detailed piping diagram showing location and type of supports, hangers, and other pertinent construction information is required for obtaining such costs. For

TABLE 9-16. MATERIALS AND LABOR COSTS FOR PIPING LAYOUT OF FIG. 9-15

Line code: 0–102				Diameter-inches*			Labor factor†				
Bill of materials	No.	Connections	Estimated cost, $		Fabrication and erection	Bolt-up of flanged ends	Fabrication and erection	Bolt-up	Man-hours	Labor cost @ $3.50 per man-hr, $	Total cost, $
			Per item	Total							
1-in. ells...........	2	4	$0.70	$1.40	4						
1-in. union.........	1	2	3.14	3.14	2						
1-in. valve.........	1	2	17.00	17.00	2						
1-in. coupling......	1	2	0.70	0.70	2						
1-in. totals........		10		22.24	10		1.30		13.0	45.50	67.74
2-in. ells...........	1	2	1.82	1.82	4						
2-in. reducer.......	1	1	‡	‡	2						
2-in. flange.........	1	1	5.20	5.20	2						
2-in. pipe..........	20 ft	0	0.57	11.40	0						
2-in. totals........		4		18.42	8		1.00		8.0	28.00	46.42
3-in. ells...........	4	8	4.70	18.80	24						
3-in. flanges........	3	3	4.80	14.40	9						
3-in. reducer.......	1	1	4.70	4.70	3						
3-in. flanged tee...	1	2	13.00	13.00	...	6					
3-in. flanged valve (VI)...	1	2	101.00	101.00	...	6					
3-in. pipe..........	118 ft	4	1.00	118.00	12						
3-in. totals........		20		269.90	48	12	1.00	0.35	52.2	182.70	452.60
Job totals.........				$310.56					73.2	$256.20	$566.76

Note: Carbon steel, schedule 40, piping materials costs from Table 9-15 and Fig. 9-16. Labor from Table 9-17.

* Number of connections multiplied by pipe diameter (fabrication and erection of welded or threaded connected pipe or bolt-up of flanged ends).

† See Table 9-17.

‡ See 3-in. price.

401

order of magnitude estimates the following costs can be used: overhead pipe racks such as shown in Fig. 9-9 range from $300 to $800; structural brackets such as shown in Fig. 9-10e range from $30 to $60; pipe hangers cost $0.25 to $0.70 per nominal pipe diameter-inch; 150-lb flange bolt and gasket sets can be priced from $0.30 to $0.40 per nominal pipe diameter-inch.

Cost of Painting and Insulation. Cost figures for these two items have been given on pages 384 to 390.

TABLE 9-17. LABOR FACTORS FOR PIPING INSTALLATIONS

A. Pipe Layout, Cutting, Welding, Threading, and Erection

Nominal pipe diameter, in.	Man-hours per diameter-inch	
	Carbon steel schedule 40	Alloy steel schedule 5-10
½–1½	1.30	1.60
2–3	1.00	1.30
4–8	0.90	1.25

B. Handling and Bolting Up Flanged Valves and Fittings

Nominal pipe diameter, in.	Man-hours per diameter-inch per end
½–1½	0.40
2–5	0.35
6–8	0.30

C. Additional Labor Factors for Various Piping Areas
1. Highly complex and confined areas: Add 20 per cent.
2. Open process towers above second floor: Add 10 per cent.
3. Elevated piping requiring temporary scaffolding: Add 15 per cent.
4. Long straight runs of piping: Subtract 10 per cent.
5. Testing of completed job: Add 5 per cent.

Total Piping Cost Estimate. The above costs are additive and must be factored for overhead and profit if contract piping costs are required. Overhead figures range from 50 to 65 per cent of direct labor and 8 to 10 per cent of materials. Profit is also a variable item, depending primarily on local business conditions; use 5 to 12 per cent of total project piping costs.

PROCESS CONTROL AND INSTRUMENTATION

Process control and automation, together with their associated instrumentation, can be considered as the mechanical brain and nerves of modern chemical processing. Variables in a process must be measured and then controlled and integrated for optimum processing conditions. Mechanical and electrical components and systems have been designed by instrumentation and control engineers to reduce labor and improve feasibility of plant operation, and to allow extensive outdoor plant construction. The economic advantage of automatic process control has been well established throughout the industry, thus accounting for the rapid growth in this area of the chemical process industry. The field is specialized and design organizations have an instrumentation and control group consisting of chemical, mechanical, and electrical engineers who have acquired education and experience in this area. This group works closely with the process engineers to accomplish their mission of incorporating the best instrumentation and control design into each new process being developed. Decisions within the group, based on a cooperative effort with other groups, must be made on the following points:

1. What are the variables that affect process operation, and to what extent do they influence the operation?
2. What physical functions should be measured, and with what degree of accuracy?
3. What physical functions should be automatically controlled, and to what extent?

Chemical engineers not specializing in process control should have a working knowledge of this field, so that they can appreciate the relative merits of process control and automation. Most schools give a required course to all chemical engineers on instrumentation for just this reason. Many excellent books and periodicals are available for those who wish to specialize in this field (see Additional Selected References). The purpose of this section of the plant design book will be to highlight the principal features of process control and instrumentation and show their relationship to chemical process and plant design practice.

Process Control

Chemical reactions are affected by a number of variables, e.g., temperature, pressure, etc. Input, output, and in-process conditions must be measured by some physical and/or chemical means to determine rates, compositions, and holdup in processing equipment and piping. Key control points are selected from process flow sheets, such as shown in Chap. 3. Instruments at these control points are assigned code numbers for identification in the design,[1] purchase, and construction phases of plant design. All the material and energy components must then be measured, compared with a desired standard, computed, integrated, and corrected, when necessary, for a smoothly operated chemical process. These are the basic functions of process control which give rise to the *closed-loop* theory of automatic control systems.

[1] See pages 74 to 82 for instrumentation symbols and flow sheets.

Figure 9-17 is a schematic diagram of a typical closed control loop. If the error detection is done by an experienced human operator, then the process is controlled manually. Manual control is used only where the required instrumentation either is not available or is too expensive from either a first cost or maintenance standpoint. Mechanical and electrical *error detecting* elements or *controllers* are almost always used in closed-loop control to give a fully automatic process control system.

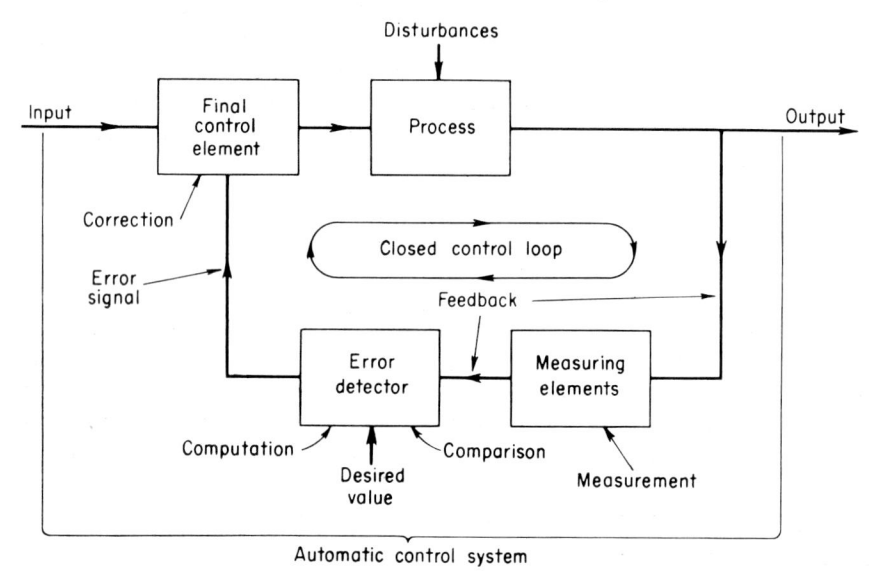

FIG. 9-17. Diagram of an automatic control system. [G. A. Hall, Jr.; by permission from "Process Instruments and Controls Handbook," Douglas M. Considine (ed.), McGraw-Hill Book Company, Inc., New York, 1957.]

Another of the essential components of the controlled loop system is the *measuring element*, often called indicating and recording instrumentation. Examples of a measuring element are a temperature indicator, a pressure recorder, or a recording flowmeter. The final component of the closed loop is the *control element*, such as a valve or switch.

Selection of Process Instrument Elements

Process design engineers should be able preliminarily to select types of instrumentation for process control. Several excellent guides for selecting instrument elements are available. In Perry's "Chemical Engineers' Handbook," 3d ed., pp. 1266–1309, there is a discussion of process variables and their measurement in terms of (1) energy variables, such as temperature, pressure, and radiation; (2) quantity and rate variables, such as fluid flow, liquid level, thickness; (3) physical and chemical characteristics, such as density, moisture content, and chemical absorption, just to men-tion a few. A more recent guide[1] is recommended for further study of process instrument elements. It is organized into 22 sections on process variables and their measurement, with 5 sections on error detection and final control elements in closed-loop automatic control systems. These guides are not intended as a substitute for manufacturers' catalogs or their recommendations; final selection should be done only by an experienced instrument engineer. Some of the factors which comprise the proper selection are: level, range and function of the instrument, accuracy required, materials of construction, possible effect on process conditions (e.g., a control valve with too large a pressure drop).

It would be advisable to look through this guide to become familiar with the concepts of process instrument elements. Their use in automatic process control systems will be explained next.

Elementary Principles of Automatic Control

It is seen that the basic functions of automatic control are (1) measurement, (2) comparison, (3) computation, and (4) correction. Thus, a mechanism used for measuring the value of a *controlled process variable* and operating to limit the deviation of this variable from a desired value or *set point* is an automatic controller. Disturbances which cause deviations in the controlled variables are classified as either *supply-load changes* or *demand-load changes*. The automatic controller regulates by making corrections to one or more *manipulated variables* of the system. These points can best be illustrated by a typical process that uses automatic control.

Example of a Process Control System.[2] Figure 9-18 shows schematically an air-operated control system applied to the control of a water heater process. The parts of this control system, arranged in block-diagram form to illustrate the basic control function that each part performs, are shown in Fig. 9-19.

Measuring Means. The measuring system of the typical controller consists of three parts:

1. Thermometer bulb (primary or sensing element). This is part of the measuring system that is directly sensitive to the controlled variable (temperature). The primary element converts energy from the controlled medium (hot water) into a measurable signal (fluid pressure).

2. Bourdon pressure element (receiving element). This part of the system evaluates the signal from the primary element, and converts it into scale readings, chart records, and actuation for the error detector.

3. The capillary tubing (transmitting means). This part of the system

[1] T. R. Olive and S. Danatos, *Chem. Eng.*, **64**(6): 288–319 (1957).

[2] The discussion in this section (through p. 408) is modified slightly from that of G. A. Hall, Jr.; by permission from "Process Instruments and Controls Handbook," Douglas M. Considine (ed.), McGraw-Hill Book Company, Inc., New York, 1957.

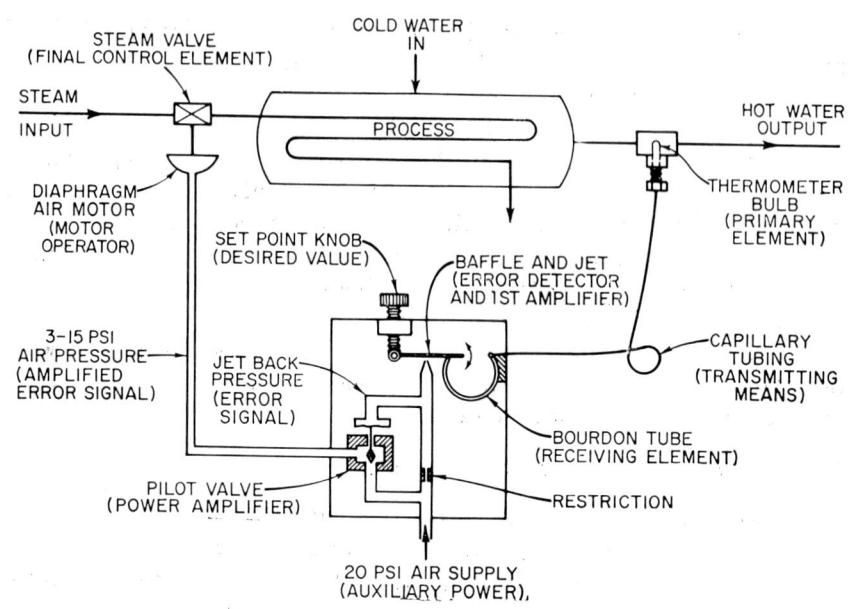

Fig. 9-18. Typical pneumatic controlled system applied to a heat-exchange process. [G. A. Hall, Jr.; by permission from "Process Instruments and Controls Handbook," Douglas M. Considine (ed.), McGraw-Hill Book Company, Inc., New York, 1957.]

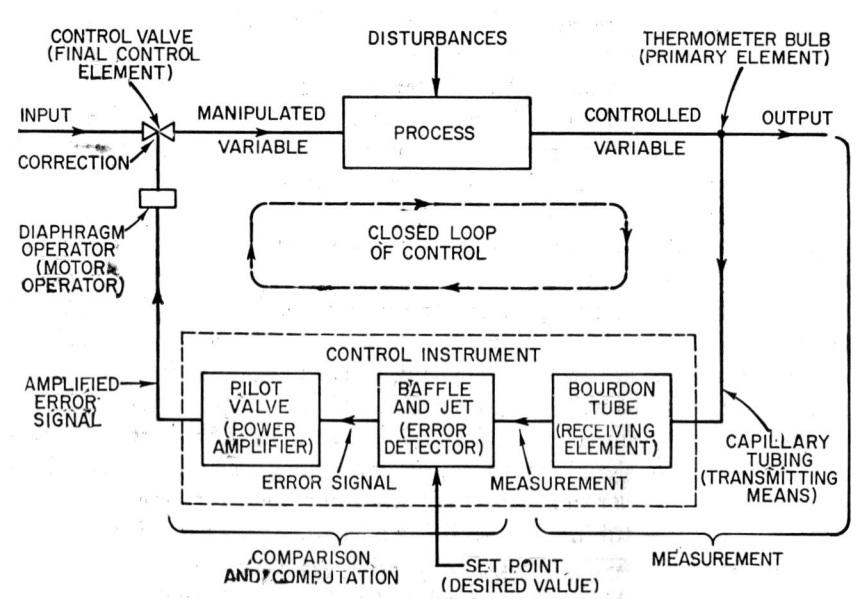

Fig. 9-19. Block diagram of the control system of Fig. 9-18. [G. A. Hall, Jr.; by permission from "Process Instruments and Controls Handbook," Douglas M. Considine (ed.), McGraw-Hill Book Company, Inc., New York, 1957.]

carries the signal from the primary element to the receiving element. In some controllers, the afore-mentioned three parts of the measuring system are combined into one or two devices.

Error-detector Elements (unbalanced detector, summing point, or primary relay). In Fig. 9-18, the error detector is the baffle and jet. The error detector compares the measured value of the controlled variable with its desired value, and produces an error signal when deviation exists. A brief description of how the error detector of Fig. 9-18 operates is given in the following paragraphs.

The *desired* value is represented by the position of the left end of the baffle, as determined by manual adjustments to the set-point knob. The *measured* value of the controlled variable is represented by the position of the right end of the baffle, as determined by the deflection of the bourdon tube. Thus, the baffle is a differential lever; the position of its center (at the jet) represents the deviation (difference) or error between the desired and measured values of the controlled variable. In order to be usable, this very weak, small deviation or error signal, represented by the position of the baffle mid-point, must be both measured and amplified. This is done (in this particular controller) by the baffle-and-jet error detector.

The jet back-pressure air system is continuously furnished with air which bleeds in from the air supply through the restriction. This air bleeds out from the jet back-pressure system through the jet to atmosphere. As the center of the baffle (representing error) is moved closer to or farther from the air jet, the resistance to the flow of air through the jet changes. This changes the back pressure in the jet air system. This back pressure is proportional to the baffled position through a small range of baffle motion. Thus, the amount of deviation or error is measured and converted into an air-jet back pressure, which is an amplified error signal. The error detector is the heart of any automatic controller, for it is the part that senses deviation and first instigates corrective action.

Amplifier. In order not to restrict the sensitivity or accuracy of the measuring system, an error detector must take very little power from the measuring system. Thus, error signals usually are very weak. In order to operate most final control elements, the error signal must be greatly amplified in power. Therefore, most automatic controllers contain a power amplifier that uses auxiliary power to increase the strength of the error signal. In the typical air-operated controller of Fig. 9-18, some amplification is obtained from the jet-and-baffle error detector, which might be called a *first-stage amplifier*. Additional power amplification is produced by the pilot valve, which might be called a *second-stage or power amplifier*. Both stages use auxiliary power furnished by the 20-psi air supply. The jet back pressure (error signal) from the error detector is applied to the diaphragm of the pilot valve. This causes the pilot-valve plug to move up and down, which delivers more or less 20-psi supply air to form an amplified error signal or *output* from the air controller that is still proportional to the error or deviation represented by the baffle midpoint.

Motor Operator. The error signal must be converted into corrections to the manipulated variable of the process. In most control systems, this requires some form of operator or motor to operate the final control element. In the air-operated control system of Fig. 9-18, the motor operator that positions the steam valve is the diaphragm air motor. The output air pressure (amplified error signal) from the pilot valve is applied to the motor-operator diaphragm.

Final Control Element. The final control element corrects the value of the manipulated variable. In the control systems of Figs. 9-18 and 9-19, the final control element is the steam valve which is in direct contact with the control agent (steam) and makes corrections to the manipulated variable (rate of steam flow).

Self-operated and Relay-operated Controllers. Some control systems obtain all power for operating the error detector and final control element from the controlled medium of the process via the primary element and measuring means. Such control systems are termed *self-operated controllers* (Fig. 9-20a).

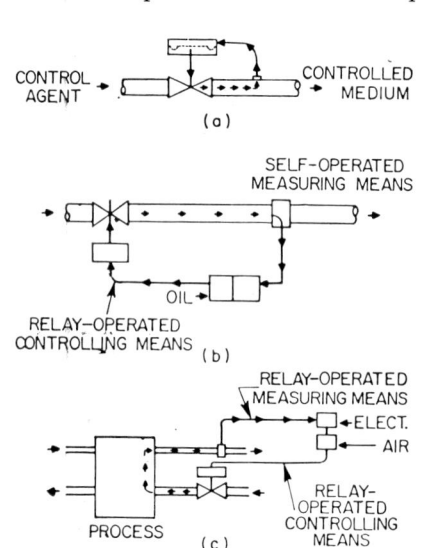

FIG. 9-20. Classification of automatic controllers: (a) self-operated controller, using energy only from controlled medium through primary element; (b) relay-operated controller with self-operated measuring means and relay-operated controlling means; (c) relay-operated controller with relay-operated measuring means and relay-operated controlling means. [G. A. Hall, Jr.; by permission from "Process Instruments and Controls Handbook," Douglas M. Considine (ed.), McGraw-Hill Book Company, Inc., New York, 1957.]

Control systems that use an *auxiliary* source of power, in addition to the power provided through the primary element, are termed *relay-operated controllers* (Fig. 9-20b and c). This auxiliary power may be introduced into the measuring system, the error detector, or one or more amplifying relays. In the typical controller of Fig. 9-18, the measuring means is self-operated. The auxiliary power is provided by compressed air supplied at the baffle-and-jet error detector and at pilot-valve power amplifier.

Dynamic Response Characteristics of Process and Control Systems.
Processes and their associated equipment have the characteristics of delayed changes in the values of the process variables. The control instrumentation also has response lags. In the previous example the following types of time lags exist:

1. Process lags
   a. Supply side capacity: heat capacity of steam coils
   b. Demand side capacity: heat capacity of water in tank
   c. Transfer resistance: heat-transfer resistance of fluid films and steam coil metal wall
   d. Velocity—distance: delay on lead time due to separation of inlet from measuring element point (depends on velocity and degree of mixing)
2. Controller lags
   a. Measuring element: resistance-capacitance characteristics of the measuring element (e.g., thermometer bulb)
   b. Controller: usually negligible
   c. Transmission: resistance capacitance of connecting pipe and diaphragm motor; keep the line length and volume of diaphragm motor as small as possible
   d. Final control element: frictional inertia; dead time results from sticking valves

Dynamic system response theory in terms of *time constants*, corresponding to each lag, has been developed by means of differential equations, while a corollary method known as *frequency response* has been developed both theoretically and experimentally for determining the dynamic response characteristics of a system or component.[1] A nonmathematical graphical approach to frequency response has been presented as an aid to engineers lacking advanced mathematics background.[2]

It is obvious that lags of any kind should be reduced to a minimum by careful design, selection, installation, and maintenance.

Modes of Control. Modern industrial control instruments are designed to produce one or several modes of control (or control responses to system disturbances) as follows:

1. On-off (two position)
2. Proportional (throttling)
3. Proportional plus reset
4. Proportional plus rate (derivative)
5. Proportional plus reset plus rate (three-term control)

These responses will be illustrated by use of the heat-transfer process of Fig. 9-18, but with different types of error detectors and compensators.

On-Off Response. Assume the hot-water output from the tank is to be held at 190°F, i.e., set point is 190°F. The error-detecting and compen-

---
[1] Stephen P. Higgins, Jr., in Douglas Considine (ed.), "Process Instruments and Control Handbook," McGraw-Hill Book Company, Inc., New York, 1957.

[2] "Principles of Frequency Response," Instrument Society of America, Pittsburgh, Pa.; film and covering booklet.

sating device, such as the on-off air-operated automatic control system pictured in Fig. 9-18, either fully opens the steam valve if the hot water is below the set point or fully closes it if the hot water is above the set point. A hunting or cyclic action results, as shown in Fig. 9-21a; the amplitude is inversely proportional to system demand capacity, and the

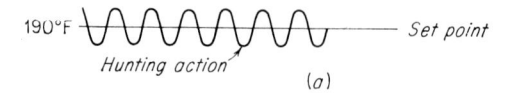

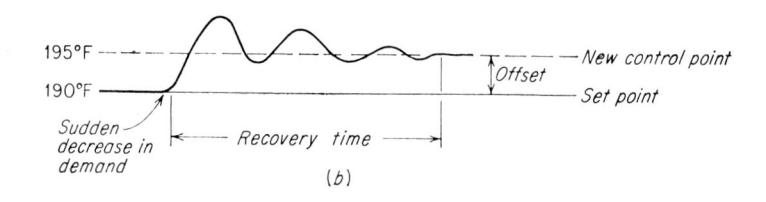

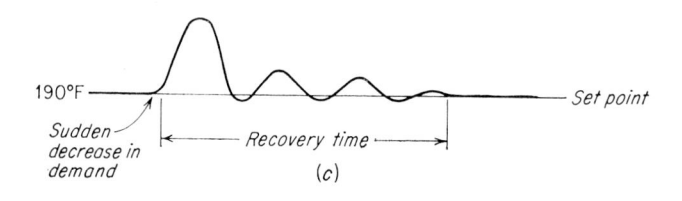

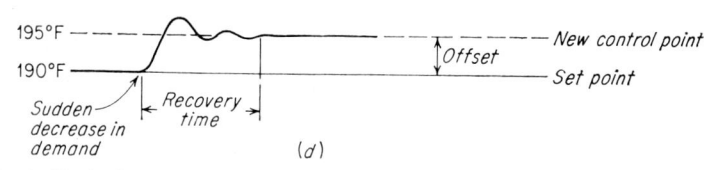

FIG. 9-21. Typical control mode characteristics. (a) On-off response (creates hunting action); (b) proportional response (eliminates hunting but creates offset except on constant preset demand); (c) proportional and reset response (eliminates hunting and offset); (d) proportional plus rate or derivative response (reduces recovery time but maintains offset unless reset is also used).

period is directly proportional to transfer resistance lag. This type of control is not generally satisfactory and should be used only for infrequent or small demand changes, high system demand capacity, and low-transfer resistance lag.

Proportional Response. To eliminate hunting, a control element movement is incorporated which is a linear function of the controlled variable.

The resulting proportional controller throttles the steam supply in proportion to the demand. Instruments are designed for a full-scale change of the controlled variable, e.g., 100°F, or from 140 to 240°F in this example. The instrument can be manually set to provide complete on to off valve action over a percentage of the full-scale range of the instrument. This percentage is termed *proportional band width.* In this example, a 20 per cent proportional band width of 20°F at a set point of 190°F would mean that the control valve would be fully closed at 200°F, fully opened at 180°F, and half open at the set point of 190°F.

This control action eliminates hunting but creates *offset or droop* with varying demand changes, such that there is a shifting control point (Fig. 9-21b). A narrow proportional band setting produces the minimum offset but maximum cycling action and recovery time. Offset can only be eliminated by manually changing the set point of the instrument.

Proportional plus Reset Response. By adding an automatic element to a proportional controller, the control point automatically returns to the set point (see Fig. 9-21c). Thus, control without offset error is possible under all load conditions, but the reset itself does not contribute to the stability of the control loop. A majority of control applications require only the proportional plus reset control. However, on systems with long process lags, such as temperature applications with long thermal lags, a more sensitive control is necessary to avoid overcontrolling and cycling, thereby reducing recovery time. This mode of control will be discussed next.

Proportional plus Rate Response. To decrease the recovery time, *derivative* or *rate response* may be added to a proportional controller. The control valve is moved at a rate proportional to the rate of change of the controlled variable with time, i.e., derivative. Thus, the valve has its greatest movement when the controlled variable is changing fastest and this type of anticipation and response reduces the recovery time of the system (Fig. 9-21d). The addition of the reset mode, giving *three-term control,* will eliminate the offset shown in this example.

These explanations serve as an introduction to the principles of process control. For further study, see the Additional Selected References. The "Process Instruments and Controls Handbook" cited previously is particularly recommended.

Process Control Center

A central control room from which all major controls are operated typifies modern chemical plants. This room houses the control equipment and operation, thus acting as the brain of the processing plant. It should be designed for comfortable and pleasant working conditions.

Within the control room, the control board and data-logging center is the key to automatic process control. Two functional control boards are used. In conventional control panel design (Fig. 9-22a), instruments are mounted closely together to centralize the working area; the operator is within easy reach of every instrument, and the control room may be reduced in size, allowing a more favorable plot location. Proponents of graphic control panels (Fig. 9-22b) have designed the panel board as a replica of the process flow diagram with instruments located in the proper position in the process. This graphic panel has the advantage of quickly impressing on the new operators the functional characteristics of each control point in terms of the entire operation. This tends to reduce errors in operating a plant. Miniature instruments designed originally for the graphic system to reduce panel dimensions are also being incorporated in conventional panel designs for the same reason.

Automatic data logging is an addition to control rooms which aids process control (Fig. 9-22b). Electronic equipment is provided which continually scans all process variables, typing on a log sheet the variables which are off their preset limits, so that operators can make adjustments when necessary. Other options include typing of a complete log of variables periodically and teletyping production rates and other information to remote points, such as the sales office.

The next step now being studied is complete automation and control of the process by means of electronic computers, without use of operating personnel. Such systems require mathematically defined knowledge of the effect of process variables, which is presently available for only a few simple processes.

To provide adequate facilities for the control room, which, incidentally, is only a small percentage of the capital investment, a few pointers are in order:

1. Provide comfortable heating and air-conditioning facilities; eliminate all flammable and toxic gases.

2. Use color schemes which are pleasing and restful; shades of green are most often specified.

3. Provide a comfortable walking surface, such as rubber tile.

4. Provide office, lunch preparation equipment, toilets, locker and shower rooms.

5. Use diffuse lighting of about 30 foot-candles intensity, except in front of and behind the panels where directional lighting of at least 50 foot-candles intensity is necessary.

6. Follow the electrical code for explosion-proof wiring if the control room is located in a hazardous area.

7. Locate control panels for operating functionality; provide at least 6 ft in the rear of control panels for instrument maintenance.

(a)

(b)

Fig. 9-22. Control room elements. (a) Example of a conventional panel control center—S. D. Warren Co., Cumberland Mills, Maine. (*Courtesy of the Foxboro Co. and S. D. Warren Co.*) (b) Example of a graphic panel and data logging control center—Sohio Refinery, Toledo, Ohio. (*Courtesy of The M. W. Kellogg Co., Standard Oil Company of Ohio, and E. M. Payne Studios, New York.*)

Control System Design

Centralization of control in one area, as described above, is possible only through the use of transmitters and telemeter devices. The signal from the measuring element located at a process point is carried back to the control room where indicating, recording, and prime controlling elements are located. The signal is then sent from this type of instrument to the final control element within the process area. Transmitters of either the electronic or pneumatic type are the signal carriers within the process area. Telemetering by telephone, wire, or microwave is used for remote operation.

The *pneumatic control system* has been the standard of the industry, but has some disadvantages which must be "designed around" to reduce their effects. Transmission lines over 300 ft should have a booster relay to speed the signal. The required instrument air is usually supplied by a separate air compression and dehydration system. The air supply must be free of oil to prevent clogging of the instrument parts, and free of water to avoid freezing. Pressure at the distribution header of about 3-in. line size is 75 psi with instrument air at 3 to 20 psi operating standard diaphragm valves. For higher-pressure systems, larger diaphragms or air relay multipliers should be specified.

The *electronic control system* is the newer of the two and has several advantages. Electric impulse signals from sensing elements, such as thermocouples or liquid-level devices, can be transmitted after suitable amplification through practically unlimited lengths of transmission lines. This provides much greater sensitivity and eliminates the difficulties caused by ice formation in air lines during extremely cold weather. The principal disadvantage is that a proportional control signal must be converted to a pneumatic valving system, since electric motor-driven valves, which could operate directly from the electric signal, are too slow and gear wear is excessive. However, an electrically operated solenoid valve can be used directly without pneumatic conversion for off-on control.

Design of Process Control Systems for High-pressure Plants[1]

The fundamentals discussed previously are applicable to plants of any pressure range. However, instrumentation of processes at pressures over 3,000 psi presents additional complicating factors which fall into one or more of the following general classifications:
1. Physical properties of process chemicals
2. Small-dimension characteristics, particularly lines
3. Safety requirements
4. Pulsating flows

[1] Y. Anderson and S. Humbla, Vulcan-Cincinnati, Inc., private communication.

There follows a brief discussion of each of these factors in relation to measurement and control of the more common variables, such as temperature, pressure, flow, and liquid level.

Physical Properties. Accurate values of process materials at high pressure are not known in most cases, nor can they be extrapolated with any accuracy. Gases are above the critical point, and liquids are compressible, so that absolute measurements are difficult to obtain. Specification of suitable instruments is difficult.

Small Dimensions. High-pressure gas lines may be quite small; $9/16$ in. OD by $5/16$ in. ID is a common size. Making flow- and pressure-measurement sensing devices to fit on small lines, particularly jacketed lines, is a real mechanical design problem.

Safety Requirements. This is not solely an instrumentation problem. Code practice should be followed with special flanges, materials, and welding procedures, specified for connection of the instrumentation elements to pressure, piping, and equipment (see Perry's "Chemical Engineers' Handbook," 3d ed., pp. 1233–1261).

Pulsating Flows. This is the most troublesome factor of all, since it adversely affects flow, pressure, and liquid-level measurements. Pulsations set up by reciprocating pumps and compressors create pressure variations superimposed on normal system pressures, along with vibration effects. Damping devices, such as snubbers or restricting orifices, should be used in the process lines. Signal damping can be specified on the instrument transmission lines, if transfer lag is not excessive. Flexible connections, in the form of U or O loop design, should be used to reduce vibration transfer to instrument lines, if the vibration cannot be removed from the process lines. The selection of proper instruments, which do not use pressure-sensing devices, is the best way to circumvent the pulsed-flow problem. Liquid level can be obtained by capacitance or radioactivity probes; flow is measured by an electromagnetic flowmeter, if the fluid is an electrical conductor.

Cost Estimation for Process Control

As cited in Chap. 6, basic instrumentation costs range from 3 to 20 per cent of installed equipment costs, depending on the completeness of automatic control. Detailed instrumentation costs are required for final cost estimates within 10 per cent. Instrumentation as listed by code numbers from the instrumentation flow sheets is set up as separate instrument items on specification sheets and sent to suppliers. Firm quotations are received after mutual agreement on the correct type and application is reached among the purchasing and manufacturing personnel. Table 9-18 lists typical instrument and auxiliary costs which can be used for estimating purposes in lieu of actual quotations.

TABLE 9-18. TYPICAL PRICE RANGES FOR INDUSTRIAL INSTRUMENTS
(1958 PRICES)*

	Estimated price range, $†		
	Low	Average	High
Temperature:			
Differential expansion thermostat	10	25	40
Indicating expansion thermometer	60	145	230
Indicating resistance thermometer	100	310	600
Indicating pyrometer	100	225	300
Pressure:			
Indicating bourdon-type gauge (noncorrosive conditions)			
0–500 psi	20	75	160
0–10,000 psi	35	115	200
0–30 in. vacuum	25	105	160
Indicating bourdon-type gauge (corrosive conditions requiring diaphragm seal)			
0–500 psi	85	140	225
0–5,000 psi	110	170	250
0–30 in. vacuum	75	150	225
Strain-gauge load cells	240	360	975
Indicating vacuum gauge—McLeod type	300	400	500
Fluid flow:			
Volumetric meter			
Nutating disk meter	200	350	500
Rotary vane meter	200	350	500
Liquid-sealed gas meter	100	200	300
Variable head			
Indicating orifice meter	180	400	595
Indicating venturi meter	110	360	700
Variable area meters			
Indicating tapered tube rotameter	30	320	950
Velocity meters			
Electric-current–turbine-wheel meter	450	730	1,000
Electromagnetic flowmeter	775	3,290	7,800
Hot wire anemometer	300	440	620
Level:			
Float type			
Gauge glass	75	150	200
Tank float with external indicator	200	810	1,590
Hydrostatic			
Differential gauge	40	210	400
Continuous capacitance	300	400	500
Radioactive-ray absorption	465	1,230	2,000
Strain gauge weighing			
50 lb–5 tons	540	600	625
10–50 tons	825	1,200	1,550

TABLE 9-18. TYPICAL PRICE RANGES FOR INDUSTRIAL INSTRUMENTS
(1958 PRICES)* (*Continued*)

	Estimated price range, $†		
	Low	Average	High
Density:			
Photoelectric hydrometer	120	550	1,000
Gas density balance	480	1,150	2,000
Radioactive-ray absorption	2,000	3,000	4,000
Viscosity:			
Continuous viscosimeter	220	2,450	4,000
Continuous consistency meter	1,500	2,300	4,160
pH	550	1,170	1,500
Electrical conductivity	200	500	800
Thermal conductivity	200	900	1,400
Combustible gas analyzers	600	800	2,000
Gas humidity	650	1,150	1,800
Spectroscopy:			
Infrared	2,000	3,000	20,000
Ultraviolet	2,000	2,500	4,000
Mass spectrometry	2,500	8,000	20,000
Radio frequency	3,000	5,000	15,000
Gas chromatography	4,500	6,500	15,000
Radioactivity measurements:			
Ionization	200	450	3,000
Scalar for proportional or G-M counting	500	1,100	1,500
Scintillation counters	900	1,400	4,000
Recording instrumentation for most of the sensing elements listed above (e.g., temperature, pressure, flow rate):			
1 point	200	500	1,200
4 points	350	800	1,400
16 points	1,000	1,500	1,900
Recording-controlling instrumentation for most of the sensing elements above (e.g., temperature, pressure, flow rate); price does not include main control valves:			
1 point	300	700	1,600
4 points	1,400	1,500	1,750
16 points	1,800	2,000	2,250

* Composite figures from eight major instrument companies.

† Add 35 per cent of the average base price for installation charges on instruments costing less than $2,000; add 20 per cent for instruments costing over $2,000.

POWER SYSTEMS

Chemical plants cannot operate without power from such sources as steam and electricity. Power requirements are shown in Figs. 6-33 and 6-34. Reliability of power sources is taken for granted by chemical plant design engineers. However, a great deal of design experience must be incorporated in such systems to guarantee economical and trouble-free sources of power. Much of the responsibility of such systems lies in the hands of electrical and mechanical engineers, but chemical engineers should have a basic appreciation of the engineering features of power systems used in chemical process plants.

Power Requirements

Power for the modern chemical plant is required for a variety of services, i.e., (1) process heat in the form of steam, electrical, nuclear, or fossil-fuel power; (2) mechanical power; (3) electric power; (4) refrigeration; (5) heating, ventilating, and air conditioning; (6) gas compression. These power requirements can be supplied in several ways, depending on economic factors. Typical power sources are (1) utility-owned hydroelectric, steam-generating, or nuclear power plants; (2) company-owned steam-generating or nuclear plants for process heat and electric power; (3) company-owned internal-combustion plants, particularly adapted to emergency stand-by operation; (4) conventional direct or indirect fossil-fuel-fired furnaces, heat exchangers, kilns, etc. A discussion of several of these types of power plants is in order.

Process Steam Power

Most chemical plants produce their own steam for a cheap, convenient heating medium, and as a direct power source for pumps, jets, and turbines, in addition to furnishing a high-temperature gaseous reactant in some chemical processes, such as catalytic re-forming of hydrocarbons. The pressure-temperature requirements of steam for the process industries vary greatly. A somewhat arbitrary division has been used in terms of low or high pressures, with 250 psia (400°F saturated) being one dividing point. Supply stations can be one or more large central boiler houses with a complex distributing system, or a number of small packaged boilers, located strategically in the process areas, depending on plant economic studies. Steam exhaust, or bleed from turbine operation, is a source of process heat. Waste-heat boilers use the available enthalpy from high-temperature gases to generate steam. An economic break-even point study on waste-heat steam generation shows that gas temperatures should be above 1000 to 1100°F.

For further study of boilers, furnaces, and waste-heat recovery units, the reader is referred to Perry's "Chemical Engineers' Handbook," 3d ed., sec. 24, pp. 1628–1652, on steam plants; sec. 23, pp. 1598–1625, on furnaces and kilns.

Nuclear Power

The development of nuclear-power devices has grown rapidly since 1950. It has been estimated that 1 lb of a fissionable fuel, such as the uranium isotope U^{235}, has an energy equivalent of 3 million lb of coal, and there is 40 to 60 times more energy potentially available from nuclear than from fossil fuel, based on known reserves. The vast amount of energy available from a small quantity of fissionable fuel found in relatively large abundance on the earth's surface, or capable of being generated by fission-fertile fuel processes, makes this type of power attractive. Steam power plants have been designed and operated with nuclear fuel as a power source. Competitive economics has limited nuclear plants to locations where fossil fuel is expensive. This situation will shift as fossil-fuel supplies are gradually reduced and allocated for small-capacity power units. Direct application of nuclear heat for chemical processing is being recommended; economic factors again limit this idea. In Chap. 10, Nuclear Chemical Plant Design, nuclear systems are discussed in more detail. The reader is referred to this chapter and its References.

Electric-power Generation

Electricity is a very important and useful form of power for the chemical industries. Electrochemical industries are based on this energy source. Other chemical plants use electricity for driving pumps, compressors, agitators, and other mechanical equipment, for process instrumentation, and for lighting.

Careful consideration must be given to the source, cost, and reliability of the primary electric-power supply. Power can be either purchased from a public or private utility, produced at the plant site by steam-driven turbogenerators or natural gas–driven engines, or purchased from an adjacent industrial plant as a by-product. An economic study is necessary to select the type or types of power supply to be specified. Important factors which are considered in such a study include:

1. Proximity to existing utility power lines. The plant location may be so remote from satisfactory public utility lines that power costs would be excessive and service undoubtedly unreliable.

2. Magnitude and type of power requirement. Many plants, particularly electrochemical plants, require large blocks of power, sometimes more than an entire municipality. Utility companies may not wish to

expand their facilities for only one user. Cooperative industrial area development would be required to interest a utility company. Frequently, peak demand and load requirements do not match the utility's output.

3. Large demand for low-pressure steam for processing and heating. Through the use of high-pressure boilers, some operating above the critical conditions for water, it is possible to extract power for prime movers, such as electric-generating steam turbines, and to obtain a cheap and plentiful source of bleed steam or high-pressure exhaust steam. This balancing of process steam and electric power is a part of the economic study which should be made if process steam requirements exceed 50,000 lb of steam per hour.

4. Availability of by-product fuel and heat. Fuel and heat sources are readily available in many chemical industries with no cost other than handling. Forms of this by-product energy differ: combustible chemical products from paper mills; combustible gases from blast furnaces, coke ovens, and refineries; hot air from smelters and cement plants.

5. Competitive capital ventures. The over-all investment picture must be considered. If competitive electric power is available from a progressive and reliable utility, the capital should not be invested in a company-owned electric generating plant, but in additional chemical plant facilities which yield higher returns.

Emergency Power

Well-designed and maintained electric-power supply systems have a very high degree of reliability. However, accidental power interruptions do occur and some stand-by system must be provided to keep essential equipment running. Certain vessels, pipes, and sumps must be drained. Refrigeration or heating may be required to prevent spoilage. Emergency lighting is important to shut down the plant safely.

A diesel- or gas-engine-driven generator set is usually provided for this purpose. These internal-combustion plants are discussed in Perry's "Chemical Engineers' Handbook," 3d ed., pp. 1652–1655. For effective action in the event of a major power failure, automatic interlock with the emergency power generator is provided. Emergency fire equipment is usually driven by direct-connected gasoline engines.

Electric System Design for Chemical Plants

There are many factors which govern the design, installation, and maintenance of industrial-plant power systems. This section will give a brief summary of many of the problems and principles considered important by the electrical group in a chemical plant design organization. The major electrical design items include:

CHAP. 9] PROCESS AUXILIARIES 421

1. Power generation or purchased power substation and switching
    a. Cost factors—investment and operating costs
    b. Operating factors—reliability, safety, voltage regulation, joint use of lines and poles
2. Distribution systems—feeders, unit substations, transformers, switchgear, and overload protection
3. Power wiring for plant equipment—motors, heaters, furnaces, welders
4. Lighting equipment—inside and outside buildings, yards, roadways, and protective lighting
5. Electrical process control systems
6. Communication equipment—intercommunication, public telephone
7. Safety equipment—fire alarm, burglar alarm, lightning and other static arresters
8. Environmental factors—excessive temperature, corrosion, and explosion hazards

Codes and Sources of Design Information. Electrical design is governed by rules and regulations of such codes and organizations as:

1. American Institute of Electrical Engineers (AIEE), 33 W. 39th St., New York.

2. Association of Iron and Steel Engineers (AISE), Empire Building, Pittsburgh, Pa.

3. American Society for Testing Materials (ASTM), 1916 Race St., Philadelphia.

4. Edison Electrical Institute (EEI), 420 Lexington Ave., New York.

5. National Board of Fire Underwriters, 85 John St., New York. (Prepares the National Electric Code.)

6. National Bureau of Standards (NBS), Washington, D.C. (Handbook 14-32 in particular contains basic safety rules dealing with the installation and maintenance of electric supply and communication systems.)

7. National Electric Code (NEC), Cl-1959 edition, with editions published triannually; available through American Standards Association, 70 E. 45th St., New York.

8. National Electric Manufacturers Association (NEMA), 155 E. 44th St., New York.

9. National Safety Council (NSC), 20 Wacker Drive, Chicago.

10. Underwriters Laboratories, Inc. (UL), 207 E. Ohio St., Chicago.

11. Local or plant codes (check with local authorities and plant engineering group).

Codes are only the minimum requirements, and with these aids a good job can be done, but experience within the electrical design group is required for incorporating the best engineering practice.

Electrical Distribution Systems. Starting at the source of power supply, the basic electrical system includes transmission, substation, transformation, switching, distribution by feeders, the items of use, and protective safety equipment. Electrical flow sheets are prepared, using symbols such as those given in Fig. 9-23. Reference to NEC code for each system item is also included. Some important facts about the principal items in this industrial plant electric system will be discussed next.

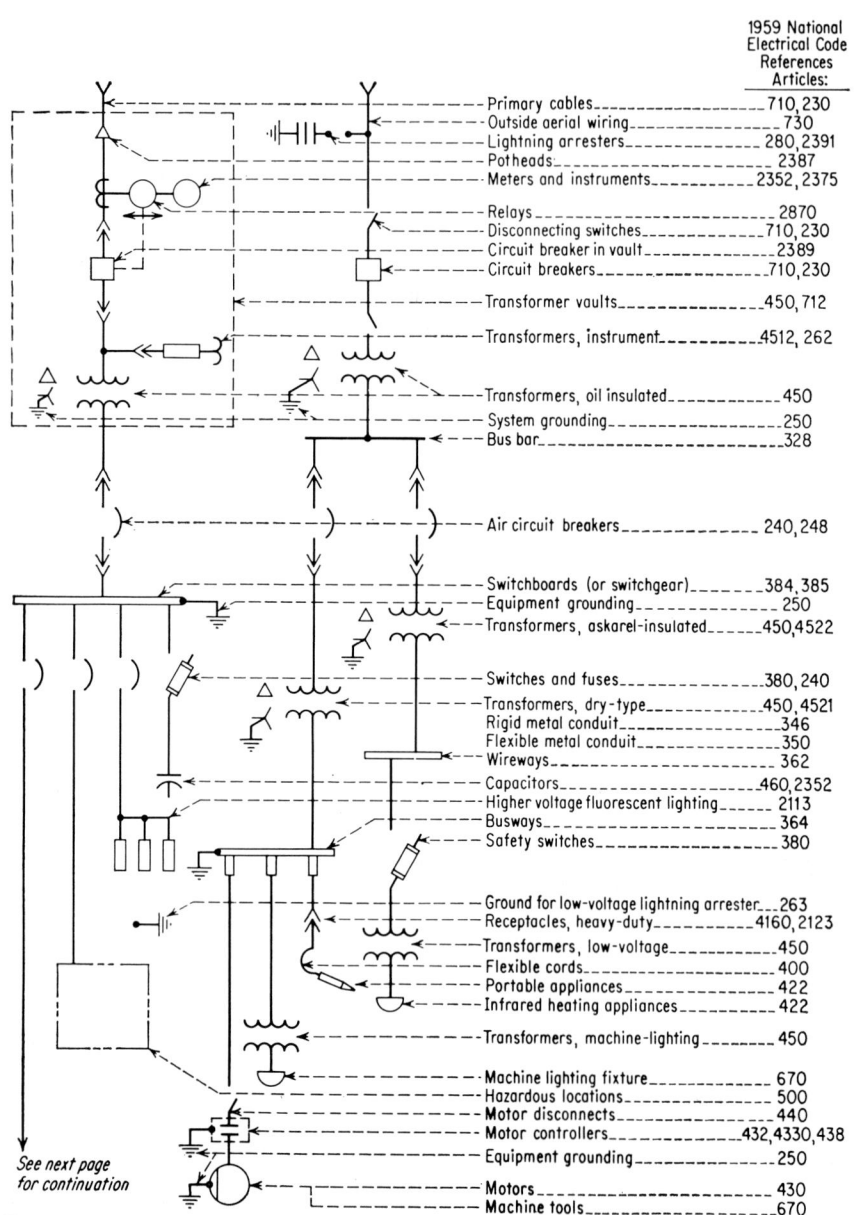

	1959 National Electrical Code References Articles:
Primary cables	710, 230
Outside aerial wiring	730
Lightning arresters	280, 2391
Potheads	2387
Meters and instruments	2352, 2375
Relays	2870
Disconnecting switches	710, 230
Circuit breaker in vault	2389
Circuit breakers	710, 230
Transformer vaults	450, 712
Transformers, instrument	4512, 262
Transformers, oil insulated	450
System grounding	250
Bus bar	328
Air circuit breakers	240, 248
Switchboards (or switchgear)	384, 385
Equipment grounding	250
Transformers, askarel-insulated	450, 4522
Switches and fuses	380, 240
Transformers, dry-type	450, 4521
Rigid metal conduit	346
Flexible metal conduit	350
Wireways	362
Capacitors	460, 2352
Higher voltage fluorescent lighting	2113
Busways	364
Safety switches	380
Ground for low-voltage lightning arrester	263
Receptacles, heavy-duty	4160, 2123
Transformers, low-voltage	450
Flexible cords	400
Portable appliances	422
Infrared heating appliances	422
Transformers, machine-lighting	450
Machine lighting fixture	670
Hazardous locations	500
Motor disconnects	440
Motor controllers	432, 4330, 438
Equipment grounding	250
Motors	430
Machine tools	670

See next page for continuation

FIG. 9-23. Graphical symbols for an industrial plant electrical distribution system. Book Company, Inc., 1950; revised by General Electric Co.]

Incoming Power Source. The characteristics of the primary power source from the main transmission line should be determined as follows:

1. Incoming line voltage and frequency. This is generally 60-cycle alternating current and 13,800 volts or higher.

2. Available short-circuit kva from the transmission line in order to size circuit breakers.

3. Main substation ownership and maintenance.

Internal Plant Power Distribution. The voltage for a distribution system is governed by several factors, including the plant area to be serviced, size of largest motors and other power equipment, voltage drop, and cost.

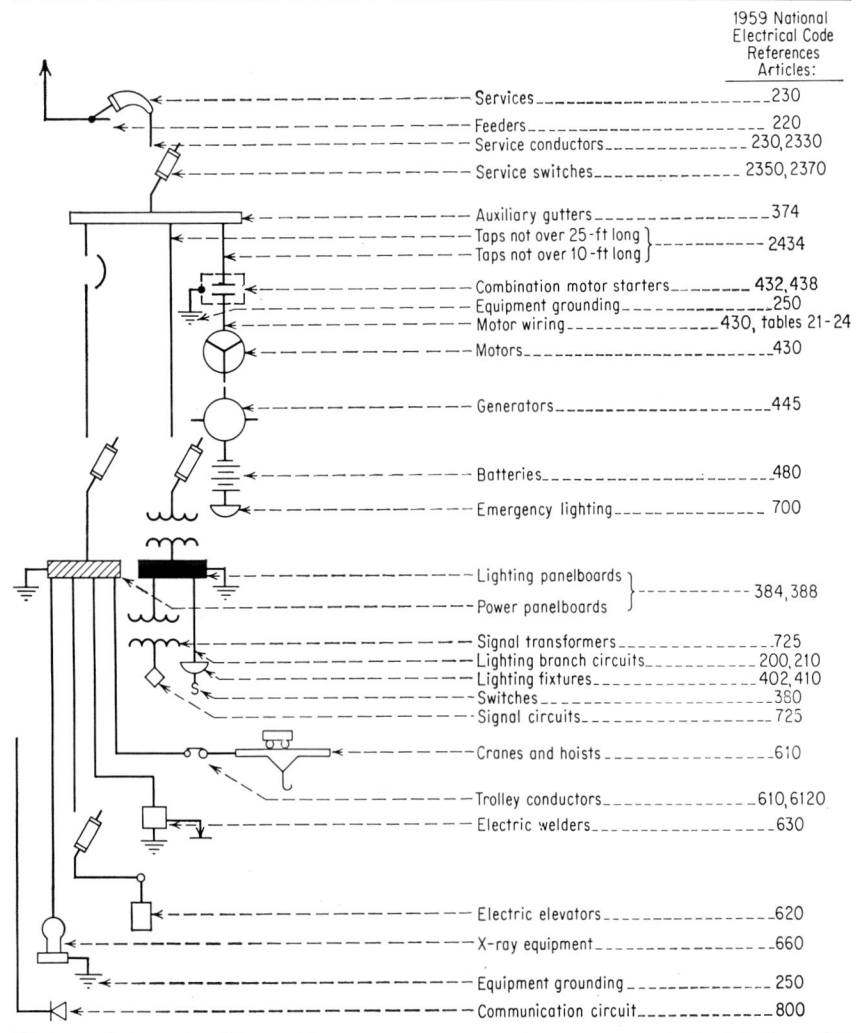

	1959 National Electrical Code References Articles:
Services	230
Feeders	220
Service conductors	230, 2330
Service switches	2350, 2370
Auxiliary gutters	374
Taps not over 25-ft long Taps not over 10-ft long	2434
Combination motor starters	432, 438
Equipment grounding	250
Motor wiring	430, tables 21–24
Motors	430
Generators	445
Batteries	480
Emergency lighting	700
Lighting panelboards Power panelboards	384, 388
Signal transformers	725
Lighting branch circuits	200, 210
Lighting fixtures	402, 410
Switches	380
Signal circuits	725
Cranes and hoists	610
Trolley conductors	610, 6120
Electric welders	630
Electric elevators	620
X-ray equipment	660
Equipment grounding	250
Communication circuit	800

[By permission from "Plant Engineering Handbook," W. Stainer (ed.), McGraw-Hill

Electrical equipment specifications are discussed in Chap. 4; in particular, see Fig. 4-5. Lighting requirements are given in Perry's "Chemical Engineers' Handbook," 3d ed., pp. 1755–1759. Voltages most commonly used are 13,800, 4,160, 2,300, 550, and 440 volts with secondary reduction to 220 or 110 volts for lighting and office fixtures.

Size of substations must be chosen with care. As size is reduced, more material is required for high-voltage feeders, transformers, and switches. However substations supplying too great an area will increase secondary circuitry costs. This is again an economic balance. Optimum power ratings for a three-phase substation are 400 to 600 kva with 208/120 or 230-volt secondaries and 600 to 1,000 kva with 460-volt secondaries.

The two general methods of running primary feeders to substations and secondary lines from there to equipment are *underground* and *overhead*. Underground distribution is generally recommended around chemical plants unless rocky ground makes the cost prohibitive or the water table is sufficiently high to flood permanently any below-ground installation. Underground feeders may be insulated neoprene, lead-covered or steel-armored cable of aluminum, or copper buried directly in the soil below the frost line. A group of cables can be run in a light-wall galvanized steel, concrete-asbestos, or compressed fiber pipe, colored red on the outside, and either buried, laid in a trench below ground, or rested on top of the ground. The best method is to use underground tunnels, accessible by manholes 100 to 250 ft apart, with sufficient room for two men to install the cable and make the necessary repairs. The advantages of the underground type of distribution are: (1) safe except to maintenance personnel under certain conditions, (2) reliable since it is not affected by severe weather conditions, and (3) does not produce an unsightly appearance.

The inherent disadvantages of the underground system are high first cost, inconvenience, and excessive cost of maintenance. *Overhead* construction is often used since it can be installed at about one-third the cost of the underground system, and repairs can be quickly made. However, such lines are exposed to impairment by weather, lightning, trucks, cranes, and corrosion, so that overhead construction should be avoided in congested areas where a reliable and safe electrical system is necessary.

Electrical Equipment Protection. Equipment must be constantly protected against excessively high currents from short circuits or faulty operation. Circuit breakers which open the circuit automatically on overload are sized and installed throughout the entire distribution system, including both primary and secondary systems. The design philosophy is to confine the interruption to the piece of equipment or circuit at fault, thereby avoiding interruption of an entire plant.

Starting devices, locally and remotely mounted to limit current overloads on power equipment, are standard installation. Manual or power disconnect switches are installed for isolating purposes when repairing or replacing equipment in various areas.

Safety Practices in Electrical Design. Damages and losses of equipment and personnel make adherence to codes a necessary requirement of design. The National Electric Code (NEC) is issued for the purpose of preventing loss of life and injury to personnel and preventing fire losses. Types of construction for electrical equipment have been dictated in this code by the degree of hazard in the surroundings and give rise to NEC area classification.

Area Classification

Class I. Hazards of combustible gas or vapor mixtures.
    Groups A through G: Order of decreasing explosion hazard atmospheres containing the following combustibles:
  Group A: Acetylene
  Group B: Hydrogen or manufactured gas
  Group C: Ethyl ether vapor
  Group D: Acetone, alcohol, gasoline, petroleum fractions, lacquer solvents, and similar vapors
  Group E: Metal dust
  Group F: Carbon black, coal or coke dust
  Group G: Grain dust
  Division 1. Location where flammable gases or vapors may exist under normal operating conditions, or under repair or maintenance conditions.
  Division 2. Location where flammable gases, vapors, or volatile liquids are handled in a *closed system* with suitable equipment, or where hazardous concentrations are normally prevented by positive mechanical ventilation; areas adjacent to Class I, Division 1, from which fumes might occasionally drift.
  Division 3. Locations where ignitable deposits of combustible finishes, such as paints, can accumulate.
Class II. Hazards of combustible dust-laden atmospheres
    Groups A through G: Same as under Class I
  Division 1. Similar to Class I, Division 1
  Division 2. Similar to Class I, Division 2
Class III. Hazards of low-ignition-point materials of construction and production
  Division 1. Manufactured areas where ignitable materials produce large surface to volume "flyings"
  Division 2. Storage areas for ignitable materials

Design for Hazardous Areas. This type of area classification tends to simplify the electrical design by directive means, but poses alternate choice problems, dictated by economics. The cost of an explosion-proof installation is as much as twice that using standard equipment. As an example of more costly electrical equipment, rigid metal conduit with threaded explosion-proof boxes and fittings are required for all Class I, Division 1, hazardous locations; either rigid metal conduit or electrical metal tubing is approved for both Class I and II, Division 2, locations.

In Class II, Division 1, locations, the wiring must be in rigid metal tubing with threaded boxes and fittings. Watertight equipment is required for damp locations; suitable materials of construction must be supplied for corrosive conditions. Everdur or aluminum is used for certain acid or alkali conditions.

Electric motor design for hazardous areas has been given in Chap. 4, and cost of motors can be found from Fig. 6-22.

Often explosion-proof equipment is not available; for example, it is difficult to buy explosion-proof motors greater than 250 hp as a stock item. One of the following methods can be used to provide low-cost protection in hazardous-area operations:

1. Remote location. Substations and switchgear can often be located outside the hazardous area. Lighting equipment can be placed outside the area, and the electric-light energy transmitted through skylights and windows. Power may be transmitted by extending sealed shafts into the hazardous area.

2. Ventilation in special rooms or chambers. Lower-cost standard motors are enclosed with light sheet metal and pressurized with uncontaminated air or other inert purge gas. A housing pressure of 1 to 2 lb is sufficient to exclude hazardous gas from the equipment enclosure. In some cases, positive-pressure ventilation of an entire room or building is recommended to allow use of less expensive standard electrical equipment throughout.

Grounding. The best-insulated and most carefully installed electrical system does not always stay perfectly insulated. Grounding, when properly installed, acts as a double safety: (1) personnel are protected from dangerous shock hazards because voltage to ground is not excessive, and (2) faults in circuitry give high current surges to ground, and rapidly open the overload switching. The two common types of grounding are permanent grounding of the distribution system and equipment grounds through the non-current-carrying metal parts. Metallic parts, which must be grounded according to code, include conduits or cable sheathing, switchgear, electric-motor frames, and similar items. Equipment placed on cement floors is not truly grounded and must have independent ground connections to water piping and the like.

Protection by Inaccessibility. Common sense dictates electric system protection, and one of the best means is to prevent unauthorized personnel from contact with electric systems, particularly systems with voltages greater than 110 volts. Outside substations are enclosed by chain fencing with barbed-wire topping, and warning signs are posted. Indoor substation rooms are also posted and locked, so that only electrical-maintenance men can have ready access. Switchgear panels are often handled in the same fashion. Color codes are generally black, yellow, and orange for overhead installations with red adopted for buried cable carrying conduit.

CHAPTER 10

Nuclear Chemical Plant Design

INTRODUCTION

Chemical engineering as applied to design in the nuclear field is basically the same as applied in other industrial realms; it is based on physical and chemical principles. Specialization within the scope of the nuclear field is termed "nuclear engineering," which involves the following problems of a chemical engineering nature: nuclear-fuel production starting from basic mineral depositions, enrichment via gaseous diffusion, nuclear-fuel "burning" and heat removal, chemical reprocessing and regeneration of spent nuclear fuel, fission-product utilization, and radioactive waste disposal.

To understand the factors which make nuclear chemical plant design different from the conventional chemical plant design discussed in previous chapters, two basic nuclear physics principles must be understood.[1] The first is the nuclear chain reaction:

Fission fuel + 1 neutron → fission fragments + 2–3 neutrons + energy

This simplified equation shows that more than one neutron is emitted for each one used in the fissioning of the fuel (U^{233}, U^{235}, or Pu^{239}). By proper design of a nuclear reactor or a chemical processing vessel containing fission fuel, the chain reaction can be controlled so that there is no net gain of neutrons available for fission with time.

The second principle of prime importance to the design engineer is that of radioactivity. In simplified form:

$$\text{Fuel and fission fragments} \rightarrow \begin{cases} \text{negative beta } (\beta^-) \text{ particles or electrons} \\ \text{positive beta } (\beta^+) \text{ particles or positrons} \\ \text{gamma rays } (\gamma) \\ \text{alpha particles } (\alpha) \text{ or helium nuclei} \\ \text{neutrons } (n) \end{cases}$$

[1] Nomenclature and nuclear constants are given on pages 469 and 470.

428   CHEMICAL ENGINEERING PLANT DESIGN   [CHAP. 10

As will be discussed under health physics, the energy represented by the products on the right-hand side can be injurious to both man and material.

Thus we see that the two novel plant design factors in nuclear chemical engineering are (1) protection of personnel and selection of materials of construction to avoid the injurious effects of radioactive energy; (2) inherent hazard of accidentally producing an uncontrolled nuclear chain reaction with the instantaneous release of vast quantities of energy and radioactivity. This chapter will emphasize the design principles involved in coping with the hazardous problems created in the nuclear engineering field.

HEALTH PHYSICS PRINCIPLES

The design engineer is responsible for protecting all people within and without the plant area from nuclear process hazards. In radiochemical processes, protection from radiation hazard is of prime importance. The human body may be exposed to radioactivity externally by direct bombardment of rays or particles or internally through breathing, ingestion, or cuts in the skin. Radiation hazard experience includes skin burns, skin cancers, eye cataracts, bone tumors, and fatalities from acute exposure. Other injurious effects reported are sterility and leukemia. Thus, it is necessary to employ good plant design practice and rigid monitoring by a plant health physics group after plant start-up.

Biological Effects

The exposure to radiation is the product of the absorbed *dose rate*, which is a rate of energy absorption, and the *exposure time*. Most of the injuries listed above are of the threshold type. The dosage received must exceed a minimum before any physiological effect is observed. Above this level, small dose rates for a long period of exposure time are less injurious than an equivalent total dosage comprised of a very-high-level dose rate for a much smaller period of time. From an engineering viewpoint, the dose rate must be expressed in quantitative units as discussed next.

Radiation Dose Units

The roentgen (r) is the unit most commonly used to define a unit of radiation energy. One *roentgen* is the quantity of X rays or gamma rays which will produce one electrostatic unit of charge as a result of ionization of 1 cm^3 of dry air at 0°C and 760 mm Hg pressure. This is equivalent to 83 ergs/g of air. The roentgen is a measure of the total energy absorption in air; as such the unit is independent of the other media, such as body tissue, in which the absorption may occur. The relationship between the roentgen and destruction of body tissue requires new units of definition.

The International Commission on Radiological Units has adopted a unit of absorbed radiation, called the *rad*, which represents 100 ergs/g of any energy absorbed by any material at the point of interest. For muscle and most soft tissue, ordinary X rays or gamma rays produce a local energy absorption per roentgen of 93 ergs/g of tissue. Under these special conditions, 1 r equals 0.93 rad. The difference between 0.93 rad and 1 rad is insignificant in specifying dosage levels for protection. Therefore, assume that 1 r of ordinary X rays or gamma rays produces a soft-tissue dose of 1 rad.

The physical damage to tissue is a function of the type of radiation, i.e., 100 ergs of X-ray absorption differs physiologically from 100 ergs of alpha

TABLE 10-1. PRACTICAL RELATIONSHIPS AMONG RADIATION UNITS

	Roentgen	Relative biological effectiveness	Radiation absorbed dose	Roentgen equivalent man
Accepted abbreviation............	r	RBE	rad	rem
Type of radiation:				
X and gamma rays..............	~ 1	1	1	1
Beta particles..................	~ 1	1	1	1
Fast neutrons...................	~ 1	10	1	10
Thermal neutrons...............	~ 1	5	1	5
Protons........................	~ 1	10	1	10
Alpha particles (internal ingestion).	~ 1	10–20	1	10–20

particles when absorbed in body tissue. A comparison has to be made on the basis of a quantity called the relative biological effectiveness (RBE), defined as

$$\text{RBE} = \frac{\text{physical dose of 200-kv X rays to produce desired effect}}{\text{physical dose of other radiation to produce same effect}}$$

Values of RBE for various types of radiation are listed in Table 10-1. The final unit of interest, the *rem* or *roentgen equivalent man*, is a true measure of the biological injury produced from various types of radiation. The dosage can be expressed in terms of units of total energy absorption multiplied by the relative biological effectiveness:

$$\text{Dose in rems} = \text{dose in rads} \times \text{RBE}$$

A summary of the relationships between the values of the r, rad, and rem for various types of radiation is given in Table 10-1.

Maximum Permissible External Exposure

To stay below the threshold levels for radiation damage, the National Committee on Radiation Protection and Measurement recommends the following standards for occupational conditions in controlled areas:

Accumulated Dose. The maximum permissible accumulated dose, in rem, at any age is equal to five times the number of years beyond age eighteen, provided no annual increment exceeds 15 rem. Thus, the accumulated MPE = $5(N - 18)$ rem, where N is the age and greater than 18. This amounts to an average of approximately 5 rem per year between ages eighteen and seventy.

Weekly Dose. The permissible weekly whole-body dose is 0.1 rem (100 mrem) or 2.5 mrem/hr on a 40-hr-week basis.

By restricting the exposure to certain parts of the body such as the skin of the hands or feet where the threshold levels for damage are greater, the MPE may be doubled. Most companies allow anywhere from 20 to 50 per cent of the 100 mrem per week value on a yearly integrated dosage basis for their personnel.

Emergency Dose. For planned short-time emergency exposures a total dose of 25 rem to the whole body is allowable, provided it occurs only once in a lifetime. Note that the lethal dosage for short-time exposure lies between 200 and 800 rem.

Maximum Permissible Internal Exposure

Radioactive material may enter the body by ingestion, by breathing, or through open wounds. Maximum permissible concentration (MPC) of various isotopes in air and drinking water have been set up by the International Commission on Radiological Protection. These are based on the average energy of radiation and an average dose rate internally of 0.1 rem per week. Threshold values of the dose are difficult to determine because of the variation of biological half life (time to reduce the radioactivity in the body to one-half its initial value). Table 10-2 lists MPC values for some isotopes likely to be encountered. Where mixtures of isotopes are encountered, multiply the MPC value for each isotope by its respective atom fraction.

Maximum Permissible Radiation Flux

The *flux* (ϕ) is the number of particles or photons flowing per unit time through a unit area perpendicular to the beam. Intensity (I) of radiation is the product of the flux and the energy (E) per particle. The dose rate (D) is the energy absorption per unit weight per unit time. It is now necessary to relate the flux and intensity terms with the biological tissue damage quantitatively through the dose rate in order to design biological shielding. The basic absorption law is

$$\phi = \phi_0 e^{-\mu x} = \phi_0 e^{-b} \tag{10-1}$$

where ϕ_0 is the flux at the front plane of the absorbing medium, ϕ is the flux at a distance x from the front plane, μ is the attenuation or absorption coefficient for the absorbing medium, and $b = \mu x$. Writing Eq. (10-1) in incremental form,

$$-\Delta\phi = \mu\phi\,\Delta x \tag{10-2}$$

Consider $\Delta x = 1/\rho$ for a unit gram mass of material with an exposed frontal plane area of 1 cm², where ρ is the density of the absorbing substance in grams per cubic centimeter. Since E is the energy in millions of electron volts (Mev) per particle or photon of radiation, then the rate of energy absorption or dose rate is

$$D \text{ in Mev/g-sec} = E\,\Delta\phi = \frac{\mu}{\rho} E\phi = \frac{\mu}{\rho} I \tag{10-3}$$

Application to Gamma Rays. One mrad $\cong$ 1 mr as shown previously. One mrad absorbed in the tissue liberates 0.1 erg/g. The mass absorption coefficient μ/ρ is 0.032 cm²/g for tissue in the energy range of 0.07 to 2 Mev. Substituting these values in Eq. (10-3) gives

$$D \text{ in mr/hr} \cong D \text{ in mrad/hr}$$

$$= \left(D \text{ in } \frac{\text{Mev}}{\text{g sec}}\right)\left(\frac{1 \text{ mrad}}{0.1 \text{ erg/g}}\right)\left(1.6 \times 10^{-6} \frac{\text{erg}}{\text{Mev}}\right)\left(\frac{3{,}600 \text{ sec}}{\text{hr}}\right)$$

$$= 5.76 \times 10^{-2} \frac{\mu}{\rho} E\phi = 1.86 \times 10^{-3} E\phi \tag{10-4}$$

If the maximum allowable dose rate for a 40-hr week is 100/40, or 2.5 mr/hr, then the maximum allowable flux is

$$\phi = \frac{2.5}{1.86 \times 10^{-3} E} = \frac{1{,}300}{E} \frac{\text{photons}}{\text{cm}^2\text{-sec}} \tag{10-5}$$

and the maximum allowable intensity is

$$I = E\phi \cong \frac{1{,}300 \text{ Mev}}{\text{cm}^2\text{-sec}} \tag{10-6}$$

Table 10-3 lists design values of flux and intensity for other types of radiation.

Radioactive Energy Sources

In previous sections, the absorption of energy in various media is discussed along with the resulting tissue damage. The source of this energy will be described next. A radioisotope is an element whose nucleus is unstable and undergoes a disintegration process in which energy is released in one or more forms of alpha or beta particles or gamma photons. The

TABLE 10-2. RADIATION PROPERTIES AND MAXIMUM PERMISSIBLE CONCENTRATION (MPC) OF ISOTOPES IN AIR AND WATER*

Substance	MPC† Air, μc/ft³	MPC† Water, μc/gal	Half-life $t_{1/2}$	Gamma (γ) rays Energy distribution, Mev (%)	y	E_{av}, Mev	Alpha (α) and beta (β) particles Energy distribution, Mev (% β unless noted)	y	E_{av}, Mev
Plutonium, Pu²³⁹ (soluble)	1.6 × 10⁻⁸	7 × 10⁻³	2.4 × 10⁴ yr	0.051 (11%), 0.013 (20%)	0.31	0.03	5.15 α	1.0	5.15
Plutonium, Pu²³⁹ (insoluble)	1.6 × 10⁻⁸		2.4 × 10⁴ yr	0.051 (11%), 0.013 (20%)	0.31	0.03	5.15 α	1.0	5.15
Uranium, natural (soluble)	2.5 × 10⁻⁷	1.3 × 10⁻¹				0.01	4.30 α	1.0	4.30
Uranium, natural (insoluble)	2.5 × 10⁻⁷					0.01	4.30 α	1.0	4.30
Uranium, U²³⁵			4.5 × 10⁹ yr	0.05 (22%)	0.22	0.05	4.18 α	1.0	4.18
Uranium, U²³⁸			7.1 × 10⁸ yr				4.5 α	1.0	4.5
Thorium, natural (soluble)	2.5 × 10⁻⁷	2 × 10⁻¹	1.4 × 10⁵ yr	0.094 (2%), 0.044 (15%)	0.17	0.18	4.82 α	1.0	4.82
Thorium, natural (insoluble)	2.5 × 10⁻⁷	7 × 10⁻⁴	1.4 × 10¹⁰ yr		0	0.050	4.10 α	1.0	4.10
Praseodymium, Pr¹⁴⁴			1.4 × 10¹⁰ yr	0	0		4.10 α	1.0	4.10
Cerium, Ce¹⁴⁴	1.6 × 10⁻⁵	10	17 min 280 days	2.18 (1%), 0.7 (1%) 0.134 (30%)	0.01 0.30	2.1 0.134	2.95 0.30 (70%), 0.17 (30%)	1.0 1.0	2.95 0.261
Praseodymium, Pr¹⁴³	1.6 × 10⁻³	100	13.7 days	0	0		0.93	1.0	0.93
Cerium, Ce¹⁴¹			33 days	0.145 (67%)	0.67	0.145	0.581 (33%), 0.422 (67%)	1.0	0.48
Lanthanum, La¹⁴⁰			40 hr	2.55 (4%), 1.6 (77%)	0.81	1.63	2.26 (10%), 1.67 (20%)	1.0	1.48
Barium, Ba¹⁴⁰	1.6 × 10⁻⁴	0.7	12.8 days	0.57 (30%), 0.50 (10%), 0.19 (60%)	1.0	0.335	1.32 (70%) 1.0 (75%), 0.4 (25%)	1.0	0.85
Cesium, Cs¹³⁷	1.6 × 10⁻³	1.6	33 yr	0.661 (92%) from 2.6-min Ba¹³⁷	0.92	0.661	1.17 (8%), 0.51 (92%)	1.0	0.56
Xenon, Xe¹³⁵	1.6 × 10⁻²	1.3	9.1 hr	0.25	1.0	0.25	0.93	1.0	0.93
Xenon, Xe¹³³	3 × 10⁻⁵	5	5.3 days	0.081	1.0	0.081	0.34	1.0	0.34
Iodine, I¹³¹	3 × 10⁻⁵	7 × 10⁻²	8.1 days	0.722 (2.8%), 0.637 (9.3%), 0.284 (87.2%), 0.163 (0.7%)	1.87	0.398	0.815 (0.7%) 0.608 (87.2%), 0.335 (9.3%), 0.250 (2.8%)	1.0	0.574
Tellurium, Te¹²⁹	3 × 10⁻⁴	12	70 min	0.080 (87.2%) 0.8	1.0	0.8	1.8	1.0	1.8
Rhodium, Rh¹⁰⁶	3 × 10⁻⁴	120	35 sec	2.42 (6%), 1.55 (3%), 1.14 (12%), 0.513 (11%)	0.32	1.17	3.55 (82%), 2.30 (18%)	1.0	3.34
Ruthenium, Ru¹⁰⁶		1.0 yr		0	0		0.039	1.0	0.039
Ruthenium, Ru¹⁰³	1.6 × 10⁻³	3	42 days	0.50	1.0	0.50	0.22	1.0	0.22
Niobium, Nb⁹⁵	7 × 10⁻⁴	300	35 days	0.745	1.0	0.745	0.15	1.0	0.15
Zirconium, Zr⁹⁵	7 × 10⁻⁵	50	65 days	0.721 (99%)	0.99	0.721	0.84 (1%), 0.371 (99%)	1.0	0.372
Yttrium, Y⁹¹	1.6 × 10⁻⁶	1 × 10⁻³	60 days	1.4 (<0.1%)	<0.001	1.41	1.54	1.0	1.54
Yttrium, Y⁹⁰			62 hr	0	0		2.20	1.0	2.20
Strontium, Sr⁹⁰	1.6 × 10⁻⁴	1 × 10⁻¹	19.9 yr	0	0		0.60	1.0	0.60
Strontium, Sr⁸⁹	7 × 10⁻³	25	54 days	0	0		1.48	1.0	1.48
Cobalt, Co⁶⁰	3 × 10⁻³	0.7	5.3 yr	1.33 and 1.17	2.0	1.25	0.31	1.0	0.31
Argon, Ar⁴¹	3 × 10⁻³		110 min	1.37 (99%)	0.99	1.37	2.5 (0.7%), 1.18 (99.3%)	1.0	1.48
Phosphorus, P³²	1.6 × 10⁻²	10	14.3 days	0	0		1.70	1.0	1.70
Sodium, Na²⁴	3 × 10⁻²	1,500	15 hr	2.76 and 1.38	2	2.07	1.39	1.0	1.39
Beryllium, Be⁷	6 × 10⁻²	250	53 days	0.479 (11%)	0.11	0.479	0	0	0
Tritium, H³			12.4 yr	0	0		0.018	1	0.018

* Based in part on recommendations of National Committee on Radiation Protection and Measurement, 1958.
† μc is a microcurie and represents 3.7 × 10⁴ disintegrations per second. For mixtures of isotopes, multiply the MPC value for each isotope by its respective atom fraction.
y = photon or particle yield per disintegration.

CHAP. 10] NUCLEAR CHEMICAL PLANT DESIGN 433

TABLE 10-3. DESIGN VALUES OF FLUX AND INTENSITIES FOR MAXIMUM PERMISSIBLE EXPOSURE (MPE)
$\left(\text{Basis: MPE} = 100 \text{ mrem/week; 40-hr work week} \atop E = \text{energy in Mev}\right)$

Type of radiation	MPE, mrad/week	Flux, ϕ, per cm²-sec	Intensity, $I = E\phi$, Mev/cm²-sec
X or gamma (γ)........	100	$1,300/E$ photons	1,300
Beta (β)...............	100	$32/E$ particles	32
Thermal neutron...........	20	650 neutrons	$650\,E$
Fast neutron ($E = 0.01$ to 10 Mev).............	10	$20/\sqrt{E}$ neutrons	$20\,\sqrt{E}$
Proton...................	10		
Alpha (internal only)......	5		

disintegration rate follows from the kinetic equation

$$\frac{-dN}{dt} = \lambda N \text{ disintegrations/unit time} \qquad (10\text{-}7)$$

where N = atoms of radioactive element
λ = decay constant
t = elapsed time when $N = N_0$

Equation (10-7) integrates to

$$N = N_0 e^{-\lambda t} \qquad (10\text{-}8)$$

A convenient measure of the rate of disintegration is the so-called half life ($t_{1/2}$), which is the time to reduce the concentration of the radioactive element to one-half its original value. Substituting in Eq. (10-8),

$$\frac{N}{N_0} = \frac{1}{2} = e^{-\lambda t_{1/2}} \qquad (10\text{-}9)$$

or

$$\lambda = \frac{0.693}{t_{1/2}} \qquad (10\text{-}10)$$

The half-life ($t_{1/2}$) values for some elements of interest are listed in Table 10-2.

It is sometimes desirable to determine the number of disintegrations per second coming from W grams of radioisotope of atomic weight A. Use of Eqs. (10-10) and (10-7) and Avogadro's number gives

$$\text{Decay rate} = (0.693)(6.03 \times 10^{23}) \frac{W}{A t_{1/2}} \quad \text{disintegrations/sec} \qquad (10\text{-}11)$$

The curie unit is in common use as a measure of disintegration rates. One curie is that quantity of any substance that gives 3.7×10^{10} disintegrations per second. Let S_c be the source strength in curies of W grams of an element of atomic weight A and half life $t_{1/2}$ in seconds. Then, by substitution in Eq. (10-11),

$$S_c = 1.13 \times 10^{13} \frac{W}{A t_{1/2}} \qquad (10\text{-}12)$$

Note that the definition of a curie involves disintegrations per second, not particles or photons per second. One must know the type of decay and the energy liberation per particle.

Example 10-1. One hundred disintegrations of I^{131} by negative beta (β^-)-particle decay produce statistically the following particles, photons, and energy (see Table 10-2):

Number	β^- particles, energy per particle, Mev	Number	Gamma photons, energy per photon, Mev
0.7	0.815	2.8	0.722
87.2	0.608	9.3	0.637
9.3	0.335	87.2	0.284
2.8	0.250	0.7	0.163
		87.2	0.080
100	$E_{av} = 0.574$ Mev	187.2	$E_{av} = 0.398$ Mev

$y = 1.0$ particle/disintegration $\qquad\qquad y = 1.87$ photons/disintegration

Typical emission data for isotopes can be found in Table 10-2, with a complete listing given in the Radiological Health Handbook, Sanitary Engineering Center, Cincinnati, Ohio, 1955.

From the example above, it is seen that the yield (y) of particles and photons per disintegration can be obtained as well as the accompanying average energy. These two variables can be inserted in Eq. (10-11) to give the source strength S_p in particles or photons per second or S_E in terms of Mev per second.

$$S_p = \frac{(0.693)(6.03 \times 10^{23})(Wy)}{t_{1/2} A} \qquad \text{particles or photons/sec} \qquad (10\text{-}13)$$

$$S_E = \frac{(0.693)(6.03 \times 10^{23})(W y E_{av})}{t_{1/2} A} \qquad \text{Mev/sec} \qquad (10\text{-}14)$$

or, with Eq. (10-12),

$$S_p = 3.7 \times 10^{10} y S_c \qquad \text{particles or photons/sec} \qquad (10\text{-}15)$$
$$S_E = 3.7 \times 10^{10} y E_{av} S_c \qquad \text{Mev/sec} \qquad (10\text{-}16)$$

To permit maximum permissible exposure (MPE) conditions or less (see Table 10-3), the source strength must be divided by an area term which is related to the space configuration between the source and the receiver. If the resulting *flux* or *intensity* is too large, biological shielding will have to be placed between the radiating source and the person or material being exposed. This brings up the very important subject of shielding design.

SHIELDING DESIGN

Chemical process plants handling radioisotopes of high-level activity will require shielding for alpha, beta, and gamma emission to avoid damage to personnel and equipment. The basic principles developed in this section will apply to all radiations but only the gamma-ray shield design is discussed in detail because the gamma rays are far more penetrating than the others. Hence, the common occurrence of β and γ radiation simultaneously requires no design for β shielding since any shielding material which will reduce the γ intensity to a safe level will automatically reduce the β intensity to a negligible value.

Plants handling heavy radioisotopes such as uranium have a problem of guarding against α-particle emission. Alpha particles of less than 7.5 Mev from stable isotopes do not have sufficient energy to penetrate the protective layer of body skin, thus requiring no external biological shielding. However, extreme precaution must be taken to prevent internal sorption as the heavy isotopes accumulate in certain sections of the body and cause serious tissue damage.

Gamma Shielding Design

The exponential law for absorbing radiation energy as given in Eq. (10-1) is not exact for gamma photons, principally because the scattering of some of the photons in the beam and subsequent return is not true adsorption. The attenuation of beam intensity is always less than predicted by Eq. (10-1) and this physical phenomenon is taken into account by modifying the exponential attenuation law to give

$$\phi = \phi_0 B e^{-\mu x} = \phi_0 B e^{-b} \tag{10-17}$$

The approximate build-up factor B for most elements up to 3 Mev is $1 + b$, but for lead, $B = 1 + b/2$. (See Fig. 10-3 for the relation between b and ϕ_0/ϕ.) In most cases, it can be assumed that $B = b$; this is known as linear build-up.

The absorption coefficient μ can be estimated from the fact that μ/ρ or mass absorption coefficient is a constant and equal to about 0.04 cm^2/g for elements from aluminum to uranium for 2-Mev photons. More exact values of μ/ρ as a function of photon energy are given in Fig. 10-1.

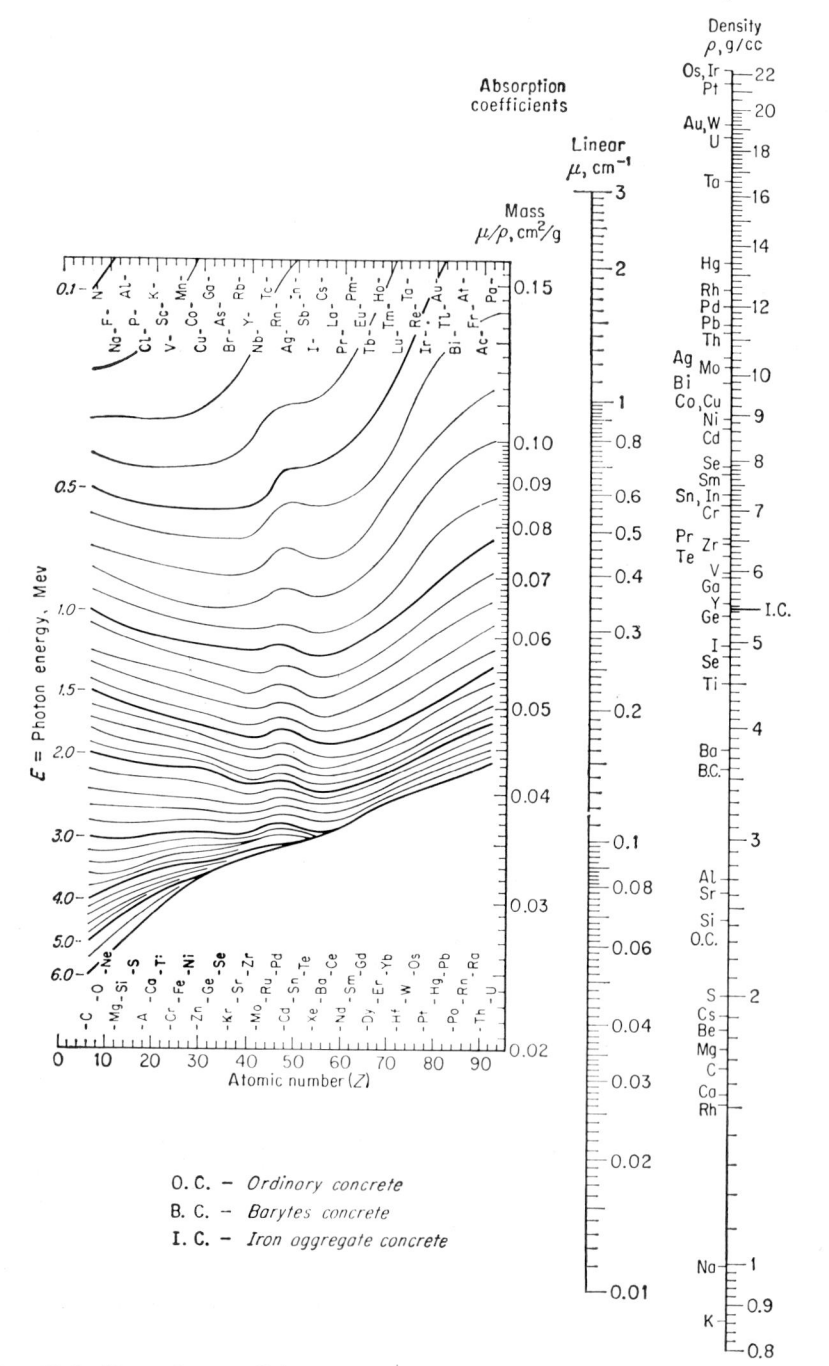

FIG. 10-1. Absorption coefficients for gamma-ray attenuation. [*Courtesy of D. G. Chappell, Nucleonics*, **14**(1): 40 (1956).]

O. C. — *Ordinary concrete*
B. C. — *Barytes concrete*
I. C. — *Iron aggregate concrete*

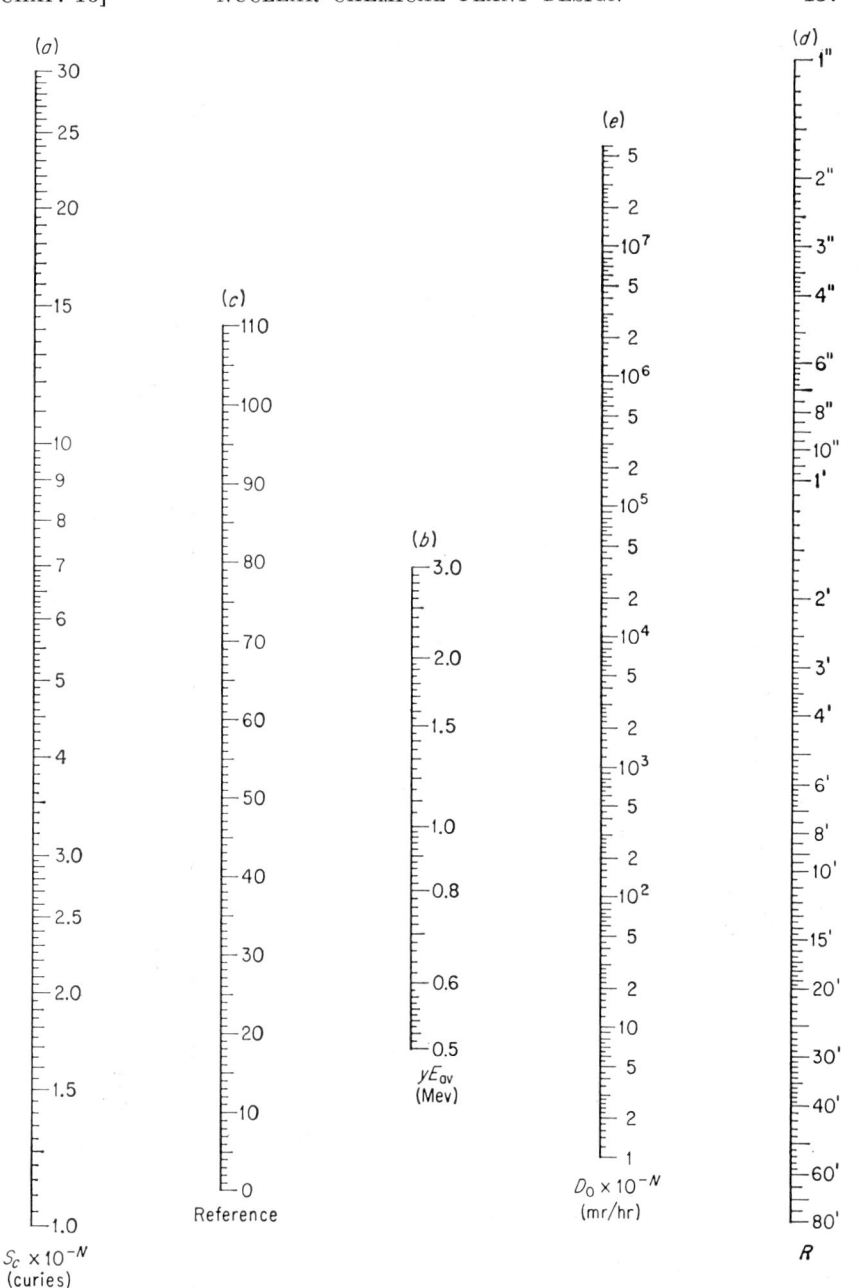

Fig. 10-2. Nomograph for point source–absorbed dose rate calculations (see Table 10-4).

438   CHEMICAL ENGINEERING PLANT DESIGN   [CHAP. 10

The flux ϕ can be converted to the health physics dose rate D by Eq. (10-4) if there were some method of relating the source strength S to the initial or frontal plane flux ϕ_0. Furthermore, if the area for absorption increases with distance as is the case for a spherical shield surrounding a

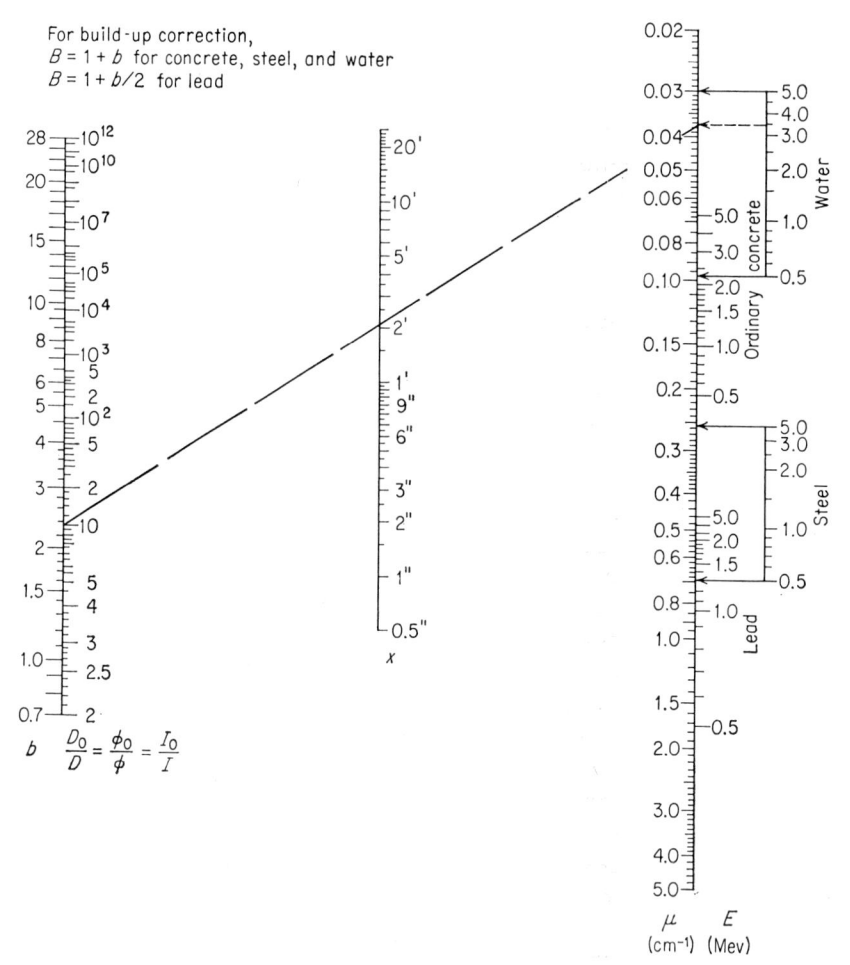

FIG. 10-3. Nomograph for shielding thickness determinations: point source or linear build-up cases (see Table 10-4).

point source or a concentric shield enclosing an internal volume source, then this area change has to be taken into account. If there is appreciable absorption of energy within the γ-ray source itself, this self-absorption complicates the picture. The complexity of such shielding calculations does not justify exhaustive treatment in a book of this scope. Modifications of the nomograph solutions of Balderston, Taylor, and

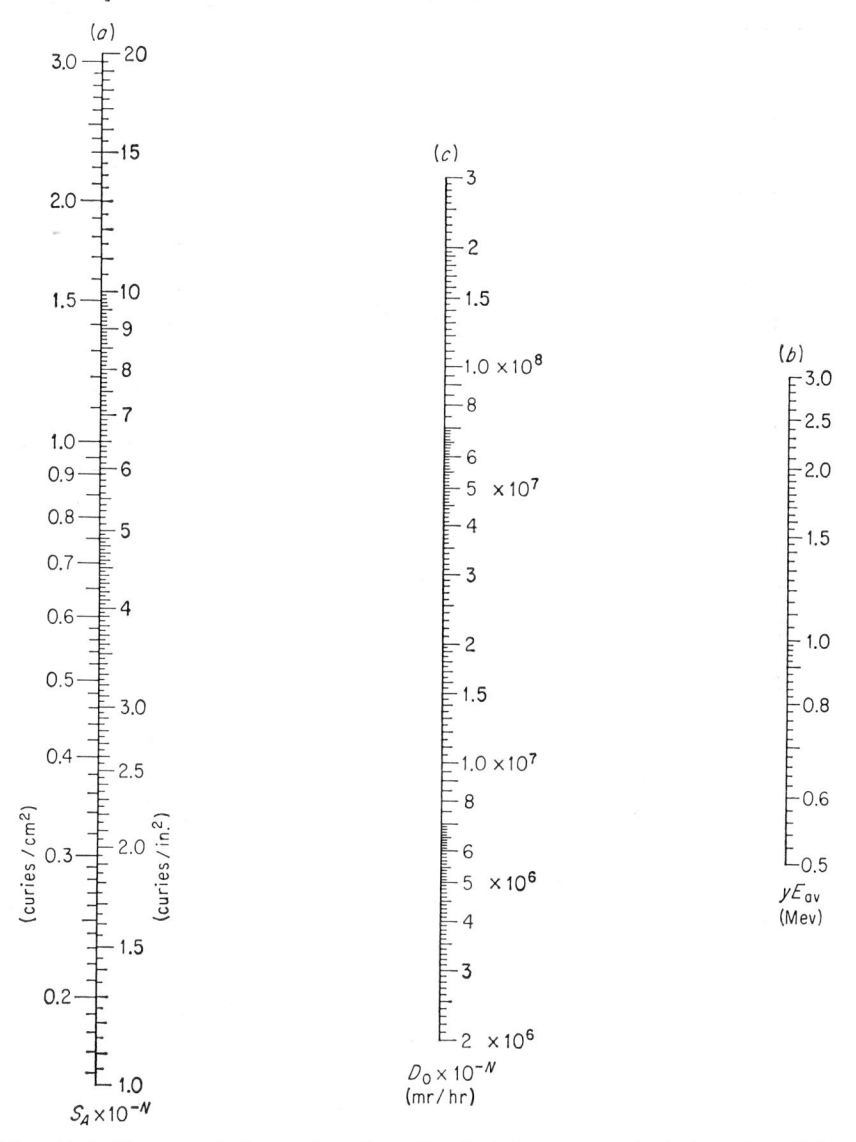

FIG. 10-4. Nomograph for surface flux–absorbed dose rate calculations (see Table 10-4).

Brucher[1] and of Chappell[2] are presented in Table 10-4 and Figs. 10-2 to 10-6 for four different cases which cover most of the geometrical configurations usually encountered in radiochemical plant design.

[1] J. L. Balderston, J. J. Taylor, and G. J. Brucher, AEC, TIS AECD-2934, 1948.
[2] David G. Chappell, *Nucleonics*, **14**(1): 40 (1956).

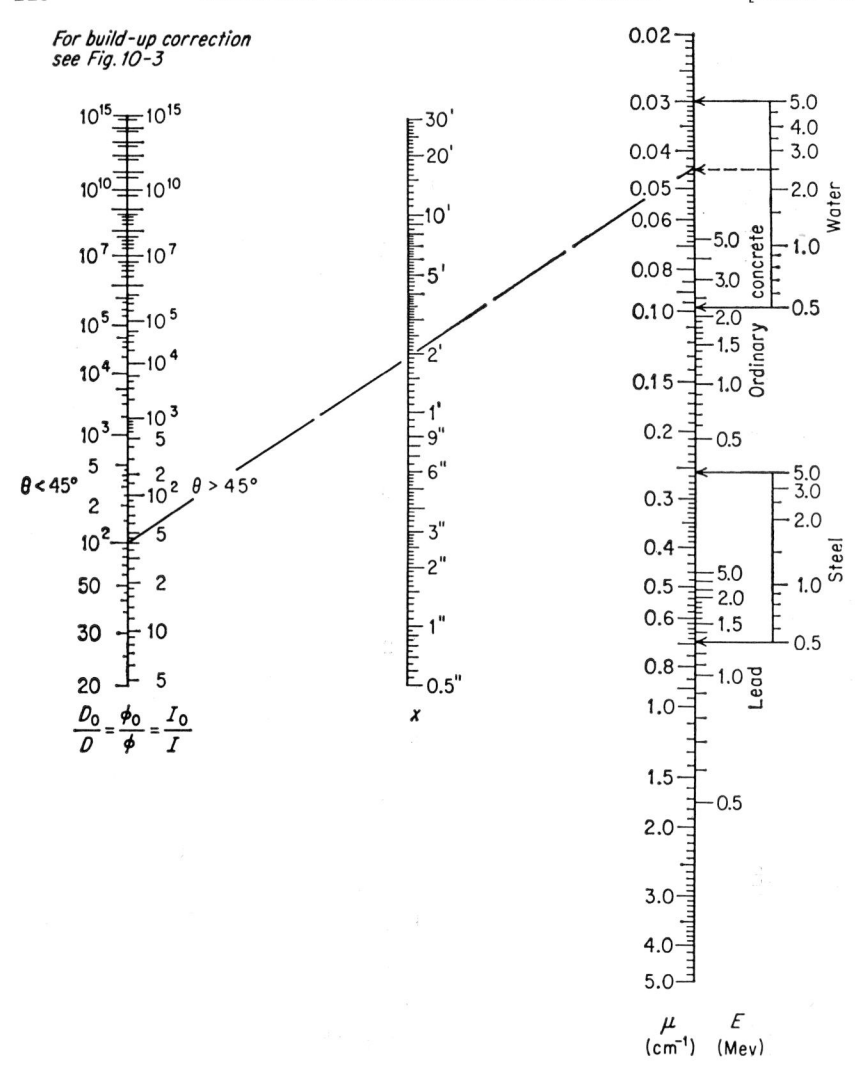

FIG. 10-5. Nomograph for shielding thickness determinations: unity ($B = 1$) and nonlinear ($B \neq b$) build-up cases (see Table 10-4).

Point Sources. (See Table 10-4, case 1.) Imagine a highly concentrated small-volume source emitting energy through a relatively nonabsorbent medium such as air. This is the no-shielding condition and the inverse-square law holds:

$$\phi = \frac{S_p}{4\pi x^2} \qquad (10\text{-}18)$$

where x is the distance from the point source to the surface where the dose

CHAP. 10]   NUCLEAR CHEMICAL PLANT DESIGN   441

rate is measured, known as the receptor plane. For the same geometry but with a gamma absorbing shield substituted for air,

$$\phi = \frac{S_p}{4\pi x^2} Be^{-b} \qquad (10\text{-}19)$$

These equations can be combined with Eqs. (10-4) and (10-15) to obtain

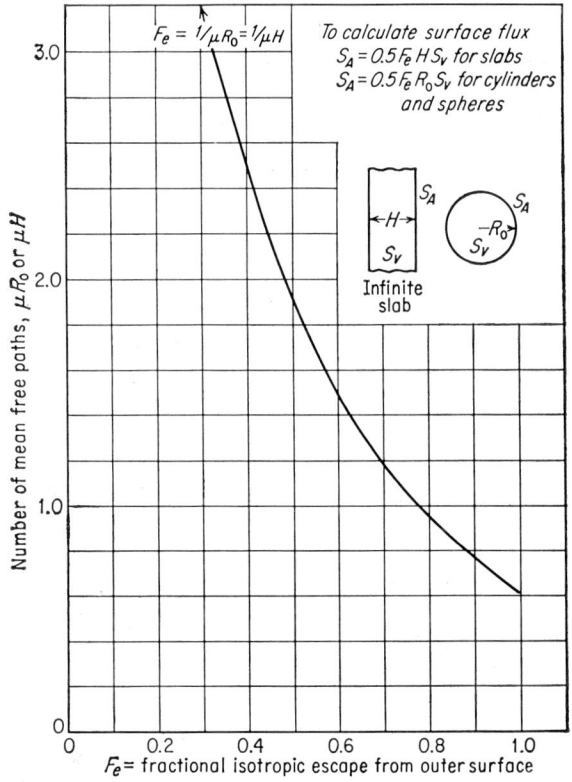

FIG. 10-6. Curve for determination of surface flux from volume source (see Table 10-4).

the dose rate for the nonshielding case:

$$D \text{ in mr/hr} = \frac{5.5 \times 10^6 S_c y E_{av}}{x^2} \qquad (10\text{-}20)$$

and for the shielding case:

$$D \text{ in mr/hr} = \frac{5.5 \times 10^6 S_c y E_{av}}{x^2} Be^{-b} \qquad (10\text{-}21)$$

These equations are plotted in nomograph form in Figs. 10-2 and 10-3 for estimating shielding or dose rate requirements. Note particularly that the shielding thickness is the same no matter whether the shield is

TABLE 10-4. SUMMARY OF SHIELDING DESIGN FOR DIFFERENT CONFIGURATIONS OF SOURCE AND RECEPTOR (Y)

Case 1. Approximate point source system, $\theta < 20°$ or $\tan \theta < 0.36$

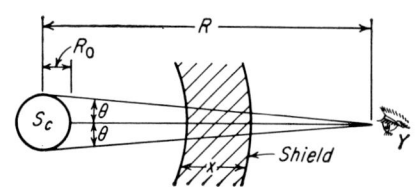

For all sources (planes, disks, lines, and small volumes) where θ is less than 20°. Figures to be used: 10-6 for self-absorption correction (multiply S_c by F_e), 10-2, and 10-3. Correct dose rate or flux ratios on left-hand scale of Fig. 10-3 as follows:

$$\frac{D_0}{D} \text{ (scale)} = \frac{D_0}{D} \text{ (actual)} \times B \text{ to obtain } x$$

$$\frac{D_0}{D} \text{ (actual)} = \frac{D_0}{D} \text{ (scale)} \times \frac{1}{B} \text{ to obtain } D \text{ at position } R, \text{ given } x \text{ and } D_0$$

Case 2. Volume source system, $20 < \theta < 45°$ or $0.36 < \tan \theta < 1$

Same as case 1, but with larger view angle θ. Shapes include slabs, short cylinders, polyhedrons, and spheres of finite volume.

Case 2l. Linear build-up ($B = b$)

Figures to be used: 10-6 for self-absorption, 10-4, and 10-3. Correct dose rate or flux ratio on left-hand scale of Fig. 10-3 as follows:

$$\frac{D_0}{D} \text{ (scale)} = \frac{D_0}{D} \text{ (actual)} \times \left(\frac{R_0}{R}\right) \text{ to obtain } x$$

$$\frac{D_0}{D} \text{ (actual)} = \frac{D_0}{D} \text{ (scale)} \times \left(\frac{R}{R_0}\right) \text{ to obtain } D \text{ at position } R, \text{ given } x \text{ and } D_0$$

Case 2b. Nonlinear build-up ($B \neq b$)

Same as case 2l except $B \neq b$.

Figures to be used: 10-6 for self-absorption correction, 10-4, and 10-5 (left-hand scale, $\theta < 45°$). Correct dose rate or flux ratio taken from left-hand scale of Fig. 10-5 as follows:

$$\frac{D_0}{D} \text{ (scale)} = \frac{D_0}{D} \text{ (actual)} \times B \times \left(\frac{R_0}{R}\right) \text{ to obtain } x$$

$$\frac{D_0}{D} \text{ (actual)} = \frac{D_0}{D} \text{ (scale)} \times \frac{1}{B} \times \left(\frac{R}{R_0}\right) \text{ to obtain } D \text{ at position } R, \text{ given } x \text{ and } D_0$$

Case 3. Long cylindrical or line sources, $\theta > 20°$ or $\tan \theta > 0.36$

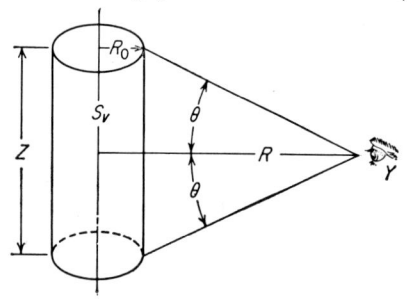

Case 3l. Linear build-up ($B = b$)
Identical to case 2l except use $(R_0/R)^{1/2}$ and $(R/R_0)^{1/2}$ in making D_0/D corrections.

Case 3b. Nonlinear build-up ($B \neq b$)
Identical to case 2b except: (1) left-hand scale of Fig. 10-5 should be used for $\theta > 45°$ or $\theta < 45°$, depending on geometry relations; (2) use $(R_0/R)^{1/2}$ or $(R/R_0)^{1/2}$ in making D_0/D corrections.

CHAP. 10]   NUCLEAR CHEMICAL PLANT DESIGN   443

TABLE 10-4. SUMMARY OF SHIELDING DESIGN FOR DIFFERENT CONFIGURATIONS OF SOURCE AND RECEPTOR (Y) (*Continued*)

Case 4. Planes or slabs, $\theta > 45°$ or $\tan \theta > 1$
Includes cylinders viewed endwise and large cubic sources.
  Case 4l. Linear build-up ($B = b$)
    Identical to case 2l except there is no D_0/D correction.
  Case 4b. Nonlinear build-up ($B \neq b$)
    Identical to case 2b except there is no R_0/R or R/R_0 correction for D_0/D.

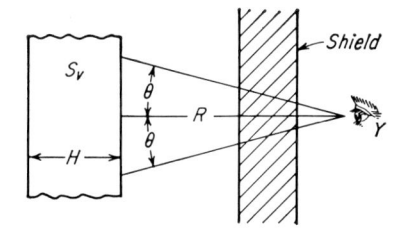

placed near the source or receptor. Minimum weight of cylindrical or spherical shielding is obtained by placing the shield close to the source.

Example 10-2. Design a water shield for a 1-kilocurie Co^{60} point source to allow a 2 mr/hr dose rate at the surface of the water.

Solution. (Use case 1.) Gamma rays of 1.17 and of 1.33 Mev energy are emitted per disintegration of Co^{60}; so $y = 2.0$ with $E_{av} = 1.25$ (see Table 10-2). From the statement of the problem, $S_c = 1,000$ curies, $D = 2$ mr/hr, $b = \mu x = \mu R$. Since $x = R$, the solution involves a trial-and-error calculation using Figs. 10-2 and 10-3. Starting with Fig. 10-2, $S_c \times 10^{-3} = 1.0$ and $yE_{av} = 2.50$, guess $R = 12$ ft and read $D_0 = 8.5 \times 10^1 \times 10^3 = 8.5 \times 10^4$. Therefore, $D_0/D = 8.5 \times 10^4/2 = 4.25 \times 10^4$. Using Fig. 10-3 with μ for H_2O at $E = 1.25$ Mev gives $x = 6.0 \neq R = 12$ ft. To estimate B, find $b = 11$ from left-hand scale when $D_0/D = 4.25 \times 10^4$. Then $B = 1 + b = 1 + 11 = 12$; so

$$\frac{D_0}{D} \text{(scale)} = \frac{D_0}{D} \text{(actual)} \times B$$
$$= (4.25 \times 10^4)(12) = 5.0 \times 10^5$$

This gives $x = 7.2$ ft when scattering is taken into account.

For a second trial, guess $R = 8.0$ ft, which gives $D_0/D = 1.9 \times 10^5/2 = 9.5 \times 10^4$ from Fig. 10-2. Using $B = 13$ from the left-hand scale of Fig. 10-3,

$$\frac{D_0}{D} \text{(scale)} = \frac{D_0}{D} \text{(actual)} \times B$$
$$= (9.5 \times 10^4)(13) = 1.23 \times 10^6$$

and $x = 7.5$ ft. This is close enough to 8 ft guessed for R; thus the design should call for 8 ft of water between the Co^{60} source and the surface of the water.

Cylindrical Sources. (See Table 10-4, cases 2 and 3). This is a commonly encountered problem in plant design involving storage tanks, extraction columns, and the designing of spent-reactor-fuel shipping containers. The case can best be discussed by an example.

Example 10-3. Cylindrical spent-fuel rods 0.5 in. in diameter and 7 ft long are removed after 3 yr from a reactor operating at a power level of 8,000 watts/lb fuel. After storage for 65 days, they are shipped in a lead-shielded, horizontally positioned cylindrical container, measuring 6.6 in. ID. The average density of the fuel including void volume is 7.84 g/cm³ with a mass absorption coefficient $\mu/\rho = 0.045$. The curie level after 65 days' cooling is 0.3 curie per operating power watt (Fig. 10-7b) and the average energy is 0.7 Mev. Calculate the thickness of the lead shielding required if the dose rate is to be 10 mr/hr at a distance 3 ft from the wall of the container.

Solution. Use case 3b because the configuration is a long right cylinder and lead shielding does not approximate linear build-up.

$$S_v = (0.3)(8,000)\left(\frac{7.84 \times 62.4}{1,728}\right) = 680 \text{ curies/in.}^3$$

$$\mu = \frac{\mu}{\rho}\rho = (0.045)(7.84) = 0.353 \text{ cm}^{-1} \qquad \mu R_0 = (0.353)(3.3 \times 2.54) = 2.95$$

From Fig. 10-6 for cylinders, $F_e = 0.330$.

$$S_A = 0.5 F_e S_v R_0 = (0.5)(0.330)(680)(3.3) = 370 \text{ curies/in.}^2$$

To obtain D_0, use Fig. 10-4 with $S_A = 3.7 \times 10^2$ and $yE_{av} = 0.7$ Mev.

$$D_0 = 1.53 \times 10^9 \text{ mr/hr}$$

$$\frac{D_0}{D} \text{ (actual)} = \frac{1.53 \times 10^9}{10} = 1.53 \times 10^8$$

In using Fig. 10-5, the shielding thickness x must be guessed to compute $\tan \theta$ and $(R_0/R)^{1/2}$ correction. Guess $x = 1$ ft. Then

$$\tan \theta = \frac{7/2}{3.3/12 + 1 + 3} = 0.81 \qquad \text{or} \qquad \theta = 39°$$

To obtain the build-up factor B, read 18 on left-hand side of Fig. 10-3 for D_0/D (actual) $= 1.53 \times 10^8$. For lead, $B = 1 + b/2 = 10$.

$$\frac{D_0}{D} \text{ (actual)} \times B \times \left(\frac{R_0}{R}\right)^{1/2} = (1.53 \times 10^8)(10)\left(\frac{3.3}{51.3}\right)^{1/2} = 3.87 \times 10^8 = \frac{D_0}{D} \text{ (scale)}$$

For a safety factor, use the absorption coefficient μ for lead at $E = 1.5$ Mev. From Fig. 10-5, read $x = 11$ in. If the less accurate linear build-up case (case 3l) had been used, D_0/D (scale) $= D_0/D$ (actual) $(R_0/R)^{1/2} = 3.87 \times 10^7$. Then using Fig. 10-3, read $x = 11.5$ in.

Example 10-4. A 500-gal spent-fuel dissolving tank is placed inside a concrete cell along with several holding tanks. This equipment is known as the "head-end" section of an aqueous spent-fuel processing plant. The dissolving tank measures 3 ft in diameter, is 10 ft high, and is wrapped with a 3-in.-thick layer of lead. The closest concrete wall is 1 ft from the outside of the tank. If the tank is full and holds a solution containing 400 curies/gal, calculate the thickness of the concrete wall required if the outside of the wall has a dose-rate tolerance of 10 mr/hr. The average energy of the radiation beam for dose-rate purposes is 0.85 Mev/curie and the shield design should allow a μ at $E = 2$ Mev for a safety factor.

Solution. Use case 3l since concrete and small thicknesses of lead are involved. $S_v = 400/231 = 1.73$ curies/in.³ From Fig. 10-1, $\mu = 0.049$ cm^{-1} for H$_2$O at $E = 2$ Mev. $\mu R_0 = (0.049)(18 \times 2.54) = 2.24$; $F_e = 0.44$ from Fig. 10-6. $S_A = 0.5 F_e S_v R_0 = (0.5)(0.44)(1.73)(18) = 6.85$ curies/in.² With this value of S_A and yE_{av} value of 0.85, the dose rate can be read from Fig. 10-4 as $D_0 = 3.4 \times 10^7$ mr/hr. To find the dose-rate ratio at the nearest distance to the inside concrete wall, use Fig. 10-3 with $E = 2$ Mev for lead. D_0/D (scale) $= 50$; $R =$ nearest distance to inside of concrete wall $= 33$ in.; $R_0 = 18$ in. (ID of tank, neglecting metal wall as shield). D_0/D (actual) $= D_0/D$ (scale) $\times (R/R_0)^{1/2} = 50 \times (33/18)^{1/2} = 67$. $D = 3.4 \times 10^7/67 = 5 \times 10^5$ at inside edge of concrete wall. Rather than guess the thickness of the concrete wall and use cylindrical attenuation, a good approximation can be made using the plane-source method (case 4l). D_0/D (actual) $= 5 \times 10^5/10 = 5 \times 10^4$. From Fig. 10-3, using $E = 2$ Mev for ordinary concrete, $x = 3.5$ ft. If this had been done by cylindrical attenuation (case 3l), guess $x = 3.3$ ft, or 40 in.

$$\frac{D_0}{D} \text{ (scale)} = \frac{D_0}{D} \text{ (actual)} \left(\frac{R_0}{R}\right)^{\frac{1}{2}} = (5 \times 10^4) \left(\frac{33}{40 + 33}\right)^{\frac{1}{2}} = 3.3 \times 10^4$$

From Fig. 10-3, using $E = 2$ Mev for ordinary concrete, $x = 3.2$ ft, which is close enough to the assumed value of 3.3 ft.

The cases and examples discussed above are typical of ones encountered in nuclear chemical plant design. More detailed analysis of shielding can be obtained by use of shielding manuals.[1,2] However, for most design work, the above-described procedures are adequate.

Heat Evolution in Shielding Design

From the discussion on absorption of radiation energy by shielding materials, it is recognized that the kinetic and electromagnetic energy must be changed to heat energy on absorption. Heat transfer from volume-heat-generating sources is sometimes an important part of shielding design. The thermal properties of the source or shield may be such as to cause damage to shield material and the source itself. For quantitative heat-transfer calculations, the rate of energy or power-rate liberation from the source material must be known.

Radiation Power Rate. For sources consisting of a known weight of a single radioisotope whose decay properties are known, the power in Mev/g-sec can be computed from Eq. (10-14) or (10-16). Conversion to watts is then made on the basis that 1 Mev/sec $= 1.6 \times 10^{-13}$ watt. The addition of other radioisotopes to the volume source requires a summation of the power value calculated for each isotope as described. The power liberation from a complex mixture of radioisotopes such as found in the fission products of U^{235} fuel is time-consuming to calculate. Figure 10-7 avoids the necessity of this by giving the β and γ power, curies and composition of the radioactive isotopes, all as a function of elapsed time after the fuel is pulled from the reactor. This elapsed time is known as the radiation *cooling period*.

Designing for Heat Dissipation. The radiation power rate must be handled by heat transfer through the walls of the container holding the isotopes in addition to the possibility of removing most of the heat by vaporization or forced convection of a fluid which surrounds or solubilizes the heat-generating material.

[1] "Reactor Handbook," vol. 1, Physics, Technical Information Service, AEC, February, 1955.
[2] T. Rockwell, "Reactor Shielding Design Manual," McGraw-Hill Book Company, Inc., New York, 1956.

Fig. 10-7. See legend on opposite page.

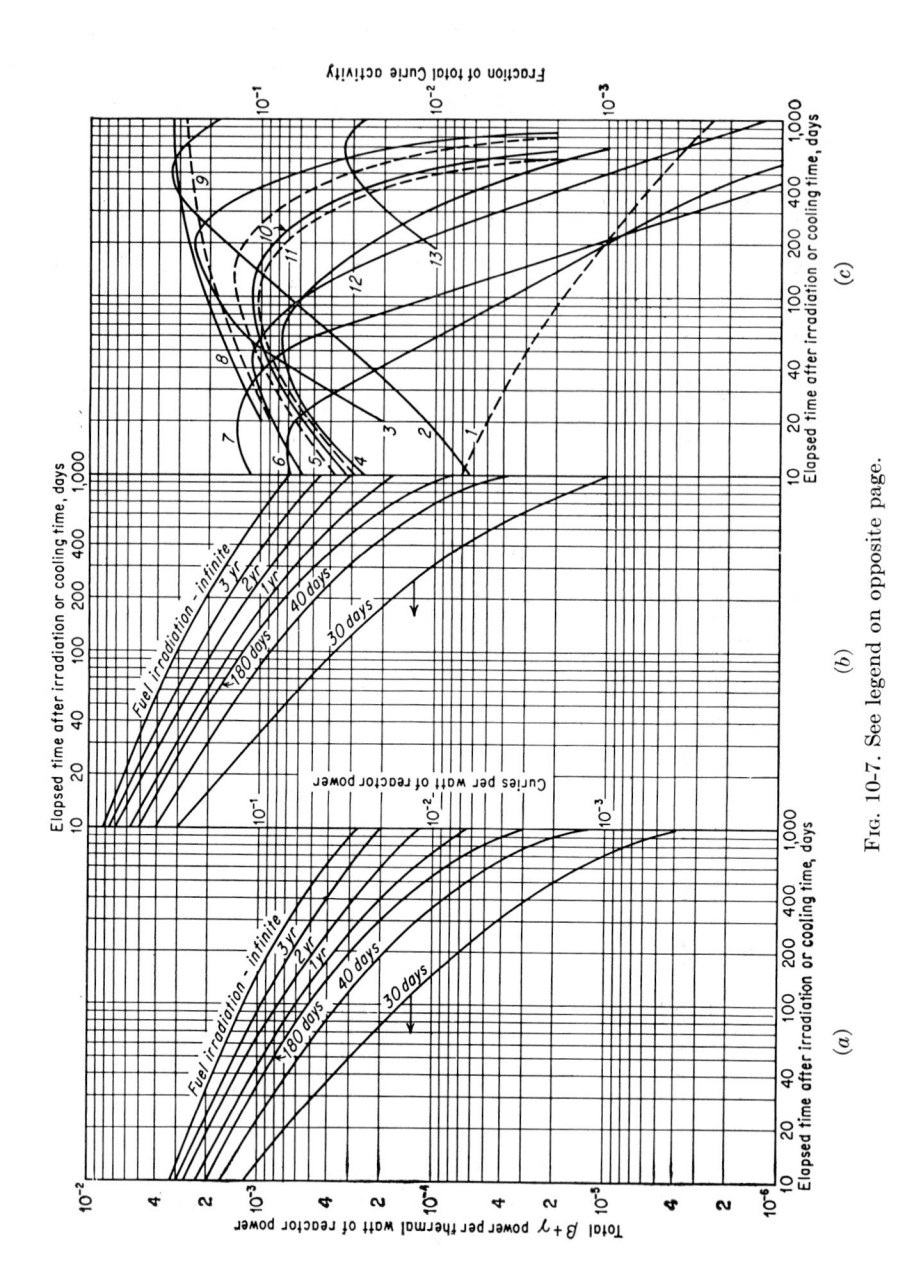

NUCLEAR CHEMICAL PLANT DESIGN

For heat dissipation only through the container wall, the exact solution of the design problem is complicated since there is heat release from the γ rays in logarithmic attenuation identical with their absorption in the shielding material. To avoid this complication, it is conservatively assumed that all of the γ power is released within the inner wall of the container and travels by conduction to the outside of the shield. The usual heat-transfer methods can then be employed.

For the cases where a major portion of the heat is to be removed by latent or sensible heat transfer to a fluid inside the container, a different method of design is employed. All the β power is transferred to the fluid but the γ power transfer depends on the γ absorption properties of the combined mixture of fluid and heat-generating substances within the vessel. Self-absorption calculations, using Fig. 10-6, are required. The design method is best illustrated by the solution of a typical problem.

Example 10-5. A chemical reprocessing plant accumulates waste solution in large storage tanks at the rate of 1,000 gal of aqueous wash per ton of fuel processed. A typical storage tank measures 9 ft in diameter and is 20 ft long. The fuel processed has a specific power rating of 5,000 watts of heat per pound of fuel over a 2-yr irradiation period. If the average density of the waste is 1.39 g/cm^3 and the tank contents represent 100 days of cooling, calculate (1) the Btu per hour heat evolution within the solution when the tank is 90 per cent full and (2) the pounds of water evaporated assuming all the internal heat is used for this process.

Solution. For the calculation of the internal heat liberation, from Fig. 10-7a at 100-day cooling and 2-yr irradiation, read 7.5×10^{-4} β-γ watts/watt of power $(7.5 \times 10^{-4})(5,000) = 3.75$ β-γ watts/lb fuel. For the calculation of F_e, the fraction of γ power which is absorbed within the solution, from the approximate mass absorption coefficient, $\mu/\rho = 0.05$, $\mu = (0.05)(1.39) = 0.0695$, $\mu R_0 = (0.0695)(54 \times 2.54) = 9.30$. From Fig. 10-6, $F_e = 1/\mu R_0 = 0.107$.

γ power absorbed in solution $= (3.75/2)(1 - 0.107) = 1.70$
β power absorbed in solution $= 3.75/2 = 1.88$
Total power liberated in solution $= 3.55$ watts/lb fuel
Volume of tank $= (0.785)(9)^2(20)(7.48) = 9,580$ gal
Pounds of fuel per gallon of waste $= 2,000/1,000 = 2$
Heat liberated at 90% of full storage capacity $= (9,580)(0.9)(2)(3.55 \times 3.415) = 209,000$ Btu/hr

Pounds of water evaporated $= \dfrac{209,000 \text{ Btu/hr}}{970 \text{ Btu/lb H}_2\text{O}} = 215$ lb H$_2$O evaporated per hour

---

FIG. 10-7. Composition and power emission from irradiated U^{235} fuel. (a) Power emission. Basis: 1 beta particle of 0.35 Mev per disintegration; 0.5 gamma photon of 0.7 Mev per disintegration; beta power = gamma power = one-half total power. (b) Curie emission. (c) Fractional curie emission; all beta emitters (except where noted γ, meaning beta *plus* gamma emission). Key to curves on (c): (1) Te-129γ; (2) Ce-144γ, Pr-144γ; (3) Nb-95γ; (4) Ru-103γ, Rh-103γ; (5) Zr-95γ; (6) I-131γ; (7) La-140γ, Ba-140γ, Pr-143; (8) Cs-137γ; (9) Sr-90, Y-90; (10) Y-91; (11) Sr-89; (12) Ce-141γ; (13) Ru-106γ, Rh-106γ. [*Courtesy of H. F. Hunter and N. E. Ballou, Nucleonics,* **9**(5): C-2 (1951); see also Table 10-2.]

CRITICALITY

This is the last important concept which must be introduced in order that the chemical engineer can have confidence in his ability to handle the hazardous design problems in the nuclear chemical engineering field. In the beginning of this chapter it was pointed out that the chain reaction with fission fuel can be controlled or prevented from occurring by proper design. What constitutes a proper design will be discussed now. The principles given apply to both chemical process vessel design and nuclear heat-generating reactors.

An uncontrolled or disastrous fission reaction occurs when an excess of neutrons, on the average, is produced with each generation of new neutrons formed. There are several basic interrelated steps which can be taken to prevent excess neutron formation.

1. *Surface Leakage of Neutrons.* The surface-to-volume ratio is controlled so that sufficient neutrons escape from the bare surface to avoid excess accumulation within the neutron-generating volume. The mass within the neutron-generating volume, generally defined in terms of the fission element only, becomes a *critical mass* just at the threshold of excess neutron accumulation. If the vessel is surrounded or jacketed by low-atomic-weight substances such as water or concrete for shielding purposes, carbon, graphite, or beryllium, these reflect or return neutrons to the fissioning volume and thereby create excess neutrons. This reactor is known as a reflected reactor and the critical mass of fission fuel and volume of the vessel is always lower than for an unreflected reactor.

2. *Absorption of Neutrons.* All materials within the neutron-generating volume of the vessel will remove excess neutrons by nonfission capture or absorption. The degree of effectiveness of removing neutrons is a fundamental property of the atom and is measured in terms of a cross section. Tables listing absorption cross sections for all the elements can be found in general reference books listed in the References for this chapter.

3. *Effect of Moderation.* Moderating materials are generally low-atomic-weight nuclei which elastically collide with high-energy neutrons and slow them down to energy levels where absorption probability is greatest. Water is a good moderator and the critical mass of a fission fuel is very sensitive to the percentage of water present. (This is often expressed as atom ratio of H/fission element.) The critical mass may decrease by a factor of 10 or more as the water content increases. However, on further dilution the critical mass will increase without limit owing to the absorption of neutrons by the excess water present.

It is thus seen that the exact critical mass of a fission element and the critical size of a process vessel depend on the nature of materials within

CHAP. 10] NUCLEAR CHEMICAL PLANT DESIGN 449

the vessel, the presence or absence of reflecting materials surrounding the vessel, and the geometry and construction layout of the vessel and its neighbors. Figures 10-8 to 10-11 give some of the relationships of these variables.

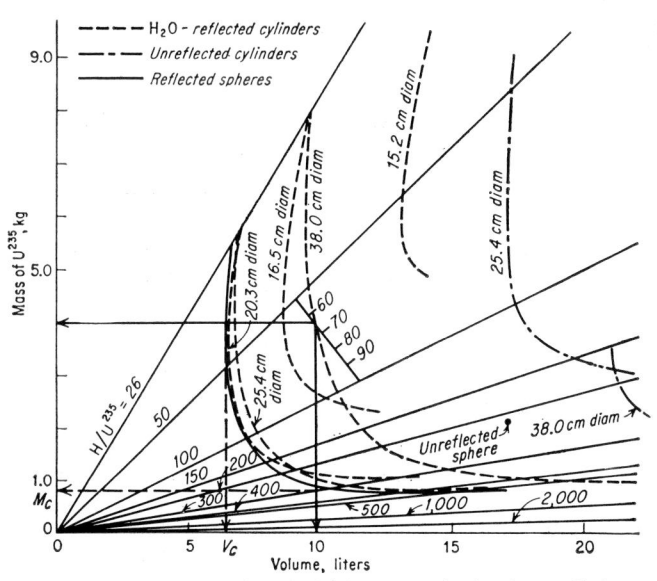

FIG. 10-8. Critical mass and volume of spheres and circular cylinders containing solutions of $U^{235}O_2F_2$ and surrounded by water. Straight lines show mass of U^{235} as function of volume for different concentrations (atomic ratio, H/U^{235}). Dashed curves intersect straight lines at critical masses for cylinders of indicated diameters. Envelope for dashed curves (solid curve at left) gives criticality parameters for spheres. Also shown are two similar curves for unreflected cylinders (no water) and a point that represents an unreflected critical sphere. [*Courtesy of D. Callihan, Nucleonics,* **14**(7): 40 (1956).]

Example 10-6. Estimate the critical mass of U^{235} and critical dimensions for a cylindrical vessel of 15 in. internal diameter completely immersed in a water shield. The vessel will be completely filled with a uranium solution containing an $H_2O:U^{235}$ weight ratio of 2.5.

Figure 10-8 will be used to solve this example.

$$\frac{H}{U^{235}} \text{ atom ratio} = \frac{(2)(2.5/18)}{(1/235)} = 65$$

Diameter of the cylindrical vessel = $15.0 \times 2.54 = 38.0$ cm. Interpolating between 50 and 100 for the H/U^{235} atom ratio of 65 on the H_2O reflected cylinder curve of 38.0 cm gives a critical mass of 4.0 kg of U^{235} and a critical vessel volume of 10 liters. The vessel height is therefore

$$h = \frac{10{,}000}{(0.7854)(38)^2} = 8.8 \text{ cm}$$

Example 10-7. Find the limiting value of critical mass and volume for U^{235} below which no chain reaction can accidentally start.

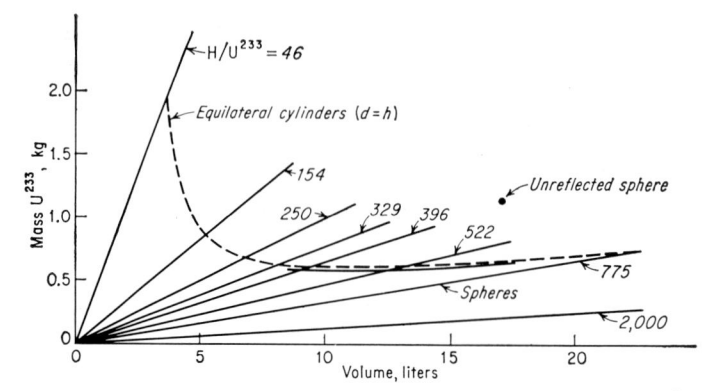

FIG. 10-9. Critical mass and volume of reflected spheres and equilateral cylinders ($d = h$) for U^{233} in aqueous solution. [*Courtesy of D. Callihan, Nucleonics,* **14**(7): 40 (1956).]

Figure 10-8 will be used to solve this example. A reflected sphere has the lowest values, so that the minimum values of the solid curve in Fig. 10-8 can be obtained. M_c is the limiting critical mass of 800 g and V_c is the limiting critical volume of 6.3 liters. These values also appear in Table 10-5.

TABLE 10-5. SAFE CRITICALITY LIMITS FOR AQUEOUS SOLUTIONS OF FISSION FUEL*

System	Restricted variable	U^{235}†	U^{233}	Pu^{239}
Any vessel, any dimension	Critical concentration, g/liter	11.6	10.9	7.8
Any vessel, any dimension	Critical mass, g	800	588	509
Spherical vessel	Critical volume, liters	6.3	3.5	5.0
	Diameter, cm	10.7	8.6	9.6
	in.	4.2	3.4	3.8
Cylindrical vessel	Diameter, cm	12.7	10.1	13.7
	in.	5.0	4.0	5.4
Slab-shaped vessel	Thickness, cm	3.5	1.2	4.8
	in.	1.4	0.5	1.9

Critical mass. Stay below this value. Divide by a safety factor of 2 to allow for process and analysis errors.

Critical volume. Stay below this value. Divide by a safety factor of 1.3.

Critical cylinder diameter. Safety is imposed by diameter alone; the vessel may be of any length and contain any mass of fission element and be surrounded by a reflector.

Critical slab thickness. Slab-type vessels of infinite extent having the above-estimated thickness values will be subcritical when reflected by water.

Critical concentration. Stay below these concentrations.

* Abstracted in part from Dixon Callihan, *Nucleonics,* **14**(7): 40 (1956).

† Approximately 90 per cent U^{235}, 10 per cent U^{238}.

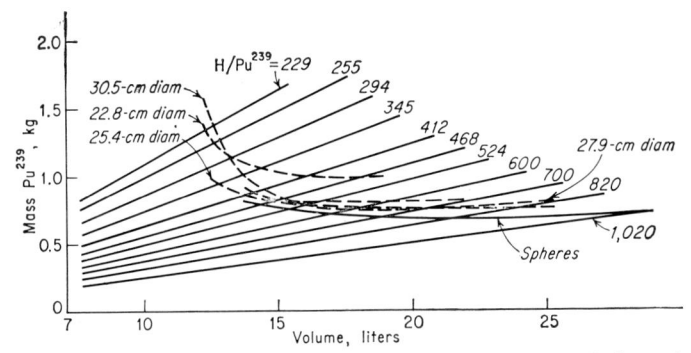

FIG. 10-10. Critical mass and volume of reflected spheres and cylinders of Pu^{239} in aqueous solution. [*Courtesy of D. Callihan, Nucleonics,* **14**(7): 40 (1956).]

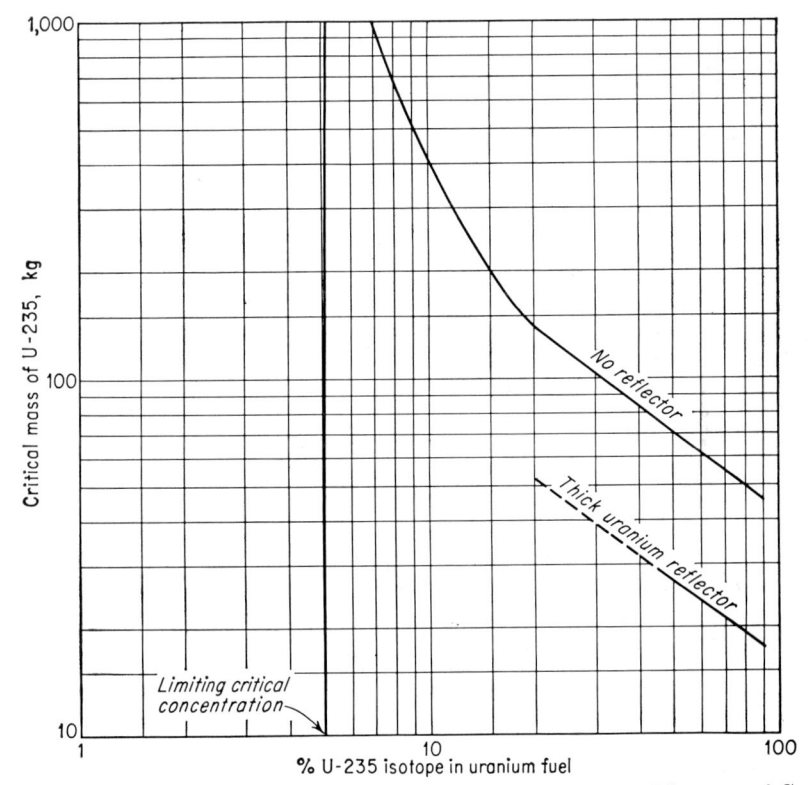

FIG. 10-11. Critical mass values for spheres of uranium metal. [*Courtesy of G. A. Graves and H. C. Paxton, Nucleonics,* **15**(6): 91 (1957).]

Table 10-5 gives the design limitations on mass, volume, container dimensions, and fuel concentration, any one of which will prohibit an accident regardless of other conditions. These values are for aqueous-type processing plants.

Dilution of U^{235} with U^{238} does not greatly affect the critical mass until the composition becomes greater than 95 per cent U^{238}. At that enrichment, the minimum mass is 1.9 kg of U^{235} and the volume of the reflected vessel is 30 liters. Further increase in U^{238} concentration increases these critical values markedly. It is impossible to make a homogeneous H_2O solution of natural uranium critical.

When fabrication and handling of *metal* nuclear fuels is contemplated, a different type of design curve is required. Moderation of fission neutrons is absent and conditions for fast-neutron fission exists. Critical values are larger under these conditions (Fig. 10-11).

Nuclear Safety Design[1]

It is possible to construct a fuel-processing plant or shipping containers in which all components are so sized that no critical accumulation can occur regardless of the quantity or concentration of material in process or the proximity of neutron-reflecting bodies. Design data would be taken from the critical-dimension data of Table 10-5. If this design is practiced, an uneconomical plant design will result, as it specifies a large number of small reactors spaced far apart and all contained in individual concrete vaults or cells. As another criterion, the mass in the process at any time can be maintained below the minimum critical mass with all other design variables economically optimized. This design will undoubtedly lead to very dilute solutions and large-size process equipment.

A practical approach is to avoid uneconomical extremes and yet be entirely safe. For example, the concentration may be low enough to permit relaxation of the stringent critical value listed in Table 10-5 by use of the experimental data plotted in Figs. 10-8 to 10-11. As another example, some solutions are automatically protected against overconcentration since some alloying agent, such as aluminum, would precipitate to a large extent before any uranium would, thereby plugging up the equipment and giving adequate warning.

PROCESS DESIGN

Chemical processes in the nuclear field relate to reactor fuels and radioactive isotopes (see Fig. 10-12). A classification can be set up as follows:

[1] N. Ketzlach, Nuclear Safety Considerations in Reprocessing Plant Design, *Chem. Eng. Progr.*, **53**(7): 357 (1957).

1. Raw materials separation and preparation of nuclear fuel
2. Nuclear fuel "burning" and heat removal
3. Spent-fuel reprocessing
4. Utilization of fission products
5. Disposal of radioactive wastes

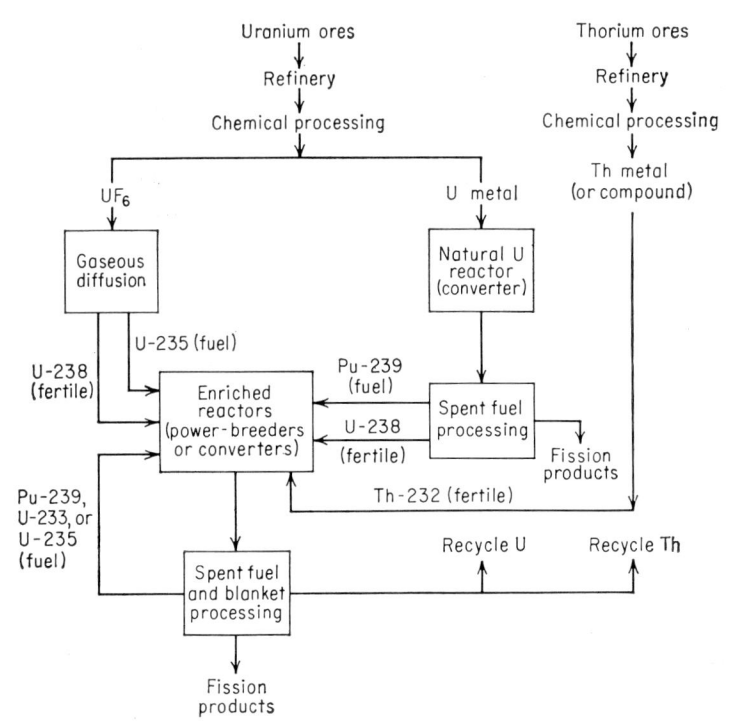

Fig. 10-12. Reactor fuel cycles. (*Courtesy of S. Glasstone, "Principles of Nuclear Reactor Engineering," p. 475, D. Van Nostrand Company, Inc., Princeton, N.J., 1955.*)

Nicholls and White[1] have stated the following essential differences between conventional chemical processes and those handling radiochemical materials under the listing above:

1. Limitations are imposed by toxic and radiation hazards and also by criticality considerations.
2. Radioactive materials such as exposed fuel elements have continually changing composition because of radioactive decay.

[1] C. M. Nicholls and A. S. White, *Chem. Eng. Progr. Symposium Ser.*, **50**(13): 129 (1954).
C. M. Nicholls, *Proc. Intern. Conf. Peaceful Uses of Atomic Energy*, Geneva, **9**: 453 (1955).

3. The systems are frequently complex and involve elements whose properties and chemical behavior are little known.

4. Materials used in fuel elements have a very high economic value.

In making a process selection, it is necessary to consider the usual factors of (a) simplicity of design and operation, (b) purity of end products, (c) safety including consideration of toxicity, flammability, chemical stability, and reactivity, (d) nonseverity from a corrosion and materials selection standpoint, (e) waste-disposal problems, and (f) economic justification. To this list must be added the factors unique to nuclear chemical processes which include (a) radiation stability of materials of construction and process reactants and products, (b) radiation safety for personnel including provision for remote or direct maintenance and decontamination of all equipment, and (c) processing and disposal of radioactive by-products or wastes. The fundamentals for handling design problems of this unique nature have been given in the previous sections on health physics, shielding, and criticality. The discussion of nuclear chemical processes which follows will be mainly to orient the reader in this field; detailed references are listed for those who desire specialized knowledge of one or more specific processes.

Raw Materials Separation and Preparation of Nuclear Fuels

All naturally occurring nuclear fuel requires concentration from basic mineral deposits by extractive metallurgical procedures. The bulk of these deposits contain uranium and thorium in concentrations under 1 per cent. Preliminary concentration, employing physical ore dressing or chemical leaching methods, is done near the mine site. Concentrated ores are then shipped to purification plants where the fuel in the form of metal or one of its chemical compounds is produced.[1,2] In cases where U^{235} is required in concentrations greater than that existing in natural uranium, it is necessary to prepare the hexafluoride salt and process it in a multistage gaseous-diffusion plant.[3] Use of the fuel in reactors other than the solution or slurry type requires special fabrication procedures to produce the shapes needed for maximum heat-transfer duty.[4] These processes are usually characterized by lack of gamma radiation hazards, but alpha radiation and the deadly toxicity of any plutonium which may be incorporated in the fuel-fabrication process require rigid health physics monitoring. Gamma radiation hazards will be encountered where

---

[1] J. W. Clegg and D. Foley (eds.), "Uranium Ore Processing," Addison-Wesley Publishing Company, Reading, Mass., 1958.

[2] W. E. Kelley, *Nucleonics*, **13**(11): 68 (1955).

[3] M. Benedict and T. Pigford, "Nuclear Chemical Engineering," chap. 10, McGraw-Hill Book Company, Inc., New York, 1957.

[4] Editorial Staff, *Nucleonics*, **13**(9): 54 (1955).

partially decontaminated fuel is to be handled by remote fabrication procedure.

Generation of Process Heat In Nuclear Reactors[1,2]

The pattern of consumption of conventional energy in the United States shows that only 20 per cent of the heat is converted to electrical energy and the balance to industrial heat, individual home space heating, and transportation. Product manufacturing by industry accounts for about 30 per cent of this nonelectrical energy consumption. The substitution of nuclear fuel for fossil fuel in this latter area is now prevalent.

Industrial process heat can be classified on the basis of heat requirements.

1. Chemical reaction heat (1500°F or higher)
    a. Preheat reactants to operating temperature
    b. Supply endothermic reaction heat for such processes as nitrogen fixation from air, coal gasification, catalytic re-forming of natural gas and petroleum, and thermal cracking
2. Process steam
    a. Low-pressure steam for evaporation and low-temperature heat requirements (50 to 150 psig and temperatures up to 400°F)
    b. High-pressure steam for special industrial process work

Many of the design problems for this type of work depend on an understanding of nuclear reactor design. This subject is beyond the scope of this text and the reader is referred to several suitable books in the Addition Selected References section.

Spent or Irradiated Fuel Reprocessing

The main objective of processes listed under this heading is to separate completely the highly radioactive fission products which act as neutron-absorbing ashes or poisons from the nuclear fuel so that the latter can be returned for further burnup. Process selection requires a consideration of these items:

1. Highly effective removal of fission products with decontamination factors as high as 10^6 to 10^7. The fission-product decontamination factor is defined as the ratio of fission-product activity per unit fuel weight entering the plant to that leaving the plant. This degree of decontamination is necessary if direct fabrication methods are used to prepare the recycle fuel. The successful development of remote fabrication methods would alleviate the requirement for a high decontamination factor.

2. High percentage recovery (99 per cent or better in most cases) of fission fuel based on its very high replacement cost.

[1] B. W. Gamson, *Chem. Eng. Progr.*, **54**(2): 74 (1958).
[2] Staff Report, *Nucleonics*, **16**(2): 62 (1958).

3. Selection of chemical processing materials which are not prone to irradiation damage and do not have objectionable corrosion properties. This latter point is very important since servicing, inspection, and replacement of plant equipment containing high levels of radioactivity is a very costly and dangerous procedure.

4. Low volume holdup to reduce shielded cell sizes and radiation damage to solvents as well as to avoid criticality conditions. Continuous rather than batch processing would favor this situation.

The various processes available for fuel reprocessing are aqueous solvent extraction, precipitation, ion exchange, fractional distillation, pyrometallurgy, and fluoride volatility.[1,2] Most of the commercial development experience has come from the solvent-extraction method for separation of uranium, plutonium, and fission products.

Utilization of Fission Products

Once the radioactive fission products are isolated by one of the separation processes, the major problem in the nuclear chemical industry must be faced since radioactivity cannot be immediately destroyed (see Fig. 10-7c for curie level of fission-product isotopes versus elapsed time after removal from the neutron source). This source of radiation energy can be employed in the food-processing industries for sterilization and in the chemical industries for such processes as hydrogenation, chlorination, isomerization, and polymerization. Design of radiation facilities to economically employ spent reactor fuel elements, composite or individually isolated fission products such as cesium 137, is one of the problems facing the design engineer in the nuclear field.

Disposal of Radioactive Wastes

Radioactive wastes come directly from nuclear-reactor-fuel reprocessing plants and from industries employing radioactivity for processing work. The dominating elements from nuclear reactor fuels are cesium 137 and strontium 90, with the latter the controlling isotope owing to low permissible concentration values (Table 10-2). Rodger[3] cites an example to illustrate the severity of the problem. In the year A.D. 2000 the installed reactor capacity on a world-wide basis is predicted to be 2.2×10^6 Mw. If this system is operated for 50 years, the Sr^{90} steady-state level (rate of production = rate of decay) would be 8.6×10^{11} curies, which would require 5 per cent of the entire world ocean volume to dilute to the maxi-

---

[1] "Processing Irradiated Fuels and Radioactive Materials," *Proc. Intern. Conf. Peaceful Uses Atomic Energy, Geneva;* 1st conf., vol. 9 (1956); 2d conf., vol. 17 (1959), United Nations, New York.

[2] E. L. Anderson, *Nucleonics,* **15**(10): 72, **15**(12): 44 (1957).

[3] W. A. Rodger, *Chem. Eng. Progr.,* **50**(5): 263 (1954).

mum permissible concentration. Furthermore, commercial utilization of the radioactivity of the products for a period of time during their early life does not in any way change the problem of final disposal.

Since natural decay and dilution processes do not appear feasible, the only alternative is to reduce the volume of waste to the minimum allowable based on β and γ heat dissipation and store in remote locations with long-term controlled containment. Methods available for bulk reduction and storage of liquid and solid wastes are given by Rodger[1] and Lieberman.[2]

1. Evaporation. This is the most widely used technique. Decontamination factors range from 10^4 to 10^6. Costs range from $0.05 to $0.70 per gallon of starting solution.

2. Calcining. Vaporization to complete dryness by high-temperature methods minimizes volume requirements and corrosion. One disadvantage is the poor heat-transfer characteristics of the heat-generating solids.

3. Adsorption and ion exchange. This method is useful for low solids concentration in liquid waste and for gaseous wastes. If clay is used for the adsorbent in liquid phase, the wet freshly adsorbed mass can be calcined and placed in the ground with insignificant release to the environment. Costs are about $0.05 per gallon of starting solution.

4. Incineration. This method is useful on combustible matter but requires particle separation from flue gas by an inertial-type settler for larger particles and a high-temperature Fiberglas filter in series. Costs range from $2.50 to $5 per cubic foot of waste.

5. Baling. Compressive baling methods are sometimes used in conjunction with above-ground disposal or burial as final storage.

6. Burial. Abandoned mines, desert areas, and deep wells are sites wherein health physics monitoring with positive control is possible. Selection of burial grounds is based on (a) characteristics of the soil to retain radioactivity, (b) rate of release of waste from the burial area, (c) the elevation at the area, (d) rate of movement of ground water from the area, (e) effect of release on contamination of ground water, and (f) distance to users of ground water.

7. Ocean disposal. Concentrated waste is mixed into concrete and the resulting forms are transported to an ocean dumping site. This method has the disadvantage of high costs because of heavy shielding during ocean transportation. A second poor feature is lack of health physics control once the containers have left the ship. Costs range from $0.30 to $1 per pound of waste.

8. Tank storage. One of the most reliable methods to date; use of corrosion-resistant materials for tank construction ensures positive control.

[1] *Ibid.*
[2] S. A. Lieberman, *Nucleonics*, **16**(2): 82 (1958).

Costs range from $0.50 to $10 per gallon, depending on tank size and corrosivity.

PLANT DESIGN

Radiochemical plants differ from the conventional type of chemical plant since no direct maintenance or process control by operating personnel can be performed within the immediate site because of the intense radioactivity when the plant is in operation. It is only possible to enter the area when the active material has been removed from the equipment

TABLE 10-6. DESIGN AND SERVICE POSSIBILITIES FOR RADIOCHEMICAL PLANTS*

Grade	Total activity present in cell	Possibility of entering cell	Design notes
1. Extremely high activity	>1,000 curies of γ or high-energy β	No entry at all until equipment has been decontaminated	No relaxing of design specification possible; all servicing during operation to be done from outside shielding by remote control
2. Very high γ activity	100–1,000 curies	Short entry (2–30 min) after extensive decontamination, assuming ease of decontamination is made a feature of the design of vessel and plant	Slight relaxing of design specification is possible if plant is required for short-time service only; for plant required to work for a long life, Grade 1 design specification is necessary
3. High γ activity	1–100 curies†	Entry possible after draining plant and a limited amount of decontamination, particularly if some local shielding is used	Careful use of some conventional plant items may be permissible
4. Low γ activity	<1 curie†	Entry may be possible during operation, particularly if local shielding is used	Conventional plant items can be used
5. α activity only	....	Entry possible at any level of activity for unlimited periods (wearing protective clothing and mask)	Owing to intense toxicity, plant should be inside "fume cupboard" or "dry-box" type of cell and should be leakproof

* C. M. Nicholls, *Proc. Intern. Conf. Peaceful Uses Atomic Energy, Geneva*, **9**: 455 (1955).

† These figures are very much dependent on the size of plant and containing cell.

CHAP. 10]   NUCLEAR CHEMICAL PLANT DESIGN   459

by drainage and cleaning procedures, also known as decontamination. Table 10-6 lists the design and service possibilities of radiochemical plants as a function of curie level. The term cell applies to a completely integrated and shielded section of the plant in which a typical portion of the process is carried out (Fig. 10-13).

This classification is only qualitative and rigid health physics monitoring must always be carried on after radioactive material first enters the process equipment. The unique features of plant designs for such conditions will be discussed next.

Plant Layout

Two basically different types of maintenance can be incorporated in the plant design. *Direct maintenance* involves personal contact service on the equipment in place and requires thorough decontamination, usually a very slow procedure, before entry into the process area. Where possible,

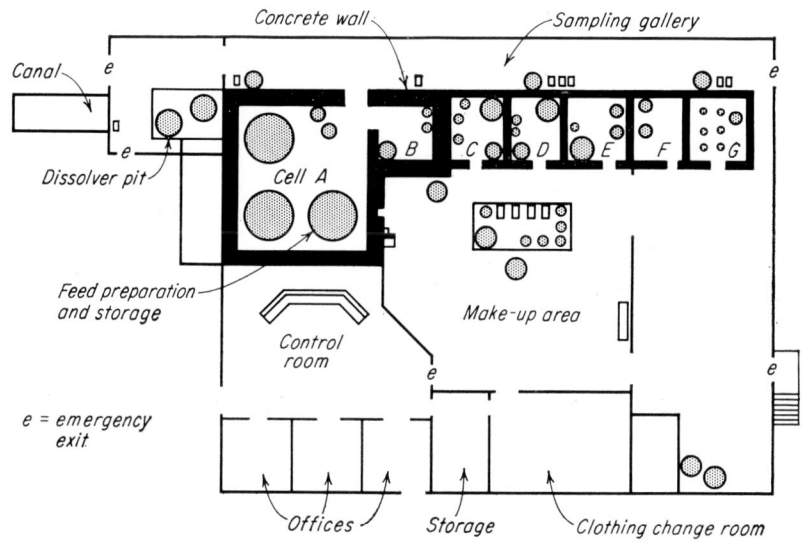

Fig. 10-13. Plan of a direct-maintenance fuel reprocessing plant. Division of equipment is according to activity level and function. [*Courtesy of H. K. Jackson and G. S. Sadowski, Nucleonics,* **13**(8): 24 (1955).]

equipment is isolated in shielded cells so that the entire plant does not have to be decontaminated. A typical direct-maintenance plant layout is shown in Fig. 10-13. *Remote maintenance* is the service of equipment by mechanical devices so that repairmen never enter the process area. A plant layout for remote maintenance is shown in Fig. 10-14. Mechanical devices are operated by a person in a shielded control cab of a crane which travels above the process equipment. Service work done in this manner involves the removal of faulty equipment by means of quick-opening connectors and setting in a replacement. The faulty equipment is either discarded in a burial pit or decontaminated in a special process area, followed by repair in a normal fashion.

The principal advantage of remote maintenance is positive and safe repair scheduling. It has several serious disadvantages: (1) design is

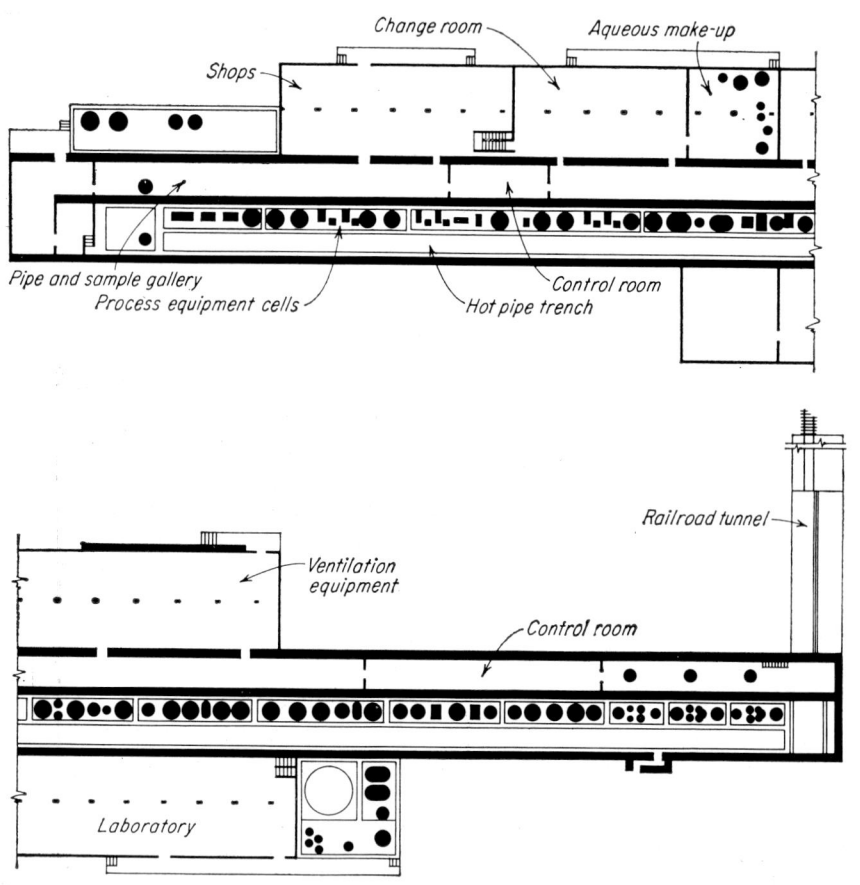

Fig. 10-14. Plan of a remote-maintenance fuel reprocessing plant. Shielded crane with operator runs above the process equipment cells to perform remote maintenance. (*Courtesy of W. M. Harty, Chem. Eng. Progr. Symposium Ser.* 50, p. 118, 1954.)

more difficult, as all layouts require access from above the equipment; (2) fabrication and construction is costly so that the initial investment is greater. On the other hand, direct-maintenance plants can be built using commercially available equipment and standard fabrication techniques. This reduces the capital cost but the operating charges per pound of material increase since off-stream time is required for decontamination, whereas this is no factor in a remote-maintenance plant. A direct-maintenance plant has the further advantage of more effectively locating equipment without regard to accessibility from the top of the cell. Thus a more compact unit cell can be designed with lower shielding costs.

Because any type of maintenance in these plants is costly, design of equipment should be aimed toward long life with no repairs. Experience has shown the following to be good practice:

1. Reduce the number of valves to a minimum.
2. Weld all pipe connections.
3. Thoroughly leak-test all equipment and fittings with Freon or other sensitive leak-detection methods.
4. Make test runs using nonradioactive simulated chemical solutions.

Critical Mass Control Requirements

As emphasized previously, critical mass accumulation must be prevented by proper process design and plant layout. Lemon and Reid[1] give a design philosophy in which each process vessel is made safe by one of the following methods:

1. Limited concentration in which the concentration of the solution is held within the range where a chain reaction is not possible. Concentration limitation can sometimes be controlled easily by low solubility of the alloying elements in aqueous-type processes.
2. Mass limitation, where the quantity of fissionable material allowed in the vessel is kept below the amount that can go critical under any possible condition.
3. Safe geometry, where the vessel diameter is such that critical conditions can never be achieved because of high neutron loss. Vessel spacing is such that interaction between vessels is minimized.

The first two methods require rigid process control except in natural or slightly enriched reprocessing plants before plutonium partitioning. Other features of the layout include a calculated pitch on the process cell floors so that a major spill will not cause the solution depth to exceed criticality. A steam jet and sump pump with alarm devices are located in a geometrically safe sump at the low point in the cell area. To avoid loss of fissionable-fuel material, there should be no gravity connections between the process tanks and the waste storage area.

Health Physics Requirements

Another unique feature of the nuclear plant layout is the requirement for special locker and shower facilities. Personnel coming from a hot area (i.e., processing cells or other contaminated areas) are generally

[1] R. B. Lemon and D. G. Reid, *Proc. Intern. Conf. Peaceful Uses Atomic Energy,* Geneva, **9**: 535 (1955).

required to remove all work clothing, shower, and have a radiation count measurement taken, particularly of the hands and feet, before leaving the plant. There should be no direct access from the street or offices to the process area except through a clothing-change room where radiation film badges and self-reading pocket dosimeters are issued. Emergency exits, opening from the inside only, are frequently provided within the process area in case of fires or explosions.

Equipment Selection

The equipment used in radiochemical plants is, for the most part, similar to that found in other industrial chemical plants. The equipment should be selected for minimum holdup, thus reducing shielding requirements and criticality hazards. For example, thin film evaporators, operating continuously, should be selected instead of pot-type vessels. The method of coupling equipment together and the remote control and handling features of equipment design may be quite different, especially for the remote-maintenance type of plant.

Mechanical devices, such as rotating shafts, are avoided if possible. Agitation with steam or gas sparging is preferred. Almost all solution transfers are made with steam jets since this method of transfer is easily adapted to remote control and is essentially trouble-free. Lack of accurate metering and handling of low flows are process disadvantages where steam jets are used. If it is absolutely necessary to use a pump, the tank-submerged type is preferred for large capacities to avoid shaft leakage. Canned rotor pumps have been used to advantage. Smaller flows can be handled with a diaphragm pump employing a remotely located head within the radioactive process area and the driving fluid pump positioned outside the shielded wall.

Air-operated bellows-sealed valves of special design are used. Metal bellows are enclosed in an all-welded housing to prevent solution leakage in the event of bellows failure.

Materials Selection

The radioactive level in most chemical processing equipment is such as to preclude nearly all types of organic compounds for structural or packing materials. Thus, the use of plastics for these functions is not recommended. The selection of metals and ceramic materials can be made on the basis of corrosion resistance under ordinary environmental circumstances, provided due allowance is made for the heat effects caused by absorption of radioactivity in the material under consideration.

Where viewing systems are necessary, the choice of materials creates unusual problems because radiation stability of the transparent properties of the materials is a prime requisite. There is a tendency toward color formation and loss of transparency with increasing radiation dosage. A list of suitable viewing materials which also act as shielding is:

Material	Density, g/cm^3
1. Water	1.0
2. 80% ZnBr in water	2.52
3. Nonbrowning lime glass	2.68
4. Corning 8362 glass	3.27
5. X-ray lead glass	4.88
6. Dense lead glass	6.20

An excellent discussion of viewing systems and material requirements is given by Stephenson.[1]

Decontamination

Materials in contact with radioactive solids, liquids, or gases will pick up activity as the result of physical or chemical adsorption. In some cases it may be impossible to do repair work, even though the radioactive contents are removed. This is true in the case of direct-maintenance plants, and materials should be selected for rapid decontamination to increase the plant on-stream time.

Materials of construction should be smooth and nonporous, nonionic in nature, with good chemical and heat-resistant properties. A listing of materials in order of increasing difficulty of decontamination includes (1) polyethylene, (2) glass, (3) stainless steel, (4) copper, (5) brass, (6) carbon steel, (7) lead, and (8) concrete. Nearly 100 per cent of the applied activity is adsorbed on a porous material such as concrete and cannot be removed. Special plastic paints and coatings which can be applied to rough surfaces and subsequently peeled off when contaminated have proved useful in building construction.

Decontamination of process equipment by chemical cleaning requires corrosion resistance to numerous solutions. A typical cleaning cycle, which may take somewhere between 3 and 30 days, employs the following chemical steps:

1. Water agitated by steam sparge
2. 10 per cent dilute HNO_3
3. 10 per cent citric acid
4. 10 per cent NaOH–2.5 per cent tartaric acid
5. 10 per cent oxalic acid
6. 0.003 M periodic acid
7. 3 per cent sodium fluoride–20 per cent HNO_3

The temperature is maintained at 75°C or above for solutions except the

[1] R. M. Stephenson, "Introduction to Nuclear Engineering," 2d ed., pp. 418–422, McGraw-Hill Book Company, Inc., New York, 1958.

last, which is left at room temperature for about 1 hr. Type 347 stainless steel has proved adequate for this service.

Process Instrumentation and Control

The radiochemical plant, which by necessity must be operated by remote control, is highly instrumented. In addition to the usual recording and automatic control instruments for flow rate, liquid level, density, and temperature, many types of commercially available radiation monitoring instruments are used. All instruments must have a high degree of precision where measurements are required for concentration control of criticality. Reliability is improved by use of transistor components in the instrument electrical circuit.

Most of the process control instruments are connected to the radioactive equipment via air purge lines which are positioned higher than the process vessel to avoid entrance of radioactive solutions into the purge lines. Both density and liquid level are determined by differential pressure between two air probes.

Sampling of liquid streams for process control and accountability of fuel poses unusual problems because of the radioactivity involved. Landry[1] describes the numerous ingenious sampling devices which are used. Most designs depend on a gas lift to recirculate the liquid to a shielded sampling box where a representative sample is removed.

Ventilation and Gaseous-waste Disposal

The principal requirement of the ventilation system is to keep radioactive gases and dusts which may be present in the process area from entering the operating-control areas. Two types of gaseous wastes are encountered: (1) ventilation gases from the process cells and laboratory hood exhausts and (2) process gases evolved from the vessel contents during chemical processing. The ventilating air is supplied to the process area after being washed, heated or cooled, and passed through the nonradioactive areas. The process area is maintained under negative pressure. The air can be exhausted to a 200- to 300-ft vertical stack by means of large motor-driven blowers if the level of radioactive materials in direct contact with this air is low.

The process gases are treated differently. All chemical vessels are vented under negative pressure up to 10 in. of water by steam-jet ejectors or stainless-steel blowers. The discharge from these blowers is generally sent to scrubbing or filter-bed towers to remove certain gases and/or entrained liquid and solid particles. The filter is made up of Fiberglas mats which have a 99.9 per cent efficiency for removal of particles over

[1] J. W. Landry, *Proc. Intern. Conf. Peaceful Uses Atomic Energy*, Geneva, **9**: 555 (1955).

1 micron (μ) in diameter at low pressure drop. The gases may be either discharged directly to a stack or passed through asbestos-fiber-paper filters if the activity of the gases from the Fiberglas filter is too great. The allowable radioactivity count of the gases discharged to the stack depends on the maximum permissible concentration of the isotopes present, the height of the stack, the meteorological conditions in the plant vicinity, and the available plant exclusion area. The motor-driven blowers for exhausting the process gases are backed up by steam-turbine drives or internal-combustion engines in case of electric-power failure.

Plant Location

The factors discussed in Chap. 7 apply equally well to nuclear power, heat, and processing plants but different factors must be stressed. The principal factors relating to problems in the nuclear field are (1) raw materials, (2) market, (3) transportation, (4) labor, (5) plant requirements, (6) power, (7) waste disposal, (8) soil structure, (9) climatic conditions, (10) ordinances: nuisance and zoning, and (11) population density or degree of isolation. A discussion of these factors for various types of nuclear plants follows:

1. Concentration of fuels from mineral deposits. The ore-treating plants are located near the raw materials because of the large tonnages to be processed for fuel concentration.

2. Conversion of ore concentrates to chemical or metallurgical fuel. The volume output is small and geographical location is unimportant. A source of chemical labor and provision for waste disposal are important considerations for this plant.

3. Separation of fission from fertile fuel. Gaseous-diffusion plants fall in this category. Requirements for large blocks of electric power and adequate cooling water dictate the location of this plant.

4. Spent- or irradiated-fuel reprocessing. The plants handling spent fuel from homogeneous power reactors where the fuel must be processed continuously will be integrated at the power reactor site. Those plants which handle solid fuels from heterogeneous reactors must be located in places where waste disposal can be adequately handled. Transportation is a consideration because the spent fuel must be shipped in shielded caskets weighing at least 15 times as much as the fuel.[1] The total exclusion area requirements will dictate the plant site within a given geographical location. Local, state, and Atomic Energy Commission ordinances may be prohibiting factors.

The above examples illustrate the fact that analysis required for plant location is the same for any industry, but the accent is placed on different factors, depending on the nature of the processes being handled.

[1] C. E. Dryden, *Nucleonics*, **14**(6): 77 (1956).

COST ESTIMATIONS

It was shown in the plant layout section of this chapter that radiochemical plants for spent-fuel reprocessing require designs which have no counterpart in the chemical industry. Shielding, criticality control, and

TABLE 10-7A. ESTIMATING FACTORS FOR RADIOCHEMICAL PLANTS*
P = process equipment costs, delivered basis

1. Process equipment installed	$1.1P$ to $1.15P$
2. Piping, process building[a]	$1.5P$ to $2.9P$
3. Instrumentation, process	$0.75P$ to $1.0P$
4. Special shielding and equipment[b]	$1.0P$ to $1.3P$
5. Process building[c]	$1.2P$ to $4.5P$
6. Laboratory and administration bldg. complete	$2.8P$ to $3.8P$
7. Reactor fuel storage bldg. complete[c]	$0.68P$ to $1.5P$
8. Gaseous waste-disposal system[d]	$1.3P$ to $1.4P$
9. Liquid waste-disposal system[e]	$1.8P$ to $2.8P$
10. Waste-disposal building[c]	$0.72P$ to $1.4P$
11. Service building complete	$0.85P$ to $1.2P$
12. Site development[g]	$1.0P$ to $1.6P$
13. Total installed physical costs	Sum of items 1 through 12
14. Construction overhead including fee[f]	40% to 50% of item 13
15. Total construction costs	Sum of items 13 and 14
16. Engineering	10% to 20% of item 15
17. Contingency	10% to 50% of items 15 and 16
18. Preoperational costs[h]	6 mo to 1 yr operating costs
19. Total capital costs[i]	Sum of items 15, 16, 17, 18

* W. G. Stockdale, *Chem. Eng.*, **63**(4): 185 (1956).
Factors are for direct-maintenance plants processing enriched spent reactor fuels.

[a] Piping costs include all labor and material for process piping, service piping, chemical drains, and pipe sleeves through concrete.

[b] Items included here are peculiar to radiochemical plants, i.e. chargers, samplers, radiation monitors, etc.

[c] Building cost is very sensitive to criticality and shielding.

[d] Includes particle and chemical cleanup systems and tall stack for final dispersal.

[e] Includes evaporator for concentration of liquid to be permanently stored as well as storage tanks. Storage tanks can be carried as operating charge but interest on investment tied up must be carried.

[f] Includes equipment rental and expense.

[g] Site development includes power lines, water supply, roads, fences, guardhouses, fire lines, etc., but not land costs.

[h] Personnel training, test runs, cold operation, manual preparation.

[i] To this cost should be added interest on construction money, working capital, land costs, and owner's expense.

remote handling are additional features which add significantly to the cost of a radiochemical plant. Stockdale has listed a range of factors useful for making estimates of direct-maintenance solvent-extraction

TABLE 10-7B. TYPICAL COSTS FOR RADIOCHEMICAL PLANTS*
For a direct-maintenance radiochemical processing plant

Capital cost distribution	Material and labor	
Process building with equipment:		
Process equipment	$678,731	
Pipe, valves, and fittings	1,734,770	
Instruments and controls	574,988	
Electrical (process)	93,766	
Special equipment	761,503	
Process building with services	2,511,009	
		$6,354,767
Waste disposal:		
Liquid waste collection and disposal	1,614,596	
Gaseous waste collection and disposal	843,942	
Waste-disposal building with services	858,892	
		$3,317,430
Administration and laboratory building		2,221,848
Fuel storage building		908,556
Service building		693,332
Yard facilities		940,985
Total labor and material		$14,436,918
Construction overhead and fee		7,001,956
Engineering, including fee		3,773,357
Total fixed capital investment		$25,212,231

Operating cost distribution (no amortization included)	Per cent
Direct cost:	
Direct operating labor	12.3
Direct supervision	4.9
Process chemicals and supplies	9.9
Utilities	20.3
Maintenance and repair	17.1
Product control laboratory	29.0
Process improvement laboratory	6.0
Engineering department	0.5
	100.0
General plant expense (per cent of direct cost):	
Overhead	11.4
Administrative	21.4
Health physics	9.9
SF accountability	3.5
Other	11.3
	57.5

* W. G. Stockdale, *Chem. Eng.*, **63**(4): 185 (1956).

plants. His method is presented in Table 10-7 together with a typical cost analysis for a plant size estimated to process around 1 ton/day of natural or slightly enriched fuel. The upper limit is based on actual plant construction experience while the lower limit is an anticipated goal resulting from experience and improvement factors. Scale-up should be computed on a 0.5 exponential rather than the usual 0.6 value described in Chap. 6 (pp. 205, 222) because of the inherently higher investment in the small plant due to shielding and remote-handling equipment.

Remote-maintenance plants will have a capitalized cost about 50 per cent higher than that shown for the direct-maintenance plant. Pyrometallurgical or fractional-distillation types of plants are in the low range of the costs estimated for direct-maintenance plants.

No data are available on cost estimating methods for gaseous-diffusion plants; ore-concentrating plants can be computed on the same basis as other extractive metallurgical plants using wet processing. Flow sheets for all processes are available in the references given in the process design section of this chapter. Equipment requirements can thus be specified from these flow sheets as a basis for a cost estimation.

LEGAL PROBLEMS

Plants handling bulk quantities of radiochemical materials must conform to regulations similar to those for nuclear reactors. The preliminary design must be approved by a safeguards committee of the Atomic Energy Commission and by local and state authorities as well as insurance groups. The construction and operation requires licensing of both the plant and its operating personnel. The design engineer is advised to first obtain the latest regulatory information from the Atomic Energy Commission and the other controlling authorities before starting on the design of a radiochemical plant.

PROBLEMS

10-1. Design the chlorination unit of the benzene hexachloride plant discussed in Chaps. 3 to 6 to utilize gamma-ray energy from typical spent-fuel elements of nuclear-power reactors and from cesium 137 isolated from fission products. Make a comparative economic study.

10-2. Develop a complete process and plant design for a chemical reprocessing plant to handle 5 metric tons per day of spent uranium fuel from nuclear process heat reactors of 20,000 Mw total heat output. The initial fuel was 1.6 per cent slightly enriched natural uranium and the average time the fuel was in the reactor at full power was 2.0 yr.

NOMENCLATURE

Note: Where units are not given, consistent units are to be used.

- A atomic weight, g
- b exponential attenuation coefficient, also linear build-up
- B shielding build-up factor, dimensionless
- D dose rate, generally at receptor position
- D_0 dose rate before shield
- E energy per particle, Mev
- E_{av} average energy of photons or particles, Mev
- H width of flat slab source
- I intensity, Mev/cm^2-sec
- N number of atoms of radioactive element
- r roentgen, photon equivalent energy
- R distance between center of source and receptor, length
- R_0 distance from center to outer edge of source, length
- S source strength
- S_A surface source strength or flux, curies/cm^2
- S_c source strength, curies
- S_E source strength, Mev/sec
- S_p source strength, photons or particles/sec
- S_v volumetric source strength, curies/cm^3
- t time
- $t_{1/2}$ half life of radioisotope
- W weight of radioactive isotope, grams
- x shielding thickness, length
- y yield of photons or particles per disintegration
- Z height of cylinder

Greek Symbols

- α alpha particles
- β electron particles, negative or positive
- γ gamma photons
- Δ incremental operator
- θ attenuation angle between source and receptor
- λ radioactive decay constant $= 0.693/t_{1/2}$
- μ linear absorption coefficient, length^{-1}
- μ/ρ mass absorption coefficient, cm^2/g
- ρ density, g/cm^3
- ϕ flux, particles or photons/cm^2-sec

CONVERSION FACTORS FOR NUCLEAR DESIGN

Length

1 foot .... 30.48 centimeters
1 inch .... 2.54 centimeters

Volume

1 cubic foot .... 28.32 liters

1 gallon	231 cubic inches
1 gallon	3.785 liters

Mass

1 pound	453.6 grams
1 kilogram	2.2 pounds
1 ton (short)	2,000 pounds
1 ton (metric)	2,200 pounds

Energy

1 watthour	3.413 Btu
1 megawatt-day (Mwd)	1.0 gram of fission fuel consumed (approx.)
1 megawatt-day	8.19×10^7 Btu
1 electron volt (ev)	1.603×10^{-12} erg
1 million electron volts (Mev)	1.603×10^{-6} erg
1 million electron volts (Mev)	1.603×10^{-13} watt-sec
1 million electron volts (Mev)	1.520×10^{-16} Btu
1 fission	$\sim$200-Mev energy release

Power

1 watt	1 joule/sec = 10^7 ergs/sec
1 watt	3.413 Btu/hr
1 watt	3.1×10^{10} fissions/sec
1 Mev/sec	1.603×10^{-13} watt

Nuclear

1 curie	3.7×10^{10} disintegrations/sec
1 barn	10^{-24} cm^2
1 radiation absorbed dose (rad)	100 ergs/g of absorbing matter
1 roentgen (r)	83 ergs/g air = 90 ergs/g soft tissue
Avogadro's number	6.03×10^{23} molecules/mole

Additional Selected References

CHAPTER 1
General

1. Barkow, C. W.: The Project Engineer, *Chem. Eng. Progr.*, **52**(3): 61 (1956).
2. Genereaux, R. P.: Engineering for Today's Chemical Plants, *Chem. Eng.*, **61**(4): 182 (1954).
3. Lobo, W. E.: Design for Tomorrow's Designers, *Chem. Eng. Progr.*, **53**(10): 6 (1957).

Computers

4. Berkeley, E. C., and L. Wainwright: "Computers—Their Operation and Applications," Reinhold Publishing Corporation, New York, 1956.
5. Bentler, J. A., and J. B. Roberts: Electronic Analogs in Reactor Design, *Chem. Eng. Progr.*, **52**(2): 69F (1956).
6. Bibliography on the Use of IBM Machines in Science, Statistics and Education, International Business Machines Corp., New York, 1956.
7. DeCarlo, C. R.: The Future of Automatic Information Handling in Chemical Engineering, *Chem. Eng. Progr.*, **51**(11): 487 (1955).
8. Johnson, C. L.: "Analog Computer Techniques," McGraw-Hill Book Company, Inc., New York, 1956.
9. McCracken, D. D.: "Digital Computer Programming," John Wiley & Sons, Inc., New York, 1957.
10. McMaster, R. C., R. L. Merrill, and B. H. List: Analog Systems in Engineering Design, *Prod. Eng.*, **24**(1): 184 (1953).
11. Reviews on Computers, Mathematics and Statistics, *Industrial and Engineering Chemistry*, annual March issue.
12. Schrage, R. W.: The Automatic Computer in the Control and Planning of Manufacturing Operations, "Advances in Chemical Engineering," vol. I, pp. 331–336, Academic Press, Inc., New York, 1956.
13. Symposium: Computers, *Chem. Eng. Progr.*, **52**(11): 449–470 (1956).
14. Symposium: Machine Computation in Petroleum Research, *Ind. Eng. Chem.*, **50**(5): 712–752 (1958).

15. Williams, T. J., R. C. Johnson, and A. Rose: Computations in the Field of Engineering Chemistry, *J. Assoc. Computing Machinery*, **4**(4): 393 (1957).

Drawing

16. Berg, R. H.: Handy Way to Scale Drawings for Flow Sheets, *Chem. Eng.*, **65**(16): 174 (1958).
17. Hoelscher, R. P., and C. H. Springer: "Engineering Drawing and Geometry," John Wiley & Sons, Inc., New York, 1956.
18. Zozzora, F.: "Engineering Drawing," McGraw-Hill Book Company, Inc., New York, 1953.

CHAPTER 2

General

1. Corley, H. M. (ed.): "Successful Commercial Chemical Development," John Wiley & Sons, Inc., New York, 1954.
2. Harper, J. I. (ed.): "Chemical Engineering in Practice," Reinhold Publishing Corporation, New York, 1954.
3. Miller, R. L., Jr.: Organization for Plant Design, *Chem. Eng.*, **63**(7): 185 (1956).

Literature Searching and Reporting

4. "Applied Science and Technology Index" (formerly "Industrial Arts Index"), The H. W. Wilson Company, New York.
5. "Chemical Abstracts," American Chemical Society, The Ohio State University, Columbus, Ohio.
6. Chemical Engineering Notation, Appendix B of this book.
7. "Chemist's Dictionary," D. Van Nostrand Company, Inc., Princeton, N.J., 1953.
8. Clark, G. L., and G. G. Hawley (eds.): "Encyclopedia of Chemistry," Reinhold Publishing Corporation, New York, 1957.
9. Crane, E. J., A. M. Patterson, and E. B. Marr: "A Guide to the Literature of Chemistry," John Wiley & Sons, Inc., New York, 1957.
10. Dyson, G. M.: "A Short Guide to Chemical Literature," Longmans, Green & Co., Ltd., London, 1951.
11. "Engineering Index," Engineering Index Co., Inc., New York.
12. Grant, J. (ed.): "Hackh's Chemical Dictionary," 3d ed., McGraw-Hill Book Company, Inc., New York, 1946.
13. Kirk, R. E., and D. F. Othmer (eds.): "Encyclopedia of Chemical Technology," Interscience Publishers, Inc., New York.
14. Kobe, K. A.: "Chemical Engineering Reports," Interscience Publishers, Inc., New York, 1957.
15. Mellon, M. G.: "Chemical Publications," 3d ed., McGraw-Hill Book Company, Inc., New York, 1958.
16. Nelson, J. R.: "Writing the Technical Report," 3d ed., McGraw-Hill Book Company, Inc., New York, 1952.
17. Rose, A., and E. Rose: "Condensed Chemical Dictionary," 5th ed., Reinhold Publishing Corporation, New York, 1956.
18. "Scientific Encyclopedia," 3d ed., D. Van Nostrand Company, Inc., Princeton, N.J., 1958.
19. Shera, J. H. (ed.): "Advances in Documentation and Library Science," vol. 1, John Wiley & Sons, Inc., New York, 1957.

20. Stephenson, H. J.: "A Dictionary of Abbreviations," The Macmillan Company, New York, 1943.
21. Ulman, J. N., Jr.: "Technical Reporting," Henry Holt and Company, Inc., New York, 1952.
22. Zimmerman, O. T., and I. Lavine: "Scientific and Technical Abbreviations, Signs and Symbols," Industrial Research Service, Dover, N.H., 1948.

Process Data Sources

23. Adams, D. P.: "An Index to Nomograms," John Wiley & Sons, Inc., New York, 1950.
24. Atack, F. W. (ed.): "Handbook of Chemical Data," Reinhold Publishing Corporation, New York, 1957.
25. Beilstein: "Handbuch der organischen Chemie," Springer-Verlag OHG, Berlin.
26. Brewer, L.: Thermodynamic Properties of Oxides and Their Vaporization Processes, *Chem. Revs.*, **52**(1): 1 (1953).
27. Chemical and Engineering Data Series, *Ind. Eng. Chem.*, biannually.
28. Chu, J. C.: "Distillation Equilibrium Data," Reinhold Publishing Corporation, New York, 1950.
29. Cushing, R.: Your Design Reference File, section VII—Physical Data, *Chem. Eng.*, **64**(10): 255 (1957).
30. Davis, D. S.: "Chemical Engineering Nomographs," McGraw-Hill Book Company, Inc., New York, 1944.
31. Doss, M. P.: "Physical Constants of the Principal Hydrocarbons," The Texas Company, New York, 1943.
32. Dreisbach, R. R.: "Physical Properties of Chemical Substances," Dow Chemical Co., Midland, Mich., 1952.
33. Dreisbach, R. R.: "Pressure-Volume-Temperature Relationships of Organic Compounds," 3d ed., Handbook Publishers, Inc., Sandusky, Ohio.
34. Egloff, G.: "Physical Constants of Hydrocarbons," vols. I–IV, Reinhold Publishing Corporation, New York, 1947–1953.
35. "Fuel Gases," American Gas Association, New York, 1941.
36. Gambell, W. R.: Process Data, *Chem. Eng.*, **64**(2): 235, (3): 271, (4): 273, (5): 263, (6): 243, (7): 263, (8): 256, (9): 267, (10): 283, (12): 261 (1957); **65**(1): 159, (3): 137, (5): 147, (7): 146, (9): 143, (11): 125, (13): 113, (17): 121 (1958).
37. Gmelin, "Handbuch der anorganischen Chemie," Verlag Chemie, Berlin.
38. Hildebrand, J. H., and R. L. Scott: "The Solubility of Nonelectrolytes," 3d ed., Reinhold Publishing Corporation, New York, 1950.
39. Hodgman, C. D. (ed.): "Handbook of Chemistry and Physics," Chemical Rubber Publishing Co., Cleveland, Ohio.
40. Horsley, L. H.: "Azeotropic Data," American Chemical Society, Washington, D.C., 1952.
41. Hougen, O. A., K. M. Watson, and R. A. Ragatz: "Chemical Process Principles, Part I—Material and Energy Balances," John Wiley & Sons, Inc., New York, 1954.
42. Hougen, O. A., and K. M. Watson: "Chemical Process Principles, part II—Thermodynamics, part III—Kinetics and Catalysis," John Wiley & Sons, Inc., New York, 1947.
43. "International Critical Tables," National Academy of Sciences and National Research Council, McGraw-Hill Book Company, Inc., New York.
44. Jacobson, C. A., and C. A. Hampel (eds.): "Encyclopedia of Chemical Reactions," vols. 1–7, Reinhold Publishing Corporation, New York, 1946–1958.

45. Kammemeyer, K., and J. O. Osburn: "Process Calculations," Prentice-Hall, Inc., Englewood Cliffs, N.J., 1956.

46. Kobe, K. A., and R. E. Lynn: The Critical Properties and Elements and Compounds, *Chem. Revs.*, **52**(1): 117 (1953).

47. Lange, N. A. (ed.): "Handbook of Chemistry and Physics," Handbook Publishers, Inc., Sandusky, Ohio.

48. "Liquid Metals Handbook," 2d ed., AEC–Dept. of Navy, 1954.

49. "Liquid Metals Handbook, Sodium-NaK Supplement," TID 5227, AEC–Dept. of Navy, 1955.

50. Maxwell, J. B.: "Data Book on Hydrocarbons," D. Van Nostrand Company, Inc., Princeton, N.J., 1950.

51. Mellan, I.: "Source Book of Industrial Solvents," Reinhold Publishing Corporation, New York, 1957–1959.

52. Perry, J. H. (ed.): "Chemical Engineers' Handbook," 3d ed., McGraw-Hill Book Company, Inc., New York, 1950.

53. Rossini, F. D., et al.: "Selected Values of Physical and Thermodynamic Properties of Hydrocarbons and Related Compounds," API Carnegie Press, Pittsburgh, Pa., 1953.

54. Sakiadis, B. C., and J. Coates: Thermal Conductivities of Liquids, *Louisiana State Univ., Eng. Expt. Station Bulls.*, Baton Rouge, La.

55. "Selected Values of Chemical Compounds," Manufacturing Chemists' Association Project, Carnegie Institute of Technology, Pittsburgh, Pa.

56. Seidell, A.: "Solubilities of Inorganic and Organic Compounds," vols. I, II and supplements, D. Van Nostrand Company, Inc., Princeton, N.J., 1941 to date.

57. Stull, D. R., and G. C. Sinke: "Thermodynamic Properties of the Elements," American Chemical Society, Washington, D.C., 1958.

58. "Tables of Chemical Kinetics—Homogeneous Reactions," *Natl. Bur. Standards Circ.* 510, Washington, D.C., 1951.

Pilot Plants

59. Clark, E. L.: Pilot Plants in Process Technology, *Chem. Eng.*, **65**(8): 155, (11): 119, (15): 119, (20): 125 (1958).

60. Fleming, R. (ed.): "Scale-up in Practice," Reinhold Publishing Corporation, New York, 1958.

61. Fragen, N., G. H. Weisemann, and K. C. Peterson: Selecting the Kind and Size of Pilot Plants, *Chem. Eng. Progr.*, **54**(8): 65 (1958).

62. Grothe, J. D.: Modern Approach to Pilot-plant Design, *Chem. Eng.*, **63**(6): 239 (1956).

63. Johnstone, R. E., and M. W. Thring: "Pilot Plant and Scale-up Methods in Chemical Engineering," McGraw-Hill Book Company, Inc., New York, 1957.

64. Jordan, D. G.: "Chemical Pilot Plant Practice," Interscience Publishers, Inc., New York, 1955.

65. Symposium: Pilot Plant Design and Construction, *Ind. Eng. Chem.*, **41**: 2011 (1949).

66. Symposium: Collection of Engineering Data on a Small Scale, *Ind. Eng. Chem.*, **50**(4): 577–603 (1958).

67. Symposium: Instrumentation for Pilot Plants, *Ind. Eng. Chem.*, **45**(9): 1836 (1953).

Safety

68. Armistead, G., Jr.: Safety and the Chemical Engineer, *Chem. Eng. Progr.*, **48**(1): 3 (1952).

69. Braidech, M. M.: Safety in Chemical Plant Operations, *Chem. Eng. Progr.*, **47**(12): 595 (1951).
70. Guelich, J.: "Chemical Safety Supervision," Reinhold Publishing Corporation, New York, 1956.
71. Kieweg, H.: Safety and Outdoor Construction, *Chem. Eng. Progr.*, **47**(7): 341 (1951).
72. Miner, H. L.: Management Viewpoints on Plant Safety, *Chem. Eng. Progr.*, **47**(12): 597 (1951).
73. Safety Workbook, *Ind. Eng. Chem.*, monthly feature.

Statistics

74. Cochran, W. G., and G. M. Cox: "Experimental Designs," 2d ed., John Wiley & Sons, Inc., New York, 1957.
75. Kempthorne, O.: "The Design and Analysis of Experiments," John Wiley & Sons, Inc., New York, 1952.
76. Lange, H. B.: Investigating Chemical Plant Process Variables, *Chem. Eng. Progr.*, **53**(6): 304 (1957).
77. Statistics Workbook, *Ind. Eng. Chem.*, monthly feature.
78. Youden, W. J.: "Statistical Methods for Chemists," John Wiley & Sons, Inc., New York, 1951.

Legal Aspects

79. Buckles, R. A.: "Ideas, Inventions, and Patents," John Wiley & Sons, Inc., New York, 1957.
80. Canfield, D. T., and J. H. Bowman: "Business, Legal, and Ethical Phases of Engineering," 2d ed., McGraw-Hill Book Company, Inc., New York, 1954.
81. Crooks, Robert: Review Patent Fundamentals, *Chem. Eng.*, **65**(4): 121 (1958).
82. Dunham, C. W., and R. D. Young: "Contracts, Specifications and Law for Engineers," McGraw-Hill Book Company, Inc., New York, 1958.

CHAPTER 3

The sources listed below furnish process information from the chemical, petroleum, and metallurgical fields. This information is usually in the form of qualitative block type or equipment flow sheets which can serve as starting points for a process and plant design problem. The student should make a thorough literature survey to understand all the chemistry and engineering aspects of the process design problem (see Reference Sources in Chap. 2).

Composite Flow-sheet Books

1. "Chemical Engineering Flow Sheets," composited from *Chemical Engineering*, 4th ed., McGraw-Hill Book Company, Inc., New York, 1944.
2. "Modern Chemical Processes," composited from *Industrial and Engineering Chemistry*, vols. 1–5, Reinhold Publishing Corporation, New York, 1958, published biannually starting in 1950.

Individual Flow-sheet Sources—Magazines

3. *Chemical Engineering*, McGraw-Hill Publishing Company, Inc., New York.
4. *Chemical Engineering Progress*, American Institute of Chemical Engineers, New York.
5. *Industrial and Engineering Chemistry*, American Chemical Society, Washington, D.C.

6. *Oil and Gas Journal*, Petroleum Publishing Co., Tulsa, Okla.
7. *Petroleum Refiner*, Gulf Publishing Co., Houston, Tex.
8. *Chemical and Process Engineering*, Leonard-Hill Limited, London.
9. *Chemical Processing*, Putnam Publishing Co., Chicago.

Individual Flow-sheet Sources—Books

10. Faith, W. L., D. B. Keyes, and R. Clark: "Industrial Chemicals," 2d ed., John Wiley & Sons, Inc., New York, 1957.
11. Kirk, R. E., and D. F. Othmer (eds.): "Encyclopedia of Chemical Technology," Interscience Publishers, Inc., New York, 1944–1957.
12. Shreve, R. N.: "Chemical Process Industries," 2d ed., McGraw-Hill Book Company, Inc., New York, 1956.
13. Riegel, E. R.: "Industrial Chemistry," 5th ed., Reinhold Publishing Corporation, New York, 1949.

Process Cycles

14. Naguv, M. F.: Material Balances in Complex and Multi-stage Recycles, *Chem. Eng. Progr.*, **53**(6): 297 (1957).

CHAPTER 4

Materials of Construction

1. ASTM Standards on Materials, American Society for Testing Materials, Philadelphia, Pa.
2. Chemicals and Materials Technical Review, *Chem. Eng.*, annually, September issue.
3. Corrosion of Engineering Materials, *Corrosion*, monthly.
4. DuMond, T. C. (ed.): "Engineering Materials Manual," Reinhold Publishing Corporation, New York, 1951.
5. Engineering Materials Reviews, *Chemical & Process Engineering*, monthly.
6. Engineering Materials Reviews, *Materials and Methods*, monthly.
7. Greathouse, G. A., and C. J. Wessel (eds.): "Deterioration of Materials," Reinhold Publishing Corporation, New York, 1954.
8. Lee, J. A.: "Materials of Construction for Chemical Process Industries," McGraw-Hill Book Company, Inc., New York, 1950.
9. Mantell, C. L. (ed.): "Engineering Materials Handbook," McGraw-Hill Book Company, Inc., 1958.
10. Materials of Construction Reviews, *Industrial Engineering Chemistry*, annually, September issue.
11. Miner, P. F., and J. B. Seastone (eds.): "Handbook of Engineering Materials," John Wiley & Sons, Inc., New York, 1955.
12. Report on Materials of Construction, *Chemical Engineering*, biannually, even years, November issue.
13. Society of the Plastics Industry, Inc.: "Plastics Engineering Handbook," Reinhold Publishing Corporation, New York, 1954.
14. Uhlig, H. H. (ed.): "Corrosion Handbook," John Wiley & Sons, Inc., New York, 1948.

Equipment Design and Selection

15. Carmichael, C. (ed.): "Kent, Mechanical Engineers' Handbook—Design and Production," John Wiley & Sons, Inc., New York, 1950.

16. "Compressed Air and Gas Institute—Compressed Air Handbook," 2d ed., McGraw-Hill Book Company, Inc., New York, 1954.

17. Cremer, H. W. (ed.): "Chemical Engineering Practice," 12 vols., Academic Press, Inc., New York, 1956 to date.

18. Cushing, R.: Your Design Reference File, *Chem. Eng.*, **64**: (1957)–(3) : 257–260 (tanks, reactors, mixers and agitators, materials handling); (4): 271–276 (piping); (5): 267–273 (pumps, water supply, power); (7): 247–253 (compressors, jet ejector and eductors, design and cost estimating); (8): 267–272 (heating, ventilating, and air conditioning); (9): 277–282 (heat transfer, heat exchangers, chemical engineering, kinetics); (10): 255–262 (materials of construction, corrosion, paints and coatings, physical data); (11): 257–264 (separation processes, distillation, evaporation, absorption, drying, air pollution, dust collection, entrainment separators); (12): 275–280 (instrumentation, refrigeration, structural engineering, illumination).

19. Davidson, R. L.: "Successful Process Plant Practices," McGraw-Hill Book Company, Inc., New York, 1958.

20. Equipment and Design Workbook, *Industrial and Engineering Chemistry*, monthly feature.

21. Eshbach, O. W. (ed.): "Handbook of Engineering Fundamentals," 2d ed., John Wiley & Sons, Inc., New York, 1952.

22. Hicks, T.: "Pump Selection and Application," McGraw-Hill Book Company, Inc., New York, 1957.

23. "Liquid Metals Handbook," 2d ed., AEC–Dept. of Navy, 1954.

24. "Liquid Metals Handbook, Sodium-NaK Supplement," TID 5227, AEC–Dept. of Navy, 1955.

25. Marks, L. S. (ed.): "Mechanical Engineers' Handbook," 5th ed., McGraw-Hill Book Company, Inc., New York, 1951.

26. Nelson, W. L.: "Petroleum Refinery Engineering," 4th ed., McGraw-Hill Book Company, Inc., New York, 1958.

27. Nielsen, C. H. (ed.): "Distillation in Practice," Reinhold Publishing Corporation, New York, 1956.

28. Reviews on Unit Operation, Unit Processes and Chemical Engineering Fundamentals, *Industrial and Engineering Chemistry*, annual March issue.

29. Scale Models Create New Equipment Design Approach, *Chem. Eng. News*, **36**: 102 (1958).

30. Staniar, W. (ed.): "Plant Engineering Handbook," 2d ed., McGraw-Hill Book Company, Inc., New York, 1959.

31. Symposium: Process Equipment Standardization, *Chem. Eng. Progr.*, **52**(4): 129 (1956).

32. Symposium: Selection of Equipment for Operating and Maintenance, *Chem. Eng.*, **65**(12): 142–179 (1958).

CHAPTER 5
General

1. Immer, J. R.: "Layout Planning Techniques," McGraw-Hill Book Company, Inc., New York, 1950.

2. Mallick, R. W., and F. T. Gandreau: "Plant Layout: Planning and Practice," John Wiley & Sons, Inc., New York, 1951.

3. Salviani, J.: "Encyclopedia of Chemical Technology," vol. 10, pp. 737–743, Interscience Publishers, Inc., New York, 1953.

4. Schwab, L., and R. G. Earnheart: Cut Repair and Costs with Alert Layout Design, *Chem. Eng.*, **63**(10): 190 (1956).

5. Smith, W. P.: How Good Is Your Layout? *Modern Materials Handling*, **9**(5): 121 (1954).

Scale Models

6. Bowen, H. J.: Scale Models, *Chem. Eng.*, **61**(8): 176 (1954).
7. Bussard, W.: It Pays to Build Design Models, *Petrol. Processing*, **12**(4): 90 (1957).
8. Cannon, R.: Models Simplify and Cut Costs on Big Revamp Job, *Petrol. Processing*, **12**(8): 48 (1957).
9. Davidson, R. L.: Can You Save Design Hours with Photo Blueprints? *Petrol. Processing*, **10**(3): 348 (1955).
10. Kershaw, H.: Design Models Are Here to Stay, *Petrol. Processing*, **12**(5): 222 (1957).
11. Kershaw, H., and A. F. Hollowell: Models—A New Maintenance Tool, *Petrol. Refiner*, **37**(1): 133 (1958).
12. Michel, A. E.: Use of Models in Design and Construction, *Chem. Eng. Progr.*, **54**(3): 86 (1958).
13. Paton, B. L.: Do Models Pay Out? *Petrol. Refiner*, **35**(11): 161 (1956).
14. Troy, W. N.: Models Can Cut Your Piping Costs, *Petrol. Processing*, **9**(2): 224 (1954).
15. Tucker, T. S.: How Photo-drawings Work with Models, *Petrol. Processing*, **12**(4): 94 (1957).

CHAPTER 6

General References

1. Aries, R. S., and R. D. Newton: "Chemical Engineering Cost Estimation," McGraw-Hill Book Company, Inc., New York, 1956.
2. Grant, E. L.: "Principles of Engineering Economy," 3d ed., The Ronald Press Company, New York, 1950.
3. Happel, J.: "Chemical Process Economics," John Wiley & Sons, Inc., New York, 1958.
4. Hur, J. J.: "Chemical Process Economics in Practice," Reinhold Publishing Corporation, New York, 1956.
5. Osburn, J. O., and K. Kammermeyer: "Money and the Chemical Engineer," Prentice-Hall, Inc., Englewood Cliffs, N.J., 1958.
6. Perry, J. H. (ed.): "Chemical Business Handbook," McGraw-Hill Book Company, Inc., New York, 1954.
7. Peters, M. S.: "Plant Design and Economics for Chemical Engineers," McGraw-Hill Book Company, Inc., New York, 1958.
8. Schweyer, H. E.: "Process Engineering Economics," McGraw-Hill Book Company, Inc., New York, 1955.
9. Thuesen, H. G.: "Engineering Economy," 2d ed., Prentice-Hall, Inc., Englewood Cliffs, N.J., 1957.
10. Zimmerman, O. T., and I. Lavine: "Chemical Engineering Costs," Industrial Research Service, Dover, N.H., 1950.
11. Zimmerman, O. T., and I. Lavine (eds.): *Chemical Engineering Cost Quarterly, Cost Engineering*, Industrial Research Service, Dover, N.H.

Reviews and Symposia

12. Annual Review of Chemical Engineering Costs, *Cost Engineering*, January issue.

13. Barnet, W. I.: Bibliography of Investment and Operating Costs for Chemical and Petroleum Plants, U.S. Bureau of Mines, October, 1949.
14. "Chemical Economics Handbook," Stanford Research Institute, Palo Alto, Calif. (quarterly reports).
15. Cost File for Preconstruction Estimating, *Chem. Eng.*, **65**(12): 187 (1958) (starts with this issue).
16. Cost Workbook, *Industrial and Engineering Chemistry*, monthly feature.
17. Cushing, R.: Your Design Reference File, section IV—Design and Cost Estimating, *Chem. Eng.*, **64**(7): 247 (1957).
18. Symposium: Economic Evaluation, *Chem. Eng. Progr.*, **52**: 399 (1956).
19. Weaver, J. B.: Chemical Cost and Profitability Estimation Reviews, *Chem. Eng.*, **61**(10): 185 (1954), **62**(6): 247 (1955); *Ind. Eng. Chem.*, **48**: 934 (1956), **49**: 936 (1957), **50**: 753 (1958).

Plant and Process Economics

20. Ashton, H. W., and G. T. Meiklejohn: Cost Estimating in Process Development, *Soc. Chem. Ind.*, **32**: 27 (1950).
21. Bauman, H. C.: Estimating Costs of Plant Auxiliaries, *Chem. Eng. Progr.*, **51**(1): 44-J (1955).
22. Beattie, R. D., and J. E. Vivian: Cost Estimation Glossary, *Chem. Eng.*, **60**(1): 172 (1953).
23. Bottomley, H.: Definitive Cost Estimating, *Petrol. Refiner*, **32**(10): 110 (1953).
24. Bottomley, H.: How to Prepare Cost Estimates, *Petrol. Refiner*, **32**(9): 211 (1953).
25. Butler, C. A., Jr.: Keep Cost Estimates Realistic, *Chem. Eng.*, **62**(1): 171 (1955).
26. Chilton, C. H.: Cost Estimating Simplified, *Chem. Eng.*, **58**(6): 108 (1951).
27. Chilton, C. H.: Six-tenths Factor Applies to Complete Plant, *Chem. Eng.*, **57**(4): 112 (1950).
28. Chilton, C. H.: What Is Cost Engineering? *Chem. Eng.*, **64**(7): 237 (1957).
29. Chilton, C. H.: What Price Process Plants? *Chem. Eng.*, **58**(5): 164 (1951).
30. Cleveland, R.: Simple Approach to Cost Estimating, *Prod. Eng.*, **24**(7): 171 (1953).
31. Duff, B. S.: Economics of Ammonia Manufacture, *Chem. Eng. Progr.*, **51**(1): 125 (1955).
32. Dybdal, E. C.: Engineering and Economic Evaluation of Projects, *Chem. Eng. Progr.*, **46**(1): 57 (1950).
33. Ferencz, P.: Statistical Analysis of Cost Estimates, *Chem. Eng. News*, **29**: 4158 (1951).
34. Gilmore, J. F.: Short-cut Estimating of Processes, *Petrol. Refiner*, **32**(10): 97 (1953).
35. Guthmann, W. S., and P. R. Inman: Cost Estimating in a Multipurpose Plant, *Ind. Eng. Chem.*, **44**: 2832 (1952).
36. Hamilton, J. P.: Economic Analysis in Petroleum Refining, *Petrol. Refiner*, **32**(10): 102 (1953).
37. Horn, A. B., et al.: Economic Analysis in Chemical Plants, *Petrol. Refiner*, **32**(10): 102 (1953).
38. Jelen, F. C.: Watch Your Cost Analysis, *Chem. Eng.*, **63**(6): 247 (1956).
39. Lang, H. J.: Cost Relationships in Preliminary Cost Estimates, *Chem. Eng.*, **54**(10): 117 (1947).
40. Lang, H. J.: Engineering Approach to Preliminary Cost Estimates, *Chem. Eng.*, **54**(9): 130 (1947).

41. Lang, H. J.: Simplified Approach to Preliminary Cost Estimates, *Chem. Eng.*, **55**(6): 112 (1948).

42. Lynn, L., and J. R. McKlveen: Simplify Your Cost Estimates by Nomographs, *Chem. Eng.*, **60**(4): 193 (1953).

43. Page, E. C.: Equipment for Small-scale Chlorination Plants, *Cost Eng.*, **3**(1): 9; **3**(2): 55; **3**(3): 85 (1958).

44. Rohrdanz, R. C.: Design for Low Construction Costs, *Chem. Eng.*, **65**(6): 133 (1958).

45. Samaniego, J. A., and C. R. Nelson: Cost Estimation in the Development of a New Process, *Chem. Eng. Progr.*, **52**(11): 471 (1956).

46. Schweyer, H. E.: How Inventory Costs Affect Your Process Economics, *Chem. Eng.*, **60**(10): 188 (1953).

47. Smith, C. A.: Cost Indexes, *Cost Eng.*, **2**(4): 110 (1957).

48. Smith, R. B., and T. Dresser: Economic Consideration in Process Design, *Chem. Eng. Progr.*, **51**(12): 544 (1955).

49. Stahl, R., and J. E. Kasch: Chemical Engineering Economics, *Chem. Eng.*, **58**(2): 270 (1951).

50. Symposia: Economic Evaluation, *Chem. Eng. Progr.*, **52**(10): 399 (1956).

51. Timpe, T. W.: Optimum Design Capacity, *Chem. Eng. Progr.*, **54**(1): 57 (1958).

52. Van Noy, C. W., et al.: Guide for Making Cost Estimates for Chemical Type Operations, U.S. Bureau of Mines, November, 1949.

53. Wells, A. J., and S. A. Senger: Predesign Cost Estimates, *Ind. Eng. Chem.*, **43**: 2309 (1951).

54. Wessel, H. R.: How to Estimate Costs in a Hurry, *Chem. Eng.*, **60**(1): 168 (1953).

55. Wilcoxon, B. H.: Unit Cost of Some Complete Plants, *Chem. Eng.*, **54**(5): 112 (1947).

General Equipment Costs

56. Behrens, J. R.: Estimating Installations of Chemical Equipment, *Chem. Eng.*, **54**(7): 106 (1947).

57. Bliss, H.: Data for Equipment Cost Estimates, *Chem. Eng.*, **54**(5): 126; **54**(6): 100 (1947).

58. Chilton, C. H.: Cost Data Correlated, *Chem. Eng.*, **56**(6): 97 (1949).

59. Happel, J., et al.: Equipment Costs and Other Items in Engineering Economics, *Chem. Eng.*, **53**(12): 97 (1946).

60. Happel, J., et al.: Estimating Chemical Engineering Equipment Costs, *Chem. Eng.*, **53**(10): 99 (1946).

61. Jandvisevits, P.: Preliminary Estimating by Selective Unit Costs, *Ind. Eng. Chem.*, **43**: 2299 (1951).

62. Kistin, H. R., C. S. Cameron, and A. P. Carter: Installed Equipment Costs per Unit of Production Capacity, *Chem. Eng.*, **60**(11): 192 (1953).

63. Molaison, H. J., et al.: Chemical Equipment Costs, *Oil Gas J.*, vol. 48, p. 159, Aug. 11; p. 107, Sept. 8; p. 299, Sept. 22; p. 91, Sept. 29; p. 343, Oct. 6, 1949.

64. Molaison, H. J., et al.: Pilot Plant Equipment Costs, *Chem. Eng.*, **57**(4): 110 (1950).

65. Reys, J.: Equipment Costs of Graphite Equipment, *Chem. Eng.*, **65**(4): 137 (1958).

66. Williams, R.: Standardizing Cost Data on Process Equipment, *Chem. Eng.*, **54**(6): 102 (1947).

67. Zimmerman, O. T., and I. Lavine: General Equipment Cost Data, *Chem. Eng. Costs Quart.*, **3**(1): 21 (1953), **4**(1): 4 (1954); *Cost Eng.*, **2**(4): 113 (1957).

Gas Handling

68. Densler, R.: Blower and Fan Costs, *Chem. Eng.*, **59**(10): 130 (1952).
69. Gerow, G. P.: High Vacuum Equipment, *Chem. Eng. Costs Quart.*, **2**(4): 80 (1952).
70. Jorgensun, R.: Fans, *Chem. Eng. Costs Quart.*, **5**(4): 84 (1955).
71. Katell, S., and J. P. McGee: Air Compressor Costs, *Cost Eng.*, **2**(1): 5 (1957).
72. Nelson, W. O.: Gas Moving Equipment, *Oil Gas J.*, vol. 48: Compressors, p. 223, June 23; Blowers, p. 91, June 30; Steam Jet Ejectors, p. 377, Nov. 17, 1949.
73. Tallman, J. C.: Ejectors Show Low First Cost, *Chem. Eng.*, **60**(1): 176 (1953).
74. Reha, T. R., and J. S. Quill: Gas Turbines and Centrifugal Compressors, *Oil Gas J.*, **51**: 113 (Feb. 9, 1953).
75. Zimmerman, O. T., and I. Lavine: Inert Gas Generators, *Chem. Eng. Costs Quart.*, **3**(1): 18 (1953).

Solids Handling

76. Arcand, H. J.: Chemical Feeders, *Chem. Eng. Costs Quart.*, **6**(4): 97 (1956).
77. Arcand, H. J.: Lime Slakers and Silica Activators, *Chem. Eng. Costs Quart.*, **6**(4): 102 (1956).
78. Fox, L. E.: Estimate of Cost of Screw Conveyers, *Chem. Eng.*, **56**(11): 128 (1949).
79. Hudson, W. B.: Cutting Costs—Materials Handling, *Chem. Eng.*, **56**(10): 102 (1949).

Pumps

80. Cramer, G. W.: Stainless Steel Centrifugal Pumps, *Chem. Eng. Costs Quart.*, **3**(4): 117 (1953).
81. Kluna, B. B.: Packless Pumps, *Chem. Eng. Costs Quart.*, **2**(4): 88 (1957).
82. Lundeen, R. W., and W. G. Clark: Cost of Installing Centrifugal Pumps, *Chem. Eng.*, **62**(8): 189 (1955).
83. Zimmerman, O. T., and I. Lavine: Cost of Stainless Steel Centrifugal Pumps, *Chem. Eng. Costs Quart.*, **3**(4): 117 (1953).

Size Reduction

84. Fattu, D. S.: Crushing and Grinding Costs, *Cost Eng.*, **3**(1): 15 (1958).
85. Hoskins, R. J.: Entolater Impact Mill and Aspirator, *Chem. Eng. Costs Quart.*, **2**(4): 100 (1952).
86. Mulcahy, P. H.: Reitz Disintegrators, *Cost Eng.*, **2**(2): 49 (1957).
87. Stern, A. L.: Size Reduction Equipment, *Chem. Eng. Costs Quart.*, **5**(4): 105 (1955); **6**(2): 50 (1956).

Mixing

88. Carlson, G. A.: Continuous Mixers, *Cost Eng.*, **3**(2): 44 (1958).
89. Carlson, G. A.: Horizontal Ribbon Type Batch Blendors, *Cost Eng.*, **3**(1): 4 (1958).
90. Diltz, J. L.: Mixing Equipment, *Chem. Eng. Costs Quart.*, **6**(2): 42 (1956).
91. Lennon, J. J.: Nettco Flomix Continuous Mixers, *Chem. Eng. Costs Quart.*, **5**(3): 65 (1955).
92. Lewis, G. E.: Your Guide to Mixer Costs, *Chem. Eng.*, **60**(1): 191 (1953).

93. Petrey, J. K.: Patterson Kelley Twin Shell Blendors, *Chem. Eng. Costs Quart.*, **5**(3): 60 (1955).

94. Zimmerman, O. T., and I. Lavine: Propeller Type Agitators, *Chem. Eng. Costs Quart.*, **3**(3): 74 (1953).

Separations

95. Apt, J., Jr.: Dorfan Impingo Filter, *Chem. Eng. Costs Quart.*, **5**(3): 75 (1955).

96. Chalmers, J. M., L. R. Elledge, and H. F. Porter: Filters, *Chem. Eng.*, **62**(6): 191 (1955).

97. Dellavalle, J. M.: Dust Collector Costs, *Chem. Eng.*, **60**(11): 177 (1953).

98. Dermody, J. L.: Cost of Colloidal Separators, *Chem. Eng. Costs Quart.*, **4**(3): 82 (1954).

99. Fleming, M. C.: Thickeners and Clarifiers, *Chem. Eng. Costs Quart.*, **1**(4): 53 (1951).

100. Flood, J. E.: Centrifugals, *Chem. Eng.*, **62**(6): 217 (1955).

101. Gery, W. B.: Thickeners, *Chem. Eng.*, **62**(6): 228 (1955).

102. Kracklauer, F. W.: Liquid Pressure Filters, *Chem. Eng. Costs Quart.*, **6**(3): 65 (1956).

103. Kriegel, P.: Filter Presses, *Chem. Eng. Costs Quart.*, **5**(2): 36 (1955).

104. Nelson, W. L.: Petrochemical Filter Press Costs, *Oil Gas J.*, **48**: 81 (June 2, 1949).

105. Samfield, M.: Dust Collecting Equipment, *Cost Eng.*, **2**(1): 106 (1957).

106. Shera, W. S.: Vibrating Screens, *Chem. Eng. Costs Quart.*, **6**(3): 94 (1954).

107. Smith, J. C.: Cost and Performance of Centrifugals, *Chem. Eng.*, **59**(4): 140 (1952).

108. Stasting, E. P.: Electrostatic Precipitators, *Chem. Eng. Costs Quart.*, **6**(2): 32 (1956).

109. Zimmerman, O. T., and I. Lavine: Hersey Reverse-jet Dust Filter, *Chem. Eng. Costs Quart.*, **4**(1): 9 (1954).

Heat Transfer

110. Bridges, F. L.: Platecoil Heat Transfer Units, *Chem. Eng. Costs Quart.*, **3**(3): 65 (1953).

111. Degler, H. E.: Cooling Towers and Air-cooled Exchangers, *Oil Gas J.*, **50**: 76 (Sept. 6, 1951).

112. Ellingen, W. E.: Condenser Costs, *Chem. Eng.*, **62**(3): 116 (1955).

113. Michell, A. M.: Shell and Tube Exchangers, *Chem. Eng. Costs Quart.*, **3**(2): 38 (1953).

114. Nelson, W. L.: Exchangers, *Oil Gas J.*, **47**: 117 (Oct. 28, 1948).

115. Rubin, F. L.. Heat Exchanger Costs, *Chem. Eng.*, **60**(10): 202 (1953).

116. Zimmerman, O. T., and I. Lavine: Cost of Salt-heating Units, *Chem. Eng. Costs Quart.*, **3**(4): 111 (1953).

Distillation

117. Bragg, L. R.: Stedman Packed Columns, *Chem. Eng. Costs Quart.*, **3**(3): 84 (1953).

118. Donovan, J.: Cost Estimation of Fabricated Plate Equipment, *Chem. Eng. Progr.*, **50**: 320 (1954).

119. Katzen, R.: Help in Picking the Right Tray, *Chem. Eng.*, **62**(11): 209 (1955).

120. Nelson, W. L.: *Oil Gas J.*—Fractionating Towers, **47**: 277 (Dec. 30, 1948), **48**: 93 (Jan. 6, 1949); Pipe Stills and Furnaces, **48**: 199 (Apr. 7, 1949), **48**: 125 (Apr. 14, 1949), **48**: 339 (June 28, 1949).

Drying

121. Crites, G. J.: Vacuum Drying, *Chem. Eng. Costs Quart.*, **6**(1): 4 (1956).
122. Lapple, W. C., W. E. Clark, and E. C. Dybdal: Drying Design and Costs, *Chem. Eng.*, **62**(8): 189 (1955).
123. Maquire, J. F.: Atms. Drum Dryers and Flakers, *Chem. Eng. Costs Quart.*, **6**(2): 37 (1956).
124. Murray, F. V., Jr.: Spray Drying Equipment, *Chem. Eng. Costs Quart.*, **6**(4): 89 (1956).
125. Russell, R. S.: Dryers and Drying Costs, *Chem. Eng. Costs Quart.*, **5**(4): 96 (1955).
126. Zimmerman, O. T., and I. Lavine: Adsorptive Dryers, *Chem. Eng. Costs Quart.*, **3**(1): 9 (1953).

Evaporation

127. Kohlins, W. D., and H. P. Englander: Cost Factors in Evaporator Design, *Chem. Eng. Progr.*, **52**(2): 45 (1956).
128. Williams, G. C.: Report on Evaporator Costs, *Chem. Eng.*, **60**(4): 156 (1953).

Reactors, Vessels, Tanks

129. Boberg, I. E., and W. R. Fickett: Relative Costs of Alternate Types of Reactor Vessel Construction, *Petrol. Processing*, **8**(5): 690 (1953).
130. Clark, W. G.: Tank Foundations, *Cost Eng.*, **1**(1): 12 (1956).
131. Cottrell, C. E.: Estimation of Vessel Costs, *Chem. Eng.*, **60**(2): 143 (1953).
132. How, H.: Short Cut Estimation of Welded Process Vessels, *Chem. Eng.*, **55**(1): 122 (1948).
133. Nelson, W. L.: *Oil Gas J.*, vol. 47, vessels, p. 113, Dec. 16, p. 81, Dec. 23; tank, p. 123, Nov. 18, p. 133, Nov. 25, 1948.
134. Plummer, F. L.: Field Erected Storage Tanks, *Chem. Eng. Costs Quart.*, **2**(3): 53 (1952).
135. Zimmerman, O. T., and I. Lavine: Haveg Equipment, *Chem. Eng. Costs Quart.*, **2**(4): 89 (1952).
136. Zimmerman, O. T., and I. Lavine: Mixing Tanks, *Chem. Eng. Costs Quart.*, **3**(3): 83 (1953).
137. Zimmerman, O. T., and I. Lavine: Cost of Cast Iron Process Vessels, *Chem. Eng. Costs Quart.*, **3**(4): 97 (1953).
138. Zimmerman, O. T., and I. Lavine: Wood Tanks, *Cost Eng.*, **2**(1): 20 (1957).

Piping, Control, and Instrumentation

(See Chap. 9.)

Building and Construction

(See Chap. 8.)

Capital Costs

139. Bauman, H. C.: Accuracy Considerations for Capital Cost Estimation, *Ind. Eng. Chem.*, **50**(4): 57A (1958).
140. Chiswell, E. G., and J. J. Merrill: Capital Costs Considerations, *Petrol. Refiner*, **33**(6): 127 (1954).
141. Jelen, F. C.: Next Time Use Capitalized Costs, *Chem. Eng.*, **61**(2): 199 (1954).
142. Lynn, L.: Make the Most of Capital Ratios, *Chem. Eng.*, **61**(4): 175 (1954).
143. Nackney, J. W.: Outline of Capital Cost Estimating, *Cost Eng.*, **1**(1): 27 (1956).

144. Neidig, C. P.: Is Capital Really Tight in the Process Industries? *Chem. Eng. Progr.*, **52**: 269 (1956).
145. Nichols, W. T.: Capital Cost Estimating, *Ind. Eng. Chem.*, **43**: 2295 (1951).
146. Schweyer, H. E.: Capital Ratios Analyzed, *Chem. Eng.*, **59**(1): 164 (1952).
147. Symposium: Capital Cost Estimation, *Chem. Eng. Progr.*, **52**: 171 (1956).
148. Tiel, R. J.: Importance of Complete and Accurate Cost Estimates, *Chem. Eng. Progr.*, **52**(5): 187 (1956).

Labor and Operation

149. Bechtel, V. R.: Inflation in Production and Operating Costs, *Ind. Eng. Chem.*, **43**: 2307 (1951).
150. Cusack, B. L.: Human Engineering and Direct Labor Costs, *Chem. Eng. Progr.*, **53**(10): 471 (1957).
151. Gropper, F.: Direct Labor Costs and Chemical Plants, *Chem. Eng. Progr.*, **53**(10): 464 (1957).
152. Nelson, W. L.: Refinery Labor, *Oil Gas J.*, **48**: 97 (1949).
153. Newton, R. D., and R. S. Aries: Preliminary Estimating of Operating Costs, *Ind. Eng. Chem.*, **43**: 2309 (1951).
154. Sweet, E. R.: Preparation of Operating Cost Estimates, *Chem. Eng. Progr.*, **52**(5): 174 (1956).
155. Wessel, H. E.: New Graph Correlates Operating Labor Data for Chemical Processes, *Chem. Eng.*, **59**(7): 209 (1952).
156. Wobus, R. S.: Estimating Direct Operating Labor for New Processes, *Chem. Eng. Progr.*, **53**(12): 581 (1957).

Maintenance, Repairs, and Engineering Costs

157. Ahliness, R. L.: Schedule Your Maintenance for Minimum Cost, *Chem. Eng.*, **60**(5): 236 (1953).
158. Cziner, R. M.: How to Control Maintenance Costs, *Petrol. Refiner*, **33**(1): 106 (1954).
159. Darling, L. A.: Maintenance Organization and Operation in Chemical Plants, *Chem. Eng. Progr.*, **48**(1): 57 (1952).
160. Darling, L. A., and H. A. Bogle: Productivity in Chemical Plant Maintenance, *Chem. Eng. Progr.*, **50**(3): 164 (1954).
161. Glauz, R. L.: Estimating Maintenance Costs in New Plants, *Chem. Eng. Progr.*, **51**(3): 122 (1955).
162. Leonard, J. L.: Maintenance Costs, *Chem. Eng.*, **58**(10): 145 (1951); **59**(4): 150 (1952); **61**(2): 206 (1954).
163. O'Donnell, J. P.: New Correlation of Engineering and Other Indirect Costs, *Chem. Eng.*, **60**(1): 188, (4): 156 (1953).
164. Pierce, D. E., and W. I. McNeill: Control of Costs in Production, *Chem. Eng. Progr.*, **50**(11): 552 (1954).
165. Sayer, J. S.: DuPont's Practice with Maintenance Data, *Chem. Eng. Progr.*, **51**(11): 492 (1955).
166. Schwab, L., and B. G. Earnheart: Cut Repair Time and Costs with Alert Design, *Chem. Eng.*, **63**(10): 190 (1956).
167. Stratmeyer, R. J.: Your Key to Maintenance Savings, *Chem. Eng.*, **62**(9): 173 (1955).
168. Whitehead, S.: Chemical Plant Maintenance, *Chem. Eng.*, **59**(8): 167 (1952).
169. Woolfenden, L. B., and R. C. Thiede: Designing for Maintenance, *Chem. Eng. Progr.*, **48**: 115 (1952).

Utilities

(See Chap. 9 also.)

170. Bauman, H. C.: Estimating Costs of Process Plant Auxiliaries, *Chem. Eng. Progr.*, **51**(1): 45J (1955).

171. Carlise, V. J.: Some Economic Factors in Waste Water Treatment, *Chem. Eng. Progr.*, **46**(7): 328 (1950).

172. Clayton, C. C.: Steam-jet Refrigerating Systems, *Cost Eng.*, **2**(2): 42 (1957).

173. Hertz, D. B.: What Does Water Cost? *Chem. Inds.*, **66**(4): 512 (1950).

174. Jacobs, H. L.: Waste Treatment—Recovery and Disposal, *Chem. Eng.*, **62**(4): 185 (1955).

175. Nordell, E.: Water Treatment, *Chem. Eng.*, **62**(10): 175 (1955).

176. Streicher, L., et al.: Demineralization of Water, *Ind. Eng. Chem.*, **45**: 2394 (1953).

Depreciation, Interest, Taxes

177. Cannon, D. T.: Depreciation Policy Changes, *Chem. Eng.*, **65**(15): 70 (1958).

178. Digest of State Laws Relating to Taxes and Revenue, U.S. Dept. of Commerce, Bureau of Census, Washington, D.C., 1954.

179. Hartogenis, A. M., and H. D. Allen: Evaluate Your Depreciation Charges, *Chem. Eng.*, **61**(2): 195 (1954).

180. Jelen, F. C.: Consider Income Tax in Cost Analysis, *Chem. Eng.*, **64**(9): 271 (1957).

181. Lawrence, A. E.: Depreciation and Amortization, *Chem. Eng. Progr.*, **51**: 227 (1955).

182. Marston, A., R. Winfrey, and J. C. Hempstead: "Engineering Valuation and Depreciation," McGraw-Hill Book Company, Inc., New York, 1953.

183. Weaver, J. B., and R. J. Reilly: Interest Rate of Return Evaluation, *Chem. Eng. Progr.*, **52**(10): 405 (1956).

Packaging, Shipping

184. LaPointe, J. R.: Freight Costs for Equipment, *Chem. Eng.*, **60**(4): 213 (1953).

185. Nelson, W. L.: *Oil Gas J.*, vol. 52, ocean shipping, p. 137, Feb. 9, p. 113, Sept. 7; shipping viscous materials, p. 108, June 29, pipeline transportation rates, p. 351, Sept. 21, 1953.

186. Smith, M. A.: Getting at Your Handling Costs, *Chem. Eng.*, **62**(2): 193 (1955).

187. Smith, S. P.: If You Ship Process Products, *Chem. Eng.*, **60**(3): 222 (1953).

188. Strong, A. K.: Economics of Shipping in Larger Loads, *Chem. Eng.*, **62**(8): 178 (1955).

189. Tighe, F. C.: Rail, Motor, and Water Transportation of Chemicals, *Chem. Eng. News*, **31**: 752, 3538, 4916 (1953).

190. Uncles, R. F., and T. L. Carter: Watch Those Hidden Packaging Costs, *Chem. Eng.*, **60**(8): 185 (1953).

Sales, Research, Administration

191. Ericsson, R. L., and L. E. Johnson: Costs in Developing Market Know-how, *Ind. Eng. Chem.*, **47**: 992 (1955).

192. Hardy, W. L.: Research and Development Funds, *Ind. Eng. Chem.*, **48**(8): 43A (1956).

193. Ladd, H. D.: Economics of Chemical Selling, *Chem. Eng. News*, **30**: 4938 (1952).

194. Research Allocations in Industry, *Chem. Eng. News*, **34**: 2236 (1956).
195. Zabel, H. W.: The Exclusion Chart, *Chem. Eng. Progr.*, **52**(5): 183 (1956).

Economic Evaluation

196. Aries, R. S.: Venture Profitability in Economic Balance, *Chem. Eng. Progr.*, **46**(3): 115 (1950).
197. Fagley, W. L., and G. W. Blum: Calculate Payout Time for Your Investment, *Chem. Eng.*, **57**(7): 116 (1950).
198. Finalyson, K.: Rate Economic Factors by Importance, *Chem. Eng.*, **65**(1): 151 (1958).
199. Happel, J.: New Approach to Payout Calculations, *Chem. Eng.*, **58**(10): 146 (1951).
200. Hicks, S. S., and L. R. Steffen: Cost Estimation and Decision Making, *Chem. Eng. Progr.*, **52**(5): 191 (1956).
201. Jelen, F. C.: Capital Costs for Comparison of Alternatives, *Chem. Eng. Progr.*, **52**(10): 413 (1956).
202. Jelen, F. C.: Combined Effect of Rate of Return, Income Tax and Inflation, *Chem. Eng.*, **65**(2): 123 (1958).
203. Krase, N. W.: Criteria for Discontinuing Operating Investments, *Chem. Eng. Progr.*, **52**: 495 (1956).
204. Newton, R. D., and R. S. Aries: Break-even Charts, *Chem. Eng.*, **58**(2): 148 (1951).
205. Roth, R. J.: Break-even Charts for the Chemical Process Industries, *Chem. Eng. News*, **30**: 5437 (1952).
206. Sandal, M., Jr.: Re-evaluate Your Capital Investment, *Chem. Eng.*, **64**(11): 231 (1957).
207. Schuette, W. A.: Break-even Charts, *Prod. Eng.*, **24**(8): 170 (1953).
208. Schwartzkopf, O.: Efficiency Doesn't Always Pay, *Chem. Eng.*, **59**(8): 140 (1953).
209. Sherwood, P. W.: How to Prepare Preliminary Cost Evaluation Reports, *Petrol. Refiner*, **31**(6): 126 (1952).
210. Ven Eck, F. M.: Venture Capital Risk vs. Opportunity, *Chem. Eng.*, **59**(2): 192 (1952).
211. Wagner, H. R.: The Appropriation Request, *Chem. Eng. Progr.*, **52**: 402 (1956)

CHAPTER 7

Plant Location

1. Aries, R. S.: "Chemical Engineers' Handbook," 3d ed., pp. 1719–1730, McGraw-Hill Book Company, Inc., New York, 1950.
2. Aries, R. S.: "Encyclopedia of Chemical Technology," vol. 10, pp. 744–753, Interscience Publishers, Inc., New York, 1953.
3. Basic Guides for Locating a Petrochemical Plant, *Petrol. Processing*, **10**(10): 1587 (1955).
4. Bierwert, D. V., and F. A. Krone: How to Find Best Site for a New Plant, *Chem. Eng.*, **62**(12): 191 (1955).
5. Bradley, W. R.: Industrial Hygiene Considerations in Plant Location and Design, *Chem. Eng. News*, **29**: 1198 (1951).
6. Competition for Plant Location, *Chem. Eng. News*, **33**: 5021 (1955).
7. Groppe, H.: Trends in the Expansion of the Chemical Industry, *Chem. Eng. Progr.*, **51**(2): 81F (1955).

8. Hurley, N. P.: Atomic Vulnerability in the Chemical Process Industry, *Chem. Eng. News*, **33**: 3654 (1955).

9. Perry, J. H. (ed.): "Chemical Business Handbook," pp. 2–76, McGraw-Hill Book Company, Inc., New York, 1954.

10. Smith, S. P.: The Selection of a Plant Site, *Chem. Eng. Progr.*, **51**(3): 134 (1955).

11. Symposium: Chemical Plant Location, *Chem. Eng. Progr.*, **45**(5): 285–322 (1949).

Market Research

12. Chaddock, R. S. (ed.): "Chemical Market Research in Practice," Reinhold Publishing Corporation, New York, 1956.

13. Strickland, J. R., and J. E. R. Carrier: Consumers—the Chemical Industries' Future, *Chem. Eng. Progr.*, **54**(1): 65 (1958).

Water

14. Gilliland, E. R.: Chemical Engineering in Augmenting Water Resources, *Ind. Eng. Chem.*, **47**(12): 2410 (1955).

15. Nordell, E.: Water Sources and Treatment, *Chem. Eng.*, **62**(9): 183, (10): 175 (1955).

16. Symposium: Re-use of Water by Industry, *Ind. Eng. Chem.*, **48**(12): 2145–2171 (1956).

17. Wright, R. L.: Let Nature Cool Your Recycle Process Water, *Chem. Eng. Progr.*, **54**(2): 99 (1958).

Waste Disposal

(See Chap. 10 also.)

18. Air Pollution and Waste Treatment Workbook, *Ind. Eng. Chem.*, monthly.

19. Blum, G. W., and O. C. Thompson: Engineering Tools and Techniques for Cleaner Air, *Chem. Eng. Progr.*, **52**(8): 332 (1956).

20. Gurnham, C. F.: "Principles of Industrial Waste Treatment," John Wiley & Sons, Inc., New York, 1955.

21. Hood, D. W., B. Stevenson, and L. M. Jeffrey: Deep Sea Disposal of Industrial Wastes, *Ind. Eng. Chem.*, **50**(6): 885 (1958).

22. How to Cope with Water Pollution Problem, *Chem. Eng.*, **65**(14): 129 (1958).

23. Jacobs, H. L.: Survey of Waste Treatment Methods, *Chem. Eng.*, **62**(4): 184 (1955).

24. Rudolfs, W. (ed.): "Industrial Wastes—Their Treatment and Disposal," Reinhold Publishing Corporation, New York, 1953.

25. Symposium: Industrial Waste Treatment, *Chem. Eng. Progr.*, **46**(7): 321–343 (1950).

26. Zimmerman, F. J.: New Waste Disposal Process, *Chem. Eng.*, **65**(17): 117 (1958).

CHAPTER 8

General

1. Austin, C. T.: Check Your Design Jobs, *Chem. Eng.*, **57**(6): 137 (1950).

2. Boast, W. B.: "Illumination Engineering," 2d ed., McGraw-Hill Book Company, Inc., New York, 1953.

3. Brown, A. A.: What Size Foundations for Your Stills and Towers? *Chem. Eng.*, **59**(10): 136 (1952).

4. Coleman, H. S. (ed.): "Laboratory Design," Reinhold Publishing Corporation, New York, 1951.

5. Cushing, R.: Your Design Reference File, Section V—Heating, Ventilating and Air Conditioning, *Chem. Eng.*, **64**(8): 267 (1957).

6. Cushing, R.: Your Design Reference File, Section IX—Structural Engineering and Illumination, *Chem. Eng.*, **64**(12): 275 (1957).

7. Dunham, C. W.: "Planning Industrial Structures," McGraw-Hill Book Company, Inc., New York, 1948.

8. Harris, E. C.: Elements of Structural Engineering, The Ronald Press Company, New York, 1954.

9. How to Plan Your 195X Plant, *Factory Management and Maintenance*, **112**(6): A2 (1954).

10. Kemp, H. S., L. T. Mullen, and A. P. Guess: "Construction of Acid Recovery Units—Indoors or Outdoors," *Chem. Eng. Progr.*, **47**(7): 339 (1951).

11. Kieweg, H.: Safety and Outdoor Construction, *Chem. Eng. Progr.*, **47**(7): 341 (1951).

12. Kidder, F. E., and H. Parker: "Architects' and Builders' Handbook," 18th ed., John Wiley & Sons, Inc., New York.

13. Lin, T. Y.: "Design of Prestressed Concrete Structures," John Wiley & Sons, Inc., New York, 1955.

14. Marshall, V. O.: Foundation Design Handbook for Stacks and Towers, *Petrol. Refiner*, **37**(5): supplement 1-16 (1958).

15. Merritt, F. S. (ed.): "Handbook of Building Construction," McGraw-Hill Book Company, Inc., New York, 1958.

16. Parker, H., and J. W. MacGuire: "Simplified Site Engineering for Architects and Builders," John Wiley & Sons, Inc., New York, 1954.

17. Perry, C. W.: Integration of Engineering Skills in Chemical Plant Construction, *Chem. Eng. Progr.*, **50**(8): 382 (1954).

18. Peurifoy, R. L.: "Construction Planning, Equipment and Methods," McGraw-Hill Book Company, Inc., New York, 1956.

19. Seelye, E. E.: "Foundations: Design and Practice," John Wiley & Sons, Inc., New York, 1956.

20. Store-bought Buildings, *Chem. Eng.*, **61**(4): 140 (1954).

21. Sutherland, H., and H. L. Bowman: "Structural Theory," 4th ed., John Wiley & Sons, Inc., New York, 1950.

22. Williams, C. D., and C. E. Cutts: "Structural Design in Reinforced Concrete," The Ronald Press Company, New York, 1954.

23. Williams, W. H.: Chemical Plant Operations and the Weather, *Chem. Eng. Progr.*, **47**(6): 277 (1951).

Construction Economics

24. Annual Report on Construction Costs, *Engineering News-Record*.

25. Bauman, H. C.: Estimating Costs of Process Plant Auxiliaries, *Chem. Eng. Progr.*, **51**(1): 45J (1955).

26. Bean, T. W.: Estimating Construction Costs in Chemical Process Industries, *Ind. Eng. Chem.*, **43**: 2302 (1951).

27. De Simone, R. E.: Costs in Construction, *Chem. Eng. Progr.*, **50**(8): 379 (1954).

28. Groseclose, C. E.: Cost Comparisons of Air Conditioning Refrigerant Condensing Systems, *Refrig. Eng.*, **62**(6): 132 (1954).

29. Kiernan, F.: Building Costs, *Cost Eng.*, **2**(2): 55 (1957).

30. Kriptow, P. E.: How to Evaluate Economy of Construction Materials, *Chem. Eng.*, **60**(1): 204 (1953).

31. Lawrence, J. C.: Cost Indices for Construction, *Ind. Eng. Chem.*, **46**(8): 65A (1954).

32. Means, R. S.: *Building Construction Cost Data*, Duxbury, Mass. (annual).

33. Minevitch, J. R., G. B. Knight, S. E. Root, and H. E. Boraks: Chemical Plant Construction Cost, Indoors vs. Outdoors, *Chem. Eng. Progr.*, **47**(8): 385 (1951).

34. Peurifoy, R. L.: "Estimating Construction Costs," McGraw-Hill Book Company, Inc., New York, 1953.

35. Rohrdanz, R. C.: Design for Low Construction Costs, *Chem. Eng.*, **65**(6): 133 (1958).

36. Weather Forecasts Save Construction Dollars, *Chem. Eng.*, **61**(3): 124 (1954).

37. What Is Lowest Cost, One Story Building? *Factory Management and Maintenance*, **112**(4): 98 (1954).

CHAPTER 9

Piping—General

1. Bigham, J. E.: Pressure Relief Devices, *Chem. Eng.*, **65**(3): 133, (7): 143 (1958).

2. Blumberg, H. S.: Steam Piping Materials for High Temperature Service, *Materials and Methods*, **45**(3): 126 (1957).

3. Crocker, S. (ed.): "Piping Handbook," 4th ed., McGraw-Hill Book Company, Inc., New York, 1954.

4. Cushing, R.: Your Design Reference File, sec. II, Piping, *Chem. Eng.*, **64**(4): 271 (1957).

5. Elliott, P. M.: Thermoplastic Materials for Pipe, *Corrosion*, **13**(10): 49 (1957).

6. Heiss, J. F., and H. C. Bromer: Cut Out Trial-and-Error in Series-Parallel Pipe Flow Design, *Chem. Eng.*, **58**(6): 112 (1951).

7. Littleton, C. T.: "Industrial Piping," McGraw-Hill Book Company, Inc., New York, 1951.

8. Kennedy, W. L., and C. C. Stueve: Sizing Crude-oil Pipe Lines, *Oil Gas J.*, **52**: 183, 264 (Sept. 21, 1953).

9. Lowenstein, J. G.: Calculate Adequate Rupture Disk Sizes, *Chem. Eng.*, **65**(1): 157 (1958).

10. McLaughlin, C. B.: Piping Materials for Chemical Processes, *Heating, Piping and Air Conditioning*, **23**(10): 85 (1951).

11. Merrimen, J. C.: Mechanical Tubing, *Materials & Methods*, **46**(7): 127 (1957).

12. Norton, F. H.: "Refractories," 3d ed., McGraw-Hill Book Company, Inc., New York, 1949.

13. Olive, T. R.: Process Piping, *Chem. Eng.*, **60**(12): 187 (1953).

14. Picardi, E. A.: How to Apply Method of Slope Deflection to Thermal Stress Analysis of Piping, *Petrol. Processing*, **8**(3): 368 (1953).

15. Piping Special Report, *Petrol. Refiner*, **37**(3): 136–161 (1958).

16. Thomas, R.: Thermal Insulation for Industrial Requirements, *Petrol. Refiner*, **31**: 1–11 (1952).

17. Weiss, W. A.: Design Your Piping to Cut Maintenance Costs, *Petrol. Refiner* **37**(1): 141 (1958).

Piping—Economics

18. Bach, N. G.: Fabricating Costs of Steel Piping, *Chem. Eng. Costs Quart.*, **5**(1): 17 (1955).

19. Bayard, R. A.: Pick Off Economic Insulation Thickness, *Chem. Eng.*, **57**(6): 142 (1950).

20. Braca, R. M., and J. Happel: Economic Pipe Sizing Brought Up to Date, *Chem. Eng.*, **60**(1): 180 (1953).

21. Dickson, R. A.: Pipe Cost Estimation, *Chem. Eng.*, **57**(1): 123 (1950).
22. Downs, G. F., and G. R. Tait: Selecting Pipeline Diameter for Minimum Investment, *Oil Gas J.*, **52**: 210 (Nov. 16, 1953).
23. Hardy, W. L.: Economical Piping Can Be Made of Stainless Steel, *Ind. Eng. Chem.*, **48**(6): 79A (1955).
24. Mattiza, D. S.: Piping and Electrical Work, *Chem. Eng. Costs Quart.*, **3**(1): 19 (1953).
25. Marvis, N. B.: Economics of Increasing Capacity of Pipe Line Systems, *Oil Gas J.*, **50**: 118 (May 17, 1951).
26. Nelson, W. L.: *Oil Gas J.*, vol. 48, clay and cement pipe, p. 99, Mar. 3; tubes and bends, p. 243, Apr. 21; pumps, p. 103, May 5; pump materials and drives, p. 121, May 12; piping and tubing, p. 109, Aug. 14; pipe line construction, p. 159, Aug. 18; refinery valves, p. 143, Aug. 25; fittings and flanges, p. 71, Sept. 1; piping, p. 137, Sept. 15, 1949.
27. Symposium: Special Report on Costs, *Petrol. Refiner*, **37**(6): 127–170 (1958).
28. Zimmerman, O. T., and I. Lavine: Cost of Pipe and Nipples, *Chem. Eng. Costs Quart.*, **5**(3): 78 (1955).

Instrumentation and Control—General

29. Automatic Plants, *Chem. Eng. News*, **35**: 42 (1957).
30. Ceaglske, N. H.: "Automatic Process Control for Chemical Engineers," John Wiley & Sons, Inc., New York, 1956.
31. Considine, D. M. (ed.): "Process Instruments and Controls Handbook," McGraw-Hill Book Company, Inc., New York, 1957.
32. Cushing, R.: Your Design Reference File, section IX–Instrumentation, *Chem. Eng.*, **64**(12): 275 (1957).
33. Eckman, D. P.: "Industrial Instrumentation," John Wiley & Sons, Inc., New York, 1950.
34. Eckman, D. P.: "Principles of Industrial Process Control," John Wiley & Sons, Inc., New York, 1945.
35. Holzbock, W. G.: "Automatic Control: Principles and Practice," Reinhold Publishing Corporation, New York, 1958.
36. Holzbock, W. G.: "Instruments for Measurement and Control," Reinhold Publishing Corporation, New York, 1955.
37. Instrumentation Workbook, *Industrial and Engineering Chemistry*, monthly feature.
38. Johnson, E. F.: Automatic Process Control, "Advances in Chemical Engineering," vol. II, pp. 34–79, Academic Press, Inc., New York, 1958.
39. Nonamaker, J. N.: The Instrument Engineer in Process Design, *Ind. Eng. Chem.*, **43**(12): 2911 (1951).
40. Reviews on Process Control and Automation, *Industrial and Engineering Chemistry*, March issue (annual).
41. Rhodes, T. J.: "Industrial Instruments for Measurement and Control," McGraw-Hill Book Company, Inc., New York, 1941.
42. Rushton, J. H.: Applications of Electronic Process Control Instrumentation, *Chem. Eng. Progr.*, **52**(11): 485 (1956).
43. Symposium: Process Instrumentation, *Ind. Eng. Chem.*, **46**(7): 1371 (1954).
44. Symposium: Instrumentation for Pilot Plants, *Ind. Eng. Chem.*, **45**(9): 1836 (1953).
45. Symposium: Instrumentation and Control, *Petrol. Refiner*, **31**(12): 97–131 (1952).

46. Wade, W. F., and E. N. Kemler: Automatic Control Bibliography, Summary Reports, Spring Park, Minn., 1955.

Instrumentation and Control—Economics

47. Considine, D. M., and J. J. Kennedy: Instrument Cost Estimation, *Chem. Eng. Costs Quart.*, **4**(2): 32 (1954).
48. Hull, J. C., and D. Tricebok: Cost of Rotameters, *Chem. Eng. Costs Quart.*, **4**(3): 68 (1954).
49. Warren, A. S., and V. A. Pards: Instrument Costs, *Cost Eng.*, **2**(1): 10 (1957).

Power Systems—General

50. Knowlton, A. E. (ed.): "Standard Handbook for Electrical Engineers," 9th ed., McGraw-Hill Book Company, Inc., New York, 1957.
51. Marks, L. S. (ed.): "Mechanical Engineers' Handbook," 5th ed., McGraw-Hill Book Company, Inc., New York, 1951.
52. Pender, H. (ed.): "Electrical Engineers' Handbook," 4th ed., John Wiley & Sons, Inc., New York, 1950.
53. Salisbury, J. K. (ed.): "Kent, Mechanical Engineers' Handbook—Power," John Wiley & Sons, Inc., New York, 1950.

Power Systems—Economics

54. Bauman, H. C.: Estimating Costs of Process Plant Auxiliaries, *Chem. Eng. Progr.*, **51**(1): 45J (1955).
55. Bayerlein, R. W.: Engines for Cheap Power, *Chem. Eng.*, **60**(7): 118 (1953).
56. Durham, E.: Cost of Steam Generating Equipment, *Chem. Eng. Costs Quart.*, **4**(2): 41 (1954).
57. Fernside, T. A., and F. C. Cheney: Fast Estimate for Power Plant Costs, *Chem. Eng.*, **60**(6): 239 (1953).
58. Katell, S., and T. A. Joyce: How to Allocate Process Steam Costs, *Chem. Eng.*, **65**(5): 152 (1952).
59. Knowlton, A. E.: Steam Station Cost Survey, *Elec. World*, **133**: 94 (Apr. 24, 1950).
60. McCabe, J. C.: Higher Costs Spark Advances in Process Steam and Power, *Chem. Eng.*, **57**(5): 121 (1950).
61. Pierce, D. E.: How to Control Costs of Kilowatts, *Chem. Eng.*, **60**(1): 195 (1953).
62. Wilson, W. B.: Should Your Plant Produce Power? *Chem. Eng.*, **60**(3): 235 (1953).

CHAPTER 10

General References

1. Benedict, M., and Thomas Pigford: "Nuclear Chemical Engineering," McGraw-Hill Book Company, Inc., New York, 1957.
2. Bonilla, C.: "Nuclear Engineering," McGraw-Hill Book Company, Inc., New York, 1957.
3. Glasstone, S.: "Principles of Nuclear Reactor Engineering," D. Van Nostrand Company, Inc., Princeton, N.J., 1955.
4. Stephenson, R.: "Introduction to Nuclear Engineering," 2d ed., McGraw-Hill Book Company, Inc., New York, 1958.
5. "Nuclear Science Abstracts," bimonthly publication with comprehensive coverage of the literature of nuclear science and engineering, Superintendent of Documents, Government Printing Office, Washington, D.C.

6. "Technical Progress Reviews," AEC quarterly publication on reactor fuel processing, technology, and materials, Superintendent of Documents, Government Printing Office, Washington, D.C.

7. *Proceedings of International Conferences on the Peaceful Use of Atomic Energy*, United Nations Publication, New York, 1956, 1959.

8. "Progress in Nuclear Energy," Series I–VIII, Pergamon Press, New York, 1956–1958. In particular, Series III, Process Chemistry; Series IV, Technology and Engineering.

9. Etherington, Harold (ed.): "Nuclear Engineering Handbook," McGraw-Hill Book Company, Inc., New York, 1958. A comprehensive coverage of nuclear engineering.

10. "Chemical Processing and Equipment," U.S. Atomic Energy Commission, McGraw-Hill Book Company, Inc., New York, 1955.

11. *Nuclear Science and Engineering*, magazine published periodically by Academic Press, New York City.

12. *Nucleonics*, monthly magazine published by McGraw-Hill Publishing Company, Inc., New York.

13. Feature articles and symposia, *Chemical Engineering, Chemical Engineering Progress*, and *Industrial and Engineering Chemistry*.

Chemical Plant Design

14. Basel, L., and J. Koslov: Pyrometallurgical Reprocessing Plant, *Nucleonics*, **15**(8): 56 (1957).

15. Bresee, J.: Damaging Effects of Radiation on Chemical Materials, *Nucleonics*, **14**(9): 75 (1956).

16. Bruce, F. R.: Chemical Processing Aqueous Blanket and Fuel from Thermal Breeder Reactors, *Chem. Eng. Progr.*, **52**(9): 347 (1956).

17. Bruce, F. R.: Nuclear Fuel Reprocessing by 1965, *Chem. Eng.*, **64**(7): 202 (1957).

18. Dryden, C. E., and J. M. Frame: Batch versus Continuous Processing, *Chem. Eng. Progr.*, **52**(9): 371 (1956).

19. Editors of *Nucleonics*: Review of Chemical Processes from Geneva Conference Papers, *Nucleonics*, **13**(9): 59–63 (1955). (1) Survey of Fuel Separation Processes; (2) Distribution Coefficients; (3) Solvent Extraction Processes; (4) Typical Plant Processes; (5) Safe Criticality Limits for Pu Solutions.

20. Graham, R. H.: Process Heat from Nuclear Reactors, *Chem. Eng.*, **63**(3): 191 (1956).

21. Hill, O. F., and V. R. Cooper: Scale-up Problems in the Plutonium Separations Program, *Ind. Eng. Chem.*, **50**(4): 599 (1958).

22. Jealous, A. C., and H. F. Johnson: Power Requirements for Pulse Generations in Pulse Columns, *Ind. Eng. Chem.*, **47**: 1159–1166 (1955).

23. Lansing, N. F. (ed.): "The Role of Engineering in Nuclear Energy Development," AEC, TID 5031. Shielding by E. P. Blizard, pp. 332–352; Radioactive Wastes by R. J. Morton and W. K. Eister, pp. 352–411.

24. Lawroski, S., and W. Rodger: New Processes Promise More Economic Fission Product Removal, *Chem. Eng. Progr.*, **53**(2): 70-F (1957).

25. Nicholls, C. M.: Criteria for Selection of Equipment for Radioactive Separation Processes, *Chem. Eng. Progr.*, **52**: 78 (1956).

26. Nicholls, C. M., and A. S. White: The Development of Radiochemical Processes, *Chem. Eng. Progr. Symposium Series*, **50**(13): 129 (1954).

27. Nuclear Engineering—A Chemical Engineering Review, *Chem. Eng. Progr.*, **50**: 217–220 (1954).

28. Ohlgren, H. A., J. G. Lewis, and M. Weech: Reprocessing Reactor Fuels, *Nucleonics*, **13**(3): 18–21 (1955).
29. Reid, D. G., and K. K. Kennedy: Direct Maintenance Fuel Reprocessing Plant Is Practical, *Chem. Eng. Progr.*, **52**(9): 394 (1956).
30. Rohrmann, C.: Processing Engineering at the Hanford Separation Plants, *Nucleonics*, **14**(6): 66 (1956).
31. Sadowski, G.: Decontamination of Processing Plants, *Nucleonics*, **15**(3): 68 (1957).
32. Technical Information Service, AEC: A Bibliography of Selected AEC Reports of Interest to Industry, part 2, Chemistry and Chemical Engineering, TID 3050.
33. Unique Uranium Reprocessing Plant, *Chem. Eng.*, **63**(1): 120 (1956).
34. Where Maintenance Is Really Tough, *Chem. Eng.*, **61**(6): 230 (1954).

Safety

35. Braidech, M. M.: The Problem of Insuring Nuclear Installations, *Chem. Eng. Progr.*, **51**(11): 513 (1955).
36. Broido, A.: Hazard Defined, *Nucleonics*, **13**(3): 82 (1955).
37. Eisenbud, M., H. Blatz, and E. V. Barry: How Important Is Surface Contamination? *Nucleonics*, **12**(8): 12 (1954).
38. Graham, R. H. (AEC): U.S. Reactor Operating History 1943–1954, *Nucleonics*, **13**(10): 42 (1955).
39. Hurwitz, H., Jr.: Safeguard Considerations for Nuclear Power Plants, *Nucleonics*, **12**(3): 57 (1954).
40. Kinsman, S. (ed.): "Radiological Health Handbook," Radiological Health Training Section, Sanitary Engineering Center, Cincinnati, Ohio, 1955.
41. McCullough, C. R.: Reactor Safety, *Nucleonics*, **15**(9): 134 (1957).
42. McCullough, C. R.: "Safety Aspects of Nuclear Reactors," D. Van Nostrand Company, Inc., Princeton, N.J., 1957.
43. McCullough, C. R.: "The Safety of Nuclear Reactors," Geneva, P/853, United Nations, New York.
44. Mesler, R. S., and L. C. Widdoes: Evaluating Reactor Hazards from Airborne Fission Products, *Nucleonics*, **12**(9): 39 (1954).
45. Taylor, L. S.: Can We Legislate Ourselves into Radiation Safety? *Nucleonics*, **13**(3): 17 (1955).
46. "International Dose Handbook," National Bureau of Standards Handbook 59, Government Printing Office, Washington, D.C.

Shielding

47. Chappell, D.: Gamma Ray Attenuation, *Nucleonics*, **14**(1): 40 (1956).
48. Chappell, D.: Gamma Ray Streaming through a Duct, *Nucleonics*, **15**(7): 65 (1957).
49. Davis, H. S.: How to Choose and Place Mixes for High Density Concrete Reactor Shields, *Nucleonics*, **13**(6): 60 (1955).
50. Dib, G.: How Flow Patterns Affect Shield Design, *Nucleonics*, **14**(11): 154 (1956).
51. Engberg, C. J.: Radiation Shields and Shielding: A Bibliography of Unclassified AEC Report Literature, TID 2032, September, 1952.
52. Fano, U.: Gamma Ray Attenuation, *Nucleonics*, **11**(8): 8; **11**(9): 55 (1953).
53. Hine, G. J., and R. C. McCall: Gamma Ray Backscattering, *Nucleonics*, **12**(4): 27 (1954).
54. Lane, J.: How to Design Reactor Shields at Lowest Cost, *Nucleonics*, **13**(6): 56 (1955).

55. Lansing, N. F.: Determining the Geometry of Thermal Shields, *Nucleonics*, **13**(6): 58 (1955).

56. Moteff, J.: Tenth-value Thicknesses for Gamma-ray Absorption, *Nucleonics*, **13**(7): 24 (1955).

57. Nucleonics Staff: Portable Radiation-shielding Materials, *Nucleonics*, **13**(5): 84 (1955).

58. Stephenson, R.: Neutron Shielding, AECD-3272.

Waste Disposal

59. Barnet, M. K., P. M. Hamilton, and F. C. Mead, Jr.: Use of Sequential Factorial Designs in the Establishment of Optimum Conditions for a Decontamination Process, Mound Labs, MLM-921.

60. Burns, R. E., and M. J. Stedwell: Volume Reduction of Radioactive Waste by Carrier Precipitation, *Chem. Eng. Progr.*, **53**(2): 93-F (1957).

61. Clouse, R. J., J. Dykstra, and B. H. Thompson: Uranium Recovery from Aqueous Wastes, *Chem. Eng. Progr.*, **53**(2): 65-F (1957).

62. Fairbourne, S. F., D. G. Reid, and B. R. Kramer: Experience of Handling Low Level Active Liquid Wastes at Idaho Chemical Processing Plant, AEC-IDO-14334, 1955.

63. Ginell, W. S., J. J. Martin, and L. P. Hatch: Ultimate Disposal of Radioactive Wastes, *Nucleonics*, **12**(12): 14 (1954).

64. Glueckauf, E.: "Long Term Aspect of Fission Product Disposal," *Proc. Intern. Conf. Peaceful Uses Atomic Energy, Geneva*, **9**: 3 (1956).

65. Hatch, L. P., J. J. Martin, and W. S. Ginell: Ultimate Disposal of Radioactive Wastes, BNL-1781, 1954.

66. Hatch, L. P., and W. H. Regan, Jr.: Concentrating Fission Products, *Nucleonics*, **13**(12): 27 (1955).

67. Hittman, F., and B. Manowitz: Progress Report on Waste Development Project. Description of Calciner Pilot Plant, BNL-323, 1954.

68. Kunin, R., and F. McGarvey: Ion Exchange, *Ind. Eng. Chem.*, **47**(3): 565 (1955).

69. Leary, J. A., R. A. Clark, and R. P. Hammond: Design and Performance of Effluent Plant for Radioactive Wastes, AECU-2818 (Los Alamos, Jan. 20, 1954); *Nucleonics*, **12**(7): 64 (1954).

70. Manowitz, B., and L. P. Hatch: Processes for High Level Waste Disposal, *Chem. Eng. Progr. Symposium Series*, **50**(12): 144 (1954).

71. Manowitz, B.: Treatment and Disposal of Wastes in Nuclear Technology, "Advances in Chemical Engineering," vol. II, pp. 82–115, Academic Press, New York, 1958.

72. Miller, H. S., F. Fahnoe, and W. R. Peterson: Survey of Radioactive Waste Disposal Practices, *Nucleonics*, **12**(1): 68 (1954).

73. "Radioactive Waste Disposal in the Ocean," *Natl. Bur. Standards Handbook* 58.

74. Ruddy, J. M.: Radioactive Liquid Waste Control, AEC BNL-2409.

75. Shannon, R. L.: Radioactive Waste Disposal—A Bibliography of Unclassified Literature, TID-374, 1950.

76. Walters, W. R., D. W. Weiser, and L. F. Marek: Concentration of Radioactive Wastes. Electromigration through Ion Exchange Membranes, *Ind. Eng. Chem.*, **47**(1): 61 (1955).

77. Wilson, E. E.: Design Consideration of Storage Tanks for Radioactive Waste, *Chem. Eng. Progr.*, **53**(3): 151 (1957).

78. Wolman, A., and A. E. Gorman: The Management and Disposal of Radioactive Wastes, *Proc. Intern. Conf. Peaceful Uses Atomic Energy, Geneva*, **9**: 9 (1956).

79. Zeitlin, H.: Economics of Waste Disposal, *Nucleonics,* **15**(1): 58 (1957); Ocean Disposal of Radioactive Waste, *Nucleonics,* **12**(12): 54 (1954).

Aerosols

80. Dennis, R., C. A. Johnson, M. W. First, and L. Silverman: Performance of Commercial Dust Collectors, Contract At (30-1)-841 (NYO-1588), Harvard School of Public Health.

81. Leary, J. A., R. A. Clark, R. P. Hammond, and C. S. Leopold: Aerosol Collection by Wetted Fiberglas Media, AECU-3072 (Los Alamos).

82. Strehlow, R. A.: *Univ. Illinois Eng. Expt. Sta., Bibliography on Aerosols,* SO-1003, 1951.

Fission-product Utilization

83. Bray, D., and C. Leyse: Food Irradiation Reactor, *Nucleonics,* **15**(7): 77 (1957).

84. Francis, W., and L. Marsden: Gamma Ray Dose and Heating from Spent MTR Fuel Elements, *Nucleonics,* **15**(4): 80 (1957).

85. Guernsey, E., and R. Ball: Reactor Irradiation for Meat, *Nucleonics,* **15**(7): 80 (1957).

86. Henley, E. J.: The Chemical Potential of Waste Fission Products, *Chem. Eng. Progr. Symposium Series,* **50**(13): 66 (1955).

87. Henley, E. R., and N. F. Barr: Ionizing Radiation Applied to Processing Industries, "Advances in Chemical Engineering," vol. I, pp. 370–441, Academic Press, New York, 1956.

88. Manowitz, B., and D. Richman: Economic Future of Fission Products for Radiation Power, *Nucleonics,* **14**(6): 98 (1956).

89. Martin, J. J.: Where We Stand in Radiation Processing, *Chem. Eng. Progr.,* **54**(2): 66 (1958).

90. Staff Article: Radiation Processing of Petroleum, *Chem. Eng. Progr.,* **53**(7): 118 (1957).

91. Young, C. A.: Utilization of Gross Fission Products—A Bibliography of Unclassified Report Literature, TID-3046, 1954.

APPENDIX A

Design Project Procedures

In the teaching of process and plant design there is concern for the development of a proper attitude of mind. The student may be well-grounded in all the fundamental science and engineering subjects he has studied, but he lacks experience in applying this knowledge to complex problems always encountered in any integrated chemical process development program.

In process design the basic teaching philosophy should be one of giving the student relatively free rein and the complete responsibility for progress and accomplishment. The students may have the option of selecting their own problem or the instructor may do this. The students should then decide on the method of approach, analyze the facts and situations, plan and carry out laboratory and calculation work, select equipment, make economic studies, and display initiative throughout.

The role of the instructor in this type of course should be that of a consultant and guide. The course may not necessarily emphasize any new scientific or engineering principles, but it should present ample opportunity to apply that which is already known. The student should be treated as if he already were a technically trained man "on the job," as if he had been adequately educated and had the essential ability to assume the entire responsibility for his success in a chemical engineering career.

Selecting the Projects

Projects may be selected in a number of ways, depending on the facilities and time available (AIChE Student Contest Problems furnish valuable ideas for projects).

1. The instructor may select a typical process or give the students the option to do so. The entire class is then organized into research, development, and engineering design departments successively as the project proceeds.

2. The instructor or students may select a chemical product and the class is divided preferentially into groups of three to six students. Each group takes one of the several processes for making the product and follows through the entire research, development, design, and cost estimating stages.

3. Each student is given the opportunity to submit several chemical manufacturing processes which he would like to study. Class discussion of each student's process ensues and one or more processes are chosen by the class for individual, small group, or entire class effort.

Search of Literature for Process Details

A search of the literature for authoritative references is undertaken. This is a library search, the work being portioned out among the students to the journals and textbooks, technical trade literature, and sources of information; conferences with other professors or instructors of the college are avoided. They are cautioned to record all information on the various methods by which the chemical commodity can be made, listing references, raw materials, conditions of preparation, and products. A review of all handbooks, trade information, and critical tables is then prepared on the chemical and physical properties of each chemical involved in the process selected. Market price curves for each commodity are then prepared using a large time span, if possible. Production figures are also required on the raw materials, products and by-products, location of present plants, and where competition can be expected. Profit margins on this basis are submitted.

Projecting the Development Work. All the data obtained are placed upon the board and tabulated, so that the class can analyze the items presented. Discussion of the various methods follows. The data must be correlated in a report emphasizing the need for development work to be carried out, including a plan of the development phase from the standpoint of experiments, basic information desired, fundamental information needed, and costs. Statistical planning of experiments can be put in practice if the process is sufficiently complex.

Small-scale Experimentation

The next step should be acquisition of experimental data through laboratory studies. In this laboratory work the students should prepare the commodity according to the process selected, for the purpose of acquiring necessary data for plant design. All thermal, chemical, and physical data not available from literature reviews must be acquired. A material balance should be made. In order to obtain yield and other general data on process reactions studied, the students should follow this order of procedure:

1. Probable costs (yield and reaction study on 1-lb batch)
2. Materials and equipment requirements
   Observations essential are:
   a. Type of reaction
   b. Quality of product
   c. Quantity of product
   d. General solubility
   e. Separation characteristics
   f. Heat considerations
   g. General operations required

Development Laboratory Experimentation

Following the laboratory- or beaker-scale experimentation when the students are satisfied as to feasibility of the preparation and have enabled themselves to acquire sufficient information to undertake a large-batch operation (5 to 10 lb), a small-scale-process laboratory study should then be undertaken. Attention should be paid to the engineering considerations involved in the production of the commodity. In order to obtain engineering data essential for the pilot plant investigation of the commodity selected, the following considerations may be important:

1. Procedure essentials
2. Raw material characteristics
3. Chemical flow sheet
4. Corrosion characteristics
5. Effect of impurities
6. Heat considerations
7. Unit operations required
8. Material handling
9. Storage
10. Engineering flow sheet

Projecting the Engineering

Although no laboratory data have yet been acquired, it is desirable for the class to make a report setting up a pattern of design for the engineering division on insufficient data. This pattern will be brought out later after confirmation has been obtained from the development report by the development laboratories; the engineering pattern will be modified later according to the requirements and changes resulting from this interim study. It is well to emphasize at this point that the engineering division requires a long time to complete its enormous task, and much time can be saved by starting the engineering immediately upon the choice of a project and by reporting changes to the engineering division as they develop. As a guide for determining what phases or items need to be considered by engineering, the following points are listed:

1. Comprehensive market surveys or locations
2. Plant sites
3. Fuel and water sources
4. Transportation facilities
5. Material supplies
6. Labor supplies
7. Laws and codes
8. Working up of process flow sheets
9. Working up design flow sheets
10. Tentative specifications of machinery
11. Tentative specifications of materials
12. Power systems
13. Calculation of heat balances
14. Calculation of material balances
15. Preliminary sizing and designing of vessels, exchanges, and process units
16. Drawing of preliminary plant layout
17. Drawing of preliminary equipment layout
18. Preparation of preliminary cost estimates

Development Laboratory Work

The students then are required to go into a development laboratory for actual experimentation work to get them acquainted with materials, materials handling, and translation of laboratory procedure into equipment, and to learn how to take exact plant data. They will also learn that there are a number of points that can be best obtained later through pilot plant operation; therefore, one objective of the development laboratory can be a pilot plant design.

The Pilot Plant Phase

Although time usually does not permit the actual setup and operation of a pilot plant, there must still not be omitted the report containing the pilot plant design and operations instructions. A pilot plant check list is given herewith:

Pilot Plant Check List

1. Flow relations:
   a. Chemical flow diagrams
   b. Breakdown into unit operations
   c. Engineering equipment flow diagrams
   d. Material and energy balances
2. Materials:
   a. Raw materials, availability, substitute raw materials, costs
   b. Impurities in raw materials and in products
   c. Corrosion, erosion, dust, fumes

APPENDIX A 501

    d. Solvents
    e. Wastes and recovery
3. Equipment and operation:
    a. Selection of equipment, elimination of obviously unsuitable equipment, or proof of applicability of standard or special types of equipment or machines such as pumps, bottles, evaporators
    b. Cost of operation
    c. Control specification
    d. Material of construction
    e. Heat transfer
    f. Mass transfer
    g. Peak loads
    h. Utility requirements
    i. Maintenance costs
    j. Instrumentation
4. Materials handling:
    a. Proper methods of handling around the plant
    b. Intermediate storage
    c. Industrial hazards (corrosion, fire, erosion, safety, health, pollution, fumes, explosions)
    d. Public nuisances
    e. Storage
5. Labor:
    a. Manpower requirements
    b. Supervision
    c. Control specifications from operator's viewpoint
    d. Process simplification from operator's viewpoint
    e. Safety from operator's viewpoint; safety requirements
    f. Saving of time and labor

The student must understand the importance of pilot plant studies. When the process finally comes out of the pilot plant, it should have been so thoroughly checked, verified, and proved that one has a practical working plan. The tentative figures that served as a preliminary basis of the process engineers will then be replaced by final and exact specifications. Definite yields and conclusive figures on all operations will be established and the most efficient process will result.

The ultimate desire is a pilot plant that will operate with the assurance that all the risks, both technical and economic, in the full-scale commercial plant have been minimized or, preferably, eliminated. It is essential in the pilot plant to determine whether improvements or changes must be made while it is yet relatively inexpensive to do so before finally establishing and fixing the design of the commercial plant.

Individual Equipment Assignments

The decision on individual assignments is best made by the instructor in order that the degree of ability of each student conforms to the requirements of each task. Group conferences are necessary to ensure coordination of operations. The intimate contact that the instructor has had previously with the individual student here aids in getting the most out of each student and a leveling-off of the work based upon unequal abilities of the members of the class. *Material and thermal balancing* of the entire project is carried out in open discussion, each student contributing that portion of the balance included in his assignment. Then calculations as to sizes and capacities are undertaken and explained. Before designs are put on paper, the individual assignments are handled mathematically; the size, capacity, number, and operational ideas relative to each unit are considered. These become specifications for the equipment. The information is presented on specification forms of the type given in Table 4-2. Preconstruction cost estimating can be initiated at this point.

Scale Models in Plant Design. Each man who makes the design calculations and who understands the complete functioning of his unit makes a model of his assignment. Scale models are an essential aid. How far to go in this is a matter of preference; some type of scale modeling will necessarily be used and the care and time will depend upon the instructor and the students. Some few students devote much outside time to modeling their units; others have a hard time scale-modeling railroads, such simple lines as yards, unloaders, or scales. Their ability to create these "tinker-toy" models and to present their cases must be taken into consideration by the instructor. A three-dimensional model is made with cardboard, plywood, or any applicable material as illustrated in Fig. 5-4. After the models of the principal equipment pieces are completed, the class assembles in a large open area where a plot plan can be laid out. Each student steps forward according to the flow diagram and places his equipment where he feels it will function best, presenting the points in favor of his decision. After much discussion and rearrangement, with due consideration to warehousing, shipping, and servicing, class agreement is finally reached.

As a final check, operating instructions are prepared and checked out on the model arrangement to pick up any flaws in equipment functioning. Over-all plan and elevation drawings can then be made with the aid of the model, followed by individual specifications for each individually assigned unit.

Commercial Unit Design

The final step in the study should be the coordination of all chemical and engineering data obtained and their translation into a definite organized unit. Access must be had to trade literature for selection of types and specific pieces of equipment. Capacities and performance should be studied. Final layouts are decided on. Organization of the equipment by means of templates and scale models will give the student a better picture of the possibilities of different layouts. After arriving at the most desirable layout, actual drawing of the plan and elevation of the assembly should be undertaken, followed by a detailed preconstruction cost estimate. In order to design a commercial unit, including housing for the production of the specified commodity, the following considerations may be important:

1. Specifications of equipment
2. Specifications of materials
3. Selection of commercial equipment
4. Plan
5. Elevation
6. Location of plant
7. Operating instruction for labor
8. Selection of personnel
9. Preconstruction costing
10. Economic evaluation

Notebooks

Notebooks must be kept with a daily log of all observations and data. Each page should have a title and date, and at the end of each period a brief résumé must be written of the day's work, signed by initials of the worker and someone who was with him in the laboratory. Notebooks should be deposited with the instructors.

Reports

Weekly reports to the class are essential, and weekly written reports should be made. The student should receive practice in presentation. Calculations and reasons for making certain decisions should be presented concisely to the class for criticism. The discussions should be informal.

Final Compilation

At the end of the year a complete report should be turned in which includes all calculations, flow sheets, material and energy balances, equipment specifications, preconstruction cost estimation figures, and economic evaluations. A carefully executed series of drawings, including plot plans, equipment layout, and elevation drawings may also be submitted if time permits.

APPENDIX B

Letter Symbols for Chemical Engineering[*]

LETTER SYMBOLS FOR PRINCIPAL CONCEPTS

Listing is alphabetical by concept within each category. Illustrative units or definitions are supplied where appropriate.

	Symbol	Unit or definition
1. General Concepts		
Acceleration	a	(ft/sec)/sec
Of gravity	g	(ft/sec)/sec
Base of natural logarithms	e	
Coefficient	C	
Difference, finite	Δ	
Differential operator	d	
Partial	∂	
Efficiency	η	
Energy, dimension of	E	Btu; (ft)(lb force)
Enthalpy	H	Btu
Entropy	S	Btu/°R
Force	F	lb force
Function	ϕ, ψ, χ	
Gas constant, universal	R	To distinguish, use R_0

[*] Adopted by American Standards Association as ASA—Y10.12, 1955. See M. Souders, Jr., *Chem. Eng. Progr.*, **52**(6): 255 (1956).

	Symbol	Unit or definition
Gibbs free energy	G, F	$G = H - TS$, Btu
Heat	Q	Btu
Helmholtz free energy	A	$A = U - TS$, Btu
Internal energy	U	Btu
Mass, dimension of	m	lb
Mechanical equivalent of heat	J	(ft)(lb force)/Btu
Moment of inertia	I	$(ft)^4$
Newton law of motion, conversion factor in	g_c	$g_c = ma/F$, (lb)(ft)/(sec)^2(lb force)
Number		
In general	N	
Of moles	n	
Pressure	p	lb force/sq ft; atm
Quantity, in general	Q	
Ratio, in general	R	
Resistance	R	
Shear stress	τ	lb force/sq ft
Temperature		
Dimension of	θ	
Absolute	T	°K (Kelvin); °R (Rankine)
In general	T, t	°C; °F
Temperature difference, logarithmic mean	$\bar{\theta}$	°F
Time		
Dimension of	T	sec
In general	t, τ	sec; hr
Work	W	Btu

2. Geometrical Concepts

Linear dimension		
Breadth	b	ft
Diameter	D	ft
Distance along path	s, x	ft
Height above datum plane	Z	ft
Height equivalent	H	ft Use subscript p for equilibrium stage and t for transfer unit
Hydraulic radius	r_H	ft; sq ft/ft
Lateral distance from datum plane	Y	ft
Length, distance or dimension of	L	ft
Longitudinal distance from datum plane	X	ft
Mean free path	λ	cm; ft
Radius	r	ft
Thickness		
In general	B	ft
Of film	B_f	ft
Wavelength	λ	cm; ft

	Symbol	Unit or definition
Area		
In general	A	sq ft
Cross section	S	sq ft
Fraction free cross section	σ	
Projected	A_p	sq ft
Surface		
Per unit mass	$A_{w,\,s}$	sq ft/lb
Per unit volume	$A_{v,\,a}$	sq ft/cu ft
Volume		
In general	V	cu ft
Fraction voids	ϵ	
Humid volume	v_H	cu ft/lb dry air
Angle	$\alpha,\ \theta,\ \phi$	
In x,y plane	α	
In y,z plane	ϕ	
In z,x plane	θ	
Solid angle	ω	
Other		
Particle-shape factor	ϕ_s	

3. Intensive Properties

	Symbol	Unit or definition
Absorptivity for radiation	α	
Activity	a	
Activity coefficient, molal basis	γ	
Coefficient of expansion		
Linear	α	(ft/ft)/°F
Volumetric	β	(cu ft/cu ft)/°F
Compressibility factor	z	$z = pV/RT$
Density	ρ	lb/cu ft
Diffusivity		
Molecular, volumetric	$D_v,\ \delta$	cu ft/(hr)(ft); sq ft/hr
Thermal	α	$\alpha = k/c\rho$, sq ft/hr
Emissivity ratio for radiation	ϵ	
Enthalpy, per mole	H	Btu/lb mole
Entropy, per mole	S	Btu/(lb mole)(°R)
Fugacity	f	lb force/sq ft; atm
Gibbs free energy, per mole	$G,\ F$	Btu/lb mole
Helmholtz free energy, per mole	A	Btu/lb mole
Humid heat	c_s	Btu/(lb dry air)(°F)
Internal energy, per mole	U	Btu/lb mole
Latent heat, phase change	λ	Btu/lb
Molecular weight	M	lb
Reflectivity for radiation	ρ	
Specific heat	c	Btu/(lb)(°F)
At constant pressure	c_p	Btu/(lb)(°F)
At constant volume	c_v	Btu/(lb)(°F)

	Symbol	Unit or definition
Specific heats, ratio of	γ	
Surface tension	σ	lb force/ft
Thermal conductivity	k	Btu/(hr)(sq ft)(°F/ft)
Transmissivity of radiation	τ	
Vapor pressure	p^*	lb force/sq ft; atm
Viscosity		
Absolute or coefficient of	μ	lb/(sec)(ft)
Kinematic	$\nu, \mu/\rho$	sq ft/sec
Volume, per mole	V	cu ft/lb mole

4. Symbols for Concentrations

	Symbol	Unit or definition
Absorption factor	A	$A = L/K^*V$
Concentration, mass or moles per unit volume	c	lb/cu ft; lb moles/cu ft
Fraction		
Cumulative beyond a given size	ϕ	
By volume	x_v	
By weight	x_w	
Humidity	H, Y_H	lb/lb dry air
At saturation	H_s, Y^*	lb/lb dry air
At wet-bulb temperature	H_w, Y_w	lb/lb dry air
At adiabatic saturation temperature	H_a, Y_a	lb/lb dry air
Mass concentration of particles	c_p	lb/cu ft
Moisture content		
Total water to bone-dry stock	X_T	lb/lb dry stock
Equilibrium water to bone-dry stock	X^*	lb/lb dry stock
Free water to bone-dry stock	X	lb/lb dry stock
Mole or mass fraction		
In heavy or extract phase	x	
In light or raffinate phase	y	
Mole or mass ratio		
In heavy or extract phase	X	
In light or raffinate phase	Y	
Number concentration of particles	n_p	Number/cu ft
Phase equilibrium ratio	K^*	$K^* = y^*/x$
Relative distribution of two components		
Between two phases in equilibrium	α	$\alpha = K_i^*/K_j^*$
Between successive stages	β	$\beta_n = (y_i/y_j)_n / (x_j/x_i)_{n+1}$
Relative humidity	H_R, R_H	
Slope of equilibrium curve	m	$m = dy^*/dx$
Stripping factor	S	$S = K^*V/L$

	Symbol	Unit or definition

5. Symbols for Rate Concepts

Quantity per unit time, in general	q	
Angular velocity	ω	
Feed rate	F	lb/hr; lb moles/hr
Frequency	f, N_f	
Friction velocity	u^*	$u^* = (g_c \tau_w \rho)^{1/2}$, ft/sec
Heat transfer rate	q	Btu/hr
Heavy or extract phase rate	L	lb/hr; lb moles/hr
Heavy or extract product rate	B	lb/hr; lb moles/hr
Light or raffinate phase rate	V	lb/hr; lb moles/hr
Light or raffinate product rate	D	lb/hr; lb moles/hr
Mass rate of flow	w	lb/sec; lb/hr
Molal rate of transfer	N	lb moles/hr
Power	P	(ft)(lb force)/sec
Revolutions per unit time	n	
Velocity		
In general	u	ft/sec
Instantaneous, local		
Longitudinal (x) component of	u	ft/sec
Lateral (y) component of	v	ft/sec
Normal (z) component of	w	ft/sec
Volumetric rate of flow	q	cu ft/sec; cu ft/hr
Quantity per unit time, unit area		
Emissive power, total	W	Btu/(hr)(sq ft)
Mass velocity, average	G	$G = w/S$, lb/(sec)(sq ft)
Vapor or light phase	$G, \bar{G}$	lb/(hr)(sq ft)
Liquid or heavy phase	$L, \bar{L}$	lb/(hr)(sq ft)
Radiation, intensity of	I	Btu/(hr)(sq ft)
Velocity		
Nominal, basis total cross section of packed vessel	v_S	ft/sec
Volumetric average	$V, \bar{V}$	(cu ft/sec)/sq ft; ft/sec
Quantity per unit time, unit volume		
Quantity reacted per unit time, reactor volume	N_R	(moles/sec)/cu ft
Space velocity, volumetric	Λ	(cu ft/sec)/cu ft
Quantity per unit time, unit area, unit driving force, in general	k	
Eddy diffusivity	δ_E	sq ft/hr
Eddy viscosity	ν_E	sq ft/hr
Eddy thermal diffusivity	α_E	sq ft/hr
Heat transfer coefficient		
Individual	h	Btu/(hr)(sq ft)(°F)
Over-all	U	Btu/(hr)(sq ft)(°F)

	Symbol	Unit or definition
Mass transfer coefficient		lb moles/(hr)(sq ft) (driving force)
Individual	k	
Gas film	k_G	To define driving force, use subscript:
Liquid film	k_L	
Over-all	K	c for lb moles/cu ft
Gas film basis	K_G	p for atm
Liquid film basis	K_L	x for mole fraction
Stefan-Boltzmann constant	σ	0.173×10^{-8} Btu/(hr)(sq ft)(°R)^4

6. Dimensionless Numbers Used in Chemical Engineering

	Symbol	Unit or definition
Condensation number	N_{Co}	$\dfrac{h}{k}\left(\dfrac{\nu^2}{a}\right)^{1/3}$; $\dfrac{h}{k}\left(\dfrac{\nu^2}{g}\right)^{1/3}$
Euler number	N_{Eu}	$\dfrac{g_c p}{\rho u^2}$, $\dfrac{g_c \rho p}{G^2}$
Fanning friction factor	f	$\dfrac{g_c \rho D(\Delta p_f)}{2G^2(\Delta L)}$
Fourier number	N_{Fo}	$\dfrac{kt}{c\rho L^2}$ or $\dfrac{\alpha t}{L^2}$
Froude number	N_{Fr}	$\dfrac{u^2}{aL}$; $\dfrac{u^2}{gL}$
Graetz number	N_{Gz}	$\dfrac{cLG}{k}$ or $\dfrac{L\bar{V}}{\alpha}$
Grashof number	N_{Gr}	$\dfrac{L^3\rho^2\beta g\Delta t}{\mu^2}$ or $\dfrac{L^3\beta g\Delta t}{\nu^2}$
Heat transfer factor	j_H	$\dfrac{h}{cG}\left(\dfrac{c\mu}{k}\right)^{2/3}$ or $(N_{St})(N_{Pr})^{2/3}$
Lewis number	N_{Le}	$\dfrac{k}{c\rho D_v}$ or $\dfrac{\alpha}{D_v}$
Mass transfer factor	j_M	$\dfrac{k_c}{u}\left(\dfrac{\mu}{\rho D_v}\right)^{2/3}$
Nusselt number	N_{Nu}	$\dfrac{hL}{k}$; $\dfrac{hD}{k}$
Peclet number	N_{Pe}	$\dfrac{Luc\rho}{k}$ or $\dfrac{Lu}{\alpha}$; $\dfrac{D\bar{V}}{\alpha}$
Prandtl number	N_{Pr}	$\dfrac{c\mu}{k}$ or $\dfrac{\nu}{\alpha}$
Prandtl velocity ratio	u^+	$\dfrac{\bar{u}}{u^*}$
Reynolds number	N_{Re}	$\dfrac{Lu\rho}{\mu}$; $\dfrac{DG}{\mu}$
Reynolds number, local	y^+	$\dfrac{ru^*\rho}{\mu}$
Schmidt number	N_{Sc}	$\dfrac{\mu}{\rho D_v}$

510 CHEMICAL ENGINEERING PLANT DESIGN

	Symbol	Unit or definition
Sherwood number	N_{Sh}	$\dfrac{k_c L}{D_v}$ or $j_M(N_{Re})(N_{Sc})^{1/3}$
Stanton number	N_{St}	$\dfrac{h}{c\rho u}$; $\dfrac{h}{cG}$
Vapor condensation number	N_{Cv}	$\dfrac{L^3 \rho^2 g \lambda}{k \mu \Delta t}$
Weber number	N_{We}	$\dfrac{L u^2 \rho}{g_c \sigma}$; $\dfrac{D G^2}{g_c \rho \sigma}$

Concept	Remarks	Superscript	Subscript

7. Modifying Signs for Principal Symbols

Concept	Remarks	Superscript	Subscript
Average value	Written over symbol	− (Bar)	
Dimensionless form	Follows symbol	+ (Plus)	
Equilibrium value	Follows symbol	* (Asterisk)	
Fluctuating component	Usually applied to local velocity	′ (Prime)	
Initial or reference value	Follows symbol		0 (zero)
Modified form	Follows symbol	′ (Prime) ″ (Double prime)	
Partial molal quantity	Written over small capitals	− (Bar)	
Sequence in time or space	Follows symbol	′ (Prime) ″ (Double prime)	1, 2, 3, etc.
Standard state	Follows symbol	° (Degree)	
First derivative with respect to time	Written over symbol	. (Dot)	
Second derivative with respect to time	Written over symbol	.. (Double dot)	

Alphabetical Index by Symbol

	Primary concept	Subscript concept
a	Acceleration Activity Area, alternative for Surface per unit volume	Acoustic Adiabatic Arithmetic
A	Absorption factor Area Helmholtz free energy	Absolute Area basis Component A
b	Breadth	Baffle Base

APPENDIX B

	Primary concept	Subscript concept
B	Heavy-product rate Thickness	Black body Boiling point Component B
c	Concentration, mass or moles per unit volume Specific heat, heat capacity	Concentration basis Contraction Conversion factor Critical Cutoff size
C	Coefficient	Component C
d	Differential operator	Discharge Disperse Drop Dry
D	Diameter Diffusivity Light or raffinate product rate	Component D Distillate
e	Base of natural logarithms	Effective Exit

E	Energy
	   Dimension of
	   In general
		Component E
		Eddy
		Entrainment

f	Frequency Friction factor, Fanning Fugacity	Film Fluid Frequency Friction
F	Feed rate Force Gibbs free energy	Feed
g	Acceleration of gravity	Gauge Gravity Vapor

G	Gibbs free energy
	Mass velocity
	   In general
	   Of vapor
		Vapor
		Vapor film basis

h	Individual coefficient of heat transfer	Heat Heated

	Primary concept	Subscript concept
H	Enthalpy Height equivalent Humidity	Heat basis Humidity Hydraulic
i		Generalized component Interface Internal or inner
I	Intensity of radiation Moment of inertia	
j	Transfer factor	Generalized component
J	Mechanical equivalent of heat	
k	Mass transfer coefficient, individual Quantity per unit time, unit area, unit driving force, in general Thermal conductivity	
K	Mass transfer coefficient, over-all Phase concentration ratio	
L	Heavy or extract phase rate Length Mass velocity of liquid or heavy phase	Liquid Liquid-film basis

m	Mass
	   Dimension of
	   In general
	Slope of equilibrium curve
		Mass
		Mean

M	Molecular weight	Mass basis Molecular
n	Number concentration Number of moles Revolutions per unit time	Generalized stage number Normal
N	Molal rate Number, in general	
o		Initial Outer
O		Over-all

	Primary concept	Subscript concept
p	Pressure	Constant pressure Particle Plate or stages Pressure basis Projected
P	Power	
q	Quantity per unit time, in general Rate of heat flow Rate of volumetric flow	Rate basis
Q	Heat Quantity, in general	
r	Radius	Radius or radial Reduced
R	Gas constant Ratio, in general Reflux ratio Resistance	Radiation Reactor volume basis Relative value
s	Distance along path Specific surface	Saturation Shape Stress Surface basis
S	Cross section Entropy Stripping factor	Cross-section basis Solid Solvent
t	Temperature Time	Tangential Terminal Transfer unit or units
T	Absolute temperature Temperature, in general	Constant temperature Total
u	Longitudinal component of local velocity Velocity, in general	Upper
U	Heat transfer coefficient, over-all Internal energy	
v	Lateral component of local velocity Nominal velocity Specific volume	Constant volume Velocity basis Volumetric

	Primary concept	Subscript concept
V	Light or raffinate phase rate Volume, in general Volumetric average velocity	
w	Mass flow rate Normal component of local velocity	Mass basis Wet bulb
W	Work Total emissive power	
x	Distance along path Fraction Mole or mass fraction in heavy or extract phase	Mole fraction basis
X	Longitudinal distance from datum plane Mole or mass ratio in heavy or extract phase	Mole ratio basis
y Y	Mole or mass fraction in light or raffinate phase Humidity Lateral distance from datum plane Mass or mole ratio in light or raffinate phase	
z	Compressibility factor	
Z	Height above datum plane	
α	Absorptivity for radiation Angle Angle in x,y plane Coefficient of linear expansion Relative distribution of two components between two phases at equilibrium Thermal diffusivity	
β	Coefficient of volumetric expansion Relative distribution of two components between successive stages	
γ	Activity coefficient, molal basis Ratio of specific heats	
∂	Differential operator, partial	
δ	Diffusivity, volumetric	Film basis
Δ	Difference, finite	

	Primary concept	Subscript concept
ϵ	Emissivity ratio for radiation Fraction voids	
η	Efficiency	
θ	Angle Angle in z,x plane Temperature, dimension of	
$\bar{\theta}$	Log mean temperature difference	
λ	Latent heat of phase change Mean free path Wavelength	
Λ	Volumetric space velocity	
μ	Viscosity, absolute	At constant viscosity
ν	Viscosity, kinematic	
ρ	Density Reflectivity for radiation	Density basis
σ	Fraction free cross section Stefan-Boltzmann constant Surface tension	
τ	Shear stress Time, alternative for Transmissivity for radiation	
ϕ	Angle Angle in y,z plane Fraction cumulative, larger than a given size Function Particle factor	
χ	Function	
ψ	Function	
ω	Angular frequency Angular velocity Solid angle	

APPENDIX C

Table of Equivalents

Chemical Equivalent. The atomic weight divided by the valence.

Heat Capacity. Specific heat is the number of Btu's required to raise 1 lb of material 1°F. The molal heat capacity is the product of the specific heat and the molecular weight. Kopp's law: molal heat capacity of *solid* compounds is the sum of the atomic heat capacities of the constituent atoms. The atomic heat capacities may be taken as follows: $C = 1.8$; $H = 2.3$; $B = 2.7$; $Si = 3.8$; $O = 4.0$; $P = 5.4$; $F = 5.4$; all others $= 6.2$. Example: that of $CaCO_3 = 6.2 + 1.8 + 3 \times 4 = 20.0$ Btu/°F. Law of Dulong and Petit: atomic heat capacity of *solid* elementary substances $= 6.2$; exceptions—C, B, P, and Si as above.

Heat of Vaporiaztion. Btu required to vaporize 1 lb of a substance. Molal heat of vaporization = heat of vaporization multiplied by the molecular weight. The molal heats of vaporization of similar liquids are approximately equal.

Humidity. Absolute humidity is the pounds of water per pound of bone-dry air.

Mass Velocity. $G = \rho V$.

Moisture Per Cent

$$\text{Dry basis} = \frac{\text{wt of water} \times 100}{\text{wt of bone-dry material}}$$

$$\text{Wet basis} = \frac{\text{wt of water} \times 100}{\text{bone-dry wt} + \text{wt of water}}$$

Units

Electrical. (For direct current only) $E = RI$. Power in watts $= I^2R$. Energy in watthours $= EIT$ when T is time in hours. 96,540 amp-sec = 1 faraday, which will theoretically deposit or liberate one chemical equivalent at an electrode.

Energy, Work, Heat. 1 Btu = 252 cal = 778 ft-lb = 0.293 watthr. 1 Pcu = 1.8 Btu = 454 cal. 1 hp-hr = 746 watthr = 2545 Btu = 1,980,000 ft-lb. 1 cal per gram, gram atom, or gram mol = 1 Pcu/lb, lb atom, or lb mol = 1.8 Btu/lb, lb atom, or lb mol.

Hydrometers

Liquids lighter than water:

Degrees Baumé $= \dfrac{140}{\text{sp gr}} - 130$

Gravity API $= \dfrac{141.5}{\text{sp gr}} - 131.5$

Liquids heavier than water:

Degrees Baumé $= 145 - \dfrac{145}{\text{sp gr}}$

Degrees Twaddell $= \dfrac{\text{sp gr} - 1}{0.005}$

Linear. 1 in. = 25.4 mm = 2.54 cm.

Mol or Mole. Molecular weight of a substance expressed in units of weight.

Power. *Rate of doing work:* 1 hp = 746 watts = 33,000 ft-lb/min = 42.4 Btu/min.

Pressure. 1 atm = 29.92 in. Hg = 760 mm. Hg = 14.7 psi = 33.9 ft of water = 1,033 g/cm².

Temperature. 1°C or K (Kelvin) = 1.8°F or R (Rankine) = 0.8°Ré (Réaumur). Freezing point of water = 0°C = 273°K = 32°F = 460 + 32 or 492°R = 0°Ré. 0°F = 460°R.

Viscosity. 100 centipoises = 1 poise = 1 g/sec-cm = 0.0672 lb/sec-ft. Kinematic viscosity $= \rho/\mu$.

Volume. 1 ft³ = 7.48 gal. 1 gal = 231 in.³ = 3,785 cm³.

Weight. 1 lb = 453.6 g = 7,000 grains avoirdupois.

General Information

Centrifugal Force. $F = 0.000341 WRN^2$ lb.

Combustion. These major combustion reactions are all reversible:

$$C + O_2 = CO_2 + 97{,}000 \text{ cal}$$
$$CO_2 + C = 2CO - 39{,}000 \text{ cal}$$
$$2C + O_2 = 2CO + 58{,}000 \text{ cal}$$
$$2CO + O_2 = 2CO_2 + 136{,}000 \text{ cal}$$
$$2H_2 + O_2 = 2H_2O + 136{,}000 \text{ cal}$$
$$H_2O + C = CO + H_2 - 39{,}300 \text{ cal}$$
$$2H_2O + C = CO_2 + 2H_2 - 39{,}600 \text{ cal}$$
$$CO + H_2O = CO_2 + H_2 - 300 \text{ cal}$$

The equilibrium of the gases above the bed of a gas producer is $0.096L = \dfrac{(CO_2)(H_2)}{(CO)(H_2O)}$, when L is depth of active fuel bed and volumes of gases are in terms of 100 volumes of dry gas.

Evaporation. The pounds of water evaporated in evaporators per pound of steam is approximately $0.85n$, where n is the number of effects.

Flow of Fluids. Venturi meter; $\sqrt{(u_2)^2 - (u_1)^2} = 0.98\sqrt{2g\Delta H}$. With a sharp-edged orifice use 0.61 instead of 0.98.

Pitot tube; $u = \sqrt{2g\Delta H}$.

Flow of Heat. By conduction $Q/\theta = U\Delta TA$.

The over-all coefficient $U = \dfrac{1}{\dfrac{L_1}{k_1 A_1} + \dfrac{L_2}{k_2 A_2} + \dfrac{L_3}{k_3 A_3} + \text{etc.}}$

When heat flows through a fluid film, substitute $1/h$ for L/k, as h is the film coefficient. The value of h for low-pressure condensing steam is often taken as 2,000. The equation for water flowing in turbulent motion in horizontal pipes (t in °F) is

$$h = \frac{0.00486(1 + 0.010t)(G^{0.8})}{D^{0.2}}$$

Gas Laws. Volume per cent = mol per cent = pressure per cent. A pound mol of any gas occupies 359 ft³ at 32°F and 1 atm (STP).

$$V_2 = V_1 \frac{P_1 T_2}{P_2 T_1}$$

The apparent molecular weight of air may be taken as 29.

Name Index

For additional names see pages 471-495

Persons

Anderson, E. L., 456
Anderson, W. T., 153
Anderson, Y., 414
Aries, R. S., 100, 195, 228, 233, 257, 336

Badger, W. L., 126
Balderston, J. L., 439
Ballou, N. E., 446, 447
Barrow, M. H., 100, 150
Bartkus, E. P., viii
Beattie, R. D., 23
Benedict, M., 454
Benenati, R., 131
Bowman, J. H., 94
Braca, R. M., 148, 149, 387
Brandt, A. D., 329
Brucher, G. J., 439
Bryant, G. A., 338, 339
Buell, W. H., 254
Bullinger, C. E., 253
Butler, W. T., viii, 75

Callihan, D., 449-451
Canfield, D. T., 94
Chappell, D. G., 436, 439
Chilton, C. H., 193-197, 222, 227, 229, 231, 235
Chope, H. R., viii
Clark, W. G., 398
Clegg, J. W. 454
Colburn, A. P., 119, 350

Danatos, S., 405
Davies, O. L., 18
DeOng, E. R., 286
Des Jardins, P. R., 143
Dolman, R. E., 147
Donovan, J. R., 41
Driesbach, R. R., 45
Dryden, C. E., 100, 465
Duggan, J. J., 31, 32
Dunham, C. W., 318
Dybdal, E. C., 224

Foley, D., 454
French, T. E., 6, 363-365
Friedman, S. J., 122
Friend, L., 19

Gamson, B. W., 132, 455
Garrett, D. E., 119, 390, 391
Geankoplis, C. J., viii
Genereaux, R. P., viii, 350
Githens, R. E., Jr., 115
Glasstone, S., 453
Goodgame, T. H., 23
Gordon, D., 15
Gore, W. L., 18
Gradishar, F. L., 43
Grant, E. L., 253
Graves, G. A., 451
Groggins, P. H., 99
Grummer, M., 131

Hall, G. A., Jr., 404–408
Happel, J., 90, 100, 148, 149, 387
Harty, W. M., 460
Hempstead, J. C., 256
Henderson, J. G., 152
Hesse, H. C., 150
Higgins, S. P., Jr., 409
Hirschberg, H. S., 269
Humbla, S., 414
Hunter, H. F., 446, 447
Hur, J. J., 256

Ireland, J. D., viii

Jackson, H. K., 459
Jordan, W. A., 390, 391

Kammermeyer, K., 90
Karr, A. E., 117
Katz, D. L., 355
Kelley, W. E., 454
Kern, D. Q., 115
Ketzlach, N., 452
Kidder, F. E., 393
Kiddoo, G., 195
Kirkpatrick, S. D., viii
Koffolt, J. H., viii
Kropf, V. J., 140

Landry, J. W., 464
Lange, H. J., 193, 195, 207
Laurent, M., 23
Lavine, I., 100, 101
Lemon, R. P., 461
Leonard, J. D., 239–242
Leva, M., 127, 131
Lieberman, S. A., 457
Lindsey, E., 124
Ludwig, E. E., 94–99

McAdams, W. H., 115
MacLean, G., 133–135
Mantell, C. L., 85
Marshal, W. R., Jr., 122
Marston, A., 256
Martin, J. C., 256
Melinat, C. H., 269
Molstad, M., 121

Nelson, W. L., 68, 201
Newton, R. D., 100, 195, 228, 233, 257, 336
Nicholls, C. M., 453, 458

Nichols, W. T., 192, 193
Norton, C. L., 381–385

O'Donnell, J. P., 40
Olive, T. R., viii, 105, 405
O'Rourke, C. E., 307
Osburn, J. O., 90
Othmer, D. L., 127
Otto, F. C., 389
Oxley, J. H., viii

Paffenbarger, R., viii
Parker, H., 393
Parkinson, R. W., viii, 319
Paxton, H. C., 451
Perry, J. H. (ed.), viii, 30, 86, 100, 106, 109, 142, 145, 147, 155, 160, 185, 309, 321, 325–330, 332, 342, 344, 349–352, 355, 363, 365, 367, 368, 404, 415, 419, 420, 424
Peters, M. S., 90, 100
Pierce, D. E., 239
Pigford, R. L., 119
Pigford, T., 454
Pollchik, M., 131
Pontius, E., viii
Powell, S. T., 274

Rase, H. F., 100, 150
Reid, D. G., 461
Riegel, E. R., 100, 101
Rockwell, T., 445
Rodger, W. A., 456, 457
Rodgers, R. D., 322
Romeo, A. P., 72, 73
Rosenbaum, G. P., 119
Rubin, F. L., 116
Rushton, J. H., 150

Sadowski, G. S., 459
Sarchett, B. R., 350
Sax, N. I., 33, 34, 329
Schiebel, E. G., 117
Schweyer, H. E., 90
Sege, G., 118
Shafer, R. J., viii
Sherwood, T. K., 119
Shockey, A. F., 94–99
Sittig, M., 127
Smith, C. A., 204
Smith, J, C., 101, 111, 113
Smith, S., 366
Souders, M., Jr., 504

NAME INDEX 521

Spitzglass, J. M., 353
Steele, A. B., 32
Stephenson, R. M., 463
Stevens, R. W., 201
Stockdale, W. G., 466, 467
Sweeney, R. J., 372, 374
Sylvander, N. E., 355
Syverson, A., viii

Taylor, J. J., 439
Thomas, R., 380
Thompson, A. R., 121
Treybal, R. E., 117
Troy, W. N., 356
Turner, W. C., viii, 380

Urquhart, L. C., 307

Valentine, K. S., 133–135
Vierck, C. J., 6, 363–365
Vilbrandt, F. C., 21, 100
Von Lossberg, L. C., 274

Walosewick, F. E., 366
Warzell, F. M., viii
Weintraub, N., 131
Wert, H. A., 366
Wessel, H. E., 221, 225, 234
White, A. S., 453
Williams, R., Jr., 205
Winfrey, R., 256
Winter, G., 307
Woodfield, F. W., 118

Zimmerman, O. T., 100, 101

Organizations and Publications

Acme Coppersmithing and Machine Co., 172, 173
Addison-Wesley Publishing Company, 454
Aluminum Company of America, 272
American Concrete Institute (ACI), 305
American Institute of Chemical Engineers (AIChE), 16, 38, 114, 497
American Institute of Electrical Engineers (AIEE), 421
American Institute of Steel Construction (AISC), 317
American Iron and Steel Institute (AISI), 342, 345

American Library Association, 269
American Oil Co., 308
American Petroleum Institute (API), 31, 135, 342
American Society of Mechanical Engineers (ASME), 31, 114, 150, 342, 358–362, 366
American Society of Testing Materials (ASTM), 31, 292, 305, 306, 316, 330, 342–345, 351, 421
American Standards Association (ASA), 31, 71, 324, 342–344, 354–366, 390, 421, 504
American Water Works Association (AWWA), 345
Artisan Metal Products, Inc., 170
Associated Factory Mutual Insurance Companies, 31
Association of Iron and Steel Engineers (AISE), 421
Atomic Energy Commission (AEC), 33, 445, 465, 468

Baker, Perkins, Inc., 171
Barnstead Still and Demineralizer Co., 164
Battelle Memorial Institute, viii
Bethlehem Steel Co., 278
Blaw Knox Co., viii, 181, 182, 366
Bureau of Explosives, 31
Bureau of Labor, 31, 204

Carbide and Carbon Chemical Corp., 277, 278
Celanese Corporation, 277
Central Soya, Inc., 310
Chemical Engineering, 3, 15, 31, 38, 40, 41, 72, 94, 99–101, 105, 111, 113, 115–118, 124, 127, 140, 141, 143, 147–149, 152, 192–197, 200, 201, 205, 207, 221, 222, 225–227, 229, 231, 234, 235, 240, 254, 277, 338, 339, 372, 374, 380–385, 387–391, 398, 405, 466
Chemical and Engineering News, 38, 226, 252
Chemical Engineering Progress, 19, 23, 38, 117, 118, 131, 132, 224–239, 262, 274, 452, 453, 455, 456, 460
Chemical Industries, 18, 234
Chemical Week, 38, 226, 262
Compass Instrument and Optical Co., 70
Concrete Reinforcing Steel Institute (CRSI), 305
Cost Engineering, 101, 204
Crane Co., 366, 374

Diamond Alkali Co., 278
Dow Chemical Co., 45
Dresser-Ideco Co., 312
du Pont de Nemours, E. I., & Company, iv

Eastern Industries, 163, 167, 169, 171, 173, 174, 176
Edison Electric Institute (EEI), 232, 421
Engineering News Record, 198, 200, 220, 336

Federal Power Commission, 232, 269
Federal Security Agency, 269
Federal Trade Commission, 269
Foxboro Co., 413

General Electric Co., viii, 423
Gump, B. J., Co., 173

Hafner Publishing Company, 18
Handbook Publishing Co., 155
Hardinge Co., 108
Harman, F. W., and Associates, 183

Industrial and Engineering Chemistry, 21, 38, 119, 121, 153, 226, 262, 350
Industrial Models Associates, 183
Industrial Nucleonics, viii
Industrial Rayon Co., 278
Industrial Research Service, 100, 101
Instrument Society of America (ISA), 74–81, 409
International Commission on Radiological Units, 429, 430
Interscience Publishers, Inc., 18
Interstate Commerce Commission (ICC), 31, 33, 136, 269

Jeffrey Manufacturing Co., 7–10

Kellogg, M. W., Company, viii, 69–71, 308, 366, 370, 372
Kennecott Copper Corp., 75

Lancaster Iron Works, 163, 167–170, 172, 175, 176
Lukens Steel Co., 88, 89

McGraw-Hill Book Company, Inc., viii, 6, 85, 90, 94, 117, 119, 127, 195, 225, 228, 253, 256, 307, 318, 322, 336, 363–366, 386, 404–409, 423, 445, 454, 463
(*See also* Perry, J. H.)
McMillen Feed Co., 310
Manufacturing Chemists Association (MCA), 31, 32, 269
Manufacturing Standardization Society of Valves and Fittings Industry, 342
Milton Roy Co., 166–169
Mixing Equipment Co., 168

National Association of Manufacturers (NAM), 269
National Board of Fire Underwriters (NBFU), 31, 150, 421
National Bureau of Casualty and Surety Underwriters, 150
National Bureau of Standards (NBS), 421
National Carbon Co., 166
National Committee on Radiation Protection and Measurement, 430, 432
National Distillers Corp., 278
National Electrical Manufacturers Association (NEMA), 138, 421
National Fire Protection Association (NFPA), 31, 135, 326, 329, 330
National Industrial Conference Board, 269
National Safety Council (NSC), 421
National Security Resources Board, 282
Nucleonics, 436, 439, 447, 449–451, 454, 455, 457, 459, 465

Ohio State University, viii, 319
Oil and Gas Journal, 38, 68, 201

Payne, E. M., Studios, 413
Petroleum Engineer, 38
Petroleum Processing, 38, 225, 356
Petroleum Refiner, 38, 366
Pfaudler Co., 87, 164–170, 174
Phillips Petroleum Co., viii
Portland Cement Association (PCA), 306
Prentice-Hall, Inc., 90

Reinhold Publishing Corporation, 10, 33, 36, 127, 131, 152, 256, 286, 329

NAME INDEX

Republic Flow Meter Co., 353
Ronald Press Company, 253

Sanitary Engineering Center, 434
Shell Development Co., 117
Solvay Process Co., 18
Standard Oil Co. of Ohio, 413
Stokes Machine Co., 172, 173

Tennessee Valley Authority, 273
Tube Turns, Inc., 342, 366, 367, 369
Tubular Exchanger Manufacturers' Association, 114

Underwriters' Laboratories, Inc. (UL), 31, 421
Union Carbide Co., 278
United Nations Publications, 453, 456, 458, 461, 464
United Scale Models, Inc., 183
United States, Bureaus, Census, 38
  Internal Revenue, 38, 243–247
  Labor, 31, 237, 238, 269, 281
  Mines, 38
  Reclamation, 279
  Standards, 31

United States, Bureaus, Weather, 294, 325
  Chamber of Commerce, 283
  Coastal and Geodetic Survey, 269
  Departments, Agriculture, 38
  Commerce, 269
  Defense, 31
  Interior, 269
  Employment Service, 281
  Geological Survey, 269
  Patent Office, 35, 38
  Public Roads Administration, 269
  Superintendent of Documents, 342
  Tariff Commission, 38
United States Navy, Bureau of Ships, 342
University of Chicago Press, 287

Van Nostrand, D., Company, Inc., 150, 453
Virginia Polytechnic Institute, viii, 17, 43
Vogt Machine Co., 171
Vulcan-Cincinnati, Inc., 414

Warren, S. D., Co., 413
Wiley, John, & Sons, Inc., 90, 100, 150, 329, 366, 370, 372, 393
Worthington Corp., 163, 168, 175

Subject Index

Administration cost references, 485, 486
Aerosols and nuclear plant design, 495
AIChE letter symbols, 504–515
Air conditioning, 329
Alloys, stainless, 86, 88, 89, 345
Aluminum pipe, 346
  relative cost of, 89
Amplifiers, instrument, 407
ASA letter symbols, 504–515
Atomic (*see* Nuclear engineering problems; Nuclear plants; Nuclear power)
Automatic control, principles of, 405–411
  of processes, 405–408
  systems diagram, 404, 406
Automation, chemical plant, 402

Batch processing, 41
Benzene hexachloride (BHC), chemical process for, 43–46
  economic evaluation of, 258–262
  equipment design calculations, 152–163
  equipment specifications, 163–176
  operating schedule, 47–50
  plant layout for, 187, 188
  process description, 42
Beta (β) radiation from isotopes, 432
  power release from spent U fuel, 445–447
Birmingham wire gage (BWG), 344
Blenders (*see* Mixers)
Brass alloy pipe, 346
Break-even chart, 256
Build-up, radiation shielding, 435

Buildings, air conditioning of, 329
  cost of, 335–339
  custom-designed, 311, 336
  design principles for, 313–334
  electrical design symbols for, 328, 422, 423
  fire code classification of, 316
  fire protection for, 326
  flooring, 321, 322
  heating of, 332
  height classification of, 313
  humidity control in, 332
  illumination for, 327
  materials of construction for, 316
  mill, 317, 318
  multistory, 314, 318, 337–339
  personnel and service facilities in, 333, 334
  prefabricated, 311, 312, 336
  reinforced-concrete, 319–321
  roof classification of, 315
  roofs for, 323, 326
  safety designs for, 30, 326
  single-story, 314, 317–318, 337–339
  types of, 310
  ventilation principles for, 329–332
  walls for, 321
By-products, cost accounting for, 223, 224, 226

Capital investment, cost references, 221, 483, 484
  fixed, cost estimates for, 191–222

Capital investment, fixed, quick estimation method for, 195, 222
  per annual ton, 196, 197
  recovery methods for, 249
  total, quick estimation method for, 222
  working, 221
Carbon pipe, 348
Cast-iron pipe, 344, 345
Centrifuges, cost of, 198
  description of, 111–113
Ceramic pipe, 347
Chemical engineer, role of, 1
Chemical engineering publications, 37, 38
Chemical processes, nuclear heat for, 455
Cladding, economics of, 88–90
Climate as factor in plant location, 280
Closed-loop theory, 403, 404
Codes, electrical systems, 421
  fire, for building design, 326
  National Electric (NEC), 425
  piping, 341, 342, 390, 391
  unfired pressure vessels (ASME), 150
Cold springing of piping, 370
Commercial plant, scheduling for, 25, 41
Compressed gas costs, 228, 233
Compressors, cost of, 144, 213
  description of, 141–144
Computers, 12–14
  applications for, 13
  for heat-exchanger calculations, 115
  references for, 471, 472
  types of, 13, 14
Concrete, definition of, 305, 306
  design principles for, 307
  precast piles of, 299
  reinforced, 307
    beams and columns, 320
    design references for, 305, 307
    form work, 320
    prestressed design, 320
  strength of, 306
Concrete flooring, 320, 321
Concrete foundations, 296–307
Construction cost references, 336, 488, 489
Construction materials (*see* Materials of construction)
Contacting columns, costs, 219
  description of, 117, 118
  design methods for, 118
Continuous processing, 41
Contracts, legal, 36
Control, of fire hazards, 31
  of pilot plants, 22
  proportional, 409–411
  (*See also* Automatic control; Process control)

Controllers, process, 404–408
Conversion factors, 469, 470, 516–518
Conveyors, cost of, 200–203
  description of, 104, 105
Copper, pipe, 346
  relative cost of, 89
Cost estimation, accuracy of, 192, 193
  for capital investment, 192–195
  diameter-inch method for piping, 398
  preprocess design, 40
Cost indices, 198–205
  Bureau of Labor, 205
  Engineering News-Record (ENR), 198
  Marshall-Stevens (MS), 200
  Nelson refining, 201
Costs, administration, 249
  auxiliaries, 194
  building, 335–339
  buildings and site development, 194, 217, 218
  centrifuges, 198
  compressed gas, 228, 233
  contacting columns, 219
  conveyor, 200–203
  crystallizer, 203
  depreciation, 243–249
  dryer, 204
  electric equipment, 194–212
  electric power, 228, 232
  electrical installation, 215
  engineering and construction, 194, 217, 220
  equipment (*see* Equipment costs)
  evaporator, 205
  as factors in design, 29, 30
  filters, 206
  fluidization process equipment, 132, 199, 218
  fuel, 228, 230
  heat-transfer equipment, 116, 117, 207
  instrumentation, 194, 209, 415–417
  insulation, 386–388
  labor, direct, 236
    indirect, 236
  maintenance and repairs, 236, 239–242
  manufacturing, 222–224
    distribution of, 225
  marketing, 249
  materials of construction, 89
  mixer, 208–211
  nuclear chemical plants, 466, 467
  outdoor vs. indoor plants, 339
  piping, 398–402
    process, 194, 208, 215
  piping materials, 398–400
  plant site, 334
  pump, 215

SUBJECT INDEX

Costs, raw material, 225, 226
  reactors, 137, 216, 217
  references, 192–205, 221–240, 252–254, 478–486
    administration, 485
    capital, 221, 483, 484
    construction, 336, 338, 488, 489
    depreciation, 247, 485
    distillation, 482
    drying, 483
    economic evaluation, 256, 486
    engineering services, 484
    evaporation, 483
    gas handling, 481
    general equipment, 200, 201, 204, 205, 207, 480
    general review and summary, 478, 479
    heat transfer, 482
    instrumentation, 491
    labor, 234, 238, 484
    maintenance and repairs, 239, 240, 484
    mixing, 481, 482
    nuclear chemical plants, 466, 467
    operating, 484
    packaging, 485
    piping, 398, 489, 490
    power systems, 231, 491
    process control, 491
    pumps, 481
    reactors, 483
    research, 485, 486
    sales, 485, 486
    separations processes, 482
    shipping, 485
    size reduction, 481
    solids handling, 481
    tanks, 483
    taxes, 252, 485
    utilities, 228, 485
    vessels, 483
  refrigeration, 232, 233
  safety, 30
  sales, 249
  size-reduction equipment, 216
  steam, 228
  summary of product, 250
  tank, 137, 216–218
  thickeners, 219
  utility, 226, 228
  water, 279
Critical mass of nuclear fuel, 448–452, 461
Criticality, nuclear systems, 448–452
Crystallization by salting out, 121, 122
Crystallizers, cost of, 203

Crystallizers, description of, 119–122
Curie, definition of, 433, 470
Curie level, of irradiated U fuel, 451
  of radioactive waste, 456
Cycles selection in process design, 41

Data, process, references for, 473–476
Decontamination of nuclear plants, 463
Depreciation, cost references, 247, 485
  definition of, 239
  equipment, allowances for, 244–247
  methods of determining, 243–249
    declining balance, 248
      double, 248
    present worth, 249
    sinking fund, 249
    sum-of-digits, 248
    unit of production, 249
Derivative control, 409–411
Design, bases for, 5
  of critical nuclear systems, 448–452
  definition of, 1
  equipment, 6–11, 150
  in equipment selection, 4
  information on, for electrical systems, 421
  of mechanical equipment, 150
  need for, 2
  patents for, 35
  of piping, 340–402
  piping stresses, 365–370
  process (see Process design)
  of radiation shielding, 435–447
  relation to sales, 3
  structural, 291
    references, 317, 318, 487, 488
Diagrams, flow (see Flow sheets)
Distillation cost references, 482
Dose units, radiation, 428, 429
Drafting procedures, instruments in, 11
  reproduction methods in, 12
  scales used, 12
Drawings, construction and installation, 11
  equipment design and, 6–11
  piping, 357–365
  for plant design, 5, 11
  plant layout, 11
  in process design, 5, 6
  references for, 6, 358–365, 472
Dryers, classification of, 122, 123
  cost references for, 204, 483
  description of, 122–124
  spray, 124
Dust collectors, cost of, 199

Earnings, new or net, 251
Economic analysis, 250-257
Economic evaluation, 189-262
  factors in, 189-191
  references, 256, 486
  report of, 257
Economic optimization references, 90
Economics, of clad steels, 88-90
  of construction projects, 336, 338, 466, 467, 488, 489
  definitions of, 28
  of piping projects, 398, 489, 490
  plant location, 268, 289
  plant and process, references to, 192, 193, 195, 479, 480
  of power systems, 491
Electric equipment, costs, 194, 212
  demand charges, 141
  installation cost of, 215
  motors, classification of, 138, 140
    description of, 137-139
  power factor of, 140
  power supply, description of, 139
Electrical hazards, 425, 426
Electrical symbols, 328, 422, 423
Electrical systems, codes for, 421
  design of, 420-426
  distribution in, 421-425
  equipment protection for, 424
  grounding of, 426
  hazards of, 425, 426
  NEC area classifications, 425
  safety practices in, 425, 426
  symbols for, 328, 422
Electrical units, 517
Electricity, cost of, 228-232
Emergency power systems, 420
Energy of nuclear particles, 430-435
Energy balance, flow sheets for, 67
  preparation of, 67-69
Engineering, cost references, 484
  report forms, 257
ENR cost index, 198-200, 220
Equipment, costs (*see* Equipment costs)
  description of, 100
  design of, 4
    mechanical, 150
    electrical, 421-425
  fabrication of, 150
  flow sheets for, 48, 49, 65, 73
  foundations, 300-305
  glass-lined, 86, 87
  for nuclear plants, 462
  references, 94, 100, 101, 105, 111, 113-119, 121, 124, 127, 131, 132, 135, 136, 141, 150
  safety designs for, 30

Equipment, selection of, 27, 81, 84, 90-176
  size-reduction (*see* Size-reduction equipment)
  specifications, comparison of, 94-99
    summary sheet for, 91
    writing of, 93
  standard vs. special, 92-93
  symbols for flow sheets, 68-71
  table of references, 100
Equipment costs, estimation of, 195
  installation, 207, 214
  insulation, 209
  methods of comparison of, 205-207
  "six-tenths" factor for, 205
  table of references, 99, 100
  (*See also* Costs)
Equipment design, development of, 150
  drawings for, 6-11
Evaporators, classification of, 125-126
  comparison of specifications for, 94-99
  cost of, 205, 483
  description of, 124
Excavation and grading, 295
Expansion joints, piping, 368
Explosive limits, 329

Feeders, solids, description of, 105-106
Filters, cost of, 206
  description of, 108-110
Fire brick, 381-383
Fire code for building design, 326
Fire hazards, prevention and control, 31
Fire protection for buildings, 326
Fire station control, 283
Fission-product utilization, references for, 495
Fission products, 456
Fixed charges, 223-249
Flooring, building, 321, 322
Flow diagrams (*see* Flow sheets)
Flow sheets, auxiliary process, 81
  discussion of, 27
  energy balance, 67-69
  equipment, 48, 49, 65-73
  instrumentation, 74-81
  material balance, 64-66
  piping, 356
  qualitative block-type, 46, 47
  references for, 40, 68, 73, 475, 476
Fluid flow, formulas for, 518
Fluid velocities in piping, 350-353
Fluidization process equipment, basic operation of, 128-130
  cost of, 132, 199, 218

SUBJECT INDEX

Fluidization processes, applications for, 130
  description of, 127–132
  design methods for, 130–132
Flux, radiation, 430, 431
Footings, 296
Foundations, 295–307
  cost of, 337
  dynamic loading and, 300
  footings for, 296
  machinery and equipment, 300–305
  mats, 297, 298
  piles for, 297–300
Frequency response in process control systems, 409
Frost line, 293, 294
Fuel, 271
  costs, 228, 230
  nuclear, 427, 448–456
    critical mass, 448–452

Gamma (γ) radiation, absorption coefficients, 435, 436
  from isotopes, 432
  maximum permissible flux for, 431, 433
  maximum permissible intensity for, 431, 433
  shielding, 435–447
Gas-handling cost references, 481
Gases, dispersal from nuclear plants, 464
  storage tanks for, 136
Glass-lined equipment, 86, 87
Glass pipe, 347
  for nuclear chemical plants, 463
Graphic panels, 412, 413
Graphite pipe, 348
Grounding of electric systems, 426

Half life ($t_{1/2}$), isotopic, 433
Hazards, chemical, 33
    handbook of, 33
  electrical, 32
  electrical design for, 425, 426
  fire, 31
  health, 34
  mechanical, 32
  radiation, 34
  ventilation control, 33
Health physics, biological effects in, 428
  MPE values for, 430, 432
  nuclear, 428–435, 461, 462
  in nuclear plant control, 461, 462
  radiation dose units for, 428, 429
Heat capacity, definitions of, 516

Heat evolution in radiation shields, 445–447
Heat exchange process, automatic control of, 405–408
Heat transfer, cost references, 482
  in fluidized beds, 132
  formulas for, 518
Heat-transfer equipment, condensers, 114
  costs, 116, 117, 207
  design procedures for, 115
  exchangers, 144
  specification sheets, 116, 117
Heating of buildings, 332
High-pressure process control systems, 414, 415
Humidity, definitions of, 516
Humidity control in buildings, 332

Illumination for buildings, 327
Income, sales, 251
Income tax, 252
Information sources, design, 36–39
Installation costs for equipment, 207, 214
Instrumentation, 402–417
  amplifiers for, 407
  elements of, 404
  error detection elements, 405–407
  flow sheet symbols, 74–81
  identification code for, 74, 75
  ISA symbols, 74–82
  nuclear plant, 464
  in pilot plants, 22
  references, 404–406, 409, 414, 464, 490
  relation to outdoor plants, 309
  relay controllers, 408
Insulation, 375–389
  characteristics of, 376–380
  costs, 209
  economic selection of, 386–388
  factors for selection of, 384, 385
  pipe applications, 385–388
  refrigeration, 388, 389
  types available, 381–385
Intensity, radiation, 430, 431
Interest charges, 251
Investment, return on, 253, 254
Ion exchange in nuclear fuel separations, 455, 456
Isotopes, chemical processes for, 452–458
  half life of, 433
  radiation from spent U fuel, 446
  radioactive, 431–435

Job evaluation, 234

530 CHEMICAL ENGINEERING PLANT DESIGN

Labor, cost references, 234, 238, 484
  costs, 236–238
  as factor in plant location, 281
  for piping installation, 400–402
  requirements for chemical plants, 234, 235
  safety precautions for, 33
  wages and material index (Bureau of Labor), 204
Layout, piping, 355–357, 364, 371
  plant (*see* Plant layout)
Lead pipe, 347
  relative cost of, 89
Legal contracts, 36
Legal problems in nuclear design, 468
Legal references, 475
Lighting of buildings, 327
Lindane (γ-BHC), economic evaluation of, 258–262
  process description, 42
  site location of plant, 286–290
Liquid metals, references, 477
Liquids, regulations for storage, 31, 135
  storage tanks for, 135, 136

Machinery, foundations for, 300, 301
Maintenance, costs, 236, 239–242, 484
  nuclear plant, 459–461
Management expenses, 249
Manufacturing costs, 222–225
Market research references, 487
Marketing expenses, 249
Markets, 26, 270, 271
Marshall-Stevens (MS) cost index, 200, 220
Mass-transfer equipment, description of, 117–127
Master plot plans, 178
Material balance preparation, 46, 51–66
Materials of construction, for buildings, 316
  clad steels, 88–90
  costs, 89
  glass-lined steel, 87
  laboratory testing of, 85, 86
  for nuclear plants, 462, 463
  piping, 345–349
  for pumps, 147
  references to, 85, 86, 100, 476
  relative prices of, 89
  selection of, final, 86, 89, 90
    plan for, 85
    preliminary, 85, 86
  for process equipment, 84–90
Materials-handling equipment, cost of, 200–203

Materials-handling equipment, description of, 103
  selection of, 103–106
Mats, foundation, 297, 298
Maximum permissible concentration (MPC), 430, 432
Maximum permissible exposure (MPE), 430
Mechanical separation equipment, 106–113
  classification of, 106
  for solids, from gases, 106, 107
    costs, 199
    from liquids, 108–113
      centrifuges, 111–113
      filters, 108–110
      thickeners, 111
    from solids, 107
Metal cutting, 151
Mixers, cost of, 208–211
  description of, 132–135
  power consumption of, 134, 135
  selection of, 135
Mixing, cost references for, 481–482
Models (*see* Scale models)
Moderators, nuclear system, 448
Modes, process control, derivative, 409–411
  on-off, 409, 410
  proportional, 409–411
Motors, electric, 137–139
  NEMA standards, 138

National Electric Code (NEC) area classification, 425
Nelson refinery construction index, 201
Nickel alloys, piping, 346
  relative cost of, 89
Nomenclature for nuclear design, 469
Nominal pipe size (NPS), 344
Nomographs, for radiation shielding, 436–441
Nonmetallic piping, 347, 348
Nuclear engineering problems, biological radiation effects, 428
  chemical plant design, 427–470
    references for, 452–458, 461, 463–467, 492
    safety in, 428, 452
  conversion factors, 469, 470
  criticality design, 448–452
    for plutonium, 450–452
    for uranium, 449–452
  fuel, 427, 448–456
  fuel reprocessing, 455, 456
  general references, 491

SUBJECT INDEX 531

Nuclear engineering problems, health physics, 428–435, 461, 462
  legal problems, 468
  plant costs, 466, 467
  plant design, 458–465
  plant equipment, 462
  plant location, 465
  plant maintenance, 461
  raw materials, 453, 454
  reactions, 427, 428
  shielding of γ rays, 435, 447
Nuclear plants, decontamination of, 463
  process control, 464
  process design, 452–458
  process heat, 455
  ventilation, 464
Nuclear power, 419

On-off control, 409, 410
Operating cost references, 484
Optimization, economic, references for, 90
Ordinances, chemical plant, 282
Outdoor plants, design features of, 307–309
  relative cost of, 339

Packaging, relation to manufacturing costs, 223, 224, 250
Painting, piping color codes, 390–391
  relation to illumination, 327
Patents, infringement of, 35
  in research and development, 35
Pay-out time, 251, 254, 255
Piles, foundation, 297–300
Pilot plants, 20–24
  cost of, 24
  instrumentation and control for, 23
  personnel for, 21, 22
  planning for, 21
  references to, 21, 23, 474
  scale-up factors, 22, 23
Pipe, aluminum, 89, 346
  brass alloy, 346
  carbon, 348
  cast, 344, 345
  ceramic, 347
  ferrous, 345
  glass, 347
  graphite, 348
  lead, 347
  nonferrous, 346
  nonmetallic, 347, 348
  plastic, 348
  rubber, 348

Pipe, seamless, 343
  steel, 345
  Transite, 348
Piping, codes for, 341, 342
  cold springing in stress relief, 370
  cost estimating, 398–402
  cost references, 398, 489, 490
  design problems in, 340
  drainage, 390–398
    capacity, 392
    codes, 390
    design, 396–398
    hardware, 394–396
    testing, 396
  drawings, 357–365
  erection and testing of, 375
  fabrication methods, 342
  flexibility design, 368–370
  flow sheets, 356
  fluid velocities in, 350–353
  hardware, 355
  installation of, 370–375
  labor factors in, 400–402
  insulation, 375–389
    (See also Insulation)
  layout, 355–357, 364, 371
    underground, 374
  material costs, 398–400
  materials, 345–349
  nominal pipe size (NPS), 344
  overhead design for, 371–374
  preinsulated, 389
  pressure relief devices for, 354, 355
  references, 342, 343, 350, 353, 355, 356, 358–367, 369, 372, 380–391
  schedule numbers, 344
  selection of, 342–355
  size selection, 351–352
  sized by internal diameters, 349–351
  steam, 352–354
  stress design, 365–370
    fluid pressure, 366
    material, 366
    thermal, 366–368
  supports, 371–373
  symbols, 358–362
Plant capacity, break-even chart for, 256
Plant design, nuclear, 458–470
Plant layout, building plans for, 185–186
  description of, 27
  equipment location in, 183, 184
  factors in planning, 177
  methods, 178, 502
  nuclear, 459
  plant expansion factors in, 185
  principles of, 183–187
  railroads and roads for, 186

Plant layout, references, 477, 478
  safety considerations in, 184
  storage location in, 183
  two-dimensional, 178–181
Plant location, 28, 265–290
  economics of, 268
  nuclear, 465
  references, 269, 486, 487
  sources of information for, 269
  summary of factors in, 266, 267
  weighted-score method for, 288
Plant site topography, 295
Plastic pipe, 348
Plot plans, unit area and master, 178
Plutonium, critical mass and volume, 449–452
Power, definitions of, 517
  electric generation of, 419, 420
    costs, 228, 232
    as factor in plant location, 272
    industry requirements for, 231
  emergency systems, 420
  nuclear, 419
  references, 231, 419, 421, 491
  requirements for chemical plants, 418
  sources of, 271
  steam, 272
Power systems, 418–426
Present worth, 255
Pressure-relief systems, 354, 355
Pressure vessels, ASME code for, 150
Price indices, types of, 198–205, 220
Process auxiliaries, control systems, 402–418
  instrumentation, 404–411, 416, 417
  piping, 340–402
  power systems, 418–426
Process control, 402–417
  center, 411, 413
  criteria for, 403
  dynamic response, 409
  electronic systems, 414
  instrumentation costs for, 415–417
  modes, 409
  for nuclear plants, 464
  pneumatic systems, 405–408, 414
  references, 404–406, 409, 414, 464, 490
  relay systems for, 408
  systems design, 414, 415
  systems example, 405–411
Process cycles, references to, 40, 41, 476
Process data sources, 43–47, 473–476
Process design, cost estimation for, 40
  nuclear, 452–458
  procedures, 40–72
  projects, 262
  selection of cycles, 41

Process design, teaching methods for, 497–503
Process development, organization for, 17, 18
  references for, 15, 21, 23, 472
  requirements for, 19
  research for, 16
  steps in, iv, 16
Process equipment (see Equipment)
Process steam piping, 352–354
Process steam power, 418
Processing, batch, 41
  continuous, 41
Product cost summary, 250
Profit, 253
Profitability evaluation, 253–257
Project, present-worth method, 255
Proportional control, 409–411
Publications, chemical engineering, 37, 38
Pumps, centrifugal, 145, 146
  cost references for, 215, 481
  description of, 144–149
  diaphragm, 145
  efficiency of, 148, 149
  materials of construction for, 147
  reciprocating, 144, 145
  rotary gear, 146
  selection of, 147–149

Radiation activity from spent U fuel, 446
Radiation dose units, 428, 429
Radiation flux, 430, 431
Radiation intensity, 430, 431
Radiation shielding, 435–447
Radioactivity in wastes, 285
Radiological Health Handbook, 434
Raw materials, costs, 226
  supply vs. location, 269
Reactors, cost references for, 483
  description of, 135–137
Refractories, thermal properties of, 381–385
Refrigeration, costs, 232, 233
  insulation, 388, 389
Regulations, shipping, ICC, 136
Reinforced concrete, buildings, 319–321
  design references, 305, 307
Repair costs, 236, 239–242, 484
Report writing, references for, 472, 473
Reports, design course, 503
  writing of, 257
Research, cost references, 485, 486
  evaluation of, 18
  expenses of, 249
  market, 26, 487
  patents in, 35

SUBJECT INDEX

Research, process, 16
    requirements for, 19
Roentgen (unit), 428, 429
Rubber pipe, 348

Safety, considerations of, 30–35
    costs, 30
    design for, in buildings, 30, 326
        in process equipment, 30
    and electrical systems design, 425, 426
    information sources, 30, 31
    and labor relations, 33
    in nuclear systems design, 428, 452
    and plant layout, 184
    references, 31–33, 474, 475
        to nuclear plant design, 449–452, 493
    ventilation standards, 329, 330
Sales, cost references, 485, 486
    income from, 251
    relation to design, 3
Scale models, 181, 183
    materials for, 183
    piping layout with, 356
    in plant design, 5
    references to, 478
    types of, 181–183
Semicommercial plant, uses for, 24
Separations processes, cost references, 482
Shielding, α and β radiation, 435
    γ radiation, 435–447
    heat evolution in shields, 445
    references, 436, 439, 445, 493
Shift schedules, types of, 41, 42
Shipping, cost references, 485
    relation to manufacturing costs, 223, 224, 250
Shipping regulations (ICC), 136
Site location, 265, 281–284
Site preparation, 291–295
    cost of, 334, 335
    references, 487, 488
Size-reduction equipment, costs, 216, 481
    description of, 101
    selection of, 101, 102
Soil, bearing load of, 293
Soil testing, 292
Solids handling, cost references, 481
Specifications, heat exchanger, 116, 117
    for process equipment, 91–99, 116, 117
Spray dryers, 124
Stainless-steel alloys, 86, 88, 89, 345
Standards, heat exchanger, 114
    motors and generators (NEMA code), 138
Statistics references, 18, 475

Steam, costs, 228
    industry requirements for, 229
    process, 418
Steel, relative cost of, 89
    structural, building design for, 316–319
Steel beams, 317
Steel piling, 299, 300
Steel pipe, 345
Steel tank supports, 301
Storage regulations for liquids, 31, 135
Stress, piping, 365–370
Structural design, 291
    references, 487, 488, 317, 318
Structures, 307–339
    enclosed, 309–339
    unenclosed, 307–309, 339
Subsurface evaluation, 292–295
Symbols, drainage piping, 397
    electrical, 328, 422
    ISA instrumentation, 74–82
    letter, AIChE and ASA, 504–515
    piping, 358–362

Tanks, cost of, 137, 216–218
    foundation supports for, 301–305
    storage, 135–137
    wooden, 216, 305
Tantalum pipe, 347
Taxes, corporation, 281
    cost references for, 252, 485
    income, 252
    local, 249
    social security, 249
Teaching methods for process and plant design, 497–503
Temperature, control of, for buildings, 332
    effects of, in piping design, 366–368
    measurement of, 405–407
    water, 276, 277
Thermal conductivity of insulating materials, 376–385
Thermal insulation (see Insulation)
Thickeners, cost of, 219
    description of, 111
Thorium, nuclear fuel cycles with, 453
Tile, piping, 392
Topography of plant site, 295
Transite pipe, 348
Transportation, of gases, 137, 142–144
    of liquids, 136, 137
    and plant location, 270

Unit area plot plans, 178
Units, engineering, 517, 518

Uranium, cooling of spent fuel, 445–447
  critical mass and volume, 449–452
  nuclear fuel cycle, 453
Utilities, cost of, 226, 228, 485

Ventilation, building, 329–331
  hazard control by, 33
  nuclear plant, 464
Vessels, cost references for, 483
  fabrication of, 151

Waste disposal, 284–286
  of nuclear plant gases, 464
  radioactive, 285, 456
  references, 456, 457, 487, 494, 495
  types of plant wastes, 284

Water, conservation of, 277, 278
  costs, 228, 279
  dams and reservoirs, 278
  industry requirements for, 227, 274
  lake sources, 278
  legal restrictions for, 279, 280
  quality of, 275
  references, 227, 274, 487
  sea, 278
  source of supply, 273
  temperatures of, 276, 277
Welding, 151
  of steel pipe, 342
Wood tank, costs, 216
  supports for, 305

Zoning regulations, 282